ANALYSIS, DESIGN AND EVALUATION OF MAN-MACHINE SYSTEMS 1995

T0226934

A Postprint volume from the Sixth IFAC/IFIP/IFORS/IEA Symposium,
Cambridge, Massachusetts, USA, 27 - 29 June 1995

Edited by

T.B. SHERIDAN
Department of Mechanical Engineering,
Massachusetts Institute of Technology,
Cambridge, Massachusetts, USA

(In two volumes)

Published for the

INTERNATIONAL FEDERATION OF AUTOMATIC CONTROL

by

PERGAMON
An Imprint of Elsevier Science

UK	Elsevier Science Ltd, The Boulevard, Langford Lane, Kidlington, Oxford, OX5 1GB, UK
USA	Elsevier Science Inc., 660 White Plains Road, Tarrytown, New York 10591-5153, USA
JAPAN	Elsevier Science Japan, Tsunashima Building Annex, 3-20-12 Yushima, Bunkyo-ku, Tokyo 113, Japan

First edition 1995

Library of Congress Cataloging in Publication Data

A catalogue record for this book is available from the Library of Congress

British Library Cataloguing in Publication Data

A catalogue record for this book is available from the British Library

ISBN 0-08-042370 1

Transferred to digital print 2008

Printed and bound in Great Britain by CPI Antony Rowe, Chippenham and Eastbourne

6th IFAC/IFIP/IFORS/IEA

SYMPOSIUM ON ANALYSIS, DESIGN, AND EVALUATION OF MAN-MACHINE SYSTEMS

Hosted by

Department of Mechanical Engineering and
Department of Aeronautics and Astronautics
Massachusetts Institute of Technology
Cambridge, MA USA
June 27-29, 1995

Sponsored by:
International Federation of Automatic Control (IFAC)
  Technical Committee on Systems Engineering
  Technical Committee on Social Effects of Automation
American Automatic Control Council

Co-Sponsored by:
International Federation for Information Processing (IFIP)
International Federation of Operational Research Societies (IFORS)
International Ergonomics Association (IEA)

International Program Committee (IPC):
Chairman: T.B. Sheridan (USA)

J.L. Alty (GB)	L. Martensson (S)
A.B. Aune (N)	P. Milgram (CAN)
S. Baron (USA)	C. Mitchell (USA)
P. Brodner (D)	N. Moray (USA)
P.C. Cacciabue (I)	L. Motus (ES)
W.J. Edwards (Aus)	L. Nemes (AUS)
J.L. Encarnacao (D)	W.K. Oxenaar (NL)
R. Genser (A)	J. Ranta (SF)
G. Giuda (I)	J.E. Rijnsdorp (NL)
B.S. Hu (PRC)	W.B. Rouse (USA)
R. Jackson (USA)	A.P. Sage (USA)
G. Johannsen (D)	G. Salvendy (USA)
K. Kawai (J)	H.G. Stassen (N)
V. deKeyser (B)	S.Q. Su (PRC)
A.H. Levis (USA)	B.G. Tamm (ES)
M. Lind (DK)	H. Tamura (J)
B. Liu (PRC)	T. Vamos (H)
N. Malvache (F)	

National Organising Committee (NOC)
Chairman: R. John Hansman, MIT

History of the IFAC Symposium on Man-Machine Systems

This is the sixth in a series of (usually) triennial symposia on man-machine systems under IFAC sponsorship. It is the first of the series to be held in the United States. The sequence has been:

1982 Baden-Baden, Federal Republic of Germany
1985 Varese, Italy
1988 Oulu, Finland
1989 Xi'an, China
1992 The Hague, The Netherlands
1995 Cambridge, Massachusetts, USA

CONTENTS

SURVEY PAPERS

Conceptual Design of Multi-Human Machine Interfaces   1
G. JOHANNSEN

Telemanipulation and Telepresence   13
H.G. STASSEN, G.J.F. SMETS

Virtual Networks: New Framework for Operator Modeling and Interface Optimization in Complex
Supervisory Control Systems   25
Y.M. YUFIK, T.B. SHERIDAN

TECHNICAL PAPERS

Decision Models

The Operator-Model Architecture and its Psychological Framework   39
S. DEUTSCH, M.J. ADAMS

Validation and Verification of Decision Making Rules   45
A.K. ZAIDI, A.H. LEVIS

Linguistic Knowledge and Fuzzy Membership Elicitation   53
JIE REN, T.B. SHERIDAN

Human Error

Evolution of Ideas regarding the Prevention of Human Errors   59
V. DE KEYSER

Prevention of Routine Errors Through a Computerised Adaptive Assistance System   65
M. MASSON

Human Recovery of Errors in Man-Machine Systems   71
T.W. VAN DER SCHAAF

Human Tunnel Vision and waiting for System Information   77
L.C. BOER

Telerobotics I

A Multiparameter Tactile Display System for Teleoperation   83
D.A. KONTARINIS, R.D. HOWE

Issues and Design Concepts in Endoscopic Extenders   89
A. FARAZ, S. PAYANDEH, A.G. NAGY

Virtual Tools for Interactive Telerobotics: Potential Fields and Terrace Following   95
D.J. CANNON, C. GRAVES, K.W. LILLY, C.S. BONAVENTURA

A Concept for Symbolic Interaction with Semi-Autonomous Mobile Systems   101
M. PAULY, K.-F. KRAISS

Human-Machine Interface for a Model-Based Experimental Telerobotics System   107
T.T. BLACKMON, L.W. STARK

Organizational Systems

Information Exchange and Team Coordination in Human Supervisory Control   113
B.S. CALDWELL

Designing Human-Machine Systems to Match the User's Needs   119
P. FUCHS-FROHNHOFEN, E.A. HARTMANN, D. BRANDT

Enterprise Support Systems: Human Interaction with Complex Organizational Systems   125
W.B. ROUSE

Information Technology in Support of Living Oriented Society   131
H. TAMURA

Aviation I: Pilot Behavior

Automated Classification of Pilot Errors in Flight Management Operations   137
S. ROMAHN, D. SCHAFER

Empirical Modelling of Pilot's Visual Behaviour at Low Altitudes   143
A. SCHULTE, R. ONKEN

Autonomy, Authority, and Observability: Properties of Advanced Automation and their Impact on
Human-Machine Coordination   149
N.B. SARTER, D.D. WOODS

Experimental Study of Vertical Flight Path Mode Awareness   153
E.N. JOHNSON, A.R. PRITCHETT

Failure Assessment I

FIXIT: A Case-Based Architecture for Computationally Encoding Fault Management Experience   159
A.J. WEINER, C.M. MITCHELL

Human Error and the Holon Cognitive Architecture   165
M.J. YOUNG

Risk Probability Estimation in Systems Using Distributed Control   171
E.J. LANZILOTTA

Fuzzy Docs: Failure Detection and Location in Large Systems   177
JIE REN, T.B. SHERIDAN

Telerobotics II

A Design for a Man Machine Interface for the Teleman Dexterous Gripper   183
A.C. VAN DER HAM, G. HONDERD, W. JONGKIND

The Importance of Movement for the Design of Telepresence Systems   189
C.J. OVERBEEKE, P.J. STAPPERS

A Laboratory Evaluation of Four Control Methods for Space Manipulator Positioning Tasks   195
E.F.T. BUIEL, P. BREEDVELD

The Development of a Man-Machine Interface for Telemanipulator Positioning Tasks   201
P. BREEDVELD

Human-Machine Interaction with Multiple Autonomous Sensors   209
S.A. MURRAY

Human-Computer Interaction I

Dynamic, Simulation-Integrated, Intelligent Visualization: Methodology and Applications to
Ecosystem Simulation   215
J. YOST, M.M. MAREFAT, JINWOO KIM

Interaction Relationships: A Paradigm for the Description of Visual Manipulation   221
M. TSCHELIGI

Information Manipulation Environments: An Alternative Type of User Interfaces   227
S. MUSIL, G. PIGEL, M. TSCHELIGI

Pattern Based Human-Machine System   233
T. VAMOS, K. GALAMBOS

Aviation II: Displays

In-Flight Application of 3-D Guidance Displays: Problems and Solutions   241
E. THEUNISSEN

Perception of Flight Information from EFIS Displays   247
R.J.A.W. HOSMAN, M. MULDER, E. THEUNISSEN

The Relationship of Binocular Convergence and Errors in Judged Distance to Virtual Objects   253
S.R. ELLIS, U.J. BUCHER, B.M. MENGES

Pilot Performance and Workload Using Simulated GPS Track Angle Error Displays   259
C.M. OMAN, M.S. HUNTLEY, Jr., S.A. RASMUSSEN

Failure Assessment II

CAB-SIM: A Computer Code to Analyze Errors of Commission for Probabilistic Risk Assessment   267
A.P. MACWAN, F.J. GROEN, P.A. WIERINGA

Method to Identify Cognitive Errors During Accident Management Tasks   273
V. GERDES

Modelling Organisational Factors of Human Reliability in Complex Man-Machine Systems   279
W. VAN VUUREN, T.W. VAN DER SCHAAF

Experimental Investigation on Mental Load and Task Effect on System Performance   285
Z.G. WEI, A.P. MACWAN, P.A. WIERINGA

Developing Safety Culture - A Never Ending Process   291
P. HOLMGREN

Highway Systems I

Fuzzy-Controlled-Driver-Simulation for Exhaust Emission Tests   295
K. PFEIFFER, R. ISERMANN

Anticipation Influenced by Different Interfaces in Simulated Bus Traffic Control Tasks 301
S. MAILLES, C. MARINE, J.M. CELLIER

Human Intervention into Automatic Decision-Making and Automatic Control 307
S. KIM, T.B. SHERIDAN

Human-Computer Interaction II

FLOWER - A System for Domain and Constraint Visualization in Computer-Aided Design 313
E. ROZSA, J.C. DILL

Toward the Application of Multiagent Techniques to the Design of Human-Machine Systems Organizations 319
E. LE STRUGEON, M. GRISLIN, P. MILLOT

Designing Support Contexts: Helping Operators to Generate and Use Knowledge 325
G.A. SUNDSTROM

Building Natural Language Interface - Its Methodology and Tools 333
H. TSUJI, Y. NAMBA, H. MASE, Y. MORIMOTO, H. KINUKAWA

Aviation III: Automation

Evaluation of Intelligent On-Board Pilot Assistance in In-Flight Field Trials 339
T. PREVOT, M. GERLACH, W. RUCKDESCHEL, T. WITTIG, R. ONKEN

Mode Usage in Automated Cockpits: Some Initial Observations 345
A. DEGANI, M. SHAFTO, A. KIRLIK

The Effect of Data Link-Provided Graphical Weather Images on Pilot Decision Making 353
A.T. LIND, A. DERSHOWITZ, D. CHANDRA, S.R. BUSSOLARI

Intelligent Tutoring Systems to Support Mode Awareness in the 'Glass Cockpit' 359
A.R. CHAPPELL, C.M. MITCHELL

Towards a New Paradigm for Automation: Designing for Situation Awareness 365
M.R. ENDSLEY

Mode Awareness in Advanced Autoflight Systems 371
S.S. VAKIL, R. JOHN HANSMAN, Jr., A.H. MIDKIFF, T. VANECK

Process Control I

A New Human-Computer-Interface for High-Speed-Maglev Train Traffic Supervision 377
J.-O. MULLER, E. SCHNIEDER

Advantages of Mass-Data-Displays in Process S&C 383
C. BEUTHEL, B. BOUSSOFFARA, P. ELZER, K. ZINSER, A. TISSEN

A Generic Task Framework for Interface Analysis and Design in Process Control 389
J.-Y. FISET

New Visualisation Techniques for Industrial Process Control 395
R. VAN PAASSEN

A Proposal to Define and to Treat Alarms in a Supervision Room 401
B. RIERA, B. VILAIN, L. DEMEULENAERE, P. MILLOT

Neuromuscular Models I

Reinforcement Learning and Dynamic Programming
A.G. BARTO   407

The Modular Organization of Motor Control: What Frogs Can Teach Us About Adaptive Learning
F.A. MUSSA-IVALDI, E. BIZZI   413

Trajectory Learning and Control Models: From Human to Robotic Arms
T. FLASH   419

Learning to Optimize Performance: Lessons from a Neural Control System
CHI-SANG POON   425

Perception of Coherence of Visual and Vestibular Velocity During Rotational Motion
H.F.A.M. VAN DER STEEN, R.J.A.W. HOSMAN   431

Human Operator Adaptation to a New Visuo-Manual Relationship
O. GUEDON, J.-L. VERCHER, G.M. GAUTHIER   437

Human-Computer Interaction III

Graphic Communication and Human Errors in a Vibratory Environment
L. DESOMBRE, N. MALVACHE, J.F. VAUTIER   443

APIIS: A Method for Analysis and Prototyping Interaction Intense Software
L.V.L. FILGUEIRAS, S.S.S. MELNIKOFF   449

Integration of Cognitive Ergonomics Concepts in Knowledge Based System Development Methodologies
M. DURIBREUX, C. KOLSKI, B. HOURIEZ   455

Human-Computer Interface Evaluation in Industrial Complex Systems: A Review of Usable Techniques
M. GRISLIN, C. KOLSKI, J.-C. ANGUE   461

Usability Evaluation: An Empirical Validation of Different Measures to Quantify Interface Attributes
M. RAUTERBERG   467

Aviation IV: Tools for Systems Design

Supportability-Based Design Rationale
G.A. BOY   473

A Probabilistic Methodology for the Evaluation of Alerting System Performance
J.K. KUCHAR, R. JOHN HANSMAN, Jr.   481

Application of the Analytic Hierarchy Process for Making Subjective Comparisons Between Multiple
Automation/Display Options
L.C. YANG, R. JOHN HANSMAN, Jr.   487

Process Control II

Intensive Task Analysis and Evaluation for Interfaces Design in Large-Scale Systems
E. AVERBUKH   493

Mediation of Mental Models in Process Control Through a Hypermedia Man-Machine Interface
J. HEUER, S. ALI, M. HOLLENDER   499

Real Time Expert System in Process Control: Influence of Primary Design Choices 505
V. GROSJEAN

Highway Systems II

Modeling Car Driving and Road Traffic 511
P.H. WEWERINKE

An Estimation of the Hazard-Controllability of Driver-Support Systems 517
Y. SATO, E. KATO, K. MACHIDA

DAISY - A Driver Assisting System which Adapts to the Driver 523
J.P. FERARIC, R. ONKEN

Car-Following Measurements, Simulations, and a Proposed Procedure for Evaluating Safety 529
S. CHEN, T.B. SHERIDAN, H. KUSUNOKI, N. KOMODA

Aviation V: Air Traffic Control

Human-Machine Organization Study the Case of the Air Traffic Control 535
F. VANDERHAEGEN

Design and Evaluation of an ATC-Display in Modern Glass Cockpit 541
G. HUETTIG, A. HOTES, A. TAUTZ

Analysis and Modeling of Flight Crew Performance in Automated Air Traffic Management Systems 547
K.M. CORKER, G.M. PISANICH

Enhanced Visual Displays for Air Traffic Control Collision Prediction 553
M. JASEK, N. PIOCH, D. ZELTZER

Controller-Human Interface Design for the Final Approach Spacing Tool 559
M.C. PICARDI

Manufacturing Systems I

Integrating the Work Force into the Design of Production Systems 565
F. EMSPAK

An Adaptive Troubleshooting Model for Complex and Dynamic System 571
H. JESSE HUANG

An Original "Human-Oriented" Assessment Approach of Design Methods for Automated Manufacturing
Systems 575
J.-C. POPIEUL, J.-C. ANGUE

From Field-Based Studies to Models to Decision Aids - An Approach for Supporting Human Decision-
Making in Advanced Manufacturing Systems 581
S. NARAYANAN, T. GOVINDARAJ, L.F. McGINNIS, C.M. MITCHELL

Neuromuscular Models II

Modeling Human Performance of Intermittent Contact Tasks 587
J. WON, N. HOGAN

Analysis of the Human Operator Controlling a Teleoperated Microsurgical Robot 593
L.A. JONES, I.W. HUNTER

A Model of the Arm's Neuromuscular System for Manual Control 599
R. VAN PAASSEN

Applying Virtual Reality to Diagnosis and Therapy of Sensorimotor Disturbances 605
T. KUHLEN, K.-F. KRAISS, M. SCHMITT, C. DOHLE

Study of Human Operation of a Power Drill 611
D. RANCOURT, N. HOGAN

Intermittency of Unimpaired and Amputee Arm Movements 617
J. DOERINGER, N. HOGAN

Design of Supervisory Systems I

Ecological Interface Design: A Research Overview 623
K.J. VICENTE

Development of Analysis Support System for Man-Machine System Design Information 629
H. YOSHIKAWA, M. TAKAHASHI, K. SASAKI, T. ITOH, T. NAKAGAWA,
K. KIYOKAWA, A. HASEGAWA

A Design Method for Incorporating Human Judgement into Monitoring Systems 635
K. TANAKA, G.J. KLIR

TOOD: Task Object Oriented Description for Ergonomic Interfaces Specification 641
A. MAHFOUDHI, M. ABED, J.-C. ANGUE

Ship Control, Manual Control

Evaluation of Two Human Operator Models of the Navigator's Behaviour 647
R. PAPENHUIJZEN, T. DIJKHUIS

Observation of Maritime Emergency Management 655
T. CLEMMENSEN

Loop-Shaping Characteristics of a Human Operator in a Compensatory Manual Control System 661
T. INABA, Y. MATSUO

Manufacturing Systems II

How Do Industry Design Assembly Systems - A Case Study 667
M. BELLGRAN, E. LAHTI

Support in Setting Feed Rates and Cutting Speeds for CNC Machine Tools Through Override Logging:
Practical Test Results with a New CNC Component 673
S. STRIEPE

Teaching Motion/Force Skills to Robots by Human Demonstration 679
S. LIU, H. ASADA

Learning

Development of Machine-Maintenance Training System in Virtual Environment 687
T. TEZUKA, K.-I. KASHIWA, T. MITANI, H. YOSHIKAWA

Industrial and Experimental Applications of Transformation Theory and Ergodynamics 693
V.F. VENDA, I.V. VENDA

Transformation Dynamics in Human-Machine Systems 699
V.F. VENDA

A New Machine Learning Method Inspired by Human Learning 705
B. YANG, H. ASADA

Design of Supervisory Systems II

The Design of Perceptually Augmented Displays to Support Interaction with Dynamic Systems 711
A. KIRLIK

A Case-Based Design Browser to Facilitate Reuse of Software Artifacts 717
J.J. OCKERMAN, C.M. MITCHELL

A Design Methodology for Operator Displays of Highly Automated Supervisory Control Systems 723
D.A. THURMAN, C.M. MITCHELL

A Designer's Associate: Software Design Support for Command and Control Systems 729
J.G. MORRIS, C.M. MITCHELL

Author Index 735

CONCEPTUAL DESIGN OF MULTI-HUMAN MACHINE INTERFACES

Gunnar Johannsen

Laboratory for Human-Machine Systems (IMAT-MMS), University of Kassel (GhK),
D-34109 Kassel, Germany

Abstract: Human-machine interfaces for cooperative supervision and control by several human users either in control rooms or in group meetings are dealt with. The information flow between the different human users and their overlapping information needs are explained. The example of the cement plant illustrates this in more detail. Cognitive science concepts for supporting visual and mental coherence as well as multi-media, hypertext and CSCW (computer supported cooperative work) technologies are discussed with respect to their usage in multi-human machine interfaces. The design process for these interfaces is outlined with emphasising the different design stages, user participation and the possibility of knowledge based design support. Some ideas for the conceptual structure of multi-human machine interfaces are also presented.

Key Words: Human-machine interface; man-machine systems; human supervisory control; cooperation; information flows; group work; multimedia; knowledge tools; cognitive systems; process control.

1. INTRODUCTION

In the past, the field of human-machine systems was mainly concerned with the interaction between a single human and the dynamic technical system, the machine (Johannsen, 1993). Particularly, the subfield of human-machine interface designs was also investigated with such a preference (Johannsen, 1995b). Real work situations of many application domains require, however, that several humans of different human user classes frequently interact with the same technical system and with each other. Some examples of such technical systems are chemical plants, cement plants, power plants, transportation systems, and discrete manufacturing systems. The state of computerisation of these systems has often been developed quite far. This has sometimes lead to what Bainbridge (1983) calls the ironies of automation. The partial failure of introducing CI - (Computer Integrated) technologies, e.g., in the case of CIM (Computer Integrated Manufacturing) can be explained to a large extent by the lack of appropriate consideration of human factors. For example, CIM needs to be complemented by HIM (Human Integrated Manufacturing) in order to have chances of success (Johannsen, 1994b).

The application domain of the cement industry shows that several persons from different human user classes need to cooperate, partially via the human-machine interfaces in the plant control room. These human user classes are control room operators, field operators, maintenance personnel, operational engineers, instrumentation and control engineers, chemists, laboratory personnel, commissioning engineers, researchers, and managers. As a task analysis showed which was performed in a Danish cement plant, the style of cooperation between the different human user classes is very flexible and communicative (Heuer et al., 1993). Therefore, this application domain is chosen as a reference testbed for the new conceptual design of multi-human machine interfaces presented in this paper.

So far, research on intelligent and adaptive human-computer interfaces as well as on intelligent and adaptive human-machine interfaces has been performed and reported in the literature (Hancock and

1

Chignell, 1989; Tendjaoui et al., 1991; Averbukh and Johannsen, 1994; Johannsen, 1994a). The intelligent interfaces contain one or several knowledge-based modules, e.g., for enhancing the dialogue and the explanation facilities of the interfaces. User modelling approaches have been pursued, particularly also for the purpose of the adaptation of several interface modules to different user classes. The adaptation of the presentation, the dialogue, and the explanation modules can be supported by an appropriate user model. The user adaptation to individual human users has not well been realised for human-machine systems. The research work in several projects is more concentrated on the adaptation to different human user classes rather than to different individual human users. Also, the dynamic adaptation to task situations has further to be developed. The common prerequisite of all these research approaches is the assumption that only one human user at a time is interacting with the machine via the human-machine interface.

The new approach suggested in this paper deals with multi-human machine interfaces for integrated user classes with overlapping skills and knowledge. It is assumed that different people from different human user classes work together and can interact via the same display screen or via several screens. The new idea is here to have also overlapping information, with respect to the different user classes with their different overlapping information needs. This overlapping information has to be implemented in the presentation, the dialogue, and the explanation facilities for one display screen with several windows or even for one window.

In the following section, the actual information flow and the overlapping information needs are explained for the example of the cement plant and for other application domains. Section 3 outlines some concepts and technologies for multi-human machine interfaces. The design process for such interfaces as well as the need for user participation and knowledge support are described in section 4. The conceptual structure of multi-human machine interfaces is exemplified in section 5.

2. INFORMATION FLOW AND OVERLAPPING INFORMATION NEEDS

2.1 The example of the cement plant

The application domain of the cement industry shows that several persons from different human user classes need to cooperate, partially via the human-machine interfaces, in the plant control room or, also, in office rooms during group meetings. It is assumed that these people from different human user classes share their overlapping skills and knowledge, and interact, at least partially, via the human-ma-

chine interface. In order to prove this assumption, additional expert analyses with unstructured interviews, walk-throughs and talk-throughs were performed by the author in the cement plant, mainly in the control room, based on the task analysis performed by Heuer et al. (1993).

One of the main results of these expert analyses is a better understanding of the information flows between different human user classes. The main interactions between these people can schematically be represented by an information flow diagram, as shown in Fig. 1. This diagram was constructed and discussed with the control room engineer and the process engineer during the expert analyses. Such information flow diagrams are independent of any hierarchical organisational structure.

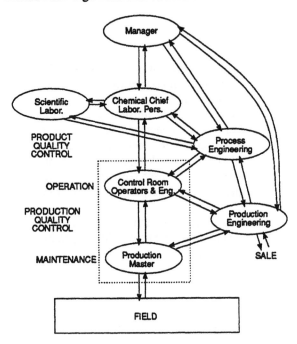

Fig. 1. Information flow diagram for some main interactions in a cement plant.

The actual information flows depend on specific tasks and problem solving needs. About 90% of the problems are small ones and can be solved online to the ongoing operation of the plant with control room personnel, production master and field operators. More complicated problems occur due to major equipment failures or with the experimentation and testing of new equipment or new products. Then, special group meetings between personnel from the control room, production engineering and process engineering are held, often only for an hour or so. The production master and, sometimes, also field operators may participate. The demand for new equipment and major problems with production quality control will be reported to the manager.

The main objective of the plant is to produce prescribed quantities of different types of cement, sometimes with frequent changes, by strictly con-

2

trolling product quality, such as fineness, strength and colour. The product quality control is guaranteed in short-term intervals of two hours by the laboratory personnel and on a long-term basis by the scientific laboratory. Problems in product quality require group meetings with people from these two laboratories together with personnel from the control room and process engineering.

All these meetings are organised in a flexible manner as soon as possible after the specific problem has occurred. Some of the meetings are held in the control room or they are held in one of the offices, e.g., in that of the process engineer. Across all human user classes, it is felt that face-to-face communication is mandatory. The group meetings are cross-organisational meetings and bring together those people from different human user classes who are specialised in the different facets of a particular problem. Thus, these people contribute their different views, skills and knowledge to the problem solution at hand. Before and during the group meetings, they have different information needs related to their particular specialisation. However, there exists also a larger area of overlapping information needs. The principle point of view of all participants of these group meetings is that they have to understand and to be interested in all information which is required by at least one member of the group.

2.2 Other application domains

In a similar way, the information flows and overlapping information needs can be identified for other application domains. In chemical plants and power plants, there is also the cooperation needed between such human user classes as operators, maintenance personnel, and engineers. The chemical plants are more similar to the cement plants with a higher degree of flexibility in the cooperation and communication between different people. In comparison, the power plant domain is characterised by less frequent changes in production.

In aircraft guidance, the crew of the pilot and the co-pilot has to cooperate permanently. Additionally, interactions occur, for example, with air traffic controllers. Generally, commercial transportation systems show the peculiarity of an additional human user class, namely the customer or passenger from the general public. This leads to special information needs which could be satisfied better in the future (Johannsen, 1995c).

3. CONCEPTS AND TECHNOLOGIES FOR MULTI-HUMAN MACHINE INTERFACES

3.1 Visual and mental coherence

Different concepts reported in the literature or currently investigated in industrial and university labo-

ratories are considered for the conceptual design of new multi-human machine interfaces. Generally, it is even not an easy task to guarantee the coherence between different windows on one screen. This problem will increase with the overlapping information for different human user classes. Concepts such as those about visual momentum, cognitive lay-outs, multilevel flow modelling, and ecological interfaces are included in the conceptual design. All these concepts are research results from cognitive science.

The concept about visual momentum was suggested by Woods (1984). During the cognitive process of perception, the attention of the human is sequentially directed towards different information sources for the purpose of finding informative areas. This information analysis is based on already existing information and knowledge as well as on their internal representation and that of possibly accessible information. Thereby, the activities of data acquisition are guided and controlled. If these processes are not appropriately supported by means of suitable display formats, the visual momentum may be low. As an example, it may be possible that the representations of a system, as presented on the display screen, do not allow the human to derive the underlying functional structure of the system. However, it depends on the available skills and knowledge of a particular human to what extent a deeper insight into relationships behind the screen can or cannot be gained. In a cooperative work situation, different human users working together may be able to derive different levels of information and knowledge from the same presentation.

This effect of different levels of insight for different human users may occur even more strongly with transitions between different pictures on one or several display screens. Generally, the visual momentum is more pronounced with parallel presentations than with serial presentations of information. Perceptual landmarks can be chosen as points of orientation for the perception and, thus, correspondingly increase the visual momentum. Also, functional overlappings between display pictures were suggested by Woods (1984) for the same purpose of increased visual momentum. These are overlappings of the information content and should not be confounded with the overlapping of windows on the screens. The overlappings of the information content have to be adapted to the different human user classes. As they may recognise different semantic relationships, it is necessary to design different perceptual landmarks and overlappings of information content for different people working together.

Another concept from cognitive science, namely the one of cognitive lay-outs, was proposed by Norman et al. (1986). The cognitive lay-out can be considered as a mental model which a human user possesses of the surface lay-out of a display screen. It is assumed that human users build up a cognitive representation or a cognitive lay-out for different information types and their relationships between the different windows and display screens. Examples for cognitive lay-outs are the linear order, the information integration, the selective attention, the processing layers, the memory storage, the zooming, and the perspective. A certain cognitive lay-out assumed by the human user strongly influences the understanding and the expectation with respect to the observable events presented by the human-machine interface. In order to achieve well designed presentations, the group formation in space and time as well as the animation are suggested. The group formation has to be accomplished in such a way that different human user classes with different overlapping knowledge always understand the coherence between a great number of pictures and windows in a complex display network.

Different human user classes working together with different overlapping skills and knowledge will probably acquire process information on several overlapping levels of abstraction. Anyone of them will manoeuvre through these levels of abstraction within goals-means hierarchies (Rasmussen, 1986). Traditional information presentations do not consider this with their topological physical structures. Additionally, the functional structures of technical systems need to be presented for the problem solving activities of the human users. Lind (1982, 1988) has suggested multilevel flow models for the presentation of higher abstraction levels.

Mass and energy balances are considered for industrial process control applications. Functional entities and contexts are described in the model by different types of functional processes. The two main ones are storage and transport processes; others are distribution, barrier etc. For all types of processes, graphical flow model symbols have been suggested as elements of a graphical representation language for functional relationships. Extensions to this graphical representation language include also symbols for sensors and controllers (Larsson, 1992).

The idea of ecological interface design was proposed by Vicente and Rasmussen (1992). The principal design objective is to present more transparent displays to the human users. It is possible to visualise the balance between the mass and the energy flows within the same display structure. The combination of both, the multilevel flow modelling structures and the ecological interface design are currently investigated for the example of the cement plant (van Paassen, 1995).

Many of the concepts described here are integrated into the graphical presentation system for the chemical process of a destillation column and its participative display design (Ali et al. 1993, 1994). Fig. 2 shows several windows within one display screen for this destillation column. The lower left window presents a dynamic goals-means hierarchy which can also qualitatively be enlarged for the upper window and interactively be operated. On the lower

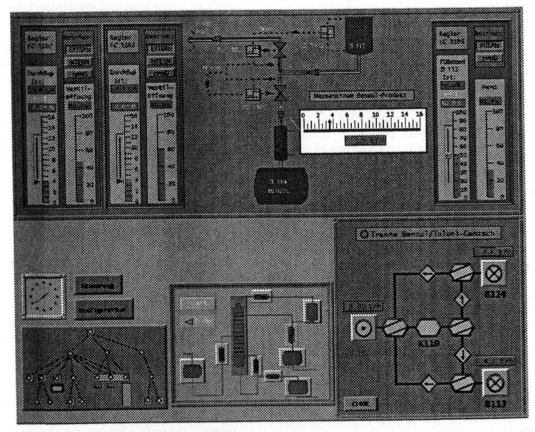

Fig. 2. Display screen for the supervision and control of a destillation column with goals-means, multi-level flow modelling, topological and functional presentations.

right side, the multilevel flow modelling representation of the highest hierarchical level is visualised. A corresponding topological component-oriented view is presented in the window between these two. The upper part of the display screen, shown in Fig. 2, presents more detailed information about controllers for the detailed topological view around the benzol tank B 114, visualised in the middle. This picture is selectable with pointing to the same tank in the lower right or lower middle overview pictures. The visual momentum between the different windows, with different graphical representations on different abstraction levels and different degrees of detail, is guaranteed by text labels and also dependent on the skills and knowledge levels of the human users. It is assumed that the cooperation between different human user classes (e.g., operators, engineers and managers) can be facilitated by the combination of several graphical representations on different abstraction levels.

3.2 Multi-media and hypertext

Relatively new communication technologies for human-machine interaction have been developed with the field of multi-media. The main idea of multi-media communication is to combine and integrate different visual and auditive media for the display and the visualisation of information about tasks to be performed with a machine or a computer. Particularly, the following media are combined with each other, namely computer-generated visual displays, video recordings, and auditive information such as recordings of noises and synthetic speech (Steinmetz and Herrtwich, 1991). Also, three-dimensional stereoscopic video scenes and stereoscopic computer graphics can be superimposed with each other (Milgram et al. 1990).

A connection of multi-media objects with an information network can interactively be used, for example, in travel agencies and libraries (Jerke et al. 1990). In this pure human-computer interaction, the querying technique is combined with the method of browsing in a hypertext environment. A hypertext system is a network of information nodes which are connected with links in a non-sequential way to arbitrary non-linear information structures (Bogaschewsky, 1992). A hypertext network has been combined with an automatic fault diagnosis module in order to support the problem solving activities during trouble-shooting (particularly, motor diagnosis) tasks of human technicians in car-repair shops (Hollender, 1995).

Generally, the human user can navigate through a hypertext network by analysing the connections between the nodes. Conceptual connections of alternative interaction procedures lead to a hypermedia information network. The separate nodes of this network are related to the separate types of information, such as text, graphic, video or audio, which are different forms of multi-media. Thus, hypermedia represent the information concept connecting several media (Begoray, 1990), whereas the multi-media encompass the combination of the different presentation media. The terms of hypermedia and of multimedia are, thus, distinguishable from each other in a similar way as the dialogue and the presentation layers in a user interface management system.

Multi-media communication systems will be introduced in the future in several application domains. Libraries and travel agencies have been mentioned already above. Also, the entertainment industry is highly interested in this new technology. Particular applications will be possible in medicine for cooperative conferencing between physicians (Kleinholz and Ohly, 1994) and for surgery. Industrial applications may be possible in the glass production, the paper and pulp industries, the chemical industries (Heuer et al., 1994), the power industry (Zinser, 1993), and the maintenance of networks such as those for distribution of electricity. A tutoring system for the latter case was suggested by Tanaka et al. (1988).

Alty and Bergan (1992) emphasise that many questions still exist with respect to the application of multi-media communication in dynamic industrial process control tasks. The general problem of the interface design remains: which information is needed by the human user, when, in which form and why. This problem becomes more severe with a larger number of technological options which increase further with the multi-media domain. Therefore, the information needs of the later human users have to be investigated by means of task analyses. This requirement becomes even more important in cases of the cooperation of several human users in cooperative work situations. The expert analyses as a kind of unstructured task analysis performed in a cement plant, as mentioned above, indicated some aspects of the cooperative work situations in that particular application domain. The face-to-face communication is absolutely mandatory. Also, the audio channel, e.g., telephone communication, is very important. Multi-media technologies can be used for integrating the video information from some of the equipment, which is now available in the control room with separate video screens. Otherwise, video observations are rejected as a spy system in this application domain.

3.3 Computer supported cooperative work

The research in Computer Supported Cooperative Work (CSCW, 1994) and groupware deals with theories, design concepts, architectures, prototype systems, and empirical results for cooperative work situations in such application domains as offices, classrooms and factories. Electronic mail systems, computer and video conference systems, office information systems, organisational knowledge bases, shared window systems and other communication and cooperation systems are investigated. Some of these systems do not allow the human users to interact and cooperate with each other directly in space and time. New approaches for the integration of action space and time have been suggested, e.g., by Ishii et al. (1992). Such support of cooperative work seems particularly to be required in industrial human-machine systems.

As the results from the expert analyses in the cement plant indicate, cooperative work should be organised as much as possible with face-to-face communication. Large projection screens are not welcome because they are very soon too much overloaded and not adaptable enough. However, they are already implemented in some other application domains, but the concept of overlapping information for different human user classes has not been taken into account with the information content of these projection screens. Thus, their use for cooperative work needs to be rethought. Display screens for group meetings in different offices and the control room are welcome as multi-human machine interfaces in the cement plant. They will also be accepted as dedicated human-machine interfaces in a network and for discussions of smaller problems at the phone.

The display screens for the group meetings may consist of one screen with four to five windows. They allow access to all pictures in the control room, rather than having just printouts, as presently available. Different most favourite pictures may be selected by different user group representatives. All selected pictures need to be seriously considered by all meeting members because cooperation rather than ego-centred views are required where each user group representative contributes. Modifications of control room pictures are foreseeable for the display screens in group meetings. Quick-change and easy-to-use editing facilities may allow to select important lines or variables from a table, qualitative zooming-in and selection of subareas of component flow diagrams, and manoeuvring or selection by sliders or text menus through different levels of abstraction. The latter range from physical form, such as scanned-in photos, e.g., from data bases or just taken of broken components (inside of a pump etc.), to goal hierarchies via multilevel flow modelling representations.

The consistency and coherence across selected and edited pictures has to be guaranteed. This will support the visual momentum which is now already available when a trend curve is selected by the cursor on a particular variable in the component flow diagram. The consistency will be increased when the computer completes the other selected pictures shown in parallel, e.g., consistent with the information reduction in the just edited picture. The information filtering, reduction and qualitative modifications may be supported by the computer or can solely be done by the group members. Further, computer-supported drawing facilities, e.g., for straight lines or for rapid prototyping (sketching) of new pictures and ideas or for modifying existing ones, are possible. However, they may be better suitable for exploratory purposes rather than for normal group meetings because these might become too long with too much computer interaction.

The overlapping information of the different user classes is already considered in the logbook, now available on a PC in the cement plant. This information has further to be implemented in the presentation, the dialogue and the explanation facilities of the human-machine interfaces, particularly of the display screen for group meetings. Thereby, the visual momentum between different windows which relate primarily to different user group representatives has to be supported.

All the suggested designs of human-machine interfaces for cooperative work have to consider also face-to-face communication. Otherwise, the social contacts will not be improved if this face-to-face contact would disappear. Tele-cooperation is often not feasible because the contact to the production will be lost, e.g., the feeling for clinker quality will disappear. Also, the work climate will deteriorate and, thus, there will eventually be no cooperation.

4. DESIGN PROCESS FOR MULTI-HUMAN MACHINE INTERFACES

4.1 Design stages and user participation

Generally, it seems to be most appropriate to combine strict systems engineering life cycle procedures with final-user participation and rapid prototyping for the design of multi-human machine interfaces (Johannsen, 1995b). Any user-oriented design of interactive software products, such as human-machine interfaces (HMI), should start with scenario definitions and task analyses in order to have a solid basis for user requirements and systems specifications, particularly also for the HMI functions specification. User requirements as specified in international standards for software development, should consider the goals-means-tasks relationship and be based on a task-oriented perspective.

The design stages with different forms of user participation are outlined in Fig. 3. This indicates that the functional specification for multi-human ma-

6

chine interfaces depends on the goals and goal structures prescribed for the human-machine system, on all kinds of technological and intellectual means for accomplishing these goals, and on the tasks to be performed by the human users with the purpose of achieving the goals by appropriate usage of the available means. The overall goals to be specified for the human-machine system are mainly (1) productivity goals, (2) safety goals, (3) quality goals, (4) humanisation goals and (5) environmental compatibility goals.

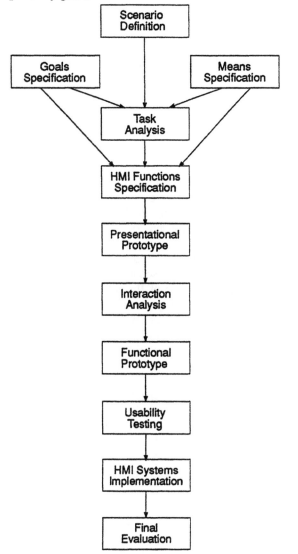

Fig.3. Design stages for human-machine interfaces with different forms of user participation.

The means for achieving the goals can be separated into pure technological means and human-related means. Most issues concerned with the design of the technical process itself but also with the degree of automation are technological means. Knowledge about the application domain as well as about human user needs and strategies can be regarded as human-related means. This may be described by structural, functional, causal, and cognitive relationships. Other human-related means are available through different

views on the application domain. These views consider the levels of abstraction, the levels of aggregation or detail, parallel versus serial presentation, navigational possibilities, and degrees of coherence. All these means are implicitly inherent in the application domain but often need to be transformed for explicit usage.

The tasks to be performed by the human users depend on certain scenarios, i. e., situations and contexts of the human-machine system which are specified together with final expert users during the scenario definition; see Fig. 3. Different experience and knowledge of the human users as well as different normal and abnormal situations and states of the dynamic technical system lead to different subjectively perceived tasks. Further, the prescribed goals and the available means for their achievements strongly determine the types of tasks to be performed in a certain scenario. These tasks may require cooperation in work situations with multi-human machine interfaces. In the special case of cooperative work, it is particularly necessary to investigate the work organisation and structure. All opportunistic and informal communication channels have also to be discovered. The real information flows need to be clearly understood, more than the formal organisation (which may or may not be so important). As has been pointed out in section 2.1, special expert analyses can be organised, either in the real application field or in a simulator environment. The results are information flow diagrams as described above. In addition, contactograms which quantify the frequency of interactions between different human user classes or single users can also be derived (Pikaar et al. 1986).

A thorough task analysis, combined with knowledge elicitation techniques, is a strong foundation for the definition of the user requirements and for the functional specifications of the multi-human machine interfaces (Johannsen and Alty, 1991; Kirwan and Ainsworth, 1992; Heuer et al. 1993). The functions of any human-machine interface should be defined in such a way that the prescribed goals can be achieved, the available means can be transformed and used appropriately, and the tasks can correctly be perceived and effectively be performed.

During later design stages, it is necessary to organise further expert meetings with the participation of different human user classes in order to evaluate intermediate prototype designs of the human-machine interfaces. The aspects of cooperative work between the different user classes have a high priority in these evaluations. The same is true for the final evaluation at the end of the human-machine interface design. In Fig. 3, this sequence of intermediate prototype designs and final HMI systems implementation, alternating with their corresponding evaluation stages, is outlined. The presentational prototype is more a surface lay-out followed by an interaction analysis with appropriate user participation in order to gather more inputs for the functional prototype. Based on the usability testing of the latter, the final design and implementation of the multi-human machine interface is accomplished followed by its final evaluation.

The analysis and evaluation techniques for all these design stages with user participation comprise scenario definition, organisational analysis, task and knowledge analyses (all with unstructured and structured interviews, walk-throughs and talk-throughs in real or simulated situations etc), less formal expert analyses, and experimental sessions with usability testing procedures.

4.2 Knowledge based design support

The designer of multi-human machine interfaces can be supported by knowledge modules, as already explained by Johannsen (1995a) for single user situations. Several knowledge modules need to be developed for supporting the designer. The example of Fig. 4 presents six knowledge modules, namely (1) the goals knowledge module, (2) the application knowledge module, (3) the users knowledge module, (4) the tasks knowledge module, (5) the human-machine interface (HMI) knowledge module, and (6) the design procedural knowledge module. These modules should not be viewed only as separate support subsystems, but also as interrelated knowledge pools.

The goals knowledge module contains information about operational design criteria with respect to efficiency and performance, safety, reliability, producibility, costs, development time, etc. Several goals-means relationships from the research litera

ture as well as from industrial experience can be included in this module.

The application knowledge module comprises knowledge about the technical process and its supervision and control (S&C) system, as well as about the knowledge-based decision support systems (DSS) of the technical system. All this structural and functional knowledge can possibly be loaded from plant and control systems data bases as well as from the knowledge bases of the decision support systems. Thus, the technological knowledge of the application domain is made available to the interface designer in a consistent form.

In a corresponding way, the users knowledge module supplies the designer with knowledge about the information needs, the cognitive strategies, and the information processing behaviour of human users. Such users knowledge is more difficult to collect than the application knowledge. However, the currently pursued combination of cognitive task analyses with knowledge-based user modelling indicates a way in which real human-centred designs of human-machine interfaces are possible in the future (Sundström and Johannsen, 1989; Sundström, 1991; Averbukh and Johannsen, 1995).

The tasks knowledge module considers the work organisation, the task scenarios and the perceived tasks. The knowledge of this module interlinks the knowledge of the two preceeding ones. Typical task scenarios may be described as time lines and causalities of events. Further, a task-related categorisation of human errors will support the designer in her or his decisions.

The human-machine interface knowledge module (HMI issues) works on the basis of the other modules which have just been described. It is the most active part of the knowlege-based designer support and contains several editors for the presentation, the dialogue and the hypertext levels, as well as related

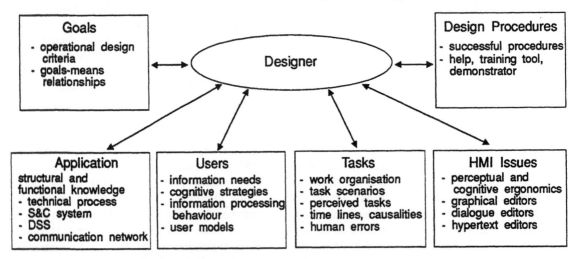

Fig.4. Knowledge modules for supporting the designer of multi-human machine interfaces.

knowledge about perceptual and cognitive ergonomics. The levels of presentation and dialogue are linked with each other through object hierarchies. The editors for the presentation level are mainly graphical editors (Fejes et al., 1993). They allow one to design different picture pyramids for different abstraction levels, different window lay-outs, different information forms (e. g., flow diagrams, functional diagrams, textual messages, analogue and digital instruments) as well as all kinds of graphical symbols and other basic elements. The editors for the dialogue level are dialogue specification tools for designing information flow and state transition networks, as well as knowlege-based dialogue assistants. The latter may be oriented towards subsystems or subtasks. Additional editors may be available for creating hypertext structures as well as other data and knowledge bases.

The design procedural knowledge module comprises knowledge about successful procedures used by experienced designers. Information about possible design failures and about how to avoid or correct them may also be included, because designers can learn not only from successful experience but from a failure, too. This module can serve as an aid, help or reminder to the design experts, as well as a training tool or a demonstrator for design novices.

5. CONCEPTUAL STRUCTURE OF MULTI-HUMAN MACHINE INTERFACES

Multi-human machine interfaces for dynamic technical systems are needed for online and off-line purposes. Fig. 5 shows that multi-human machine interfaces in control rooms allow different human user classes to interact in an online way with the technical system. The off-line usage of multi-human machine interfaces is mainly required in office

meetings, as mentioned in section 3.3. Any kind of information needed in these office meetings with different human user classes may be selected from the different libraries of the technical system or from other data and knowledge bases of the plant, such as design data bases. The information from the libraries of the technical system may contain detailed interaction data from previous situations, particularly also critical ones such as those with human errors or technical equipment failures. Also, production and product quality data can be provided. In addition, actual online data from the technical system may be observed via computer networks in the office meetings, and snapshots can be frozen or printed when requested. Thus, a much higher flexibility of interactions can be supported with new technologies and, particularly, appropriate user-oriented conceptual structures of the multi-human machine interfaces.

A possible architectural concept of such multi-human machine interfaces is outlined in Fig. 6. The flexible and creative interaction of several human users with the different functionalities of the multi-human machine interface and the technical system are better supported with distributed architectures, with respect to flexible knowledge processing and creative usage of knowledge. In the case of online usage, the real-time constraints are particularly critical. High-speed computing allows distributed architectures also here when the dominant time constants of the technical system are not too small, i.e., the technical system is not too fastly responding. A data-driven approach can then be applied, particularly, when shared knowledge bases are provided in a blackboard architecture, as suggested in Fig. 6; see also Fabiano et al. (1994).

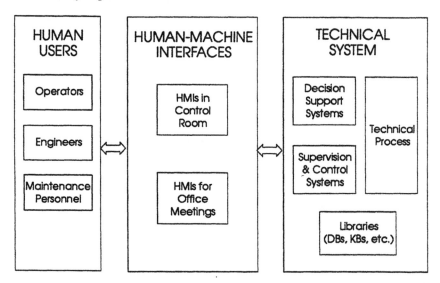

Fig.5. Human-machine interfaces in control rooms and for office meetings between different human user classes and the subsystems of the technical system.

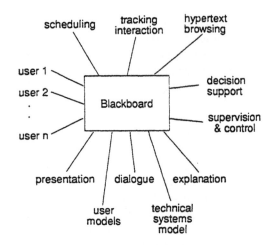

Fig. 6. Blackboard architectural concept of multi-human machine interface.

The architectural concept of Fig. 6 illustrates the interaction of *n* human users with the decision support and the supervision & control systems of the technical system via the blackboard. This interaction is accomplished and controlled by means of several functionalities of the multi-human machine interface. These functionalities have also permanent access to the blackboard. They can be differentiated in information content related functionalities and control functionalities. The information content related functionalities are presentation, dialogue, explanation, user models and technical systems model. The control functionalities include scheduling, tracking interaction and hypertext browsing. Many of these functionalities have to be adapted to the special requirements of multi-human machine interfaces.

6. CONCLUSIONS

The design of multi-human machine interfaces for dynamic technical systems is a newly developing subarea of the field of human-machine systems. The reality of work situations in the interaction with dynamic systems is considered more appropriately with this approach. Many disciplines contribute to this new approach, such as single user human-machine interfaces, human-computer interaction, cognitive sciences, work and organisational psychology, multi-media and hypertext technologies, computer supported cooperative work, and a few more. It is particularly necessary to pursue the further development of multi-human machine interfaces in a task and user oriented manner. As task and expert analyses indicate, new high technologies do not guarantee high user acceptance across all cooperating user groups. Their particular information needs, task objectives and overall systems goals have to be strongly considered for successful cooperative work, with users satisfaction, face-to-face communications

and good social work climate as important prerequisites for safe, economic and efficient results of human-machine systems production.

ACKNOWLEDGEMENTS

The author highly appreciates the stimulation of this work through the support by the Aalborg Portland Cement Factory in Aalborg, Denmark, and by two research projects which are currently carried out in the Laboratory for Human-Machine Systems at the University of Kassel. These projects are supported by the Deutsche Forschungsgemeinschaft (German research foundation), Bonn, and by the Commission of the European Union, Brussels under the BRITE/EURAM Programme in Project 6126 (AMICA: Advanced Man-Machine Interfaces for Process Control Applications). The consortium partners of the AMICA-Project are: CISE (Italy), ENEL (Italy), FLS Automation A/S (Denmark), University of Kassel (Germany), and Marconi Simulation and Training (Scotland). The author is very thankful to all industrial partners as well as to several colleagues in his research group.

REFERENCES

Ali, S., J. Heuer, M. Hollender and G. Johannsen (1993). Participative design of human-machine interfaces for process control systems. In: *Adjunct Proc. INTERCHI'93*, Amsterdam, pp. 53-54.

Ali, S., J. Heuer, M. Hollender and G. Johannsen (1994). Partizipative Bediengestaltung von Oberflächen für Prozeßleitsysteme. In: *Proc. Leitwarten-Kolloquium*, Köln.

Alty, J.L. and M. Bergan (1992). The design of multimedia interfaces for process control. In: H.G. Stassen (Ed.), *Analysis, Design and Evaluation of Man-Machine Systemes* (Preprints, The Hague). Pergamon Press, Oxford.

Averbukh, E.A. and G. Johannsen (1994). Intelligent human-machine communication and control for industrial processes. In: *Proc. First Asian Control Conference ASCC*, Tokyo Vol. 1, pp. 49-52.

Averbukh, E.A. and G. Johannsen (1995). Human performance models in control (Special Issue on Human Interaction with Complex Systems). *IEEE Trans. Systems, Man, Cybernetics*, Vol. 25 (under review).

Bainbridge, L. (1983). Ironies of automation. (Special Issue on Control Frontiers in Knowledge Based and Man-Machine Systems). *Automatica*, Vol. 19, pp. 775-779.

Begoray, J.A. (1990). An introduction to hypermedia issues, systems and application areas. In: *Int. J. Man-Machine Studies*, Vol. **33**. pp. 121-147.

Bogaschewsky, R. (1992). Hypertext-/Hypermedia-Systeme: Ein Überblick. *Informatik-Spektrum*, Vol. **15**, pp. 127-143.

CSCW (1994). *ACM Conference on Computer Supported Cooperative Work*, Proceedings Chapel Hill, NC.

Fabiano A.S., C. Lanza and J. Kwaan (1994). *Toolkit Architectural Design*. Topical Report, D4-2, Issue 1, BRITE/EURAM AMICA-Project.

Fejes, L., G. Johannsen and G. Strätz (1993). A graphical editor and process visualisation system for man-machine interfaces of dynamic systems. *The Visual Computer*, Vol. **10** (1), pp. 1 - 18.

Hancock, P.A. and M.H. Chignell (Eds.) (1989). *Intelligent Interfaces: Theory, Research and Design*. North-Holland, Amsterdam.

Heuer, J., S. Ali, M. Hollender, J. Rauh (1994). Vermittlung von Mentalen Modellen in einer chemischen Destillationskolonne mit Hilfe einer Hypermedia-Bedienoberfläche. Hypermedia in der Aus- und Weiterbildung. *Dresdner Symposium zum computerunterstützten Lernen*, Dresden.

Heuer, J., S. Borndorff-Eccarius and E.A. Averbukh (1993). *Task Analysis for Application B*. Internal Report IR 1-04, BRITE/EURAM AMICA-Project. Lab. Human-Machine Systems, University Kassel.

Hollender, M. (1995). *Elektronische Handbücher zur Unterstützung der wissensbasierten Fehlerdiagnose*. Doctoral Dissertation, University of Kassel.

Ishii, H., M. Kobayashi and J. Grudin (1992). Integration of inter-personal space and shared work space: Clearboard design and experiments. In: *Proc. CSCW'92-Sharing Perspectives*. ACM Press, Toronto, pp. 33-42.

Jerke, K.-H., P. Szabo, A. Lesch, H. Rößler, T. Schwab and J. Herczeg (1990). Combining hypermedia browsing with formal queries. In: D. Diaper, D. Gilmore, G. Cockton, B. Shackel (Eds.), *Human-Computer Interaction - INTERACT'90*. North-Holland, Amsterdam, pp. 593-598..

Johannsen, G. and J. L. Alty (1991). Knowledge engineering for industrial expert systems. *Automatica*, Vol. **27**, pp. 97-114.

Johannsen, G. (1993). *Mensch-Maschine-Systeme (Human-Machine Systems, in German)*. Springer, Berlin.

Johannsen, G. (1994a). Design of intelligent human-machine interfaces. In: *Proc. 3rd IEEE Int. Workshop Robot and Human Communication*, Nagoya, pp. 18-25.

Johannsen, G. (1994b). Integrated systems engineering: The challenging cross-discipline. In: G. Johannsen (Ed.), *Integrated Systems Engineering* (Preprints IFAC Conference Baden-Baden). Pergamon Press, Oxford, pp. 1-10.

Johannsen, G. (1995a). Knowledge-based design of human-machine interfaces. *Control Engineering Practice*, Vol. **3**, pp. 267-273.

Johannsen, G. (1995b). Computer-supported human-machine interfaces. *Journal of the Japanese Society of Instrument and Control Engineers SICE*, Vol. **34**, No. 3, pp. 213-220.

Johannsen, G. (1995c). Human-machine interfaces for cooperative work. In: *Proc. 6th Int. Conf. on Human-Computer Interaction*, Yokohama.

Kirwan, B. and L. K. Ainsworth (Eds.) (1992). *A Guide to Task Analysis*. Taylor and Francis, London.

Kleinholz, L. and M. Ohly (1994). Supporting cooperative medicine: The Bermed project. *IEEE MultiMedia*, Winter, pp. 44-53.

Larsson, J.E. (1992). *Knowledge Based Methods for Control Systems*. PhD thesis. Department of Automatic Control, Lund Institute of Technology.

Lind, M. (1982). Multilevel flow modelling of process plants for diagnosis and control. *Int. Meeting Thermal Nuclear Power Safety*, Chigaco, IL.

Lind, M. (1988). System concepts and the design of man-machine interfaces for supervisory control. In: L.P. Goodstein, H.B. Andersen, S.E. Olsen (Eds.), *Tasks, Errors and Mental Models*. Taylor and Francis, London, pp. 269-277.

Milgram, P., D. Drascic and J. Grodski (1990). A virtual stereographic pointer for a real three dimensional video world. In: D. Diaper, D. Gilmore, G. Cockton, B. Shackel (Eds.), *Human-Computer Interaction-INTERACT'90*. North-Holland, Amsterdam, pp. 695-700.

Norman, K.L., L.J. Weldon and B. Shneiderman (1986). Cognitive layouts of windows and multiple screens for user interfaces. *Int. J. Man-Machine Studies*, Vol. **25**, pp. 229-248.

van Paassen, R. (1995). New visualisation techniques for industrial process control. This IFAC Symposium on Analysis, Design and Evaluation of Man-Machine Systems.

Pikaar, R.N.,T.N.J. Lenior and J.E. Rijnsdorp (1986). Control room design from situation analysis to final lay-out: Operator contributions and the role of ergonomists. In: G. Mancini, G. Johannsen and L. Mårtensson (Eds.), *Analysis, Design and Evaluation of Man-Machine Systems* (IFAC Proceedings), Pergamon Press, Oxford, pp. 299-303.

Rasmussen, J. (1986). *Information Processing and Human-Machine Interaction*. North-Holland, New York.

Steinmetz, R. and R.G. Herrtwich (1991). Integrierte verteilte Multimedia-Systeme. *Informatik-Spektrum*, Vol. **14**, pp. 249-260.

Sundström, G.A. and G. Johannsen (1989). Functional information search: A framework for knowledge elicitation and representation for graphical support systems. In: *Proc. 2nd European Meeting on Cognitive Science Approaches to Process Control*, Siena, pp. 129-140.

Sundström, G.A. (1991). Process tracing of decision making: An approach for analysis of human-machine interactions in dynamic environments. *Int. J. Man-Machine Studies*, Vol. 35, pp. 843-858.

Tanaka, H., S. Muto, J. Yoshizawa, S. Nishida, T. Ueda, T. Sakaguchi (1988). ADVISOR: A learning environment for maintenance with pedagogical interfaces to enhance student's understanding. In: H.-J. Bullinger, E.N. Protonatorios, D. Bouwhuis, F. Reim (Eds.), *INFORMATION Technology for Organisational Systems (EURINFO'88)*. North-Holland, Amsterdam, pp. 886-891.

Tendjaoui, M., C. Kolski and P. Millot (1991). An approach towards the design of intelligent man-machine interfaces used in process control. *Int. J. Industrial Ergonomics*, Vol. 8, pp. 345-361.

Vicente, K.J. and J. Rasmussen (1992). Ecological interface design: Theoretical foundations. *IEEE Trans. Systems, Man, Cybernetics*. Vol. 22, pp. 589-606.

Woods, D.D. (1984). Visual momentum: A concept to improve the cognitive coupling of person and computer. *Int. J. Man-Machine Studies*, Vol. 21, pp. 229-244.

Zinser, K. (1993). Integrated multi-media and visualization techniques for process S&C. In: *Proc. IEEE International Conference on Systems, Man and Cybernetics*. Le Touquet, pp. 367-372.

TELEMANIPULATION AND TELEPRESENCE

Henk G. Stassen' and Gerda J.F. Smets*

'*Faculty of Mechanical Engineering and Marine Technology*
Mekelweg 2, 2628 CD Delft, The Netherlands
* Faculty of Industrial Design Engineering
Jaffalaan 9, 2628 BX Delft, The Netherlands
Delft University of Technology

Abstract: Telemanipulation is studied by many disciplines; however, independently. This plenary paper intends to contribute to the integration of these disciplines by bridging the gap between the control engineers in the man-machine systems field, the computer scientists in the field of human-computer-interaction, and the biomedical researchers in the field of severely bodily disabled persons.
Three major problems in telemanipulation are recognised, i.e. the lack of tactile and touch information, the lack of information for depth perception since visual displays only generate 2D data, and the existing time-delay in the human operator - telemanipulator - control loop.

As for control theory the importance of the Internal Model for obtaining optimal filtering and control is highlighted and for perception theory the choice for the so-called ecological approach is explained.
Control theory as well as perception theory point to the importance of an increase in feedback modalities, for instance by pointing to the importance of proprioceptive feedback for 3D-perception and by designing two-handed controls paying attention to the asymmetry in the effectivities of both hands.
It will be argued that combining proprioceptive feedback and 3D-perception is a need for good telemanipulation and for telepresence. Also a short note on compensation methods by predictive displays will be given. Finally the traditional approach will be compared with the ecological approach.

Keywords: Human Computer Interaction, Human Machine Interface, Manual Control, Robotic Manipulators, Telemanipulation.

1.INTRODUCTION

In, and just after World War Two, the research on and development of industrial manipulators and robots started slowly. Primarily the attention was focussed on the use of robots in the production of goods. Those robots can highly repetitively perform predictive tasks in a controlled environment. They work with high speed accurately and execute boring and sometimes very human unfriendly tasks. The main motive for the development of this type of robotics is efficiency, economics, reliability and availability, and product quality. Note that in general these robots are programmed; there is no human individual in the control loop.
In the early seventies the interest in another application grew; teleoperation and telemanipulation came in the picture, because in particular situations in which the tasks to be performed were too dangerous or even impossible to execute directly by

human operators. Areas to be mentioned are:
- Space and deep undersea applications
- Nuclear power plants and radio active hot cells, toxic waste clean up
- Military and police operations; life saving
- Telediagnosis and inspection tasks in hazardous environments
- Medicin: Assistive devices and prostheses for the disabled; endoscopy used for inspection of the gastrointestinal tract; and minimal invasive surgery.

A very important difference between the industrial robot and the telemanipulator is that the human operator is always in the control loop, whether the task is manual or supervisory control. The human operator makes the decisions and defines the control strategies for the manipulation of the teleoperator at distance, mostly he has to execute many different and varying tasks. In particular, the design of telemanipulation is the topic of this plenary paper.

In order to avoid misconceptions, it will be sensible to define the concepts *Telemanipulation* and *Telepresence*. We will follow the definitions as formulated by Sheridan [1992].

Telemanipulation is defined as the extension of human sensing and manipulation capabilities by coupling it to remote artificial sensors and actuators.

Telemanipulation, sometimes used as synonym for teleoperation, can be used in the generic sense to control by a human using joystick or a master-slave positioning device.

Telepresence means that the operator receives sufficient information about the teleoperator and the task environment, displayed in a sufficiently natural way that the operator feels physically present at the remote side.

In telemanipulation tasks several serious problems are inherently present; they are:

- The lack of tactile and touch information and consequently the mismatch with the proprioceptive feedback. Special techniques are proposed to overcome this problem [Sheridan, 1992; van Paassen, 1994]; in fact as early as 1969 Simpson already mentioned the concept of Extended Proprioception in the design of pneumatically powered arm prostheses for the Softenon babies [Simpson, 1969].

- The lack of information for depth perception, since visual displays generate 2D data. Many informational concepts are generating possibilities to reconstruct 3D pictures from these 2D data [Sheridan, 1992; Breedveld, 1995; Overbeeke et al, 1995].

- Due to the fact that at distance manipulations or operations have to be performed, there is always a time delay in the human operator-telemanipulator control loop, which may yield severe problems with reference to the stability of the control loop. Attempts to introduce predictive displays are promising [Sheridan, 1992], and with the present state of the control theory it can be expected that most of the problems in this area can be solved.

This plenary paper aims to show how the design of telemanipulators can be optimised by multidisciplinary research where the study of perception and action is integrated with measurement and control theory. In particular the choice of a suitable perception theory is of major importance, and it is not an accidental occurence that the perception-action loop is considered as a control loop in itself. Experimental psychological research about depth perception, fitting in with control theoretic concepts, leads to insights in experiences with teleoperation. It illustrates where the present available telepresence systems fail, and where assistive devices are added that are not helpful.

2. THE KNOWLEDGE OF A HUMAN OPERATOR

Human control of telemanipulators can only be achieved if a human operator possesses knowledge of the system to be controlled, of the task to be executed and of the statistics of the disturbances acting on the telemanipulator. This knowledge is called the *Internal Representation* of the human operator [Stassen, et al, 1990]. By analogy with the internal representation it has been proven that in optimal filter theory [Conant, et al, 1970; Kalman, et al, 1961] and in optimal control theory [Francis, et al, 1975] this concept is called the *Internal Model*. These authors showed that filtering and control only can be achieved optimally if, and only if, the internal model describing the three types of knowledge just-mentioned, is available. In this way we may state that a human performance model should be based in the internal model, representing the internal representation of the human operator [Stassen, 1989]. Hence the presentation of information by the display serves at least two purposes.

Firstly, the naive human operator has to learn to understand and to interpret the dynamics of the teleoperator, its environment and the possibly resulting disturbances. Furthermore, the operator should be able to interpret the tasks to be executed. Via displays, visual, auditory and/or tactile, he is able to get that information as well as he is able to see the concequences of actions to be taken. So the goal to be achieved refers primarily to the learning and training period. In fact, we may say that the operator at this stage is building up a correct Internal Representation. Here it is important to mention that the learning process is highly encouraged when the information is displayed by more than one modality. Experiences with the Softenon children, children who do not possess upperextremities due to the use of a sleeping drug used in an early stage of the pregnancy of the mother, have indicated that those children had severe problems in 3D-perception [Verkuyl, 1964]. It was considered that the fact that no tactile information was available, was the main source of the problem; visual information alone turned out to be insufficient. This insight had changed the rehabilitation concept drastically; now-a-days children without upperextremities are fitted with a very simple prosthesis at a very early age, and the problem in 3D perception could be avoided. Hence the integration of the different modalities is of a vital importance in understanding the process of building up the Internal Representation.

Secondly, the well trained human operator needs to be informed about the actual state of the teleoperator under control; that means that on the basis of the internal representation the human operator has to interpret the displayed information in order to keep the internal representation up to date. In particular this is important when system dynamics may vary and when disturbances may occur. For the human operator the entire system, that is the system to be controlled including both the Human Machine Interfaces has to be known (Fig.1) during manual and supervisory control. As mentioned before, the major problem is the lack of 3D-perception and tactile, proprioceptive information, or even the combination of visual and proprioceptive feedback [Verkuyl, 1964; Akamatsu, et al, 1994, Stassen, 1989]. It is therefore that we want to focus at a suitable theory of perceptual motor skills in order to understand how a human operator is able to control complex technological systems.

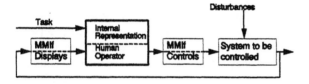

Figure 1: Block diagram of the control loop human operator, Human Machine Interface and System.

3. CHOOSING A SUITABLE THEORY OF PERCEPTION

The primary role of a perceptual theory in telemanipulation is in augmenting its functionality [Bruce, et al, 1990; Gordon, 1989]. A suitable theory can enhance the functionality of a telemanipulator by specifying what perceptual/motor information is necessary to fulfil a certain task, and what information adds little or no user capability. Most perceptual theories, however, highlight either the physiological or cognitive roots of perceiving. For engineering problems this is the wrong level of description. Here an experiental perceptual theory is needed, that specifies what visual input conveys a specific perceptual meaning. This is the main concern of the so-called *ecological theory of perception* propounded by Gibson [1966, 1979]. The term *ecological* refers to the importance of the relationship between the observer's body, its environment and the task. An example. One's eyeheight corresponds to the visual horizon. The latter always cuts objects on the ground at the same height, eyeheight, regardless of their distance; this is probably used to perceive the size of objects, such as chairs, doors, etc. [Sedgwick, 1973; Haber

et al, 1992]. There is no need for distance estimates. The inclusion of a horizon in a display is, when size estimates of objects are required, a simple and effective means to improve task performance.

The fit between the observer's body, its environment and the task holds for a moving observer as well. Gibson accentuates the fact that, in daily life perceivers are acting, and that the observer gains important information out of this perception/action coupling. For a moving observer the retinal projection undergoes a continuous serial transformation that is unique to both the movement pattern and the environment. Depth differences are specified by movement parallax. Parallax refers to the fact that when two objects are at a different distance from a moving observation point, the viewing angles shift at different rates. One speaks of *movement parallax* when the movement of the observation point is self-generated, and of *motion parallax* when this motion is imposed. Movement parallax offers potential information for the perception of absolute distances, since the parallax shifts on the retina and the observer's movement are tightly and unambiguously interlocked. Grafting movement parallax patterns on a flat display therefore realises depth television; it allows for valid depth estimates. This will be explained in the next section. Motion parallax, the shift that occurs in ordinary TV broadcasting, does not contain absolute depth information.

Furthermore, Gibson considers perceiving as performed by a *perceptual system*, accentuating how different sense modalities are interdependent [Simpson, 1969; Akamatsu, et al, 1994].

Last but not least, the theory points to the importance of irregularities in the perception/action loop. Learning a skill, i.e. achieving an optimal perception/action coupling, is impossible without, at times, occuring disturbances [Stassen, et al, 1995]. This is important for telemanipulation based on human hand movements. The shape of a gesture is not fixed, determining which gesture is meant requires advanced, spatial, pattern recognition techniques. Some advances have been made by modelling the gesture patterns with non-linear differential equations, as done by Altman [Voss, 1995], where each gesture is represented by the attractors of the set equations.

The ecological theory of perception is relatively recent. It was introduced in the computer community by Norman [1988, 1992] who implemented it in order to optimize the human computer interaction and by Gaver [1993] implementing it in multimedia research at Xerox EuroPARC. Yet, it seems that, as Flach [1995] points out, till now, hardware engineers have not fully detected its implementation potential. However, the Gibsonian experimental study of the perception action loop fits in with control theory. It specifies the perceptual relevance of the three aspects of the internal representation [Stassen, et al, 1995]. The term representation has led to furious debates in psychology and philosophy. Gibson states that the perceptual-motor system is direct. It needs no internal representation. This still is a controversial assertion, implying that perceptual-motor knowledge can be processed and stored as such and needs not be translated in a cognitive, symbolic format. There is no contradiction, however, with Stassen's [et al, 1990] internal representation, since he uses this concept in a purely operational sense, as the knowledge of the system to be controlled, of the task to be executed and of the statistics of the disturbance acting on the telemanipulator, without implying how this knowledge is processed or stored. Next some implications of the study of the perception-action loop for television and telemanipulation will be discussed.

4. TELEVISION

As Sheridan [1992] points out: *The whole world has been exposed to television, and the technology is inexpensive and highly developed. In resolution, contrast, frame rate and colour quality it is quite adequate for most teleoperation tasks. However, there remain some problems which, while not critical in ordinary TV, are crucial for teleoperation.* Depth perception continues to be a major reason why performance of direct manipulation is not matched by that of telemanipulation [Sheridan, 1992]. Let us see whether Gibson's theory of perception can be of help here.

Depth until now: Apart from learning or memory, pictorial cues and binocular disparity are the most popular answers for explaining depth perception, with motion parallax being a good third. The movement of the observer is thought to be irrelevant for visual perception. Experiments about visual perception, even those about parallax effects, are often designed so that the subject cannot move, the head being fixed by a chin rest, for example. As a consequence, depth in teleoperation generally is reproduced by high resolution stereoscopic images of the tele-environment. Two images are obtained from horizontally separated viewpoints and presented to the corresponding left and right eyes. As far as motion is concerned, it is restricted to presenting parallax shifts on the stereo display as a consequence of performances in the tele-environment. Movement of the human operator is important to control the mobility of the effectors of the workstation, but it is not considered to be important for visual perception as such. The human operator

provides largely symbolic commands, concatenations of typed symbols or specialised key presses, to the computer. Some fraction of these commands is analogical, body control movements are isomorphic to the space-time-force continuum of the physical task, since *they are difficult for the operator to put into symbols* [Sheridan, 1989]; yet this analogy is not considered to be important to enhance teleperformance nor telepresence.

A stereo display typically employs two broad band colour television channels. The transmission requirements are clearly high for such systems. The danger of overloading the communication system between workstation and operator is reduced by, e.g. transmitting a HI-res B&W image along one television channel and a LO-res colour image along another and combining the images at the operator's station in a colour stereo image.

Depth in the ecological approach: The ecological approach does not question the validity of reproducing depth using binocularity and high resolution. However, it offers alternatives that, depending on the task, offer excellent performance for less money. Movement parallax offers potential information for the perception of absolute distances, and suggests that depth television can be created by continuously updating the display output to match the observer's movements in front of the screen. This is accomplished by sensing the observer's head position and moving a camera in the remote site, accordingly. The screen then does not act as a screen anymore, but as a window, where what lies behind it forms a rigid whole with what is before it. Experiments with the Delft Virtual Window System [Smets, et al, 1987, 1988, 1990; Overbeeke, et al, 1987; Overbeeke, et al, 1988] indicate a perceptual advantage of an active observer, whose head movements steer the camera, and a passive onlooker, whose movements are not coupled to the display output (Fig. 2). Both active and passive observers receive identical output on their monitor screens, yet their perceptual input differs. The former receives movement parallax information, the latter motion parallax. The test used involved aligning wedges on local and remote objects. Telemanipulations of the active operator almost match direct manipulation performance: errors are consistently small and their variance does not differ significantly from the variance of errors in real life condition. The teleperformance of the passive onlooker, on the contrary, is not consistent, although the average error size remains small. The variance of error size for passive observers is significantly greater than that of active observers and than that obtained in the real life setup. Results indicate that the active condition, where movement parallax information was provided on the display, even allows for things apparently projecting in front of the screen. In this case, however, perceived depth is somewhat less compelling, than in real life set-up. There is a systematic underestimation in aligning a wedge virtually projecting from the screen with a real wedge mounted in front of the display, yet the variance remains relatively small. Movement parallax allows for telemanipulation, motion parallax does not. Although several head-slaved teleoperator systems are mentioned in the literature [e.g. Schwarz, 1986; Merritt, 1987], none of them reflects on the fact that movement parallax shifts in themselves are sufficient for depth perception.

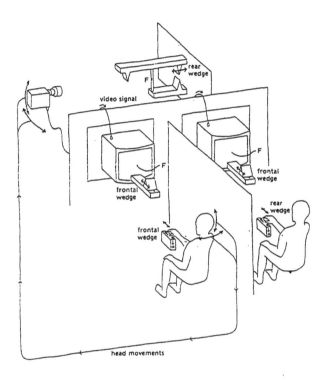

Figure 2: Experimental situation: Comparison between active and passive observer.

Conclusion: The choice for stereoscopy or movement parallax for television depends on the task. Stereoscopy is good in static situations, where the observer cannot change viewpoints. Stereoscopic images can be looked at by several observers at once, movement parallax intrinsically is individually bound. Movement parallax is good for tasks where the observer has to detect things by changing viewpoints, such as a small box that is hidden behind a larger one. Furthermore, movement parallax requires only one camera. Therefore, it is ad-

vantageous in situations where minimal invasive tools are required, such as in medical or industrial endoscopy. Stereoscopic object recognition always requires high pixel resolution. Object recognition using a head-slaved camera does require far lower pixel resolution images than using an ordinary static or moving camera. Therefore bandwidth requirements in television can be limited by the trade-off among pixel resolution and movement, frames per viewing angle, [Smets, et al, 1993, 1995]. Some of these trade-offs were already studied by Ranadive [1979] in the context of master-slave manipulation using a video-display, but without a head-slaved camera.

Movement parallel effects hold for other sense modalities than the visual as well. Everyone has experienced that stereophonic sound enhances the experienced width of a sound recording. Coupling head movements to delay times of left and right headphones differentially, extends the spatial impression further: Outside the head [Overbeeke, 1989].

5. TELEMANIPULATION

Human operator behavior can be classified according to the three-level model of Rasmussen [1983]. Rasmussen distinguished the following behaviors (i) *Skill-Based Behavior, SBB*, reacting on *signals, (ii) Rule-Based Behavior, RBB*, acting on *patterns and signs*, and (iii) *Knowledge-Based Behavior, KBB* reacting on *symptons and symbolic information*. In particular, the cognitive capabilities and limitations are contributing to human performance at the KBB level, however, also the two lower levels, RBB and SBB, are of vital importance. Therefore any well-designed Human-Machine Interface, HUMIF, has to support the human operator at all three levels [Johannsen, et al, 1994]. A relevant difference between RBB and SBB at one side, and KBB at the other, is the fact that the operator at the RB- and SB-level has a correct internal representation of system, task and disturbances at his disposal, whereas at the KBB-level he might be uncertain about the actual internal representation. In the latter case, a strong plea has to be made on the intelligence and creativity of the human operator.

Since one manipulates the physical world most often and most naturally with the hands, there is a desire to apply the skills and naturalness of the hand directly to teleoperation. However, in general, telemanipulation relies heavily on the human cognitive capabilities. This can be attributed to the three major problems as stated in the introduction: (i) The lack of tactile and touch information or the shortcomings in handedness, (ii) the lack of depth perception and limited field of view, and finally (iii) the mostly unknown but existing, large time delay between control action and per/received response. Table 1 classifies these problems in terms of SBB, RBB and KBB. Hence, although cognitive skills are very important also rule-based and skill-based skills can and will contribute to better telemanipulation. In fact one has to support the operator in teleoperations at all three levels of behavior.

behavior / problem	skill based	rule based	knowledge based
handedness	*		*
3 D- perception		*	*
time delay		*	*

Table 1: Classification of the three major problems in telemanipulation in terms of SBB, RBB and KBB.

It is often forgotten that solutions in *Computer-Human Interaction*, CHI, might also serve this purpose, since recently CHI is described more and more in manual terms than in cognitive terms. This is illustrated by Walker [Rheingold, 1991]. He made the way one interacts with the computer, the basis for a taxonomy of computer evolution. In the forties one controlled computers by plugs, ten years later by punched cards, another decade later by a keyboard, and now most people use computer menus. The next step ought to be *direct manipulation*. Usually, as in Walker's case, direct manipulation still is a *point and click* manipulation, using a mouse controlled cursor on a screen, or a finger on a touch-sensitive screen. Gibson's theory of perception and action points to more radical solutions, since it starts from the fit between the observer's body, its environment and the task. This means that perceptual and action problems are attacked on the perceptual-motor level or the SBB, and not on a cognitive level, as is done in many other theories of perception [Wallace et al, 1993].

What are the solutions to tactile and touch information, depth information and time delay offered by CHI? Firstly, each problem will be stated, then the perception-theoretical aspects will be discussed, and finally, the CHI solution given and the implementation to telemanipulation will be discussed.

5.1 Handedness

Problem: Some 40 years ago, in man-machine system research most of the attention was focussed on *Manual Control, MC*. Then, around the mid-eighties the interest shifted from MC to *Supervisory Control, SC*. This shift yielded a significant change in the tasks to be performed by the operator and in the responsibility and authority between the operator and the management [Sheridan, 1985; 1980; Stassen, et al, 1995] due to the increasing complexity of supervised plants [Wieringa, et al, 1995]. It included also a certain lost of interest in MC problems; new areas as Neural Networks, Fuzzy Set Theory and all kinds of operator decision support-systems were asking all the attention. However it is recognized that in telepresence and telemanipulation manual control problems remain essential.

With the development of modern technology for CHI, especially of computerized interfaces, human cognitive skills have been greatly supported, but here also manual skills have been neglected at first. All the subtilety of precision tools had been replaced by a single type of button, distinguished only by its label: The cognitive skills also take over the work of the manual skills. These systems can greatly help the manually impaired, [Stassen, et al 1989], but people with manual skills, have to work with one hand tied on their back, sometimes almost literally. Especially spatial tasks which are easily performed in real life, such as picking up a jar and unscrewing the top, require mind-boggling decompositions into elementary motions, each under 1-dimensional control through a button, lever, or knob.

Handedness in the ecological approach: Most research into *Gestures* is aimed at communicative gestures, like pointing at things or recognizing symbols from sign language. A *gesture* is defined as a motion of the body that contains information. Beckoning with your index finger is a gesture, but handwriting is not a gesture because the motion of the hand expresses nothing; it is only the resultant words that convey the information. [Kurtenbach, et al, 1990]. Communicative gestures are very interesting for the use in telecommu-nication, but less for telemanipulation, because they are aimed at conveying a symbolic meaning to an anlooker, and not for performing an act on an object. Examples of work on manipulative gestures can be found in pure research [Guiard, 1987], in Human Computer Interface applications [Buxton, 1990] and in the field of rehabilitation [Balkom, et al, 1994] Guiard and Baxton find that much of the power of manual skills lies in the cooperation between two hands playing complementary roles. Guiard [1987] defines a taxonomy of human hand movements. He distinguishes three classes: (i) Asymmetric unimanual activities like teeth brushing, (ii) asymmetric bimanual activities like playing the violin, (iii a) symmetric bimanual activities where the two hands are in phase as in weightlifting, and (iii, b) symmetric bimanual activities where the two hands are out of phase as in rope climbing. Asymmetric bimanual tasks are the most common in daily life. However, there is little or no research on bimanual movements and none at all on asymmetric bimanual movements. Few researchers seem to realise that the latter require *two qualitatively differentiated contributions* of the hands, with one exception, i.e. those researchers who are involved in the research on the rehabilitation of severely disabled persons. In the designs of arm prostheses this fact was already very well-known; it is recognized that the functionality of an arm prosthesis for uni-lateral amputees is totally different for those with a bilateral amputation [Stassen, 1989]. Also the tests and training in the Activities of Dayly Living, ADL, by the occupational therapist is based on the different functions of both the upperextremities [Soede, 1980]. Based on the experimental research Guiard concluded that three high-order principles appear to determine the hand's asymmetry. Note that the conclusions are stated for a right dominant hand, whereas for left-handers it is just the opposite:

(i) The right hand typically finds its spatial references in the results of the motion of the left hand. The movements of the left hand provide steady states for the movements of the right. When writing, for example, the left hand positions the paper, thus allowing the right hand to do it's job.

(ii)The right hand is capable of producing finer movements than the left, i.e., they operate on a different temporal scale, micrometric for the right and macrometric for the left. The right hand is more manipulative, the left one more postural. [Stassen, et al, 1972].

(iii) The contribution of the left hand to global bimanual performance starts earlier than that of the right. Human manipulation prefers going from macro- to micrometric.

This taxonomy and the research based on it has consequences for the design of actuators; these will be discussed in the section application to teleoperation.

Developments in HCI: An interface that uses hand gestures must have a way to register these. To do this, the system must measure movements of the hand, and relate these to meanings carried by these movements. First one will consider how to get the

movements, then how to come from these movements to gestures, and finally the influence of the feedback will be discussed. Much of this work has been executed in the context of Virtual Reality research, VR-research. Here it should be noted that it is interesting to see that the manufacturers of arcade video games seem to recognize something that the majority of main-stream computer systems ignore: That users are capable of manipulating more than one device at a time in the course of achieving a particular goal [Buxton, 1990]. Buxton gives several implementations for ordinary CHI, of which the most promising one is the Toolglass [Bier, et al, 1994], combining the use of a trackball for the non-dominant hand with the cursor steering through a mouse by the dominant hand.

The first landmark was the development of datagloves. The first one with a firm market place was the VPL-dataglove [1987]. Users wearing this glove in a VR-application see a graphic hand that follows the motions of their own hand in the simulated environment. By pantomiming reaches and grabs, the user causes the graphic hand to reach and grasp objects in the simulated environment [Sturman, et al, 1994] . Some of these datagloves include touch feedback, by small balloons mounted into the gloves fingers, that inflate when an object is touched upon. A survey of glove-based input devices is given by Sturman, et al, [1994] who also mention glove interfaces in teleoperation and robotic control for a facile, dexterous control of the remote hand. Gloves have the advantage that they let one apply the manual dexterity to the tool. However, they are cumbersome too, because of the fact that they do not fit most hands and that they are often rather stiff. Therefore alternatives were developed, and one can see an evolution from recording the positive form of the hand gloves, through recording its *negative form,* the surface where the hand touches an object, to non-intrusive recording techniques, using optical recognition methods, and fixing no devices to the hands themselves [Voss, 1995]. Examples of the second category are a thick joystick that fills the hand and has a pressure sensor for each finger, e.g. the Virtuality Flexor, and a soft cube, somewhat like a child's toy, which contains sensors that register deformations that are applied to it [Murakami, et al, 1994]. An example of optical recognition techniques is found in the work of Gershenfeld who developed a method to separate the user's hands from a video image, and recognize its form using pattern recognition methods [Voss, 1995].

Hand movements enter the interface as sequences of numerical values, but what really do these values mean? The problem of classifying movement patterns into a particular meaning has a long and troublesome history, but hope is found in new techniques such as AI, fuzzy sets and neural networks. Key problem is the variability of gestures, both within and between users, and also the changes of gestures during the learning of a skill. Ideally, one is looking for adaptive user interfaces which fit theirselves to their users, adapt to their individual differences, and learn as they co-evolve with their users. Some advances have been made in tuning neural networks to recognize patterns of sign language made by different people, as shown in Fels and Hinton's speaking glove. [Kalawsky, 1993, p. 304]. Although results until now are limited, e.g., dealing with fixed, pre-determined categories in a supervised learning paradigm, trends in research on adaptive systems are promising.

Application to Teleoperation: Hand movements and gesture recognition are obvious candidates for inclusion in teleoperators. Hands serve specific skills, and Guiard's analysis has implications on how to develop the manual interface. One wants to use two hands, but not in identical ways, because their effectivities are not interchangeable. The asymmetry in the effectivities of the two hands implies that one may want two types of hand input tools: One precise and detailed for the dominant hand, and one more crude, and for support for the non-dominant hand. Similarly, the affordances are essential: It should be directly visible how things can be grasped, held, turned, moved, bent, etc.
Guiard's analysis certainly strikes a different noise in the field of HCI. However, it is interesting to see that for many years this asymmetry in the effectivities of both the hands was very well recognised in the area of the rehabilitation of severely bodily disabled persons with defects of the upper-extremeties. In fact the different functions of the dominant and non-dominant hand are basis of the training programs in ADL by the occupational therapist [Soede, 1980]. It was also the basic design philosophy for arm prosthesis for uni-lateral amputers [Stassen, 1989]. In addition, an extensive field evaluation of the use of arm prostheses in The Netherlands proved the importance of the recognition of the different function of dominant and non-dominant hand [v.Lunteren, et al, 1983].
Not all gestures are equally important for telemanipulation. Communicative gestures, like sign language and pointing, may be reduced to movement patterns, but in manipulative gestures, like picking something up, one touches and applies force to the thing one handles, and senses the reaction forces. For these tasks, force feedback is important because it strengthens the perception-action loop. In some tasks it can be replaced by other modalities, for example, a sound can tell you whether the

object is gripped or free, but gradual and direction-sensitive spatial manipulations have to rely on the strengths of the proprioceptive perceptual force feedback system. The development of force feedback devices follows an evolution similar to that of hand measurement devices to a minimal device approach, only a few strings pulling at fingertips [Ishii, et al, 1994].

5.2. Field of View, Depth information

Problem: People have a wide field of view, of more than 180°, which head-mounted screens have difficulty to emulate. Most VR systems have a restricted field of view, typically 60°, resulting in the feeling of looking at a scene instead of participating in a world, even when it allows for interaction.

Field of View in the ecological approach: A restricted field of view hampers telemanipulation, because it disturbs:
- The overview of the environment and the observer's position in it: Human position in space is largely determined by the perception of continuous texture gradients in the environment. Every visible surface is textured, i.e., it reflects the light in a non-uniform way because of its coarseness. Texture gradients depend of the point of view of the observer because close to him the texture is coarser than further away. A restricted field of view, as in head-mounted does not allow for continuous texture gradients.
- The link between our body and its environment: Manipulating is a lot easier when you can see your hand(s). Seeing other parts of the body as well, enhances orientation skills and the feeling of telepresence. Parts of the body, for example part of the nose, are always in the field of view when one looks around. The body also throws a shadow. And although one is not aware of it, recent research [Smets, et al, 1988] suggests that this helps us to know where one is. The absence of a shadow hampers the feeling of presence.
- Peripheral vision: Human vision consists of central and peripheral vision. Movements in the peripheral field makes the head orientates in the direction of the movement. The central field is more sensitive for detecting details, acuity, and the peripheral field is more sensitive to movement or potential danger. It also serves orietation and the perception of self-motion [Warren, et al, 1990].

Implementation in CHI and telemanipulation: An evident solution is to make the screens larger, but a cheaper solution is to use those screens with HI-res in the centre and LO-res, HI-freq in the periphery. The optimal size of the field of view depends on the task. When the task demands orientation, it should be large; when the task centers on acuity and fine manipulation a limited field of view might be an advantage, as it heightens the operator's concentration. If, however, the task is hampered by the limited field of view, the information, salient for the task, should be made more obvious [Breedveld, 1995]. This can be done by introducing or accentuating depth information like perspective, horizon, texture and gradients. Other sense modalities might be exploited as well to help orientation, for example: Auditive feedback when moving. The actuator hand should be visible for the operator. The field of view can be increased by means of head slaved virtual window systems allowing the operator to scan the scene. Another solution is providing the operator with multiple views of the same scene, e.g., one global and one detailed to emulate normal vision with two screens. For telecommunication the usefulness of this solution was tested by Gaver [Gaver, et al, 1993].

5.3. *Time delay*

Problem: As stated in the introduction, telemanipulation is needed in situations where at great distances tasks have to be performed, such as operation in the hostile environment of a nuclear plant or in deep sea and space environments. This yields that in the human operator-telemanipulator-control loop a time delay will always occur. For instance in the control of space manipulators [Breedveld, 1995] the time delay may vary from three to seven seconds, depending on the distances to be covered and the information density of the transmitting station. This time delay causes that the human operator is always too late in the control and perception of the telemanipulator, with the consequences that the operator will not be able to build-up an accurate internal representation, hence no telepresence can be achieved.

Ecological approach: In order to realise a situation in which telepresence can be achieved one has to find ways to compensate for the varying time delays. Then, and only then, one may expect that the human operator *feels* what is going on at far distances, i.e. telepresence. It is also the only way for the human operator to update, in time, his internal representation of the telemanipulator and its environment. Not too many practical compensation methods are available, but a well known method is the application of predictive displays [Veldhuijzen, et al, 1977; Sheridan, 1992]. With such a display it is possible to predict the state variables of the

manipulator in advance. As a consequence, via a simulation with a predictive display, telepresence in an actual environment can be achieved, be it at a certain time delayed.

5.4. *Implementation*: The basic idea of the introduction of predictive displays in order to achieve telepresence is a correct one. However, such an implementation is not easy, since it yields at least two essential problems.

To start with, one should be able to describe the dynamics of the telemanipulator, even for the case that the telemanipulator is a very flexable robot as for instance in space applications [Breedveld, 1995]. In terms of control theory it means that the system is highly non-linear and time-varying. Moreover, there is the fact that one should be able to model the environment in which the telemanipulator has to work. Modelling a non-linear system or an approximation of it is difficult, but with the nowadays available computer capacity it is certainly possible. Modelling the environment is a much more crucial topic, since often this environment will change in time, and sensors to detect such a change will not always be available. Hence, simulation by a predictive display of a known environment is very well possible, but for an unknown environment it is almost impossible. Only with advanced pattern recognition techniques solutions may become available in future.

A second problem is the time-scaling of the predictive display. When the human operator is controlling a telemanipulator by means of a predictive display, one wants directly a feedback on the results of an action in order to achieve telepresence, so one wants to compensate for time delays. Such a display requests a high computational performance since the system to be controlled is complex and non-linear. In addition one wants to have a 3D-presentation of the tasks to be performed by the telemanipulator. All together, the performance of the computer system prescribes the possibilities, and is to a certain extent the limitation for simulation.

At this moment a Human-Machine Interface including a predictive display, as given in Fig. 3 can be realised. The predictive display can be about 100 to 500 time faster than the real world. So in controlling the system via the predictive display one has a direct feedback of the control action. Hence, telepresence is simulated. From the control of the predictive display the human operator can make a decision on the actual control of the telemanipulator by transmitting the required control actions. In this way one has achieved a compensation of the time delays in the system. Good results have already been achieved in the control of very large crude carriers [Veldhuijzen, et al, 1977], be it with a well-known environment. In cases where telemanipulation in unknown environments is required, however, not much research is executed. This area is a very challenging research topic.

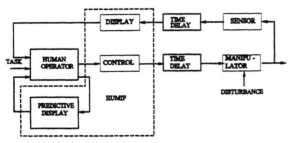

Figure 3: The application of a predictive display in the control of a telemanipulator.

6. CONCLUDING REMARKS

In drawing concluding remarks, it is important to realise that the authors have tried to bring three different disciplines together. However, the enormous amount of literature in these fields made that only a fraction of all published papers are covered. Nevertheless some interesting conclusions and remarks can be made.

- It is amazing to recognise that researchers in the three fields of man-machine systems, human-computer interaction and the rehabilitation of severely bodily disabled persons are not aware of each others research activities; often almost independently they work on the same problems. A possible reason for this might be the fact that each field has developed its own vocabulary, methods and approaches, and has its own journals.
- Most of the literature reports on one of the three major problems on telemanipulation, i.e. proprioceptive feedback, 3D-perception and time delays. Very little research is focussed on the consequences of an integration of methods to solve the total problem. This can lead to specifying the trade-off between the amount of sensory feedback, e.g. resolution, the control of it, e.g. by a moving observer, and the ability to modify our environment, e.g. by an input device allowing for direct manipulation. Such a trade-off is suggested already in Sheridan's [1992] telepresence model. In telemanipulation the concept of telepresence is essential, and all three problems are of equal importance.
- The ecological approach offers a new approach to solve problems and may lead to a better insight in human behavior in the control of complex technological systems. In particular, new 3D-concepts may originate from this approach.
- The idea of a two-handed control based on the

different functions of the dominant and non dominant hand are promising, and should be further exployed. Much can be learned from the analyses of the Activities of Daily Living in rehabilitation.

- Telepresence certainly will be increased by using multi-modality sensory feedback, a concept that is used in human-computer interaction and rehabilitation. It would be nice and advisable, if the man-machine systems society could contribute with their research to this interesting topic.

7. REFERENCES

Akamatsu, M. and S. Sato (1994). Multimodal Mouse: A Mouse type Device with Tactile and Force Display Presence. *Teleoperators and Virtual Environments*, **Vol. 3**, No. 1, pp. 73-80.

Balkom, H.v. and M. Welle-Donker (1994). *Kiezen voor co mmunicatie* (Choise for communication). Intro publ. Nijkerk, NL, 350 pp. ISBN 90-55-740047.

Bier, E.A., M.C. Stone, K. Fishkin, W. Buxton and T. Baudel (1994). A taxonomy of See-Through Tools. In: *Proc. Conference of InterCHI '94*, pp. 358-364.

Breedveld, P. (1995). The Development of a Man-Machine Interface for Telemanipulator Positioning Tasks. In: *Proc. 6th IFAC/IFIP/IFORS/IEA Symposium*, Cambridge, MIT, 7 p.

Bruce, V. and P. Green (1990). *Visual perception: Physiology Psychology and Ecology*. Erlbaum, Hillsdale NJ., 2nd ed.

Buxton, W.A.S. (1990). There's more to interaction than meets the eye: Some issues in manual input. In: *User centered system design: New perspectives on Human-Computer Interaction*. Norman, D.A. and S.W. Draper (Eds). Lawrence Erlbaum Ass., Hillsdale, NJ, pp. 319-337.

Conant, R and W.R. Ashby (1970). Every good regulator of a system must be a model of that system. *Int. J. Syst. Sci*, **Vol.1**, pp. 89-97.

Flach, J.M. (1995). The ecology of human machine interfaces: A personal history. In: *The ecology of human machine systems*. Flach J.M., P.A. Hancock, J. Caird and K.J. Vincente (Eds). Lawrence Erlbaum Ass., Hillsdale NJ. in press.

Francis, B.A. and W.M. Wonham (1975). The internal model principle of linear control theory. In: *Proc. of 6th IFAC World Congress*, Boston, MIT MA, paper 43.5.

Gaver, W.W., A. Sellen, C. Heath and P. Luff (1993). One is not enough: Multiple views in a media space. In: *Proc. Conference InterCHI93*, Amsterdam, pp. 335-341.

Gibson, J.J. (1966). *The senses considered as perceptual sys tems*. Houghton Mifflin, Boston MA.

Gibson, J.J. (1979). *The ecological approach to visual percep tion*. Houghton Mifflin, Boston MA.

Gordon, I.A. (1989). *Theories of Visual Perception*. John Wiley & sons, Chichester.

Guiard, Y. (1987). Asymmetric division of labor in human skilled bimanual action: The kinematic chain as a model. *Journal of Motor Behavior*, **Vol. 19**, No.4, pp. 486-517.

Haber, R.N. and C.A. Levin (1992). The perception of object size is independent of the perception of object distance. In: *Proc. 33rd Annual Meeting of the Psychonomic Society*, St. Louis, MO, pp

Ishii, M. and M. Sato (1994). A 3D spatial interface device using tensed strings. *Presence*, **Vol. 3**, No.1, pp. 81-86.

Johanssen, G., A.H. Levis and H.G. Stassen (1994). Theoretical problems in Man-Machine Systems and their experimental validation. *Automatica*, **Vol. 30**, No. 2, pp. 217-231.

Kalman, R.E. and R.S. Bucy (1961). New results in linear filtering and prediction theory. Trans. on ASME, *J. of Basic Eng.* **Vol. 23**, pp. 95-107.

Kalawsky R.S. (1993). *The science of virtual reality and virtual environments*. Wokingham, Addison-Wesley.

Kurtenbach, G. and E.A. Hulteen (1990). Gestures in human-computer communication. In: *The art of human computer design*. Laurel, B. (ed.) Addison Wesley, London, pp. 309-317.

Lunteren, A.v., G.H.M.v.Lunteren-Gerritsen, H.G. Stassen and H.J. Zuithoff (1983). A field evaluation of arm prostheses for unilateral amputees. *Prosthetics and Orthotics*, **Vol. 7**, No.3, pp. 141-151.

Merritt , J.O. (1987). Visual-motor realism in 3-D Teleoperator display systems. In: *Proc. of SPIE*, Vol. 761, True Imaging Techniques and Display Technologies, pp

Murakami, T. and N. Nakajima (1994). Direct and intuitive input device for 3-D shape deformation. In: *Proc. Conference of InterCHI '94*, pp. 465-470.

Norman, D.A. (1988). *The Psychology of everyday things*. New York, Basic Books.

Norman, D.A. (1992). *Turn signals are the facial expressions of automobiles*. Reading, M.A., Addison-Wesley.

Overbeeke, C.J. (1989). Special hearing: Coupling perception and action. In: *Proc. ICEPA5*. Miami Univ., Oxford, OH, USA.

Overbeeke, C.J. and P.J. Stappers (1995). The importance of movement for the design of telepresence systems. In: *Proc. 6th IFAC/IFIP/IFORS/IEA Symposium*, Cambridge, MIT., 7 p.

Overbeeke, C.J., G.J.F. Smets and M.H. Stratmann (1987). Depth on a flat screen II. *Perceptual and Motor Skills*, **Vol. 65**, No. 120, pp.

Overbeeke, C.J. and M.H. Stratmann, (1988). Space through movement. Delft, DUT, 207 p. PhD thesis.

Paassen, R. van (1994). Biophysics in Aircraft Control. Delft, DUT, 277 p. PhD-thesis.

Ranadive, V. (1979). Video resolution, framerate, and gray scale tradeoffs under limited bandwidth for undersea teleoperation. Boston, MIT, MA., p. MS-thesis.

Rasmussen, J. (1983). Skills, Rules and Knowledge: Signal, signs and symbols, and other distinctions in human performance models. *IEEE Trans. on SMC*, SMC-133, pp. 257-267.

Rheingold, H. (1991). *Virtual Reality*. London, Secker & Warburg.

Schwartz, A. (1986). Head tracking stereoscopic display.In: *Proc. of Society for Information Display*. Vol. 27, No 2, pp. 133-137.

Sedgwick, H.A. (1973). The visible horizon: A potential source of visual information for the perception of size and distance. Cornell University. University of microfilms no. 73-22530, p. PhD thesis.

Sheridan, T.B. (1989). Telerobotics. *Automatica*, **Vol. 25**, No. 4, pp. 487-507.

Sheridan, T.B. (1992) *Telerobotics, Automation and Human Supervisory Control*. MIT Press, Cambridge, M.A., 393 p.

Sheridan, T.B. (1985). 45 Years of Man-Machine Systems: History and Trends. In: *Proc. 2nd IFAC/IFIP/IFORS/IEA Conf. on Analysis, Design and Evaluation of Man-Machine Systems*. Varese, It. Invited papers, pp. 5-13.

Sheridan, T.B. (1980). Computer control and human alienation. *Technology Review*, MIT Press, Cambridge MA, pp. 60-76.

Simpson, D.C. (1969). An externally powered prosthesis for the complete arm. *Bio-Medical Engineering*, **Vol.4**, No. 3, pp. 106-119.

Smets, G.J.F. and C.J. Overbeeke (1993). Trading-off spatial versus temporal resolution: About the importance of actively controlled movement for visual perception. In: *Visual Search 2*, D. Brogan, A. Gale and K. Carr (Eds.). pp. 389-400, London, Taylor and Francis.

Smets, G.J.F. and C.J. Overbeeke (1995). Visual resolution and spatial performance; the trade-off between resolution and interactivity. *Proc. VRAIS '95, IEEE Virtual Reality Annual International Symp.*, Irvine, North Caroline.

Smets, G.J.F., C.J. Overbeeke and H.H. Stratmann (1987). Depth on a flat screen. *Perceptual and Motor Skills*, **Vol. 64**, pp. 1023-1034.

Smets, G.J.F., H.H. Stratmann and C.J. Overbeeke (1988). Method of causing an observer to get a three-dimensional impression from a two dimensional representation US. Patent Nr. 4, 757, 380.

Smets, G.J.F., H.H. Stratmann and C.J. Overbeeke (1990). Method of causing an observer to get a three-dimensional impression from a two-dimensional representation. European Patent Nr. 0189233.

Soede, M. (1980). On the mental load in Arm Prosthesis Control. TNO-NIPG, Leiden/Delft, 237 p. PhD thesis.

Stassen, H.G., G. Johannsen and N. Moray (1990). Internal Representation, Internal Model, Human Performance Model and Mental Load. In: *Automatica*, **Vol. 26**, No. 4, pp 811-820.

Stassen, H.G. (1989) On the modeling of manual control tasks. In: *Application of Human Performance Models to System Design*, R. Grant et al (Eds), Plenum Press, NY. pp. 107-122.

Stassen, H.G., A. Macwan and P.A. Wieringa (1995). Man-Machine System Studies and Human Reliability Analysis. To be published in: *Proc. of Workshop on HRA: Theoretical and Practical Challenges*. Stockholm, Sw, 16 p.

Stassen, H.G., R.D. Steele and J.A. Lyman (1989). A Man-machine system approach in the field of rehabilitation: A challenge or a necessity. In: *Proc. 4th IFAC/IFIP/IEA/-IFORS Conf. on MMS. Analysis, design and Evaluation*, Xi'an, PRC, pp. 88-97. ISBN 0-08-035743-1.

Stassen, H.G. (1989). The rehabilitation of severely disabled persons. A man-machine system approach. In: *Advances in Man-Machine Systems Research*, W.B. Rouse (Ed.). Vol. 5, JAI Press Inc.-, pp. 153-227. ISBN 1-55938-011.

Stassen, H.G., A. van Lunteren. J.S.M.J. van Dieten et al (1972). Progress report Jan. 1970 until Jan 1973 of the Man-Machine Systems Group: Ch.9. Man Machine problems in externally powered prosthesis/orthosis for upper extremities. Delft THD, pp.254-256. WTHD 55.

Sturman, D.J. and D. Zelter (1994). A survey of glove-based input. In: *Proceedings of the 1994 IEEE Computer Graphics & Applications*, pp. 30-39.

Veldhuyzen, W. and H.G. Stassen (1977). An application to modelling human control of large ships. *Human Factors*, **Vol. 19**, Nr. 4, pp. 367-380.

Verkuyl, H. (1964). Personal communication. Rehab. Center "De Hoogstraat", Leersum, NL

Voss, D. (1995). Are you waving at me? *New Scientist*, 1962, pp. 36-39.

Walker, J. (1988). *Through the looking glass*. Sausilito, CA, Autodesk, Inc.

Wallace, M.D. and T.J. Anderson (1993). The stagnation of theory in HCI. Research Report, Department of Information Systems, University of Ulster.

Warren, R. and A.H. Wertheim, (1990). Perception and the control of self-motion. In: *Recources for Ecological Psychology*. Lawrence Erlbaum, Hillsdale, NJ.

VIRTUAL NETWORKS: NEW FRAMEWORK FOR OPERATOR MODELING AND INTERFACE OPTIMIZATION IN COMPLEX SUPERVISORY CONTROL SYSTEMS

Yan M. Yufik[1] and Thomas B. Sheridan[2]

[1]*Institute of Medical Cybernetics, Washington, D.C.*
[2]*Massachusetts Institute of Technology, Cambridge, MA.*

Abstract. A fairytale tells about a gigantic genie who could turn himself into a smoke pole that would gradually enter a small bottle, to be miraculously reassembled there back into a fully functional genie, although of a much smaller size. Cognitive learning (vs. conditioning) is no less miraculous. When learning to supervise a complex system, one memorizes procedures and other fragments of operational knowledge gradually, one piece at a time. Miraculously, these fragments become integrated into a unified and coherent memory structure. What is the mechanism of such integration? Models based on the computer metaphor of cognition do not answer the question satisfactorily. We offer a different approach to cognitive modeling based on the concept of memory self-organization.

Keywords: supervisory control, cognitive model.

INTRODUCTION.

Development of increasingly more sophisticated technology creates a pressing need to understand better the powers and, importantly, the limitations of cognitive mechanisms responsible for the performance of complex control tasks. Cognitive modeling has proven to be a difficult and elusive problem, and will probably remain so in the foreseeable future. The difficulties can hardly be attributed to the lack of data. On the contrary, vast amounts of data have been accumulated in industry and academia. It is the magnitude of the problem and the deficit of theory that hamper progress.

Most designers will probably agree that thus far efforts in cognitive modeling have produced useful but limited results. That is, various models have accounted for restricted subsets of supervisory function exercised under simplified and constrained experimental conditions. However, the problem with such results is that they do not carry over to less restrictive circumstances, and do not scale up to larger control tasks. In short, no broadly applicable models and principles have been advanced to inform the design process and provide assurance that final products would stimulate rather than overwhelm their users. Why is it so?

In our view, the tendency in human-machine studies has been to conceptualize human performance in terms of computational metaphors borrowed from control theory, signal detection theory, and, most prominently, computer science. Is it possible that cognition cannot be adequately understood in those engineering terms alone? Our question is a practical one: what is gained and what is lost by following the lead of computer engineering? This paper offers an answer, and suggests some alternative approaches.

Section I formulates the role of cognitive models in the design of supervisory control. Section II presents a brief review of modeling methods motivated by the computer science metaphor. The review trades precision for simplicity so that the common source of apparently diverse models is not obscured by the technical detail. Section III outlines a new modeling framework called Virtual Networks. (VN). Finally, section IV uses the framework to derive new methods of interface optimization and operator training and decision aiding in complex tasks. The VN framework is motivated by the following three critical questions:

1. How do diverse experiences acquired in training and practice become integrated into unified and coherent memory structures?

2. How are those structures influenced by the operator's psychological conditions (e.g., stress) ?
3. What memory mechanisms underlie cognitive complexity?

Our proposal is admittedly incomplete and needs further validation. Moreover, it is centered around factors not lending themselves easily to objective study, such as memory structures, goals and values. We believe, however, that those factors are critical determinants of supervisory performance and need to be accounted for in the analysis, design and evaluation of human-machine systems.

I. COGNITIVE MODELS IN THE DESIGN OF SUPERVISORY CONTROL.

I.1. Supervisory control.

Supervisory control comprises five generic functions (Sheridan, 1992, p.14):
a) Planning what task to do and how to do it.
b) Programming plans into the computer.
c) Monitoring the computer's execution of plans.
d) Intervening in emergencies or under abnormalities.
e) Learning from experience new ways to do planning.

The supervisory console usually contains either a panel with dedicated input and output displays and controls, or multifunction displays (MFDs), or a combination of both. Other types of interface devices (e.g., eye-controlled input) appear in experimental prototypes but are not common. A generic supervisory problem requires the operator to respond to changing goals and conditions while maintaining many critical interdependent parameters within some target boundaries, e.g., safety limits. Task planning includes goal prioritization and prediction of the future system states. Monitoring involves choice, vigilance, and deliberate re-direction of attention. Intervention requires diagnosis, retrieval of standard procedures, and on-line composition of new control sequences to counter unforeseen eventualities. Learning improves timing and error rate in the performance of all cognitive functions.

In the design of supervisory control systems input from the operator's cognitive model is needed throughout the entire design life-cycle but is most critical during the conceptual design and prototype development stages when it serves the following objectives (Executive Summary, Army-NASA Aircrew/Aircraft Integration Program, 1990):

a. Determine control panel and MFD layout that satisfy engineering constraints while facilitating learning and intuitive interface manipulation.
b. Identify excessively complex interactions between the supervisor, tasks, equipment, and environment.
c. Modify the design to reduce complexity without compromising system functionality and reliability.
e. Evaluate training ramifications of design features.

In view of these objectives critical questions listed in the introduction can be narrowed to the following:
a. How are standard control procedures learned and retrieved on demand?
b. How are non-standard procedures planned in real time to satisfy conditions and goal priorities?

In this paper we will comment on the limitations of some prototypical answers to these questions offered by the computer metaphor, leaving room for a more constructive part of the discussion, with the primary focus on procedural learning.

II. COMPUTER METAPHOR OF COGNITION.

II.1. Genesis of the computer metaphor.

In our view, cognitive modeling has been in great part motivated by the theory of effective computability. A brief excursion into the theory will help to articulate this contention.

A function $F(x_1, x_2,...x_n)$ is effectively computable if there exists a "mechanical procedure" for determining the value of $F(k_1, k_2, ... k_n)$ when the arguments $k_1, k_2, ... k_n$ are given. There is no precise definition for the concept of "mechanical procedure" other than "a process that requires no ingenuity for its performance" (Mendelson, 1984, p.221). The history of artificial intelligence demonstrates keen interest in exercising the effective computability agenda by first isolating examples of intelligent behavior and then seeking systems of "mechanical procedures" for their approximation. Importantly, those AI attempts were undertaken initially without "any prejudice toward making the system simple, biological, or humanoid" (Minsky, 1982, p.7). Cognitive modeling has pursued a similar agenda, although with the prejudice of giving humanoid interpretations to concepts formulated originally in a purely computational context. In this way, cognition was not only approximated by computation but reduced to it, typically, in the following three steps:
a. Identify samplers of cognitive functioning. e.g., observe that the operator makes control decisions $D(k_1, k_2, ... k_n)$ given plant conditions $k_1, k_2, ... k_n$.
b. Construct a function $F(k_1, k_2, ... k_n)$ approximating the decision pattern $D(k_1, k_2, ... k_n)$.

c. Equate mechanisms of decision making to computation F, that is, equate stages of the decision process to computational steps, equate memory organization to data organization, equate cognitive complexity to computational complexity, equate acquisition of decision skills to improved approximation, equate units of cognitive activities to elementary computations, etc.

The remainder of the section takes issue with this form of reductionism. We suggest that current cognitive models reflect mathematical rather than psychological realities. The common conceptual foundation of many of those models is a psychologically infeasible mathematical abstraction called Markov's machine.

II.2. Markov's machine.

Effective computability research seeks universal means for approximating complex functions by combinations of the simplest possible operations. Markov discovered circa 1947 that computational simplicity can be attained in the form of what he called "substitutions." Skipping the detail, Markov's ideas can be summarized as follows (Markov, 1954). Let A be a finite alphabet, and let P and Q denote words in a language L on A comprising all possible words and expressions. Markov's machine includes an ordered list V of all the allowable simple substitutions P-> Q, some of which are marked as terminal. Given any legitimate input P, the machine initiates a chain of iterative substitutions (called normal algorithm $\Im(P)$) until the list is exhausted or a terminal substitution is encountered, transforming input into some output R = $\Im(P)$. If the process does not terminate, the machine cannot resolve P.

A remarkable result obtained by Markov (normalization principle) demonstrated that broad classes of functions can be approximated by a normal algorithm. (More precisely, any partial recursive function is partially Markov-computable, and any recursive function is Markov-computable.) That is, for every function F (P) in L there exists a normal algorithm $\Im(P)$ such that $\Im(P) \approx F(P)$.

These findings have established what seems to be the Holy Grail of AI and cognitive science: a coveted but yet elusive goal of formulating a self-contained method for modeling cognition using nothing other than the samples of cognitive behavior. Consistent with that goal, the premise has been that if the model provides a good enough approximation to behavior, then its algorithm is a good enough representation of the mechanism responsible for behavior. Granted, hardware that carries out the algorithms is different in the computer and in the organism, but that in principle cannot be construed to invalidate the premise. Making such use of Markov's theory has been difficult since the theory asserts the existence of algorithms but does not tell how to construct them. Examples of such construction are listed next.

II.3. Instances of effective computability paradigm.

A. Production-based models.
Productions are a form of substitutions expressed as IF-THEN statements, or rules (i.e, IF p_i THEN q_i). Production systems architecture includes three generic modules called Long Term Memory (LTM), Working Memory (WM) and Control Structure (CS) interacting iteratively. LTM is an unstructured set of productions each containing some data in the right-hand side and conditions for transferring this data into WM in the left-hand side. Transfer decisions are made by comparing conditionals p_i in the LTM productions against the input P = $(p_1, p_2, ...p_n)$. Successful matching moves the corresponding data to WM, until the WM contents match some terminal Q posed by the user as a goal. If no match is found, the system creates a subgoal to resolve the impasse, with each subgoal potentially giving rise to a cluster of lower subgoals. After the impasse is resolved the cluster (chunk) is added permanently to LTM. Accumulation of chunks in LTM implements learning (Newell, 1992).

As apparent from the earlier discussion, production-based systems are a straightforward realization of Markov's machine. As such, they inherit its power but also its limitations. Specifically, these systems face the problem of an exploding search space as the number of productions increases with the growth of the problem size. Formation of chunks cannot completely resolve combinatorial difficulties and can even exacerbate them due to the accumulation of auxiliary items in the production list.

The normalization principle establishes a mathematical fact, not a psychological one. In mathematics memory can be viewed as a sort of teenager's closet, i.e., one heap in a large container from which unrelated articles are fished out one by one when needed. In psychology this view has been long challenged as untenable: "There is no manifold of coexisting ideas; the notion of such a thing is a chimera. Whatever things are thought in relation are thought from the outset in a unity, in a single pulse of subjectivity, a single psychosis, feeling, or state of mind" (James, 1890, v.2, p.278). The key concept here is that of integration and dynamically sustained unity of memory structures. In a sense, an automatic closet organizer is needed. A formalization of this concept is offered in Section III.

B. Schema-based models.

Schema theory acknowledges the role of "units," that is, cohesive structures populating memory, although leaving unclear the mechanism by which they emerge and dissolve (Shallice, 1988; Norman & Shallice, 1980). The theory represents the contents of LTM as a set of schema units triggered by environmental inputs. Once triggered the unit runs its course until completion. Each unit is a fixed chain of substitutions, some of them being complex, i.e., calls to other chains. The model includes a separate module to resolve conflicts between concurrent schemes, while an attentional module is invoked when conflict resolution fails and when the schemes need to be interrupted or modified. The mechanism of coordinating such modifications across the schema set remains largely unspecified.

The model portrays memory organization as being essentially static. Inputs exercise the already existing schemes but have no active and continuous role in their construction: the latter is accomplished via a central controlling and schema coordinating agency (i.e., deliberate attention). This architecture can hardly scale up from restricted tasks to memory coordination problems of realistic complexity because of an extreme computational burden placed on the central control. Section III proposes a mechanism of automatic memory coordination.

C. Neural networks.

Neural net models comprise a variety of function approximation heuristics (Hornik et al, 1989). Given the input vector $\mathbf{P}$ and the desired output vector $\mathbf{Q}$, the input/output relationship is approximated by a composition of concurrent summations $q_i = \sum_{j=1}^{n} p_j w_{ij}$, where p_j and q_i are components of input and output vectors correspondingly, and w_{ij} are numeric coefficients. Summation terms are called "neurons" (note that the term is a form of complex substitution). Bi-directional Associative Memory (BAM) (Kosko, 1988) illustrates a prototypical neural net model.

BAM consists of two interconnected and reciprocal layers of units, having no connections inside the layers. The following simple algebraic exposition is not entirely faithful to the BAM concept but helps to appreciate its computational contents. Consider a system of equations $\mathbf{W} \mathbf{P} = \mathbf{Q}$ defining one of the input/output mapping directions in BAM:

$$w_{11}p_1 + w_{21}p_2 + ...+ w_{m1}p_m = q_1$$
$$w_{12}p_1 + w_{22}p_2 + ...+ w_{m2}p_m = q_2 \qquad (1)$$
$$\overline{}$$
$$w_{1m}p_1 + w_{2m}p_2 + ... + w_{mm}p_m = q_m.$$

Integrating "neurons" are rows in the system, i.e., outputs q_j from the "neurons" are obtained as the sum of inputs p_1, $p_2...p_m$ multiplied by the corresponding "synaptic strength" values w_{ij}. BAM seeks coefficients w_{ij}, given inputs $\mathbf{P} = \{p_1, p_2...p_m\}$ and the desired outputs $\mathbf{Q} = \{q_1, q_2,...q_m\}$. That is, BAM seeks approximation of function $\mathbf{Q(P)}$ given samplers of function operation. By contrast, a conventional algebraic problem seeks "input" $\mathbf{P}$ (roots) given the coefficients w_{ij} and "output" $\mathbf{Q}$. If matrix $\mathbf{W}$ is symmetrical, it equates to a product of two mutually transposed triangular matrices $\mathbf{W} = \mathbf{V}^1 \mathbf{V}$, transforming system (1) into two systems: (1a) $\mathbf{V}^1 \mathbf{R} = \mathbf{Q}$ and (1b) $\mathbf{V} \mathbf{P} = \mathbf{R}$. Solution of the conventional problem can then be obtained in two steps: a step forward computes $\mathbf{R}$ in 1a, and a step backward computes $\mathbf{P}$ in 1b. BAM reverses the conventional problem. Accordingly, BAM multiplies the transpose of the input vector and output vector for every input/output pair, and constructs the weight matrix $\mathbf{W}$ by summing up the products.

Note that if the operation of biological neurons is indeed just that of input summation, then a "hidden" computer (and programmer) are needed in addition to neurons to carry out BAM's algorithms. BAM is an algebraic process that integrating "neurons" (summators) cannot implement on their own: system (1) has no way of producing its own solution. The same is true of most other neural net models.

A three layer neural network can be visualized as two systems of equations such that the output of one system serves as input to the other. As in BAM, the problem is, given samplers of input/output mapping, find coefficients yielding an approximation of that mapping. This problem is not quite tractable. As a result, the required mapping is obtained iteratively via computing first the output vector given the input and some arbitrary coefficients, determining the output error, and then adjusting the coefficients in both systems so that the compounded error is reduced. The algebraic process of iterative error reduction assumes the name of "network training."

As function approximation methods, neural nets offer computational advantages in some classes of constraint satisfaction and pattern recognition problems. Irrespective of their practical utility, attention to neural nets can be attributed mostly to the claim that they represent biological neurons and "brain-like computation," and have properties that "correspond closely to the characteristics of human memory and are exactly the kind of properties we want in any theory of memory" (Rumelhart, 1989, p.148). We disagree with this assessment.

Taking into consideration that biological neurons have on the average 10,000 synapses whose internal mechanisms are not quite known (e.g., Sakman, 1994), appeal to biological plausibility seem to oversimplify the reality. "Brain-like" computational features such as tolerance to input noise are not specific to neural nets, but are inherent in any approximate computation. Like other instances of the effective computability paradigm, neural nets do not scale up. "There is good experimental evidence that training large neural networks is extremely time-consuming, and there is good theoretical evidence that the difficulties are intrinsic" (Judd, 1990, p. 39). Evidence in this case is based on neural net implementations counting on the average several hundred units. Presence of about 10^{11} neurons in the brain makes it unlikely that higher cognitive functions can be understood with the help of neural net models. A list of memory properties that neural nets have been designed to represent (Rumelhart, 1990, p. 147) will help to articulate our alternative proposal in the next section:

a. A familiar pattern is amplified.
b. An unfamiliar pattern is dampened.
c. Missing parts of familiar patterns are "filled in."
d. A pattern similar to a stored one is distorted toward it causing the system "respond strongly to the central tendency of similar patterns, although the tendency itself was never stored."

To relate to the list, consider, e.g., the state of BAM after matrix W has been computed for some maximum size pattern set. To learn a new set of patterns BAM must compute a new matrix W^{new} to replace the previous one W^{old}, or an altogether different net is needed. The two matrices and/or the two nets can have nothing to do with each other. The list implies no relation between the nets and no continuity between the learning cycles, so that a pattern "familiar" to one net can be totally "alien" to another one or even to the same net at some later point in time.

Apparently neural nets conceptualize memory as conglomerations of isolated episodes rather than as an integrative process. The resulting structure can only be a heap of fragments, not a unified and coherent whole. The concept of memory in neural net models turns out to be similar to that implied by the schema-based and production-based models, pointing back to the common computational paradigm from which the models emanate.

In our view, however, the term "memory" connotes an evolving global organization of the mind rather than an arithmetic sum total of its contents. Consequently, we would want a theory of memory to explain how traces of different experiences, both similar and dissimilar, become integrated into a global structure, and how this structure adjusts to and influences subsequent memorization. In the next section we will outline a different proposal. The gist of it is this: to the extent computation is independent of its purpose, origin and material embodiment, there is no computation in cognition.

III. VIRTUAL NETWORK (VN) MODEL OF MEMORY INTEGRATION

III.1. Basic assumptions.

The VN framework for modeling integrative memory is based on a number of assumptions. This paper will elaborate only some:
1. Learning and other higher cognitive functions involve the entire nervous system and the body, not a few neurons.
2. LTM is a self-organizing system of unlimited capacity where traces of experiences form a unified associative structure. The strength of each association depends on the frequency and subjective significance of the corresponding experience.
3. LTM is being continuously self-partitioned into clusters, or packets which are internally cohesive and externally weakly coupled.
4. Working Memory (WM) and Short Term Memory (STM) are limited capacity systems where traces of earlier experiences are matched against the present goals. A packet is a unit of STM/LTM exchange.
5. LTM stabilization is attained via spontaneous and deliberate changes of association strengths. Local changes may cause restructuring throughout LTM.
6. LTM comprises several levels (planes) with the packets in each higher level reflecting significant connections among some packets in the lower level. Thus the associative structure in each LTM level is a minimal description of the lower level topology.
7. Cognitive complexity of a task is determined by the minimal size of its description in LTM. Cognitive complexity of supervisory control is determined by the minimal LTM description of the entire range of control procedures.

III.2. Cognitive complexity.

We start the discussion by introducing measures of cognitive complexity (item 7), and than proceed with the rest of the list in order to justify those measures within the overall VN modeling framework.

Imagine a device having only ten control variables, each represented by a single button. Potentially, the device admits at least 10! = 3,628,800 control procedures. Partition the device into three independent modules having 3, 3 and 4 buttons. The number of procedures is now reduced to only 3! + 3! + 4!= 36. If those simple procedures have to be mutually coordinated, the number of combinations will start growing again towards the 10! ceiling. However, if modules are designed to keep the traffic between them at minimum, modularization will produce maximum combinatorial reduction.

We believe that procedural learning produces associative structures in LTM that are being continuously partitioned into minimally interdependent and maximally cohesive modules, or packets. In general, dynamic binding of environmental variables into such packets is the most efficient mechanism the nervous system can employ to accommodate input regularities. It is a fact that memorization takes advantage of the input invariants, whether one becomes consciously aware of those invariants or not (e.g., Reber, 1989). Due to this fundamental memory property disjoint fragments of experiential knowledge acquired in the consecutive stages of learning become integrated into simultaneous memory structures. Those structures are dynamically partitioned into internally cohesive and weakly coupled packets depending on the co-occurrence of input variations throughout the entire learning history. The VN model offers an account of cognitive learning based on the concept of a dual integration/partitioning memory process.

To visualize complex procedural learning in supervisory control, imagine a large control panel with multiple control instruments, and trace control procedures across the panel. Assign weights to the traces based on the procedure frequency and relative criticality. The result is a weighted network such that the weight of each link (instrument-to-instrument connection) is a function of the weights of all the procedures sharing the link. To simulate evolution of memory structures in the course of learning, account for integration of gradually acquired procedures into a connected weighted network, and partition the net into packets computed as minimal cutsets, that is, subnets such that the summary weight of the internal links in each subnet exceeds the summary weight of the external links connecting the subnet to the rest of the structure (Yufik, 1989). Structures capable of such self-organization are called Virtual Nets.

We have proposed (Yufik et al, 1994; 1992; 1989) that complexity of interactive tasks be measured by Kolmogorov's entropy K_h of the partitioned virtual network evolving in the course of task learning. K_h is determined as the minimum information sufficient for describing memory packets and their interrelations. To appreciate the concept, take into account that packets are highly cohesive, that is, well remembered, while connections among them are less certain. Retrieval from within the packets is effortless in comparison to retrieval from the rest of the structure. To account for this difference when describing the structure, collapse every packet to a single node. The adjacency matrix of the resulting condensed net describes its topology, while the size of the matrix is the minimal informational measure K_h of such description. We will return to this concept at the end of this section.

What processes are responsible for packet formation in biological memory, and how are packets stabilized if, according to packet definition, any new experience can trigger dissolution of old packets and creation of new ones throughout the network? These questions bring us to items 1 through 3 in the previous list.

III.3. Biological memory.

The mechanism of information integration in the nervous system is not known (Sakman, 1994), although the overall principles of the system's functional organization have become understood better over the last several decades due to (Luria, 1973), and more recently (Edelman, 1987; Grosssberg, 1987) and others.

The nervous system architecture contains three inextricably intertwined functional units: I[st], or cortical tone-setting unit, IInd, or gnostic unit, and IIIrd, or goal-setting unit (Luria, 1973).

The first unit includes the activating reticular formation, and is responsible for energy distribution (arousal) in the cortex, entailing preferential amplification of biologically significant stimuli.

The second (gnostic) unit is located in the lateral posterior areas of the neocortex and includes visual (occipital), auditory (temporal) and general sensory (parietal) regions. All the areas work in concert. "Human gnostic activity never takes place with respect to one single isolated modality; the perception - and still more, the representation- of any object is a complex procedure, the result of polymodal activity, *originally expanded in character, later concentrated and condensed.* Naturally, therefore, it must rely on the combined working of a complete system of cortical zones" (Luria, 1973 p. 73). The primary projection areas of the second unit are formed mainly of the highly specific neurons of the afferent layer, responding to restricted singular features of visual stimuli. Secondary cortical zones surrounding the primary areas consist of less specific associative neurons superposed over the previous afferent layer and allowing the "incoming excitation to be combined into necessary functional patterns, and thus subserve a synthetic function" (p. 69). Tertiary cortical zones are responsible for *"integration of excitation arriving from different analyzers"* (p.73). The majority of neurons in these zones are multimodal. Their principal role is "connected with the spatial organization of discrete impulses of excitation entering the various regions and with the *conversion of successive stimuli into simultaneously processed groups*" (p. 73).

Finally, the third unit is located in the frontal lobes and anterior regions of the cortical hemispheres. Its overall function is general regulation of all cognitive activities and their subordination to plans and goals. *"The frontal lobes are in fact a superstructure above all other parts of the cortex"* responsible *"for the most complex forms of behavior"* (Luria, 1973, p.91). Summarily, formation of goals, their prioritization and attentional control are performed by the IIIrd unit.

Based on the above notions, we proceed now to formulate our model. Fig. 1 suggests a grossly simplified diagram of cortical interactions. Fig. 1a indicates that the memory process is shaped dynamically by the cortical interactions and feedback from the motor-kinesthetic (muscular-skeletal) system. There is no room in this paper to elaborate this suggestion, and Fig. 1b focuses instead on only one aspect of cortical interaction, that is, formation of associative links in LTM.

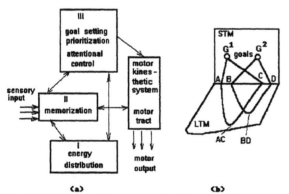

Fig. 1. (a) Interaction among the functional units in the nervous system, and (b) formation of associative connections n LTM due to STM/LTM interaction.

As Fig. 1b illustrates, elementary reactions (e.g., elementary motor responses A, B, C, D) become associated in LTM when A, C and B, D are co-selected by the low capacity serial STM in conjunction with the current goals G^1 and G^2, correspondingly. The strength of each such connection depends on its frequency and relative priority of goals that caused the association. (More generally, association strength reflects contiguity, persistency and intensity of the stimuli, and their significance to the organism.) The emerging fragmentary nets are integrated and partitioned into packets. Summarily, Fig. 1 conceptualizes the main function of cognition as active mobilization and progressively more efficient allocation of the organism's resources in the service of intrinsic goals and environmental changes. By contrast, other models see recognition as the preeminent cognitive function thus reducing the role of cognition to contemplation.

What mechanism is responsible for the stabilization of evolving memory packets vis-a-vis changing conditions and new goals and experiences? Fig. 2 explains another critical assumption of our model.

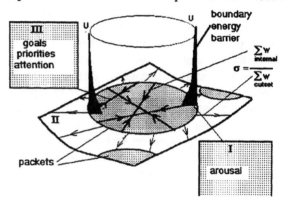

Fig. 2. Interplay of associative forces and interunit influences determining the stability/plasticity balance of LTM structures.

The decisive stabilizing influence is exerted by energy barriers formed at the packet boundaries. The relative height U of such barriers depends on the ratio of summary strength of the internal and external connections, and on the local cortical arousal T reflecting energy distribution in the cortex via the I[st] unit. Association strength is a function of its relative significance established by the IIIrd unit. Eq. (2) and Fig. 2 define combined influence of the functional units on memory processes.

$$U = \sigma - T\,(d\sigma/dT) \qquad (2)$$

Eq. 2 is equivalent to the Gibbs-Helmholtz thermodynamic stability law, where $\sigma = \sum_{internal} W / \sum_{external} W$ is the packet stability factor.

Subjective significance of associations and, consequently, the value of σ depend on the learner's psychological state. Hence, σ(T) and dσ /dT. The height U of the energy barrier determines the cognitive effort required for crossing packet boundaries and for packet dissolution. Packets dissolve when the summary strength of external associations pulling the packet apart exceeds the summary strength of the internal associative forces keeping it together.

The higher the frequency of associations within the packet and the higher their relative subjective significance, the stronger the associations and higher the packet stability. The resulting increased height of the boundary energy barrier U makes it more difficult to deviate in future decisions from the overlearned connections. By contrast, the higher the arousal, the lower the barrier and, consequently, the easier it is to lose the focus and the higher the probability of crossing the barrier in pursuit of the less relevant associations.

Compelling although indirect evidence in support of the packet concept comes from the violations of memory processes in pathology (Luria, 1973). One major form of violation causes an incessant flood of associations and pathological distractibility by irrelevant stimuli. For example, asked to repeat a phrase the patient repeats the first word, which associates with whatever comes to his mind momentarily in conjunction with that word, and so on., so the initial phrase is never repeated. An opposite dysfunction manifests as complete rigidity and fixation on whatever stimuli have been presented. For example, asked to repeat two sentences, the patient would keep repeating only the first sentence, sometimes inserting a word from the other one. Both forms of pathology can be explained in terms of the model as resulting from abnormally high or low cortical tone level T, correspondingly. High tone decreases the U barrier causing excessive distractibility and pathologically easy transition from one associative packet to another. Abnormally low level, by contrast, makes such transition impossible. However, "in an abnormarly inhibitory state ...the law of strength is upset and weak stimuli begin to evoke the same reactions as strong" (Luria, 1973, p. 157). According to Eq.(2), such inversion of the stability factor σ can cause a decrease of U and, consequently, "a flood of equally probable possibilities" (p. 157). Clinically these dysfunctions can be attributed to lesions in the first and/or in the third cortical units.

Brain pathology amplifies variations in memory processes experienced normally under adverse performance conditions. For example, in a supervisory control task, an operator's excessive arousal might be manifest in high distractibility but also in greater than usual flexibility, ease of associations, and "openness" to different views. Stress, fear, and fatigue increase U thus causing fixation, "tunnel vision," and vacillation over decision alternatives. In pathology those normal memory trends extend to the extreme.

A massive experimental study conducted in the Russian Space Biology Institute and spanning more than three decades, has investigated memory strategies in learning complex stimuli as a function of the degree of stimuli overlap (Vlasova, 1989). Based on the study results, highly successful training techniques were developed for cosmonauts and pilots improving procedural learning, target recognition, and fast categorization of new stimuli. In a typical experimental setup the subjects had to identify and learn interconnections between buttons and indicators variously located in generic control panels, as exemplified in Fig. 3. In the learning phase the subjects were demonstrated control sequences including either simple or overlapping complexes, as illustrated in Fig. 3a and 3b. During testing the task was to change the state of a few select indicators in the shortest time and in a minimum number of trials. Results of the study were summarized in a theory of procedural learning emphasizing organization of associative memory structures caused by the overlapping and non-overlapping (simple) interconnections in the stimuli.

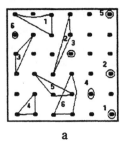 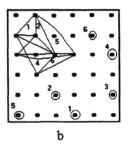

a   b

Fig. 3. Experimental control panels in (Vlasova, 1989). Buttons are circled. a. Simple indicator groups. b. Overlapping indicator groups.

A general conclusion was (Vlasova, 1989, p.45) that learning involves formation of neuronal "substrates common to the new and already acquired interconnections, simultaneous functioning of such substrates with common elements in the primary and secondary projection areas, local redistribution of excitation in the interacting structures, and change of afferent/efferent synthesis processes under the influence of verbal instruction." It was determined that in a rapid succession of presentations (150 msec intervals) repetitive overlap among the complex components impedes memorization (the phenomenon was called "associative inhibition").

That is, longer learning periods and larger numbers of training demonstrations were demanded by overlapping control sequences in comparison with the simple, i.e., procedurally and spatially non-overlapping ones.

However, under slower presentation speed (over 4 sec intervals) both simple and overlapping complexes were learned equally well. Importantly, it was found that administering of small doses of caffeine prior to rapid stimuli presentation invariably reduced the consequences of associative inhibition or eliminated them altogether. A special experimental technique was developed to assess selective influence of caffeine on the degree of arousal (excitation) in different components of the subjects' memory structures evolving in the course of learning. Fig. 4 illustrates typical experimental findings.

Subjects had to respond to standard probe signals emitted in different phases of testing, e.g., after retrieval of overlapping vs. non-overlapping stimuli components. Changes in the response delay to the same probes in the context of different stimuli and under different measurement conditions were used to assess the relative excitation characteristics of the evolving associative memory structures. As Fig. 4 illustrates, the caffeine-induced degree of excitation in the overlapping components exceeds that in the non-overlapping ones.

In terms of the VN model, the overlapping components constitute memory packets. Accordingly, a preferential increase of excitation T inside the packets by caffeine lowers energy barrier U, thus facilitating interpacket transitions and reducing the impact of associative inhibition.

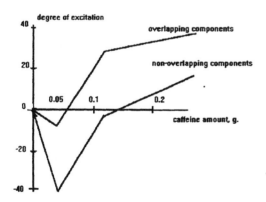

Fig. 4. Different response to caffeine by overlapping (packets) and non-overlapping (outside packets) components of associative memory structures (Vlasova, 1989).

The main finding was that procedural learning can be improved by instructions directing the subjects' attention to the overlapping procedural components.

Training methods were developed that consistently demonstrated improved long-term retention and substantial reduction of error rates, even under adverse performance conditions such as stress, fatigue, restricted mobility and sensory deprivation. Our present research findings are consistent with these results, and generalize them further in the form of training and design aids and interface optimization tools.

A phenomenon similar to associative inhibition was studied recently in experiments involving the so-called "repetition blindness" (Park & Kanwisher, 1994). It has been observed that subjects who viewed a rapid serial visual presentation (RSVP) of stimuli (words or pictures) had difficulty in reporting stimuli components that appeared twice in the series. For example, in their report of two presented sentences the subjects would omit occurrence of a word in the second sentence if it was present in the first one. Omissions can occur even within a single sentence, such as "It was work time so work had to get done." A typical report after RSVP would be: "It was work time so had to get done" (Park & Kanwisher, 1994). We think that the second appearance of the stimuli component requires the already formed group to dissociate and form jointly with the subsequent stimuli a new associative structure. According to the VN model, such dissociation is an energy consuming process requiring energy influx in the amount equal or exceeding U. If the energy supply available within the presentation interval is insufficient, the initial group will not dissolve and another group will be formed independently of the first one. As the time of exposition increases, reorganization can be supported and repetition blindness duly disappears.

An interesting phenomenon somewhat analogous to "repetition blindness" has been observed in the disturbances of visual synthesis. A visual dysfunction called "simultaneous agnosia" makes the patient unable to perceive two objects simultaneously regardless of the object's size. Luria, 1973 obtained improvement of this condition in a patient with a bilateral wound in the visual region by administering caffeine into the disturbed area, which allowed the patient to perceive several objects simultaneously for the duration of the caffeine action. Apparently, caffeine injection caused local increase of arousal T, thus changing packet dynamics toward lower stability and allowing packet dissociation and merger as required for visual synthesis.

Increased frequency of and/or attention to the external connections in integrated associative structures increases their weight, which can upset the stability balance and trigger a wave of restructuring propagating through memory. The extent and area of such restructuring are strictly optimal with respect to stability criteria (2), although no computation whatsoever is needed to coordinate and optimize the process.

Fig. 5 illustrates the local restructuring in a peculiarly unstable structure. Redirection of attention changes the weight balance, alternately causing merger and dissociation in the "vase/profiles" figure.

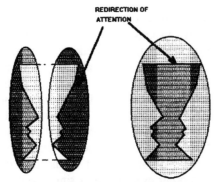

Fig. 5. Alternate merger and dissociation of packets due to redirection of attention and redistribution of weights. Peculiar topology causes low packet stability and perceptual ambiguity.

To summarize, Fig. 1 defined the dynamics of a cognitive self-organization, allowing the organisms to respond to new challenges based on the automatically maintained packets of relevant old experiences. The reorganization is limited by the persistence and significance of those experiences in the past, and the present goals and conditions. In the context of complex system control, each packet comprises actions that in the past proved to be mutually cooperative with respect to the control goals. Packets reorganize in the course of training and practice.

As stated earlier, neuronal mechanisms of learning are not precisely known, entailing a variety of coexisting theories. The Theory of Neuronal Group Selection (TNGS) in (Edelman, 1987) makes a breakthrough departure from the "computational" views by emphasizing learning via competitive formation of cohesive and externally weakly coupled neuronal groups. We think that our model is congruent with this important insight and the overall approach of the TNGS.

III.4. Learning dynamics in virtual nets.

Consider an infinite memory "plane" occupied by a virtual net subject to continuous self-partitioning. Any new experience either changes the weight of

some already existing links, or adds new weighted links. In both cases, new packets can be formed and old packets can be modified. In this way the virtual network can keep a globally organized residual "record" of an unlimited variety of experiences.

Let each packet in the net contain a combination of control actions that have been used jointly and cooperatively in the past throughout the history of control decisions. Due to self-partitioning, the contents of each packet are continually optimized vis-a-vis all other packets in the net. As a result, each packet automatically maintains an optimal combination of actions that proved to cooperate successfully in the past. Chances to improve goal satisfaction degrade steeply as soon as the packet boundaries are crossed. Strong associative connections within the packet lead directly from one action to another. Thus no search is needed to form a response within a packet.

If performance demands change and the goal is no longer satisfied, new connections are formed while previously less frequent intergroup connections get exercised and acquire greater strength, causing changes in the packet contents and relative stability. Again, the self-partitioning process automatically optimizes those changes across the entire structure.

To expand further on the model, allow packets to stabilize, and place another network above the first one. Let each node in the upper level net correspond to a packet below. Subject the upper network to the same self-integration and self-partitioning process as the net underneath, so that the upper packets correspond to the persistent patterns of path traversal in the lower net. The upper network then becomes a map of the net below, as Fig. 6 illustrates.

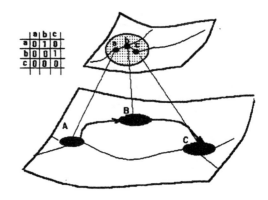

Fig. 6. Efficient multistage grouping of variables in a hierarchy of virtual networks in LTM.

It is easy to see that memory structures evolving in the course of procedural learning might encompass several levels, with the lowest level organization reflecting the order in which specific devices are operated, the next level above reflecting the predominant order in which the lower level packets have been exercised., etc. In this way a hierarchy of virtual networks acquires an unlimited capacity for integrating diverse control experiences into a global unified structure containing a residual record of the entire interaction history. Most importantly, the record of the past is optimally organized for its efficient use in the future. As can be seen in Fig. 6, connections in the higher level packet implement an adjacency matrix for a set of interconnected packets in the level below (1 if the corresponding low level connection is strong, and 0 if otherwise).

LTM is self-descriptive since network structure in each level describes the topology of some network domain underneath. Kolmogorov's complexity measure captures this self-descriptive aspect of evolving memory structures.

Working Memory (WM) is a "vertical" (Wickelgren, 1979) relatively low capacity system comprising a number of interrelated packets in different LTM layers. WM facilitates synergism among the layers manifest as competent dynamic construction of complex integral responses from a repertoire of well-learned alternative fragments at different layers in the hierarchy.

For example, in the development of motor skills such competence allows dynamic formation of "integral kinesthetic structures or kinetic melodies when a single impulse is sufficient to activate a complete dynamic stereotype of automatically interchanging elements" (Luria, 1973, p. 176). STM is a subset of WM, in a sense, a mobile small "window" in the vertical WM structure operating within individual packets and moving along the vertical and horizontal interpacket connections. Fig. 7 summarizes the architecture of the VN model.

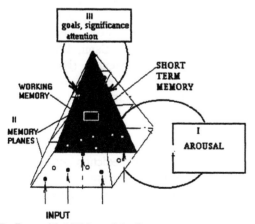

Fig. 7. Pyramidal VN model of memory.

WM is represented as a vertical cross section in the pyramid. Packets emerge and stabilize in horizontal

LTM layers under the dual influence of input regularities and system propensities. The packet network in each higher plane reflects the topology of the network below. WM is comprised interrelated packets in several ascending planes. The overall pyramidal structure represents interrelationships between the Long Term and Working Memory, and other functional units. Our pyramidal model upholds an insightful idea of horizontal and vertical memory structures advanced in (Wickelgren, 1979), although our conceptualization of those structures and their dynamics is different and in some respects orthogonal to that in (Wickelgren, 1979).

The external-input/internal-propensities interplay has been strongly emphasized in the concept of "adaptive resonance" underlying the theory of competitive learning (Grossberg, 1987). The term "adaptive resonance" captures the predominant characteristic of a dynamic self-organizing system continuously seeking inputs that satisfy its intrinsic propensities and values. The Theory of Neuronal Group Selection also conceptualizes learning as a "value driven" process. Substantiation of this concept in the VN model is beyond the scope of this paper. We would note at this point, however, that the now rather dated operant conditioning theory of B.F. Skinner, with its reinforcement-driven specialization of behavior ("differentiation") and gradual association with other learned patterns ("generalization") called for (but did not specify) some mechanisms similar to an extent to formation and dissociation of packets in the VN model.

To summarize, this section has defined criteria of stability operating autonomously in the memory planes. In each plane a virtual network is formed gradually based on the cross-correlation of inputs from the plane below. Once the network assumes a certain organization, it persists in maintaining that organization until the packets are pulled apart by the growing strength of interpacket associations causing the network at some instance to switch to another organization. There exists an energy threshold U at each packet's boundary that needs to be overcome for the packet to dissolve. Increase of cortical tone T lowers the threshold, thus increasing packet plasticity. The number of planes in the IInd unit corresponds to the number of cortical layers. Thus, networks in the first, second and third VN planes represent environmental variables, simultaneous associative patterns, and cross-correlations among the patterns. Accordingly, violations of intralayer processes and interlayer synergism are manifest in disturbances of simultaneous syntheses and inability to carry out complex intellectual operations.

Memory integration and partitioning constitute the innermost mechanism of learning. Orienting response and spontaneous and deliberate investigative activities integrate disconnected memory fragments. Memory self-partitioning and gradual consolidation of packets underlay conversion of concrete experiences into abstract thinking. Basic aspects of this process can be accounted for within the VN formalism (Yufik & Sheridan, 1995). However, this account and its practical applications need further development and validation.

IV. OPERATOR TRAINING AND INTERFACE OPTIMIZATION IN COMPLEX SYSTEMS.

The VN model of memory processes has led to innovative and potentially powerful methods for optimizing operator training and the design of control panels, multifunction displays (MFDs), menus and other forms of human-machine interface (Yufik, 1988; Yufik & Hartzell, 1989; Yufik & Sheridan, 1991). The model, as well as analysis of available pertinent psychological data and interface design practices (Yufik & Sheridan, 1991; Yufik et al, 1992) have allowed us to isolate two factors determining the cognitive complexity of supervisory control. The first factor is the extent to which the associative memory structures that capture regularity patterns in control procedures lend themselves to organization into packets. The second factor is the extent to which partitioned memory can be mapped onto the interface arrangement. Thus, to design a good interface one needs to assess the degree of memory organization allowed by the control procedures, and the degree of isomorphism between the memory structure and interface arrangement allowed by the design constraints. Those considerations were translated into an interface optimization methodology implemented as a software toolkit (Yufik & Sheridan, 1991; 1994).

In the prototype toolkit (called Interface Design Aid, or IDA) control variables (information sources) are represented as nodes in a network. Accordingly, control procedures become represented as partially overlapping network paths. Links in the network are weighted based on the relative frequency of their traversal during procedure execution, and on the relative significance of operational goals causing such traversal (see section III.3.) The optimization process produces network partitioning into minimally interacting subnetworks, with the intensity of interaction between the subnets determined by the weight of the cutset separating each subnet from the rest of the structure (Yufik & Hartzell, 1989). In the design of control panels,

cognitive complexity is reduced when control instruments corresponding to each such subnet are placed in close proximity to each other, and at sufficient distance from the other subnets. In the design of MFDs, subnets are placed on separate pages. Minimization of cross-reference among the pages reduces complexity (Yufik, 1990), as Fig. 2 and Eq. 2 have explained. Menu design is subject to the same complexity minimization principle.

Translation of the interface optimization problem into the problem of weighted network partitioning represents a critical insight stemming from the VN model of memory. IDA supports two stages of interface optimization. Preliminary partitioning results in subnets (blocks) that satisfy user defined limitations (e.g., minimum and maximum number of instruments allowed in the subnet) in such a way that the summary weight of the cutsets in the entire network is minimized in comparison with the summary weight of all the blocks. In the second stage a more precise near-optimal partitioning is obtained, maximizing stability factor σ. In this stage, in accordance with Eq. 2, the summary weight of each subnet (packet) is required to exceed the weight of the corresponding cutset by some threshold value Δ, that is, for each packet $\sum_{int ernal} W - \sum_{external} W > \Delta$. Also, variable compatibility, mandatory preservation of some of the intrapacket connections, size limitations and the degree of spatial overlap among packets, and other constraints need to be taken into consideration.

The problem of packet partitioning is considerably more demanding than that of block partitioning. A more extensive theoretical analysis resulted in a number of proprietary algorithms allowing fast processing of sufficiently large networks. Analysis of our basic underlying algorithm indicated computational complexity $O(N \log(N))$, where N is the number of network elements. This result is a tangible improvement over earlier partitioning methods (e.g., Luccio & Sami, 1969). Networks of realistic size and complexity (hundreds of nodes) can be partitioned in seconds, thus satisfying the precondition for making IDA useful in design practice.

Packet computation in IDA can be used in two ways. First, it allows prediction and diagnosis of the operators' memory dynamics in the course of training and subsequent practice. Second, it recommends to the designer an interface layout congruent with the learner's memory organization and processes. As a result, learning time and effort can be reduced and performance characteristics improved, including performance under adverse conditions (Vlasova, 1989). IDA supports intuitive graphical and textual entry of control procedures, and provides tools for viewing and analyzing the resulting structures prior to and after partitioning. Images of procedural nets can be superposed on the control console layout which helps the designer to validate entry and interpret IDA recommendations. The designer can then vary constraints and other parameters and observe layout changes without appreciable delay. A special MFD design component in IDA automatically maps packets onto button sets in MFD pages. The designer can easily vary MFD format including the layout and other characteristics, each time testing impact of those variations on performance.

Experiments were conducted at the psychology department of The Johns Hopkins University, using the IDA tools and training method. 29 students were engaged in two training and test sessions of about 90 min duration in each session, requiring subjects to learn altogether 16 abstract procedures presented on a simulated MFD. 10 subjects used a conventional display layout, 10 others were presented with IDA optimized displays, and 9 employed an optimized training process and optimized layout. During training, correct sequences were highlighted. During testing, subjects reproduced the learned sequences by pressing the MFD buttons, with each such action recorded and its correctness automatically identified. As depicted in Fig. 8, the averaged error/correct response ratio was 0.182 in the first group, 0.151 in the second group, and 0.117 in the third one.

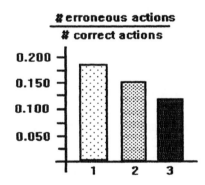

Fig. 8. Testing of procedural learning efficiency in MFD. 1 - conventional layout, 2 - optimized layout, 3 - optimized layout and training process.

These results, although consistent with (Vlasova, 1989), present as yet no convincing evidence of the merits of the approach. Three hours of abstract learning were already taxing our subjects' interest and capability to focus on the task. At the same time, we believe that results of at least 40- 50 hr. of conventional and optimized learning need to be compared in order to assess reliably the degree of improvements yielded by the optimization. Such extended assessments would require experiments with dedicated professionals engaged in real-life learning assignments. Discussions and initial studies are presently under way involving experiments with pilots and cosmonauts learning to control remote manipulators installed on board of a space craft. IDA will be used to optimize the configuration of the training simulator and training processes on the ground. Preliminary results look encouraging. Research is presently under way to extend application of VN concepts (Yufik, 1981) to the design of adaptive displays capable of self-reconfiguration based on the usage patterns and user-specified goals and priorities. A U.S patent is pending for this and other applications.

To conclude, this paper has proposed a Virtual Network (VN) model of learning and memory accounting for the dynamic self-partitioning of associative memory structures into flexible modules that can merge, dissolve and reorganize depending on the changing goals and input history. Current network models of learning recognize the need for memory modularization but can offer only fixed modules with predefined interactional constraints (e.g., Murre, 1992). The VN model suggests a solution to the plasticity/stability dilemma (Grossberg, 1987) concerning the nervous system's capability to remain plastic in view of significant events, while showing stability when encountering the irrelevant ones. The VN model combines this capability with that of dynamic memory integration. In our view, such integration and automatic maintenance of optimal plasticity/stability balance in the memory constitute the foundation of intelligent behavior, and is a unique property of our model.

REFERENCES.

Grossberg, S. 1987. *Competitive Learning: From Interactive Activation to Adaptive Resonance*. Cog. Sci. 11, pp. 23-63.

Edelman, G. M. 1987. *Neural Darwinism: The Theory of Neuronal Group Selection*. Basic Books, NY.

James, W. 1890 *The Principles of Psychology. 1950*. Dover, N.Y.

Judd, J.S. 1990. *Neural Network Design and Complexity of Learning*. A Bradford Book. The MIT Press. Cambridge, MA.

Hornik, K., Stinchcombe, M., White, H. 1989. *Multilayer feedforward networks as universal approximators*. Neural Networks, 2, pp.359

Kosko, B. 1988. *Bidirectional Associative Memories*. IEEE Trans.SMC. pp. 49-60.

Luccio, F., Sami, M. 1969. *On the decomposition*

of networks into minimally interconnected nets. IEEE. Trans. Circuit Theory, CT-16, 184-188.

Luria, A.R. 1973. *The Working Brain. An Introduction to Neuropsychology.* Basic Books, Inc. N.Y.

Markov, A. 1954. *The Theory of Algorithms.* Mat. Inst. Steklov. Moscow.

Mendelson, E 1979. *Introduction to Mathematical Logic.* D. Van Nostrand

Minsky, M. 1968. *Semantic Information Processing.* The MIT Press.

Murre, M.J. 1992. *Learning and Categorization in Modular Neural Networks.* LEA, New Jersey.

Newell, A. 1992. *Unified theories of cognition.* In: Anderson, J.R. (Ed.) Soar: A cognitive architecture in perspective. LEA, New Jersey.

Norman, D.A, Shallice, T. 1980. *Attention to action: Willed and automatic control of behavior.* Cog.Sci. Ins. UC San Diego.

Park, J., Kanwisher, N. 1994. *Determinants of repetition blindness.* J.Exp.Psychology: Human Perception and Performance. 20, 3, 500-519.

Reber, A.S. 1989. *Implicit learning and tacit knowledge.* J.Exp. Psy.: General, 118, 219.

Rumelhart, D.E. 1990. *The Architecture of Mind: A Connectionist Approach.* In: Posner, M.I. (Ed.) Foundations Of Cognitive Science. A Bradford Book. MIT Press. pp.133-159.

Sakman, B. In: Thomas Bass. *Reinventing the Future.* Addison-Wesley. p. 153-171.

Shallice, T. 1988. *From neuropsychology to mental structure.* Cambridge. U.K.

Sheridan, T.B. 1992. *Telerobotics, Automation, and Human Supervisory Control.* The MIT Press. Cambridge, MA

Vlasova, M. 1988. *Mechanisms for recognition of the unknown. Space Biology.* Moscow

Wickelgren, W.A. 1979. *Chunking and consolidation: A theoretical synthesis of semantic networks, configuring in conditioning, S-R versus cognitive learning, normal forgetting, the amnesic syndrome, and the hippocampal arousal system.* Psy. Rew. 86, 1, 44-60.

Yufik, Y.M., Sheridan, T.B. 1995. *Assessment of cognitive complexity: helping people to steer complex systems through uncertain and critical tasks .* Report. NASA Ames. CA.

Yufik, Y.M. 1994. *Virtual Network formalism.* Trans. IEEE SMC, 981-987.

Yufik, Y.M., Sheridan, T.B. 1993. *Framework for measuring cognitive complexity.* In: Smith,

M.J., Salvendy, G. (Eds.) Human- Computer Interaction. Elsevier p. 587- 592.

Yufik, Y.M., Sheridan, T.B., Venda, V.F. 1992. *Knowledge measurement, cognitive complexity and cybernetics of mutual man-machine adaptation.* In: Negoita,C.S.(Ed.) Appllied Systems and Cybernetics. Marcel Dekker, pp. 187-238.

Yufik, Y.M. Sheridan, T.B. 1991. *Interface design optimization.* Interim Report. NASA Ames Research Center. Moffett Field. CA.

Yufik, Y.M. 1990. *Application of cognitive modeling and knowledge measurement in diagnosis and training of complex skills.* In: Diaper, D. (Ed.) Human-computer interaction. INTERACT-90, Elsevier, 887-892.

Yufik, Y.M., Hartzell, J. E. 1989. *Design for trainability: Assessment of operational complexity* In: Salvendy, G, Smith, M.J. (Eds.) Designing and Using Human- Computer Interfaces. Elsevier Science Publishers.

Yufik, Y.M. 1990. *Computational and cognitive models of configurative synthesis.* Proc. IEEE International. Conf. on Intelligent Control. Philadelphia, PA. 89-96.

Yufik, Y.M., Sheridan, T.B. 1986. *Hybrid knowledge-based decision aid for operators of large scale systems.* Large Scale Systems, North Holland, 10, 133-146.

Yufik,Y.M., Sheridan,T.B., 1985. *Intelligent decision and training aid with optimization based inference engine.* Proc. Aerospace Applications of Artificial Intelligence, Dayton, Ohio, 277 - 286.

Yufik,Y.M., Sheridan,T.B. 1983. *A framework for the design of operator planning/decision aids.* Trans. American Nuclear Society, 45, 360 - 362.

Yufik, Y.M. 1981. *Simulation of cognitive strategies and systems with virtual structure.* CSI Paper. Cog. Sci. Inst, U.C. San Diego, CA.

This research was supported in part by contract NAS2-1370 to IMC, Inc. Kevin Corker was Technical Monitor and collaborator. Barry Smith, Michael Shafto and other colleagues at NASA Ames provided support and encouragement. Steven Yantis of Johns Hopkins University helped in experimental design and made other valuable contributions. Early support by Jim Hartzell of NASA Ames made this project possible.

The Operator-Model Architecture and its Psychological Framework

Stephen Deutsch
Marilyn Jager Adams

Bolt Beranek and Newman Inc.
10 Moulton Street
Cambridge, MA 02138
(617) 873-3168
sdeutsch@bbn.com

Abstract: The operators of complex equipment are frequently members of a team who must manage their control functions across numerous interruptions. They succeed, in part, because of their multi-tasking skills. The OMAR System includes a suite of representation languages as the basis for constructing models of these human multi-tasking behaviors. Prior to the development of the computational languages, a psychological framework was developed that attempts to identify key elements of the computational foundation for these behaviors. The psychological framework and the design of the representation languages for developing models of human multi-tasking behaviors are described.

Keywords: Human supervisory control, Man-machine interaction, Simulation languages, Knowledge representation

INTRODUCTION

In investigating human operator performance in the management of complex equipment one must recognize that, it is often the case that, the operators are members of a team executing their tasks in the presence of frequent interruptions. Exploring this performance through simulation requires new levels of fidelity in human performance modeling. The Operator-Model Architecture (OMAR) is a new computational framework designed to support the development of simulation models of human agents interacting with other human agents, both simulated and real, in executing these complex tasks. While OMAR is intended to support the development of a broad range of perspectives vis-à-vis modeling methodologies, its development has been guided by a particular psychological framework. The development of this framework has been based on recent research in experimental psychology, cognitive science, neuropsychology, and computer science. It reflects the requirement to portray the human operator's complex capability to intermix thoughtful and automatic behaviors in addressing proactive and reactive situations.

THE PSYCHOLOGICAL FRAMEWORK

In developing a psychological framework, it is important to identify and quantify the sources of human performance in a manner that makes it possible to develop a computational model of these behaviors. The literature on experimental psychology reinforces the basic premise that human performance capability is rooted in the ability to manage multiple "simultaneous" tasks, but raises many questions with respect to the structure of multi-task behaviors and automaticity. These are the factors that shape both novice and skilled performance, and are the sources of human error at each skill level. The neuroscience and computer science literatures have also been relevant. Gerald Edelman's reentrant and global maps have provided a neuroscience basis for the procedure network that is described later in this paper for modeling multi-task performance and automaticity. The representations of goals and plans, the data-flow model of procedure activation with data arriving from multiple sources, and the *race/join* semantics for specifying the parallel execution of network procedures have been adapted from computer science.

Theoretical Orientation

In the formulation of the psychological framework, the Memory Module effectively comprises a large number of process or event memories of varying degrees of complexity and works with a variety of data. These memories communicate with one another via a complex of excitatory and inhibitory interconnections. The strength of any given complex of connections reflects its frequency of use as well as its current level of activation, such that the strength of its response to a message is a joint function of learning, expectation or priming, and the compatibility of the message with its structure.

Complementing the Memory Module, a Cognitive Module is proposed. Within the Cognitive Module, the activities and interests of oneself as well as the impinging world are represented in terms of means-ends frames. This is consistent with Kant's (1787/1965) arguments that neither superficial similarities nor contiguities in space and time are enough to support rationality in an associative model—a person's ability to sort and order experiences of the phenomenal world depend additionally on an understanding of causality. The knowledge structures of the Cognitive Module are envisioned to consist, in part, of goals with plan structures similar to those developed in the cognitive science and artificial intelligence literatures.

The work of the system is carried out through the interarticulation of the Memory and Cognitive Modules. Memory messages trigger means-ends frames of the Cognitive Module. Unresolved means-ends frames seek out relevant data from the Memory Module, selectively activating prior knowledge or priming the system for relevant messages from the environment. In addition, with particularization and resolution, the frames direct action and response by invoking and, hence, coordinating and monitoring the effects of the corresponding action knowledge in the Memory Module. Thus, it is the transactions between the Cognitive and Memory Modules that govern the system's receptivity while weaving its

fabric of cognitive coherence and behavioral integrity.

Neuroscience and the Psychological Framework

In developing the psychological framework, the potential for the contribution of recent work in the neurosciences could not be overlooked. Edelman (1987) discusses the psychological functions of "development, perception (in particular, perceptual categorization), memory, and learning" and how they relate to the brain. Edelman (1989) extends his analysis to consider "perceptual experience—the interaction of memory with the present awareness of the individual animal," that is, perceptual awareness and conscious experience.

Reentrant Maps, Global Maps, and Degeneracy: Neural maps refer to the ordered arrangement and activity of groups of neurons as distinct from single-neuron connections. They are highly and individually variant in their intrinsic connectivity. Changes in the behavior of the network are the result of changes within particular *populations* of synapses. "These structures provide the basis for the formation of large numbers of degenerate neuronal groups in different repertoires linked in ways that permit reentrant signaling" (Edelman, 1987, p. 240) where, in *degenerate* systems, functional elements in a repertoire may perform more than one function and a function may be performed by more than one element (Edelman, 1987, p. 57). Reentry is a basic mechanism suitable for synchronizing the neuronal activity across the mappings at diverse hierarchical levels. *Global mappings* have a dynamic structure that reaches across reentrant local maps and unmapped regions of the brain to account for the flow from perception to action. Motor activity, an essential input to perceptual categorization, closes the loop.

A Neuroscience View of Memory as Active: In contrast to the view of memory as storage—as a data base—Edelman, as did many who preceded him, views memory as *process*. For him, memory is the "*ability* to categorize or generalize associatively" (Edelman's italics, 1987, p. 241). Categorization occurs at the level of a global map and is degenerate. Edelman is well aware of the distinctions between declarative and procedural memory, but he is also quick to point out that these distinctions may be less than generally assumed. He suggests that there may be a procedural base supporting declarative memory.

In Edelman's view of memory as process, perception, categorization, generalization, and memory are closely linked. "Memory is a form of recategorization based upon current input; as such, it is transformational rather than replicative" (Edelman, 1987, p. 265). Memory is an active process of classification leading to recategorization and, thus, a partitioning of the world that is presented as one "without labels." Storage, to the extent that it exists, is one of procedures for mapping inputs to responses; hence, full representations of objects are neither stored nor required: *It is the complex of capacities to carry out a particular set of procedures (or acts) leading to recategorization that is recollected"* (Edelman's italics 1987, p. 267). This view contrasts sharply with memory cast as data residing in a data base, where content is passive, references are made to it, it may fade with time, and in the case of short-term memory, new memories may reinforce or replace existing memories. In such schemes, something processes memory as data, reinforcing some of it and degrading other parts of it.

Related theories, focusing on the dependence of learning on the learner's actions or responses, have posited that even ostensively static, declarative knowledge is represented in memory only as it is embedded in the larger context, including the actions and events through which it was learned and is used (e.g., Neisser, 1976; Yates, 1985). That is, procedural knowledge is distinctly active. People have processes or procedures for doing things which they constantly modify and adapt to use in their everyday being. They take inputs from the world, process them and effect changes in the world around them. The processes work more or less well with each other depending on circumstances. In this sense, memory is procedural knowledge and, in particular, has an active component. Accepting that "declarative" knowledge is embedded in procedural knowledge, semantic memory also becomes active and procedure-like. Short-term memory is a cluster of actors re-enforcing or suppressing one another, modulated by conflict-resolution that is itself based on the class structure of the memory processes. Long-term memory and explicit and implicit memory are variations on this theme of memory as process. In this spirit, Edelman's (1987, 1989) reentrant and global mappings are, in part, a data-flow network of processes.

Multi-tasking, Automaticity, and the Psychological Framework

Human performance models are expected to represent human behaviors in the operation of complex computer-supported equipment. The range of behaviors to be modeled have perceptual, cognitive, and motor components with varying degrees of complexity in each category, and there are frequently occasions in which operators are required to execute several tasks "simultaneously." In these models, it is also necessary to represent a range of human behaviors from novice to expert along one dimension and from errorful to error-free along another dimension. The domain is complex, but there is a wealth of knowledge on these issues in the experimental psychology literature. Because the richness of the psychology literature also represents a diversity of views, the first task was one of establishing a psychological framework in which the more important threads in this research may be represented.

Dual-Task Performance: Psychological theories of task interference can be divided into two categories: In one, processing draws on resources in a graded fashion; in the other processing resources can only serve one task at a time. To be sure, many theories are hybrids, of which the following are representative examples:

The Resource Model (Wickens, 1984) may be characterized by:
- processing that relies on resources, where the resources may be used in graded quantities, with greater allocation of resources producing more efficient or faster processing; and
- processing resources that may be divided into separate resource pools, where tasks that can proceed simultaneously do so with fewer resources available to each task and, hence, each task proceeds at a reduced rate.

In the Postponement Model (Pashler and Johnston, 1989):
- cognitive operations of each task demand simultaneous access to a processor or processors, where some processors can only service one task at a time;
- while a task is occupying a single-user processor or cluster of processors other tasks are postponed; and
- the individual stages of processing that constitute the single-channel bottleneck might be:
  - perceptual identification,
  - decision and response selection, or
  - response initiation and execution.

Toward reconciling such differences, note that the single-user processors of the Postponement Model may be considered "resources" that are bimodal; that is, a task gets all of the resource or none of it. In adopting this perspective in the psychological framework, it is posited that contention among particular classes of tasks is based on dynamically computed metrics.

Automaticity: Automatic processes typically differ from those of nonautomatic processes in the following ways (Logan, 1988; Shiffrin and Schneider, 1977): They are fast, effortless, autonomous or obligatory, able to operate in parallel, consistent or stereotypic, and unavailable to conscious awareness. Significantly, some theorists have stressed that some of these are *relative* characteristics (Cheng, 1985). Indeed, it is largely by probing the *relative* expression of these characteristics across experience and training that psychologists have begun to disentangle their underlying nature. These explanations may be divided into two categories. In the first category, increases in automaticity are attributed to the increased efficiency of processes per se. In the second, they are ascribed to more comprehensive and complete memory support for practiced tasks.

Within the view of automaticity as a product of processing efficiency:
- automatic processing is processing without attention;
- the development of automaticity represents the gradual withdrawal of attention from the process itself; and
- attention is viewed as a resource, the demand for which diminishes with practice.

Within the view of automaticity as a product of memory support:
- automaticity is rooted in the memory;
- performance is automatic to the extent that it depends on single-step, direct-access retrieval of solutions from memory;
- automatization reflects the build-up of relevant traces in memory; and
- this build-up of traces underlies expert performance.

Importantly, the memory-based view of automaticity follows from the procedural view of memory that has been adopted and, further, does so in a way that explains the attention-based view of automaticity.

THE OMAR ARCHITECTURE

In the psychological framework, a set of functional human capabilities were identified. OMAR (Deutsch *et al.*, 1993) was designed to facilitate the implementation of human performance process (HPP) models with these capabilities (see Young, 1992). OMAR encompasses a suite of representation languages, graphic language editors and browsers, and a simulation environment. The Simple Frame Language (SFL) provides the traditional role of declarative knowledge representation and simultaneously forms the bridge to the object-oriented implementation based on Common

Lisp. A graphic editor provides a network view of the object definition hierarchy and a table view of the slots of individual objects. Appropriate editing functionality is available in each view.

Agent behaviors are represented in the Simulation Core (SCORE) language whose implementation is based on ACTORS (Agha, 1986) semantics. SCORE facilitates the representation of an agent's goals and plans and the networks of procedures that implement the behaviors. The rule language was included to represent novice through intermediate decision-making and problem-solving skills. The ability of tasks to wait for and to signal events provides the data-flow interactions discussed earlier. The development of SCORE procedures is aided by the graphic browser for individual procedures, and for the network formed by the set of procedures for an agent. Network views are available across the procedure caller/callee and generate-signal/await-signal perspectives. The SCORE simulator efficiently executes agent behaviors represented as compiled goals, plans, and procedures. Scenarios may be developed and executed with recorders and a traveler (Manning, 1987) available to provide on-line and post-run insight to agent performance through task and event timelines.

Attributes of HPP Models and OMAR Tools

The psychological framework identified particular aspects of human behaviors that are necessary for HPP model development. Features of the representation language that are essential to express these agent behaviors are the focus of the following discussion.

Proactive Behaviors: The psychological framework requires a representation of goals as the basis for proactive human behaviors. A *goal* expresses the conditions necessary for its achievement and includes a *plan* of subgoals directed toward the achievement of the goal. The leaves of the plan are distinguished by their invocation of and response to one or more procedures or agent actions. A goal expresses what is to be accomplished, and a plan outlines the steps to achieve the goal. The actions or procedures direct agent behaviors. SCORE, a procedural language, provides the representation of goals, plans, and procedures as defined here.

Multi-tasking and Task Contention: The modeling of multi-task behaviors requires a formalism to represent the interactions between tasks contending to execute. The psychological framework suggests basing this competition on the accrual of activation credits vis-à-vis other tasks. As tasks are developed, they naturally form *contention* classes—classes within which the competition to execute will need to be modeled. The unique feature identified here is the classification of tasks by the way in which they compete with other tasks to execute. Depending on their internal structure and processing requirements, tasks may compete with other tasks within their own class or with tasks from different classes. SFL, used to define the basic objects of the OMAR environment, is used as the basis for specifying the class membership of SCORE tasks. SCORE tasks are SFL objects.

Given the classification of tasks, it was necessary to provide a basis for developing the protocols for implementing the explicit contention among tasks. Generic functions were developed for managing the decision to suspend or resume a task, as well as the actual suspension and resumption of individual tasks. The accrual of activation credits for a task is typically driven by events in the environment or by a cognitive action that promotes the activation of a particular task. Additionally, provisions were required to enable a suspended task to dictate its own behavior during the period for which it is suspended—perhaps by simply adjusting the level of activation of the task over time. A task may also impact the activation of tasks of a given class operating in parallel with it. As the activation levels of tasks are adjusted over time, these functions provide the basis for a task asserting its right to execute. A loss in activation level for an active task will open a window of opportunity for competing tasks, while a gain in activation level by a nonexecuting task will enhance its opportunity to execute. The means to establish restart points for resuming a task is also provided. In this environment, there is no explicit scheduler. Instead, agent behaviors emerge from the dynamics of the activation levels among competing tasks.

Reentrant Map Semantics: The psychological framework is realized, in part, by a network of procedures. Emerging agent behaviors are the result of goal-directed actions and data-driven events. Task activation, with inputs arriving from multiple sources, results in a data-flow architecture that resembles Gerald Edelman's (1987) *reentrant* and *global maps* as described in his Theory of Neuronal Group Selection. Here, procedures correspond roughly to reentrant map elements that are driven by multiple inputs and that generate outputs for multiple consumers. The data flow in the network imposes sequential execution of procedures within a data-flow path, while enabling parallel execution of procedures along parallel paths.

In the data-flow architecture, all arguments need not come from a single source; that is, one procedure may provide some of the arguments to a subsequent procedure, but additional arguments from another procedure may be required before the subsequent data-flow node may execute. In a breaking with a traditional data-flow architecture, not all the arguments need necessarily be present for a procedure to execute. The psychological framework does not endorse a pure data-flow architecture.

A typical SCORE procedure is much like a Lisp function in that it may return values to its calling procedure—but this event simply does *not* happen in data-flow networks. Instead, procedures in the data-flow network typically use the SCORE *loop-forever* macro, sending out their results using *signal-event* as they near the completion of an iteration and return to the top of its *loop-forever* to await its next activation. The top of the iteration cycle is typically a wait for the arrival of one or more signals. A set of SCORE forms, *asynch-wait*, *with-signal* and *with-multiple-signals*, are provided for this purpose. As this and other issues were addressed, the implementation has evolved into *reentrant-map* semantics that support the implementation of the psychological framework.

Parallel Execution: The parallel execution of procedures must be handled, both at the language level and in the simulator. At the language level, parallel execution may be expressed at two levels of abstraction. At the procedure level, the forms *race*, *join*, and *satisfy* manage the execution of subprocedures in parallel. Parallel procedures operating under *race* all complete when the first one completes. For *join*, every subprocedure must complete before subsequent procedures execute. Within these parallel forms a failing sub-procedure will cause the entire form to fail immediately. The *satisfy* form traps sub-procedure failures and completes execution of the form when the first successful sub-procedure completes. In the plan for a goal the same forms are used to manage parallel subgoal execution. The SCORE compiler expands the parallel forms into a run-time form emulating parallel execution as outlined here.

Rule-Based Behaviors: While it is likely that expert knowledge formulated as rules will not lead to expert performance (Holyoak, 1991), there are performance levels at which human behaviors are rule-based (Rasmussen, 1983; Dreyfus and Dreyfus, 1986). To model these levels of performance, a rule-based system was required as part of the OMAR tool set. Since the reasoning that is modeled frequently involves decisions on actions to be taken, the rule system may reference SCORE-based goal, plan, and procedure definitions, just as it may reference any SFL object. Rules are assembled in rule packets that focus on particular domain issues. A procedure that invokes a rule set defines the context in which the rule set operates—a function that must usually be covered by additional *if*- clauses.

Skill Levels in Human Performance: Human skill level has a significant impact on measured performance. HPP models developed in the OMAR environment must be capable of exhibiting performance associated with selected skill levels. Across a range of skill levels, there will be qualitative as well as quantitative changes in the actions that an agent will take to complete a particular task. In modeling these actions as procedures, the quantitative changes will reflect the efficiency with which the tasks are carried out. The transition from thoughtful deliberate action to automatic behavior is one of the more notable qualitative changes in performance with improved skill levels. The performance of a given task at these very different levels of performance will be represented by distinct sets of procedures, frequently with very different structure.

AN ATC SCENARIO

An air traffic control scenario was the first modeling effort undertaken using OMAR. HPP models were developed for the controllers and the two-person flight crews of each aircraft. In the scenario the controllers manage the airspace through radio conversations with the flight crews and telephone conversations with neighboring controllers. The flight crews attend to controller directives and maintain a dialogue to coordinate flight deck operations.

ACKNOWLEDGMENTS

The authors wish to thank Michael Young the USAF Armstrong Laboratory and Carl Feehrer of BBN for their continued support in this effort. The research reported on was conducted under USAF Armstrong Laboratory Contract No. F33615-91-D-0009.

REFERENCES

Agha, G.A. (1986). *Actors: A model of concurrent computation in distributed systems.* Cambridge, MA: MIT Press.

Cheng, P.W. (1985). Restructuring versus automaticity: Alternate accounts of skill acquisition. *Psychological Review, 92,* 414-423.

Deutsch, S.E., M.J. Adams, G.A. Abrett, N.L. Cramer, and C.E. Feehrer (1993). *RDT&E Support: OMAR Software Technical Specification,* AL/HR-TP-1993-0027. Wright-Patterson AFB, OH.

Dreyfus, H.L., and S.J. Dreyfus (1986). *Mind over machine.* New York: Free Press.

Edelman, G.M. (1987). *Neural Darwinism: The theory of neuronal group selection.* New York: Basic Books.

Edelman, G.M. (1989). *The remembered present: A biological theory of consciousness.* New York: Basic Books.

Holyoak, K.J. (1991). Symbolic connectionism: Toward third-generation theories of expertise. In K.A. Ericsson and J. Smith (Eds.), *Toward a General Theory of Expertise: Prospects and Limits.* Cambridge, MA: Cambridge University Press.

Kant, I. (1787/1965). *The critique of pure reason.* New York: St. Martin's Press

Logan, G.D. (1988). Toward an instance theory of automatization. *Psychological Review, 95,* 492-527.

Manning, C. (1987). *Acore: The Design of a Core Actor Language.* Masters Thesis, Massachusetts Institute of Technology.

Neisser, U. (1976). *Cognition and reality: Principles and implications of cognitive psychology.* San Francisco: W.H. Freeman.

Pashler, H., and J.C. Johnston (1989). Chronometric evidence for central postponement in temporally overlapping tasks. *Quarterly Journal of Experimental Psychology, 41A,* 19-45.

Rasmussen, J. (1983). Skills, rules, and knowledge; signals, signs, and symbols; and other distinctions in human performance models. *IEEE Transactions on Systems, Man, and Cybernetics, SMC-13,* 257-266.

Shiffrin, R.M., and W. Schneider (1977). Control and automatic information processing: II. Perceptual learning, automatic attending, and a general theory. *Psychological Review, 84,* 127-190.

Wickens, C.D. (1984). Processing Resources in attention. In P. Parasuraman and R. Davies (Eds.), *Varieties of attention.* New York: Academic Press.

Yates, J. (1985). The content of awareness is a model of the world. *Psychological Review, 92,* 249-284.

Young, M.J. (1992). *A Cognitive Architecture for Human Performance Process Model Research.* AL-TP-1992-0054, Wright-Patterson AFB, OH.

VALIDATION AND VERIFICATION OF DECISION MAKING RULES*

Abbas K. Zaidi
Alexander H. Levis

Center of Excellence in C3I
George Mason University
Fairfax, VA 22030

Abstract: A methodology for the validation and verification (V&V) of decision making rules is proposed. The methodology addresses the general problem of detecting problematic cases in a set of rules. The rules are expressed as statements in formal logic. The definition of decision rules in formal logic makes the problem general in terms of application domains, and also provides an analytical base for defining errors. The approach is based on viewing a rule base as an organization of information that flows from one process (rule) to another. Since Petri Nets provide a powerful modeling and analysis tool for information flow structures, the methodology transforms a set of decision rules into an equivalent Petri Net representation. The static and dynamic properties of the graph are shown to reveal patterns of Petri Net structures that correspond to the problematic cases. The tools and techniques presented in this paper are based on theory and are supported by software tools.

Keywords: Decision Making, Rule-based Systems, Petri Nets

1. INTRODUCTION

The need for the Validation and Verification (V&V) of decision rules arises from the fact that often processes an organization has to perform consist of sets of rules. An example of such an organization could be one that gets inputs from a set of sensors, identifies a task based on these inputs, and on the basis of attributes of the task determines a response. The set of decision rules may have been derived from a theory (normative or prescriptive) or they may have been obtained from empirical studies (descriptive) including knowledge elicitation from domain experts. The problem of dynamic *task allocation* in team decision making requires the partitioning of these rules across the individual decision makers. The manner in which these rules are obtained and the partitioning of the rule base across the decision making entities (human and machine) can introduce inconsistencies, incompleteness, redundancies, as well as problems in coordination.

In an organization, even if its physical and coordination structures are feasible, the presence of such problems in the set of rules assigned to decision making entities can result in poor performance and unreliable system response.

Several techniques and methods for the V&V of rule bases have been reported in the literature, especially by the Expert Systems (ES) community. Boehm (1984), Botten et al. (1989), Lee et al. (1994), and Zaidi (1994) provided condensed bibliographies of the related work in

verification and validation of expert systems. A collection of some of the recent approaches in V&V of expert systems is compiled in Gupta (1990) and Ayel (1991). However, all the reported cases have at least one of the following weaknesses: a) they are based on running test cases; b) they are restricted in terms of implementation, representation of rules, types of problems handled, and constraining requirements for the applicability of the approach; c) Errors defined only between pairs of rules or among rules in a subset of the entire rule base; d) combinatorial enumeration required to solve problems.

This paper proposes a new approach to the general problem of detecting problematic and erroneous cases in a set of rules, that overcomes the limitations of the existing methods. The methodology requires that the rules be expressed as statements in formal logic. An algorithm is then used to transform these conditional statements to an equivalent Petri Net (PN) representation. The transformation is based on a mapping between the logical operations of conjunction ($\wedge$), disjunction ($\vee$), and implication and the notions of synchronization, concurrency, choice, etc., of Petri Net theory. This visualization of a rule base in terms of an organization of interacting processes opens a wealth of analytical tools and techniques that have been developed by researchers and system analysts to perform structural and dynamic analyses on PNs. Once a rule base is transformed into a PN, the solution to the problem becomes a direct application of these analytical tools of Petri Net theory. The V&V of the rule base is done by first exploiting the structural properties of the Petri Net

---
*The research was conducted with support provided by the Office of Naval Research under contract numbers N00014-90-J-1680 and N00014-93-1-0912.

representation and then by constructing the Occurrence graph directly from the Petri Net representation.

In the next section, an example is presented, which illustrates the issues involved in task allocation in team decision making. The problem of task allocation is presented by first considering a hierarchical organization of five Decision Makers (DM). An initially "correct" rule base is then decomposed and partitioned so that the rules can be assigned to the individual DMs. The decomposed rule base is then checked for inconsistencies, incompleteness, and redundancies by the proposed methodology. The Petri Net transformation of the rule base is presented in Section 3. The results of the static and dynamic analyses performed on the Petri Net representation are presented in Sections 4 and 5, respectively.

2. PROBLEM DEFINITION

The example set of organizational decision rules illustrated in this section was motivated by the "Message Puzzle Task (MPT)" of Wesson and Hayes-Roth (1980). The MPT involved a game-like environment in which words and phrases move about in a two-dimensional grid that resembles a puzzle board. A group of players, each of whom can see a portion of the grid, must communicate among themselves to identify the moving items and eventually solve the puzzle. In the illustration presented in this section, a 4×3 grid (consisting of 12 sectors) representing the puzzle board is considered (Figure 1). Contrary to the MPT experiment, the messages on the grid do not move but appear on certain sectors in the grid. The messages consist of letters from an input alphabet of integers, where each integer represents an event. Based on the appearance of these messages in certain sectors, the set of rules infer a sequence of events out of a possible three sequences in which these events can occur.

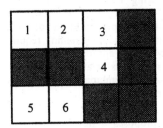

Figure 1 An Instance of the Grid

The set of all events E is given as: $E = \{1, 2, 3, 4, 5, 6\}$, where each integer represents the occurrence of an event. The appearance of an integer (from set E) in one of the sectors of the grid is considered as the basic input to the set of decision rules. The following 12 basic inputs are identified:

$$U = \{P1, P2, P3, ..., P12\} \tag{1}$$

where the proposition symbols P1-P12 represent the following information from the grid:

P1:   Integer 1 in Sector 1
P2:   Integer 2 in Sector 2
P3:   Integer 3 in Sector 3
P4:   Integer 1 in Sector 4
P5:   Integer 2 in Sector 5
P6:   Integer 3 in Sector 6
P7:   Integer 4 in Sector 7

P8:   Integer 4 in Sector 8
P9:   Integer 5 in Sector 9
P10:   Integer 6 in Sector 10
P11:   Integer 5 in Sector 11
P12:   Integer 6 in Sector 12

Based on these inputs, the rule base RB1 tries to interpret the inputs in terms of three possible outcomes (sequence of events), which are characterized as the main concepts of the rule base:

$$\Psi = \{A, B, C\} \tag{2}$$

where

A: Sequence of Events is 4, 5, 6, 3, 2, 1
B: Sequence of Events is 1, 2, 3, 4, 5, 6
C: Sequence of Events is 6, 5, 4, 3, 2, 1

Rule Base, RB1

Rule1: $P1 \wedge P2 \wedge P3 \rightarrow Q1$
Rule2: $P1 \wedge P2 \wedge P6 \wedge P7 \rightarrow R2$
Rule3: $P4 \wedge P5 \rightarrow Q3$
Rule4: $P3 \wedge Q3 \rightarrow R3$
Rule5: $P4 \wedge P5 \wedge P9 \wedge P12 \rightarrow R3$
Rule6: $R2 \wedge R4 \rightarrow C$
Rule7: $P7 \wedge P9 \wedge P10 \rightarrow Q2$
Rule8: $R3 \wedge R1 \wedge P8 \wedge P9 \rightarrow A$
Rule9: $P8 \wedge P11 \wedge P12 \rightarrow R4$
Rule10: $Q1 \wedge Q2 \rightarrow B$
Rule11: $P6 \wedge P7 \rightarrow R1$

where

Q1: Partial Sequence of Events is 1, 2, 3
Q2: Partial Sequence of Events is 4, 5, 6
Q3: Partial Sequence of Events is 2, 1
R1: Partial Sequence of Events is 4, 3
R2: Partial Sequence of Events is 4, 3, 2, 1
R3: Partial Sequence of Events is 3, 2, 1
R4: Partial Sequence of Events is 6, 5

The objective of the organization design problem is to decompose RB1 into five (possibly disjoint) sets of rules, where each set can be assigned to a DM in the organizational hierarchy presented in Figure 2, which also shows the possible interactions among DMs. It is assumed that the physical architecture of the organization provides these DMs with the means to communicate with each other whenever required by the rules. At first, the decomposition of the rule base is done in a *vertical* manner (Mesarovic et al., 1970); the decision rules are decomposed into three *layers* of sub-rules of increasing complexity. In the decomposed rule base RB2 rules of the form   "Ri $\wedge$ Rj $\wedge$ ... $\rightarrow$ _" represent the rules assigned to DM1. The Ri's represent the responses of the lower level decision makers communicated to DM1. Similarly, the set of rules defined at the intermediate layer is assigned to DM2, where the rules are of the form "Qi $\wedge$ Qj $\wedge$... $\rightarrow$ _". Finally, the set of rules at the lowest layer of the rule base is further decomposed *horizontally* (partitioned) into three sets and is assigned to three decision makers, DM3, DM4, DM5, where the rules are of the form "Pi $\wedge$ Pj $\wedge$... $\rightarrow$ _". The decomposition of the original rule base is done by taking into account the fact that the set of basic concepts U can be divided into the following three subsets (not necessarily disjoint), where each set represents information from a different sector (area of awareness) assigned to a decision maker (DM3, DM5, and DM4 respectively).

$$U_1 = \{P1, P2, P3, P4, P5\}$$
$$U_2 = \{P6, P7, P8, P9, P10\}$$
$$U_3 = \{P8, P9, P10, P11, P12\} \qquad (3)$$

The rules in RB2 correspond to the decomposition of the following rules in RB1: Rule2, Rule3, and Rule4 in RB2 correspond to the decomposition performed on Rule2 of RB1; Rule6, Rule7, and Rule8 in RB2 represent decomposition of Rule4 of RB1; Rule8, Rule9, Rule10, and Rule11 in RB2 correspond to Rule5 of RB1; Rule14, Rule15, Rule16, and Rule17 in RB2 correspond to Rule8 of RB1; Rule18, Rule19, and Rule20 in RB2 represent decomposition of Rule9 of RB1; and finally, Rule21 and Rule22 of RB2 correspond to Rule10 of RB1. The rest of the rules are left as they were; Rule1, Rule5, Rule12, Rule13, and Rule23 in RB2 correspond to Rule1, Rule3, Rule6, Rule7, and Rule11 in RB1, respectively.

From now on, any reference to the rules implies the rules in RB2 unless otherwise stated.

Rule Base, RB2

Rule1: $P1 \wedge P2 \wedge P3 \rightarrow Q1$
Rule2: $P1 \wedge P2 \rightarrow \neg Q1$
Rule3: $P6 \wedge P7 \rightarrow Q4$
Rule4: $\neg Q1 \wedge Q4 \rightarrow R2$
Rule5: $P4 \wedge P5 \rightarrow Q3$
Rule6: $P3 \rightarrow Q10$
Rule7: $Q10 \wedge Q3 \rightarrow R3$
Rule8: $P4 \wedge P5 \rightarrow Q3$
Rule9: $P9 \wedge P12 \rightarrow Q5$
Rule10: $P9 \rightarrow Q6$
Rule11: $Q3 \wedge Q5 \rightarrow R3$
Rule12: $R2 \wedge R4 \rightarrow C$
Rule13: $P7 \wedge P9 \wedge P10 \rightarrow Q2$
Rule14: $P8 \wedge P9 \rightarrow Q5$
Rule15: $R1 \wedge R3 \rightarrow Q3$
Rule16: $Q3 \wedge Q5 \rightarrow R5$
Rule17: $R5 \rightarrow A$
Rule18: $P11 \wedge P12 \rightarrow Q2$
Rule19: $P8 \rightarrow Q8$
Rule20: $Q2 \wedge Q8 \rightarrow R4$
Rule21: $Q1 \wedge Q2 \rightarrow R6$
Rule22: $R6 \rightarrow B$
Rule23: $P6 \wedge P7 \rightarrow R1$

where
Q4: Partial Sequence of Events is 4, 3
Q5: Partial Sequence of Events is 5, 6
Q6: P9 Observed
Q8: P8 Observed
Q10: P3 Observed
R5: Sequence of Events is 4, 5, 6, 3, 2, 1
R6: Sequence of Events is 1, 2, 3, 4, 5, 6

For illustration purposes, the following two sets of mutually exclusive propositions are also defined:

$$\mu_1 = \{P8, P10\}$$
$$\mu_2 = \{R1, R5\} \qquad (4)$$

The decomposition process by no means ensures the fact that the addition of new rules representing the replaced ones has not introduced errors. The effect of these new rules on the rest of the rule base can not be determined unless they are checked against the entire set of rules. This is where the methodology for the detection of problematic and erroneous cases comes into the picture: one started with a correct rule base, decomposed it and assigned different rules to different decision makers in a decision making organization, and as part of this process the resulting rule base no longer holds the property of being correct (at least it can not be claimed as correct.). Since problematic cases may involve rules across rules assigned to individual DMs, the rule base RB2 is considered as a whole.

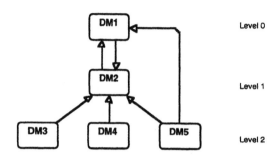

Figure 2   Organizational Hierarchy

3. PETRI NET REPRESENTATION OF RULES

An individual rule of the form "$P1 \wedge P2 \ldots \wedge Pn \rightarrow Q$" is transformed to a PN with a single transition with n input places, each representing a single input proposition Pi, and an output place representing the assertion Q (For a general and more detailed description of this technique, see Zaidi, 1994). The labels of the places and transitions correspond to the propositions and rules they represent. The rule of inference is implemented by the execution mechanism of PN theory. A token in a place represents the truth assignment of a proposition. If all the places in the preset of a transition representing the premises of a rule being satisfied are marked, the rule (transition) is enabled and can execute (fire) making all the consequents (output places) valid (marked). The entire set of decision rules is then obtained by *unifying* all the individual rules. The process represents the causal relationship among the rules and the facts of the knowledge base. This method unifies the rules by merging all the places with identical labels into a single place. The PN obtained as a result of applying this technique to RB2 is shown in Figure 3. In Figure 3, the basic inputs and main concepts defined for the rule base are shown aggregated into virtual inputs and outputs, P_{in} and P_{out}, with transitions T_{in} and T_{out}, respectively. The parts of the net in the figure that are drawn by a broken line represent this aggregation and do not represent a rule in the rule base.

It can be observed in Figure 3 that the Petri Net representation of a rule base does not provide an explicit relation between a predicate p and its negation ¬p. This lack of syntactic relation between predicates and their negations results in the absence of an implicit representation of the following axiom (tautology).

$$((\alpha \rightarrow \beta) \rightarrow (\neg \beta \rightarrow \neg \alpha)) \qquad (5)$$

where α and β are well-formed formulae of Propositional Calculus.

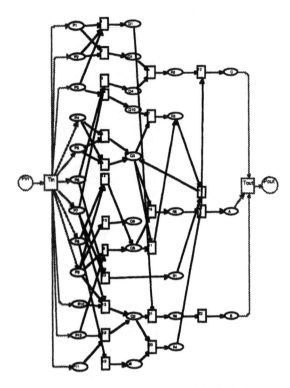

Figure 3 Petri Net Representation of the Rule Base

An obvious solution to this problem is to include all the implications of the decision rules to RB2, before transforming it to an equivalent Petri Net representation. The advantage of putting all the implications in the set of decision rules is that it will make the analyses independent of the form of the original rule base. The corresponding Petri Net will be called Enhanced Petri Net (EPN). However, such an approach will double the size of the rule base, and correspondingly with the Petri Net representation (not necessarily with the same ratio). Zaidi (1994) presented a graph-based approach that *normalizes* the EPN by removing certain *unnecessary* nodes present in the net. The parts of the EPN which help reveal the problematic cases are the only structures required and the rest of the net can be considered unnecessary. The algorithm that normalizes the EPN requires the use of two algorithms called *FPSO* and *FPSI*, two variants of an earlier FindPath algorithm (Jin, 1986). The *FPSO*(p) algorithm, when applied to a node p in a PN, collects all the nodes that have directed paths to p and returns a subnet whose nodes are the ones collected by the algorithm. The *FPSI*(p) algorithm, on the other hand, collects all the nodes to which p has a directed path and returns a subnet composed of those nodes. The normalized EPN represents that part of the rule base that is influenced by the valid inputs and influences the output of the system. The rules in the rest of the net are considered unnecessary for the obvious reason. The net presented in Figure 3 is a normalized net.

4. STATIC ANALYSIS

4.1 Detection of Incomplete Rules

Definition *Incompleteness* (Levesque, 1984)
   A knowledge base is incomplete when it does not have the information necessary to answer a question (*appropriately*) of interest to the system.

A . *Rules with Ambiguous Conditions (Ambiguous Rules)*
A rule base is incomplete if, given a valid input, the system can not interpret it in terms of applicable conditions in order to arrive at a conclusion. A cause of such incompleteness is the presence of at least one proposition in the premise which can not be explained or defined in terms of the basic concepts, U.

B. Rules with Useless Conclusions (Useless Rules)
Given an input, if the conclusion derived is not the main concept, the set of rules applicable to this input is incomplete.

C. Isolated Rules
A rule is an isolated rule if and only if
   • All the Propositions in its premise can not be explained in terms of the basic inputs, U, and
   • its conclusion is neither the main concept nor there exist(s) rule(s) taking R to an assertion α, where $\alpha \in \Psi$.

The incompleteness, as defined in this paper is determined by the lack of certain connections in the PN representation.

The isolated rules can be easily detected by a mere inspection of the normalized EPN; a rule in RB2 will be an Isolated rule if the transition t, representing the rule, and the transition t', representing the implication obtained through expression (5) are both absent in the net in Figure 3. In the illustration, no isolated rules are found.

On the other hand, some of the obvious incomplete cases can be detected by simply searching the net for dangling places — places with either inputs or outputs. The following algorithms provide a comprehensive approach to detect and identify incompleteness.

Algorithm for Ambiguous Rules
   • Apply *FPSI* to P_{in}, and compare the output of $FPSI(P_{in})$ with the original net. Those places of the original net that do not appear in the output of $FPSI(P_{in})$ identify Ambiguous rules.

Algorithm for Useless Rules
   • Apply *FPSO* to P_{out}, and compare the output of $FPSO(P_{out})$ with the original net. Those places of the original net that do not appear in the output of $FPSO(P_{out})$ identify Useless rules.

The application of these algorithms resulted in Rule10 identified as a Useless rule, since Q6 (P9 Observed) appeared as a dangling places with only input arcs. The transition in the preset of this assertion identify the corresponding Useless rule; the rule's consequent represent the assertion that the system might infer (from the basic inputs to the system) but can not interpret in terms of the required outputs.

4.2 S-Invariant Analysis

The S-invariant analysis looks at all the directed paths in the PN and searches these paths for problematic cases. The analysis is shown to reveal certain patterns of PNs that correspond to circular and inconsistent rules.

The analysis calculates the *minimal* S-invariants of the PN after it is transformed to an equivalent *Marked graph* representation in which every place has exactly one input and output. The algorithm to convert a PN into a Marked Petri Net (MPN) is given as follows:

- Merge the virtual input and output places, P_{in} and P_{out}, into a single external place P_e.
- $\forall p$ so that $|{}^\bullet p| = m$ (>1) and $|p^\bullet| = n$ (>1), create $m \times n$ copies of p, denoted by p^i, where $i = 1, 2, ..., m \times n$. Create n *links* from each of m transitions, $t_i \in {}^\bullet p$, to all of the n transitions, $t_j \in p^\bullet$, through these copies of the place p. The process will create $m \times n$ links between the input and output transitions of place p. (Figure 4 illustrate the process.)

A place p is said to constitute a link from a transition t_i to another transition t_j if ${}^\bullet p = \{t_i\}$ and $p^\bullet = \{t_j\}$. The process of converting a PN to a MPN preserves the connectivity of the original PN in the sense that if two nodes are directly connected in PN, they will remain connected in the MPN.

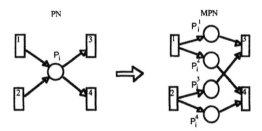

Figure 4 Process of Converting a PN to MPN

The net in Figure 3 is converted to an equivalent MPN representation, and the S-invariants are calculated for it. The following theorem characterizes the *S-components* corresponding to the *minimal Support* S-invariants of the MPN.

Theorem 1 (Hillion, 1986)
   The minimal S-components of a marked Petri net are exactly its directed elementary circuits.

Based on the results from Theorem 1, the following Proposition characterizes two different types of circuits in the MPN.

Proposition 1
   Let <Xi> be the minimal Support of the calculated S-invariant Xi;
- If <Xi> contains the external place P_e, then the S-component associated with <Xi> represents a directed path from a basic input to one of the main concepts.
- If <Xi> does not contain the external place Pe, then the S-component associated with <Xi> represents a loop inside the Petri Net structure .

The following Proposition interprets the minimal Supports calculated for MPN in terms of circular and possible inconsistent rules. The notation $p^{(\cdot)}$ used in the proposition indicates the fact that the place p might get replicated during the process of converting PN to MPN. The notation illustrates that the algorithm only

identifies the predicate p with no regard to the superscript assigned to it during the conversion process.

Proposition 2
- If there exists a <Xi> which contains places $p^{(\cdot)}$ and $q^{(\cdot)}$, where p and q $\in \mu$ (set of mutually exclusive concepts), then <Xi> indicates the presence of inconsistent rules in the rule base. The S-components associated with <Xi> and identifies the set of inconsistent rules
- If there exist <Xi> and <Xj>, where $p^{(i)} \in$ <Xi> and $q^{(j)} \in$ <Xj>, p and q $\in \mu$, and <Xi> $\cap$ <Xj> $\neq \emptyset$, then <Xi> and <Xj> indicate the presence of *possible* inconsistent rules in the rule base. The S-components associated with <Xi> and <Xj> identify the set of inconsistent rules.
- If there exists a <Xi> which does not contain the external place Pe, then <Xi> indicates the presence of circular rules in the rule base. The transitions in the S-component associated with <Xi> identifies the set of circular rules.

The first case presented in Proposition 2 corresponds to rules where it is possible to reach from a predicate "p" to its negation or to a mutually exclusive concept "q", which is not permissible in formal logic. The second inconsistent case presented corresponds to a rule which requires conflicting assertions to be simultaneously true in order to execute.

The application of this analysis performed on PN representing RB2 resulted in the following problematic cases:

- Circular Rules
   Rule15 and Rule7;
   Rule15 and Rule11

Figure 4 shows the reported circular cases in the PN. In one of the reported circular cases (Rule15 and Rule7) the system infers R3 (Partial Sequence of Events in 3, 2, 1) through Q10 (P3 Observed) and Q3 (Partial Sequence of Events in 2, 1) as illustrated by Rule7. On the other hand, Rule15 requires Q3 to infer R3. Similarly in the second case (Rule15 and Rule11), Rule11 infers R3 through Q3, where Q3 is required to be inferred through R3 in Rule15.

The analysis of RB2 reveals that the reported circular cases were introduced during the decomposition of rule 4, 5, and 8 of the original rule base RB1.

- Inconsistent Rules
   Supports <Pe, P6, R1, Q3, R5, A>, and <Pe, P7, R1, Q3, R5, A> indicate the presence of an inconsistent case, since both have R1 (Partial Sequence of Events is 4, 3) and R5 (Sequence of Events is 4, 5, 6, 3, 2, 1) present in them, where R1 and R5 are defined as mutually exclusive concepts. The following rules correspond to this inconsistency:
   Rule15 and Rule16.

The inconsistent rules, Rule15 and Rule16 $(R1 \wedge R3 \rightarrow Q3$ and $Q3 \wedge Q5 \rightarrow R5)$, were introduced in the rule base as a result of the decomposition of Rule8 of the original rule base RB1.

   Supports <Pe, P8, Q8, R4, C>, and <Pe, P10, $Q2^2$, R4, C> indicate the presence of another inconsistent case. The S-components corresponding to these Supports reveal the fact that Rule20 requires both P10 and P8 to be present simultaneously in order to fire. P10 and P8 are defined as mutually exclusive concepts, and can be true at the same time. Therefore, the following inconsistent rules are found:

Rule13, Rule19, and Rule20

Rule 19 and Rule 20 (P8→ Q8 and Q2∧Q8→R4) were added to the rule base as a result of the decomposition of Rule 9 of the original rule base, while Rule13 already existed in the rule base.

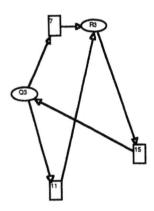

Figure 4 Circular Cases

5. DYNAMIC ANALYSIS

The dynamic analysis of the PN is performed on the Petri Net to explore the behavioral properties of the graph. The specific technique presented in this section is the Occurrence graph (OG) analysis. This analysis considers all the possible states of a net that can be reached from an initial state (marking) after a finite number of firings of transitions in the net.

With a given input, the analysis can generate an Occurrence graph for the PN and the Occurrence graph can be searched for states that can be reached from more than one distinct paths in the OG—*redundant cases*—, and for conflicting states reachable from a single input—*conflicting cases*. But, with no knowledge of the feasible input domain one is left with 2^n possible combinations of input vectors, (n = |U|). The construction of these many OGs is simply too large a problem. An approach is presented to overcome this problem.

The redundancies present in a rule base appear in the PN representation as places with multiple inputs representing the redundancy in arriving at a conclusion, and with places with multiple outputs representing several paths originating from the same input. This heuristic is used to determine whether the PN representing RB2 should be analyzed by the OG analysis for the redundant cases or not.

The heuristic can also be used to narrow down the search for redundancies to only those parts of the PN with such places. The following steps, performed on the PN, extracted the problematic parts of the original PN.

- For the PN construct a set, A;

  A = {p | p ∈ set of places in PN and |•p| > 1 }.

- From A select a place p_i (starting from the elements of set Ψ), and apply *FPSO*(p_i). The result is denoted as PN_i.
- If PN_i contains places with multiple output arcs, put PN_i in a set RN. Otherwise, declare PN_i free of redundant cases.
- Remove from A all the places that are present in PN_i. Repeat the process as long as there are elements in A.

The technique is applied to the PN representing RB2. Figure 5 shows the elements of RN.

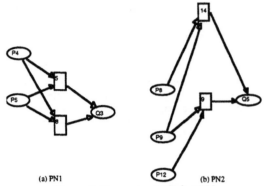

(a) PN1   (b) PN2

Figure 5 Elements of the Set RN

The only thing remaining is the determination of initial markings before any OGs can be constructed and searched for states that correspond to the problematic cases. The solution to this problem reverses the direction of all the arcs in each PN_i. The idea is to scan the feasible valid input space from a known and a much smaller set of outputs. The resulting net is denoted as PN_i' and defined as $PN_i' = (P, T, O, I)$, where PN_i is given by the tuple (P, T, I, O). The modified net is then initialized by putting a single token in the place p_i (Note, p_i is the place on which the FPSO algorithm was applied that resulted in the determination of PN_i). Since all the nets in RN are Y-nets, every transition in the PN_i' will have only one input, a single token in place p_i is enough to ensure that every place in the PN_i' will appear as marked at least once in the Occurrence graph constructed for PN_i'. For PNs that are not Y-nets, an algorithm (Zaidi, 1994) is used to convert them to an equivalent Y-net representation. It has been shown that the OGs constructed for the modified nets PN' exhibit the behavioral properties that can be interpreted in terms of the properties of the original net PN (Zaidi, 1994). The following propositions now states the results of the OG analysis applied to the nets in RN.

Proposition 3 *Redundant Rules*

Let PN_i ∈ RN;

Distinct firing vectors corresponding to multiple paths between two nodes of OG_i, or paths from one root node to two or more nodes having the same set (or a subset) of places as marked, identify redundant rules. The transitions whose corresponding components in the firing vectors are positive represent the rules that are redundant.

The algorithm proceeds by reversing the arrows of the nets PNi in set RN and initializing the modified nets, termed as PNi', by putting a token in source places. The Occurrence graphs OGi are constructed for each PNi'. The Occurrence graphs so obtained are searched for redundancies. The corresponding Occurrence graphs are shown in Figure 6. The following problematic cases are found:

• Redundant Rules
Rule5 and Rule8 are redundant rules, since they represent two copies of a single rule present simultaneously in RB2.

An investigation of the original rule base and the decomposition process reveals that the rule $P4 \wedge P5 \rightarrow Q3$ already existed in the original rule base, and another copy of this rule was added to the set of decision rules as a result of the decomposition of Rule5 of RB1.

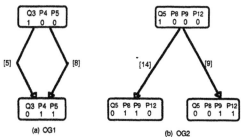

Figure 6 Occurrence Graphs

It turns out that an inconsistent case, in which conflicting assertions can be made through a single input, can be detected by a similar methodology used for the redundancies. The procedure is described as follows:

• For a set of mutually exclusive concepts μ_i present in the rule base, merge all the places in μ into a single virtual place p_i.
• Apply *FPSO*(p_i). The result is denoted as PN_i.
• If PNi contains places with multiple output arcs, put PN_i in a set CN. Otherwise, declare PN_i free of conflicting rules.
• Repeat for all sets of mutually exclusive concepts defined for the system in hand, except for those sets whose elements have already been identified in the conflicting rules reported by the static analyses.

The technique extracts the subnets of the PN with possible inconsistencies. Figure 7 presents the net obtained as a result. The subnet is then used as an input to the approach applied for detecting redundancy. The following Proposition characterizes the conflicting cases that can be detected by the analysis.

Proposition 4 *Conflicting Rules*
Let $PN_i \in CN$ and let there not exist any redundant case inside the decision rules represented by PN_i. Then, application of the techniques used for redundancy and results from Proposition 3 on Occurrence graph OG_i, identifies conflicting rules.

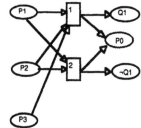

Figure 7 Element of the Set CN

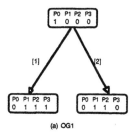

Figure 8 Occurrence Graphs for the Element of CN

The OG in Figure 8 identified the following inconsistent case in RB2:

• Inconsistent Rules
Rule1 and Rule2 are inconsistent, since OG1 in Figure 8 represents a redundant case. Rule1 infers Q1 by observing inputs P1, P2, and P3. However, Rule2 falsifies Q1 by observing a subset of the information that inferred the truth of Q1, thus creating an inconsistency.

6. CONCLUSION

The results of the example successfully illustrated the types of errors that might find their way into a set of decision rules during the decomposition of the rule base for assignment to several decision makers. The types of inconsistencies introduced during the decomposition of rules are common especially when different rules are decomposed by different individuals; different people tend to interpret identical inputs differently (sometimes oppositely) based on their assumptions about the problem and experience biases. The maintenance of consistency over a longer period of time becomes more and more difficult especially when the size of the rule base is also growing rapidly. The illustrative example started with a smaller rule base and resulted in a much larger set of decision rules. Some of the problematic cases were intentionally introduced to illustrate the capabilities of the methodology, but the resulting reported cases included cases that were not apparent during problem formulation.

The methodology for uncovering problematic rules in a rule-based system presented stands out among all the reported V&V methods due to its generality in terms of the constraints imposed on the representation of the rule base, flexibility as it can be applied at any stage of knowledge acquisition. A unique feature of the approach is that the inconsistencies are defined both *syntactically* and *semantically* (Ginsberg, 1988), however, semantic inconsistencies are defined with respect to user provided set of mutually exclusive concepts.

The approach presented in this paper can easily be extended to rule bases in first-order predicate calculus (Zaidi, 1994). The approach uses the Colored Petri Net formalism to solve the problem.

REFERENCES

Ayel M. and Laurent J. (1991), eds., "Validation, Verification and Test of Knowledge-Based Systems", John Wiley & Sons, Chichester.

Boehm B. W. (1984), "Verifying and Validating Software Requirements and Design Enhancement," IEEE Software, January, 1984.

Ginsberg A. (1988), "Knowledge-Base Reduction: A New Approach to Checking Knowledge Bases for Inconsistency and Redundancy", Proc. of 7th AAAI Conference.

Gupta U. G. (1991), ed., "Validating and Verifying Knowledge-Based Systems", IEEE Computer Society Press, Los Alamitos CA.

Hillion H. P. (1986), "Performance Evaluation of Decisionmaking Organizations using Timed Petri Nets," LIDS-TH-1590, Laboratory of Information and Decision Systems, MIT, Cambridge, MA.

Jin V. Y. (1986), "Delays for Distributed Decisionmaking Organizations," LIDS-TH-1459, Laboratory for Information and Decision Systems, MIT, Cambridge, MA.

Lee S. and O'Keefe R. M. (1994), "Developing a Strategy for Expert System Verification and Validation," IEEE Transactions on Systems, Man and Cybernetics, vol. 24, no. 4, pp. 643-655.

Levesque H. J. (1984), "The Logic of Incomplete Knowledge Bases", in *On Conceptual Modeling: Perspective from Artificial Intelligence, Databases and Programming Languages*, M. L. Brodie, J. Mylopoulos and J. W. Schmidt (eds.), Springer-Verlag, New York, pp. 165-189.

Mesarovic M. D., Macko D., and Takahara Y. (1970), "Theory of Hierarchical, Multilevel Systems," Academic Press, New York.

Wesson R. and Hayes-Roth F. (1980), "Network Structures for Distributed Situation Assessment," R-2560-ARPA, Rand Report, Santa Monica, CA.

Zaidi S. A. K. (1994), "Validation and Verification of Decision Making Rules," Technical Report, GMU/C3I-155-TH, Center of Excellence in C3I, George Mason University, Fairfax, VA.

LINGUISTIC KNOWLEDGE AND FUZZY MEMBERSHIP ELICITATION

JIE REN* and THOMAS B. SHERIDAN *

*Massachusetts Institute of Technology, Human-Machine Systems Laboratory, Cambridge, MA 02139, USA

Abstract. Linguistic rule bases provide an alternative or complement to analysis where traditional mathematical models do not exist and human expertise is widely available. A linguistic term inherited three types of inexactness (generality, ambiguity and vagueness) which can be dealt with by membership functions. This paper discusses imprecision of human knowledge and the process of membership elicitation to establish linguistic rules. A Random-Sampling-Without-Replacement technique is proposed to maximize the efficiency of knowledge elicitation from human experts. A fuzzy rule network (FRN) structure for representing linguistic knowledge is implemented, and a method for adaptive rule calibrations is presented to maximize overall rulebase performance.

Key Words. Fuzzy Modelling; Linguistic Variables; Fuzzy Systems; Calibration

1. Introduction

Approximate reasoning plays a central role in human thinking as well as in the inquiry and meaning of various soft or empirical sciences. All human-in-the-loop systems are more or less affected by subjective expertise which is not the form of a simple boolean-type logic. Human logic theory seeks to classify the varieties of subjective reasoning and the codification of rules, to assess degrees of belief of conclusions and to investigate rule rationality.

In the knowledge elicitation process, Szolovits and Pauker (1978) found that experts often refuse to give precise numerical estimates of outcomes and tend to make verbal judgments. Kochen (1992) further concluded that precision and clarity beyond a certain point may be as ineffective in representing the truth as vagueness.

Imprecision arises from a variety of sources, such as incomplete knowledge, inexact language, ambiguous definitions, inherent stochastic characteristics, measurement errors. It has been quantified primarily by means of probability theory. Several authors have emphasized the need for differentiating among the sources of imprecision underlying particular assumptions or items of evidence. The linguistic knowledge base, or the fuzzy knowledge system, is a promising approach. It draws conclusions by measuring linguistic rules and the consistency of facts in terms of possibility, and thereby forms a special kind of knowledge system.

For a linguistic term, there exist three kinds of inexactness: *generality*, in which a word applies to a multiplicity of objects in the field of reference; *ambiguity*, which occurs when a finite number of alternative meanings have the same phonetic form; and *vagueness*, in which there are no precise boundaries to the meaning of a word.

This inexactness in linguistic rules can lead to disaster. For example (Behn, 1982), in 1961, President Kennedy ordered a review of the CIA's plan for an invasion of Cuba by expatriates. The general in charge concluded that the chances of overall success were *fair*, by which allegedly he meant that they were thirty percent. But the report to the president did not mention the percentage; instead the report stated: "The plan has a *fair* chance of ultimate success". The rest is history. Later the general recalled: "We thought other people would think that *a fair chance* would mean *not too good*".

The membership function in a fuzzy set can be used to overcome such linguistic inexactness. Generality occurs when the portion of the universe of discourse where the membership value equals one is not just one point; ambiguity occurs when there is more than one local maximum of the membership function; and vagueness occurs when the function takes values other than just 0 and 1.

The essential idea of a fuzzy rule base is to make use of human expertise with linguistic *if-then* rules. The rule base emulates the behavior of human experts to derive proper actions. However, there is no consensus on how best to represent linguistic variables (such as *cold, warm, short, tall, et cetera*) as membership functions. It sometimes is concluded that there are two attributes that generally describe fuzzy membership functions: the crossover point and the support (Fleischman, 1992). The crossover point is a point whose degree of membership is 0.5, and the support is the set of points having a degree of membership greater than 0. The crossover point and support are subjective and therefore membership elicitation is an essential procedure for defining linguistic terms with the corresponding expertise.

2. Membership Elicitation

The process of generating fuzzy membership functions can be done by using clustering algorithms (Bao, 1984) (Takagi, 1991). With this method, a user defines the number and relative order of linguistic terms as they apply to a particular attribute. The clustering algorithm determines membership functions based on sample data.

It is necessary to generate enough sample data for analysis in the clustering process. The objective of designing a membership elicitation experiment (with human subjects) is to minimize the number of samples or maximize the credibility of the resulting membership function.

Random Sampling Without Replacement

Membership elicitation experiments used to be designed in such a way that the subjects were prompted with inputs generated by a computer-supplied random generator. While it is true that the computer-supplied random generator theoretically has the property of uniform distribution, the assumption is based on *infinitely large* samples. For any real experiment with subjects, however, the number of samples is limited and therefore the property of uniform distribution can only be approximated.

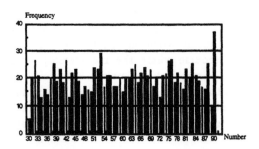

Fig. 1. Histogram from the Compiler-supplied Random Generator

For instance, we have a series of random values

from 30 to 90 and set the total number of samples as 1220. For a desired uniform distribution, each value within the range should appear exactly 20 times. However, the histogram of limited samples from a compiler-supplied random generator shows that the actual frequency varies from 13 to 27 as shown in Figure 1, where the ordinate is each random number and the abscissa is the frequency. It is seen that if we use this *random* data set in experiments without a complex compensation algorithm, the resulting membership function will certainly be distorted.

To solve this problem, a method called *random sampling without replacement (RSWOR)* is proposed. The main idea is to create a number pool. First, a fixed number of samples for each discrete value in the range are thrown into the pool. Then one number is picked at random from the pool until all numbers are exhausted. In this way, the entire sample will be guaranteed a strictly uniform distribution and the resulting membership will not be distorted due to uneven sample distributions.

Figure 2 shows the resulting histogram of a random generator using the *RSWOR* method. The distribution is strictly uniform.

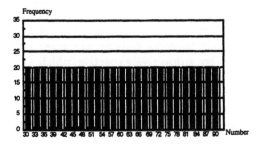

Fig. 2. Histogram from the RSWOR Method

Linguistic Term Evaluator

A computer program, *Linguistic Term Evaluator*, has been developed to elicit fuzzy membership functions using the *RSWOR* method. Figure 3 shows an example screen display for evaluating the linguistic variable *Driving-Speed-on-the-Highway*. Driving speed was chosen for demonstration because of the easy availability of experts (other linguistic variables could be evaluated using the same method).

The speed ranges from 30 mph to 90 mph, which is shown by a grey rectangle. It has 61 discrete possible values. For each value, the sampling frequency is set to 5 and hence there are total $61 \times 5 = 305$ numbers in the pool. Five linguistic terms, *Very Slow, Slow, Normal, Fast and Very Fast*, are evaluated.

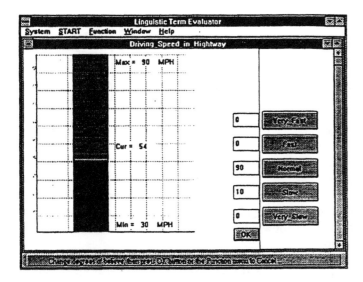

Fig. 3. Membership Elicitation Experiment

The subject is prompted with a random *RSWOR* speed, and is asked to evaluate each speed by clicking a mouse on the appropriate linguistic buttons on the right side of the screen. Upon clicking the mouse, a row of degree of belief from 0 to 100 (percentage) appears on the left of these buttons. As a default, the degree of belief for the button clicked is 100 and the others are 0. However, the subject has the option to modify the numbers according to his/her judgment (the degrees of belief for five terms are constrained to add up 100 for consistency). For instance, in Figure 3 the subject was prompted with a random speed of 54 mph. He evaluated this speed as *Normal* with a degree of belief of 0.9 and as *Slow* with a degree of belief of 0.1.

A weighted histogram of all the degrees of belief is constructed after the *RSWOR* samplings are completed. Our experiments show that all the memberships for five linguistic terms have reasonable ranges and move to the left for more conservative drivers.

Figure 4 shows a typical histogram, without any modifications, that was obtained by using the methods discussed above. The ripples partially reflect the imprecision of human judgment.

A trapezoidal membership form (with four unknown parameters) is used to smooth the curves in Figure 4, because of its simplicity of being linear. Figure 5 shows the membership function for *Driving-Speed-On-the-Highway* after data fitting.

The linguistic rules are further calibrated by an adaptive learning mechanism based on the elicited

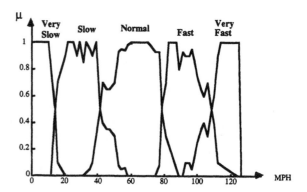

Fig. 4. Histogram of Driving Speed

membership functions.

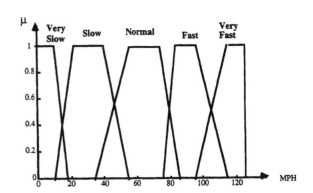

Fig. 5. Membership Function of Driving Speed

3. Linguistic Rule Calibration

In contrast to modern control theory which has a design scheme such as pole-placement or state-space design approach, the design of a fuzzy rule base depends entirely on the knowledge and experience of experts, or the physical sense as well as intuitions that are not well structured.

The issue of recursive learning is therefore of particular interest for a fuzzy rule base. Instead of designing a *perfect* rule base, the strategy is to allow it to *learn* proper input-output relationships. Mamdani (1975) proposed a performance index table for updating rules with some success, but the generation of the performance index table itself is not an easy job. Barto, Sutton and Anderson (1983) used two neuron-like elements to solve the cart-pole problem. Lee and Berenji (1989) extended the concept further by incorporating neural network architecture in their work. However, these structures suffer from a lack of generality and may be difficult to apply for larger scale systems due to the fact that developing mathematical functions for the trace function and the credit assignment are not trivial.

Later, various types of fuzzy-neural-network knowledge bases have been proposed, such as the associative memory neural networks which controls an experimental helicopter by Yamaguchi, Goto and Takagi (1992). These approaches adapt the neural networks' learning algorithm as well as the weight structures.

Fuzzy Rule Network (FRN)

A fuzzy rule structure – Fuzzy Rule Network (*FRN*), which has the capability of adaptive supervisor learning based on teaching of input-output pairs, is implemented. The structure of a *FRN* is shown by an example in Figure 6 and Equation 1.

$$IF\ x_1\ is\ \tilde{X}_1^i\ and\ x_2^k\ is\ \tilde{X}_2\ THEN\ y\ is\ w_{ik} \quad (1)$$

where both $\tilde{X}_1$ and $\tilde{X}_2$ have the universe of discourse of {*small, large*} and w has the universe of discourse of {w_1, w_2, w_3, w_4}.

A *FRN* has three layers. The input layer, or fuzzification layer, includes a fuzzifier whose task is to match the values of the input variables i to the linguistic terms k used in fuzzy rules. μ_{ik} is the degree of belief for the corresponding partitioned linguistic terms. The hidden layer, or fuzzy logic layer in this network corresponds to the rules and the decision-making logic which is character-

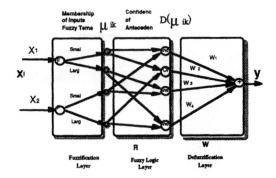

Fig. 6. Fuzzy Rule Network (FRN)

ized by the triangular T-norms and T-conorms[1]. $D(\mu_{ik})$ is the membership over all the antecedents of a rule.

The output layer (defuzzification layer) maps a space of fuzzy actions, defined over an output universe of discourse into a space of nonfuzzy value. The commonly used method is the center-of-area defuzzification method which states:

$$\hat{y} = \frac{\sum_{i=1,k=1}^{m,n} [D(\mu_{ik}) \cdot w_{ik}]}{\sum_{i=1,k=1}^{m,n} D(\mu_{ik})} \quad (2)$$

where w_{ik} is the right-hand value of a rule.

Equation 2 can be rewritten as:

$$\hat{y} = \sum_{i=1,k=1}^{m,n} (r_{ik} \cdot w_{ik}) \quad (3)$$

where

$$r_{ik} = r_{ik}(x_1, x_2) = D(\mu_{ik}) / \sum_{i=1,k=1}^{m,n} D(\mu_{ik}) \quad (4)$$

The above equations show that the output layer in the *FRN* combines fuzzy inference and defuzzification into one single procedure. This approach enables us to estimate what really is in a human subject's mind by identifying the w_{ik} from the input-output pairs in a way similar to neural network learning.

In contrast to a neural network, which has no systematic way to determine the number of neurons and in which the weights have no physical meanings, the *FRN* has a meaningful structure and the identified parameters have physical meanings. In the *FRN*, the number of nodes and their distributions are closely related to the fuzzy variables

---

[1] T_{norm} and T_{conorm} are fuzzy logic operators.

and their partitions. The number of nodes in the fuzzification layer depends on the number of input variables as well as the number of partitions in their corresponding universes of discourse. The fuzzy logic layer represents the antecedent of the linguistic rule base R and forms fuzzy sets $\tilde{X}_1 \times \tilde{X}_2$ in the Cartesian space of $U \times V$. The number of nodes in the fuzzy logic layer equals the number of elements in the Cartesian space. Each element has the value of $\mu_{ik} = \mu_{(X_{1i} \ and \ X_{2k})}$. The number of nodes in the defuzzification layer equals the number of outputs. Each learned parameter between the fuzzy logic layer and the defuzzification layer is the decision value for the corresponding rule.

If we rewrite Equation 3 in a matrix form, it is:

$$\hat{y} = R \times W \tag{5}$$

where

$$R = \begin{bmatrix} r_{11} & r_{12} & \cdots & r_{1n} \\ r_{21} & r_{22} & \cdots & r_{25} \\ \cdot & \cdot & & \cdot \\ \cdot & \cdot & r_{ik} & \cdot \\ \cdot & \cdot & & \cdot \\ r_{m1} & r_{m2} & \cdots & r_{mn} \end{bmatrix} \tag{6}$$

and

$$W = \begin{bmatrix} w_{11} & w_{12} & \cdots & w_{1n} \\ w_{21} & w_{22} & \cdots & w_{25} \\ \cdot & \cdot & & \cdot \\ \cdot & \cdot & w_{ik} & \cdot \\ \cdot & \cdot & & \cdot \\ w_{m1} & w_{m2} & \cdots & w_{mn} \end{bmatrix} \tag{7}$$

Recursive Learning of an FRN

The weight matrix W is used to calibrate an FRN as shown in Figure 7.

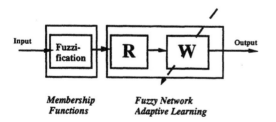

Fig. 7. Fuzzy Recursive Learning

It is noticed that R, which is a non-linear function, is the relation matrix and that W, which is a linear function, is the fuzzy network weighting matrix. For the given input (x_1 and x_2 in Figure 6), matrix R is fixed when the membership functions are fixed. W is an adjustable matrix

in which each element can be identified in such a way that the output $\hat{y}_i$ (computed from Equation 5 with the input values x_i) agree as closely as possible with the measured output y_i. That is, the cost function

$$J(W) = \frac{1}{2} \sum_{n=1}^{N} \epsilon_k^2, \tag{8}$$

is minimized, where

$$\epsilon_i = y_i - \hat{y} = y_i - \sum_{i=1, k=1}^{m,n} (r_{ik} \cdot w_{ik}) \tag{9}$$

$$i = 1, 2, ..., N \tag{10}$$

Because the relation between the input and output of W is linear, we can recursively update the weighting matrix W based on the teaching data using the least-squares method.

4. Conclusion

In contrast to classical mathematics and classical logic which demand precise and quantitative data, a human expert possesses the ability to reach reasonably good decisions based on imprecise, qualitative data. It has been shown that a linguistic rule base could generate a design that achieves satisfactory results much faster than the classical approach. Since it is found that experts often refuse to give precise numerical estimates of outcomes, and that excessive precision and clarity may be as ineffective as excessive vagueness, a fuzzy rule base is a good candidate to represent these *fuzzy* rules.

A membership function in fuzzy set theory can be used to overcome linguistic inexactness. The shapes as well as the support of a membership function are critical to a linguistic rule. To maximize the efficiency of knowledge elicitation from human experts, a Random-Sampling-Without-Replacement technique is proposed. An example of Driving-Speed-on-the-Highway is discussed and the result shows that it agrees well with a human driver's expertise. Based on the fuzzy relation and fuzzy defuzzification, this paper implements a network-type fuzzy knowledge structure with a weight matrix subject to calibration from teaching data. Each element in the weight matrix has its physical meaning in the rules and can be used as an indication of convergence. This adaptive approach reduces design time, and improves performance, accuracy and reliability at lower cost.

5. REFERENCES

Bao, S. (1984). *Pattern Recognition, Application to Large Data-Set Problems.* Marcel Dekker, Inc.

Barto, A.G., Sutton. R.S., and Anderson, C.W. (1983). Neuronlike adaptive elements that can solve difficult learning control problems. In: *IEEE Transactions on Systems, Man, and Cybernetics*, V13, p834-846.

Behn, R.D. and Vaupel, J.W. (1982). *Quick Analysis for Busy Decision Makers.* New York: Basic Books, Inc.

Fleischman, R. M. (1992). *Supporting Fuzzy Logic Selection Predicates on a High Throughput Database System.* M.S. Thesis, MIT, Cambridge, USA.

Kochen, M. (1979). Enhancement of coping through blurring. In: *Fuzzy Sets and Systems*, V2, p37-52.

Lee, C.C. and Berenji, H. R. (1989). An intelligent controller based on approximate reasoning and reinforcement learning. In: *Proc. of IEEE Int. Symposium on Intelligent Control*, Albany, NY.

Mamdani, E. H. and Assilian, S. (1975). An experiment in linguistic synthesis with a fuzzy logic controller. In: *International Journal of Man-Machine Studies*, 7(1), p1-13.

Szolovits, P. and Pauker, S.G. (1978). Categorical and probabilistic reasoning in medical diagnosis. *Artificial Intelligence*, V11, p115-144.

Takeda, H. and Asai, K. (1984). Fuzzy solution in fuzzy linear programming problems. In: *IEEE Transactions on Systems, Man and Cybernetics*, SMC–14, p325-328.

Takagi, H. and Hayashi, I. (1991). Artificial neural network driven fuzzy reasoning. In: *Int. J. of Approximate Reasoning*, 5(3), p191-212.

Yamaguchi, T., Goto, K. and Takagi, T. (1992). Intelligent control of a flying vehicle using fuzzy associative memory system. In: *IEEE International Conference on Fuzzy Systems*, p1139-49, San Diego, CA.

EVOLUTION OF IDEAS REGARDING
THE PREVENTION OF HUMAN ERRORS

V. De Keyser *

University of Liege, Faculty of Psychology and Educational Sciences
Work Psychology Department, Bd. du Rectorat, 5 (B32)
B-4000 Liege (BELGIUM)

Abstract : Human errors have proved to bear severe economical, social and environmental consequences when breaking down the safety barriers of high risk industrial systems, such as systems for energy production, and this problem is maybe still more worrying in Eastern & Central European countries than in the EEC.
This paper introduces a formal session organized around an European Network on 'Prevention of Human Errors in Systems for Energy Production and Process Industry' supported by the CEC. It deals with the possibility of transfer of research on human error between Western and Eastern European countries.

Keywords : Human reliability, transfer of research, error prevention, error management, East-West European countries

1. INTRODUCTION

All over the industrialised world, the consequences of human errors and of system breakdowns can be dramatic. They have been revealed to a large public in the most spectacular and tragic manner by catastrophes like Chernobyl, Three Miles Island, Bhopal, Zeebrugge, Teneriffe, Challenger, Airbus, etc.

Human errors have proved to bear severe economical, social and environmental consequences when breaking down the safety barriers of high risk industrial systems, such as systems for energy production, and this problem is maybe still more worrying in Eastern & Central European countries than in the EEC.

Some impressive numbers are reminded :
- As it is well known, the most serious technological catastrophe of human history - the Chernobyl Nuclear Power Plant accident - was clearly connected with a series of severe operator's errors and violations. See the *All-Russian Programme for Liquidation of Consequences of the Chernobyl NPP Accident*, under the auspices of the Government of the Russian Federation, Prof. A.B. Leonova and B. Levichkovsky, Moskow State University.
- Human factors also contributed to Three Mile Island NPP incident in the USA[1]
- According to official data, there were 118 reactor scrams at NPPs of former Soviet Union in 1989, including 55 caused by human errors and 63 due to equipment failures[2].
- In Romania, the numbers of incidents in the thermoelectric power plants are the following ones : (percentage of grave incidents are given between parentheses) : 1987 : 2688 (47.7 %), 1988 : 2541 (51.2 %), 1989 : 1588 (44.4 %). Numbers provided by Prof. Gh. Iosif of the Institute of Psychology of the Romanian Academy.

---

[1] Moll, T.H. & Sills, D.L. (1981). *The Three Mile Island Nuclear Accident : Lessons and Implications.* New-York, NY : Academy of Sciences.

[2] Munipov, V.P. (1991). New technology and culture : a macroeconomic analysis of the Chernobyl NPP accident. In Roe, R. & coll. (Eds). *Technological Change Process and its Impact on Work*. Budapest : European Network of Organizational and Work Psychologists.

* Coordinator of a Human Capital Mobility (HCM) network of the European Communities, a network for the prevention of human error in risk industries between Western and Eastern Europe.

- British Steel looses L 10K per hour if the steel mill is shut down. In a competitive steel market, quality of product is important. One roll costing L 30K may have to be rejected. Numbers provided by Prof. K. Duncan of the School of Psychology of the University of Wales in Cardiff.

To prevent continuation of these losses in both human and economic terms, join efforts of Eastern and Western researchers are needed in order to better assess the conditions for human errors and incidents, especially regarding cognitive and organisational aspects, and to ensure both prevention and error management.

Research on human error in Europe and in the United States has been widely developed over recent years. Three currents of analysis can be distinguished, which appeared successively, but which today coexist. The first is a *quantitative* current, on Human Reliability Assessment. This current gives to error the status of a technical failure, evaluating the probability of its appearance. The perspective can be reductive, but it allows the inclusion of human error within the evaluation of risks associated with the design and choice of technical systems. It opens the door to the idea of debate on the *level of acceptable risk*[3] in the society- a democratic debate indeed. But this debate relies on the possibility of arranging errors into categories that are exclusive of each other and to post them; it foresees a standardization of the method of analysis. This condition is not fulfilled in the facts. There are numerous classifications, having different advantages, and any comparison of the results of one research to another is practically impossible[4]. The second current is qualitative; it gives to human error its own status. This latter is no longer a product derived from technical reliability, but rather a characteristic *of the limits of cognitive functioning*- which is explored in its modalities of attentional control as automatic. This current also insists on heuristics, on shortcuts of human reasoning and their adaptability to uncertain and dynamic situations[5]. *The accent placed on the situation*-or the context- will be the distinctive sign of the third research current. The concepts of "situated cognition" of "man in situation" emerge[6]. The error is studied empirically; the conditions in which it appeared are analyzed. Taxonomies of risk situations are developed; the complexity, the risk and the dynamism of these situations are brought out[7]. The methods of analysis privilege, in this case, the process of error and accentuate the chain of factors that can lead to an accident.

When one makes an assessment of these three currents, it is very positive.. In the high-risk industries [8] - the collection and analysis of human errors reoriented the design of technical systems, training and organization of work[9]. Furthermore, some sectors that were formerly removed from these preoccupations are now opening themselves up to them. Human reliability in anesthesia, for example, borrows its models from aeronautics and continuous processes, and full-scale simulators are appearing in the United States and in Europe[10].

Taking into account this evolution, it was natural to think that a transfer of knowledge in the area of human error was foreseeable between Eastern and Western Europe. Even if the impact remained modest, the spread of the results of these studies seemed to be able to contribute to a greater safety in

---

[3] Fischhoff, B., Lichtenstein, S., Slovic, P., Derby, S.L., Keeney, R. L. (1981). *Acceptable risk*. Cambridge : Cambridge University Press.

[4] Leplat, J. (1985). *Erreur humaine. Fiabilité humaine dans le travail*. Paris : A.Colin

[5] cf. notably :
- Faverge, J.M. (1970). L'homme, agent de fiabilité et d'infiabilité du systeme technique. *Ergonomics*, 13, 3, 301-329.
- Faverge, J.M. (1970). *Les bases heuristiques de l'ergonomie*. Bruxelles : Office Belge pour L'Accroissement de la Productivité.
- Tversky, A. & Kahneman, K. (1974). Judgement under uncertainty : Heuristics and biases. *Science*, 185, 1124-1131.
- Reason, J. (1992). *Human Error*. Cambridge : Cambridge University Press

[6] cf. notably:
- Sternberg, R.J. & Wagner, R.K. (1994). *Mind in context. Interactionist perspectives on human intelligence*. Cambridge : Cambridge University Press.
- Klein, G.A., Orasanu, J., Calderwood, R. & Zsambok, C.E. (1993). *Decision Making in Action : Models and Methods*. Norwood, N.J. : Ablex Publishing Co.
- Hollnagel, E., Mancini, G. & Woods, D.D. (Eds) (1988). Cognitive Engineering in Complex Dynamic Worlds. London : Academic Press.

[7] De Keyser, V. (1994). Le temps dans la recherche en ergonomie : développements récents et perspectives. Papier de position. Proceedings of the 12th Congress of the International Ergonomics Association. Toronto (Canada), 15-19 août 1994. To be published in *Ergonomics*.

[8] such as nuclear electricity production, aeronautics, chemistry and petrochemistry.

[9] De Keyser, V. (1991). La prise en compte du facteur humain dans les technologies à risque. In N. Witkowki (Ed.). *L'Etat des Sciences et des Techniques*. Paris : Editions La Découverte.

[10] cf. notably :
- Gabba, D., Fish, K. & Howard, S. (1994). *Crisis Management in Anesthesiology*. New York : Churchill Livingstone.
- De Keyser, V. & Nyssen, A.S. (1993). Les erreurs humaines en anesthésie. *Le travail humain*, numéro spécial d'hommage à Jacques Leplat, 56, 2-3, 243-266.

the East. A research network, grouping Russian, Rumanian, Hungarian, German, Dutch, French and Belgian researchers was thus created, under the auspices of the Commission of the European Community, and within the framework of the program of Human Capital Mobility (HCM). The reflections that follow, and especially the diverse presentations of this symposium, form an initial assessment of these exchanges. They urge caution, since the weight of political, economic and cultural factors is heavy where prevention is concerned. We will successively discuss three points: the relationships between reliability and safety, between prevention and recovery, and functioning in a deteriorated mode.

2. HUMAN RELIABILITY AND SAFETY

As early as the 1970s in Europe the ECSC (Economic Coal and Steel Community) research program, on safety in the mines and in industry, placed less emphasis on *safety* than on *reliability* of man-machine systems[11]. Safety was only connected with accidents; reliability accounted for all the malfunctions of the man-machine system: breakdowns, errors, incidents and accidents. The source or this semantic displacement was the discovery that the same risk factors produced both one and the other. Among these risk factors, those relative to training, the the organization of work, to the design and to the maintenance of technical devices stand out in a particular way; they are often latent in the work situation- that is to say easily spotted outside of any incidental or accidental context. It is only when combined with others that they are activated and link up to lead to the accident. In this fatal spiral, human errors very often appear as fortuitous elements, that contributed to the accident, but on a turf already mined by latent risks. And it is onto these latter that prevention should first be directed.

Twenty years later, Reason[12], with a totally different progression, arrived at the exact same conclusions. Starting from laboratory research, and examples drawn from everyday life-with a cognitivist paradigm- he emerged onto industrial prevention. He targeted the importance of work organization and of the design of installations- and brought out the existence of latent risk factors that human error can activate suddenly.

The common conclusions of these two research currents illuminates the importance of methods of analysis that are not limited to error but encompass the situation in its entirety. For example, procedures based on the logical and chronological

outcome of the incidental situation belonging to the family of fault trees. They allow the detection, in an incidental process, of the the role of factors that would otherwise be overshadowed by the accident, or by the focus on of human errors. In the HCM network mentioned, the team of the University of Liège, and that of the Eindhoven University of Technology, used techniques of this type to bring out the interrelations between latent risk factors and errors.

But the start-up and the impact of these techniques depend essentially on the culture and the organizational management of the firm. The willingness of the direction, the fluidity of communications, the organizational coherence and the social climate are just some of the factors that can be the driving force of prevention; this must be based on a social dynamism. Szekely and De Keyser[13] compared the use of some simplified fault tree method - 'l'arbe des causes' - in four different locations of the same firm. For one year, in each location, a team of workers analyzed the work accidents. In the location where safety was the worst, and the latent risk factors the most difficult to modify, the powerlessness of the team was translated into a progressive impoverishment of the fault tree. As time passed, the workers gave up including within the diagrams the factors linked to the design of machines and to the organization of work, and eventually limited themselves only to human error. This insistence on the human factor did not appear in the locations where the directors and the union organizations closely followed research, and each in his own way weighed the adoption of prevention measures. A view that privileges human error too often creates, within the firm, a powerlessness to overcome other factors.

The Eindhoven University of Technology undertook the treatment of errror cases arising in thermo-electric plants in Rumania by the method of near-miss accidents, close to the fault tree[14]. The application itself poses hardly any problems. But in order for it to lead to prevention measures one would need an orientation with regards to safety, and an infrastructure that does not exist in this country. An active attitude, integration of safety at all levels of the organization and the installation of rapid decision-making circuits leading to prevention measures - just some of the factors absent at this time. Moreover, it can be said that within the most dynamic firms, like the SMF (Small and Medium-sized Firms) that tend to be created in the East, the integration of safety into the management has not

[11] Faverge, J. M. (1972). *Psychosociologie des accidents de travail*. Paris : Presses Universitaires de France

[12] Reason, J. (1992). Op. cit.

[13] Szekely, S. & De keyser, V. (1980). La sécurité dans deux usines de textile. Paris : ANACT.

[14] Iosif, G. (1994). *Multiple causality in incident and accident genesis*. Paper presented at the 23rd International Congress of Applied Psychology, Madrid (Spain), 17-22 July, 1994.

been foreseen at this time. The old conflict of production vs. safety can only be resolved in a middle- term perspective - safety being a beneficial investment for productivity. The fact is, however, the current economic climate focuses on the short term.

3. PREVENTION AND RECOVERY

It is often advanced today that, given the complexity of developed systems, all errors can not be predicted. Consequently, rather than talk of prevention, it is useful to focus attention on the management and recovery of errors. This is the thesis developed notably by the University of Giessen[15], forming part of the HCM network. The idea is intriguing as much for equipment designers as for scientists. For example, in aviation, the number of human errors is almost stationary- in spite of all the efforts being made in the area of safety for almost ten years[16]. Today it is generally accepted that top priority must be given to efforts directed towards devices allowing the detection and recovery of errors, whether they be technical, human or organizational. In the recent air catastrophe of the A 320 in Strasbourg, where the pilot in all likelihood confused two modes of airplane descent, the absence of the cross-checking of the co-pilot into the introduction of the descent data, and equally the absence of a remote detector on the ground of the GPWS type (Ground Proximity Warning System) were the decisive elements in the development of the catastrophe[17].

This philosophy, however, contains some trappings. *The first consists of focusing on recovery while neglecting an anthropocentric approach from the outset.* We know that one of the objectives of ergonomics is to introduce the human factor as early on as the drawing board-that is to say to endow the designers with good models of man, of the future task and of the environment[18]. First the

[15] Frese, M. (1994). *Error Management : An alternative concept to error prevention in organizations and in technical system design.* Paper presented at the 23rd International Congress of Applied Psychology, Madrid (Spain), 17-22 July, 1994.

[16] Statistical Summary of Commercial Jet Aircraft Accidents Worldwide Operations 1959 - 1991. Seattle, Washington : Boeing Commercial Airplane Group (1992).

[17] METT (1993). Rapport de la Commission d'enquête sur l'accident survenu le 20 janvier 1992 près du Mont Saint Odile (Bas-Rhin) à l'Airbus A 320 immatriculé F-GGED. Exploité par la compagnie Air-Inter. ISSN n° 1148-4292. Paris : Ministère de l'Equipement, des Transports et du Tourisme.

[18] Wickens, C. (1992). *Engineering Psychology and Human Performance.* New York : HarperCollins (pp. 11-12).

development and experimentation of prototypes followed by the collection of incidents, accidents or errors provide useful information for minor corrections. The essential must be thought out ahead of time. But for the designers, taking these models into account, and adapting them to each particular case, is a heavy constraint. It is tempting to make light of it by thinking that in any case the safety features will limit the potential consequences of errors. We can not overemphasize the fact that recovery must not be focused on residual errors- and that any latent risk linked to the design of equipment must be eradicated from the outset. *The second trap involves underestimating the conditions of recovery.* Certain authors[19] have shown that the nature of errors has an influence on self-recovery; the slips are more easily recovered than the mistakes. But other characteristics intervene: the sometimes collective character of error, the delay between the moment it is committed and the initial appearance of its consequences, the disassociation between the one who commits the error and the one who undergoes the consequences are some of the factors that can halt recovery. Indeed, most studies of recovery start from a very simple paradigm: someone commits an error and has a feed-back that allows him, in principle, to recover it. Reality is very different. In situation, the error is often more diffused, the product of decisions involving multiple actors, with consequences that are only observable in the long term. A good example is the contaminated blood incident in France. It can be seen there that what halted recovery was the whole social organization. Decisions of an exclusively economic nature were made without any regulation loop intervening to redirect or control the play- for example, without the main parties directly interested by the transfusions having any say in the decisions made by the medical authorities. Indeed, the error can only be recovered in systems where all actors have a power of initiative, and where all the multiple rationalities are admitted and negotiated at the same time. This is far from being the case at this time in the eastern countries. On the contrary, the priority given to rapid capitalist development is pushing the economic sector to autonomise and free itself from any political or social regulation. This situation, no doubt transitory, is shared by a number of developing countries.[20] But it can endanger any vague recovery impulse, and leave the shadow of

[19] cf. notably :
- Woods, D.D. (1984). *Some results on operator performance in emergency events.* Institute of Chemical Engineers Symposium, Series n° 90.
- Rizzo, A. (1988). Human error detection processes. In E. Hollnagel, G. Mancini, D.D. Woods (Eds). Cognitive Engineering in Complex Dynamic Worlds. London : Academic Press.

[20] Touraine, A. (1994). Faut-il, encore ou à nouveau, parler de développement ?. Public lecture given at the University of Liege, Liege, November 18, 1994.

latent risk factors hovering over the environment. Hence the design of nuclear plants in the East, which according to experts is less safe than those of the West, becomes worrisome in a period of transition seeking new rules for social functioning. *The third trap involves confusing error and responsibility.* Doing this prevents the establishment of an atmosphere of trust that allows for the expression of errors. Their collection and analysis is only possible in an environment where everyone believes in the prevention goal pursued. Present studies in anesthesia encountered this difficulty and overcame it. For the anesthetists, the willingness to give top priority to patient safety won out over worries in the face of public opinion and a legal apparatus more and more attentive to possible medical errors. These physicians made error into a topic for reflection and study, favoring their collection and their analysis in an interdisciplinary framework and seeking paths of prevention[21]. Without a doubt, such examples can also be found in the East, but as a general rule, the aftermath of bureaucratization linked to the controlled economy more often incites workers to camouflage anything they could be reproached for rather than bringing it out into the open. The crisis of confidence inherited from a long history of truncated information and of fudged statistics is clearly not a favorable terrain on which to play the transparency card.

4. FUNCTIONING IN A DETERIORATED MODE

In the numerous studies carried out on the technology transfers in developing countries, Wisner[22] relentlessly insisted on the particular characteristic specific to transfer- the economic, cultural, social and political conditions of the country in which the transfer is inscribed are of paramount importance. He also brings out the existence of an industrial functioning in a *deteriorated mode* : the machines do not function at full capacity, the incidents are numerous and the raw materials themselves are not always of the required quality. And the workers, undernourished, poorly trained and travelling long distances before arriving at work, are not giving their best performance. The economic resources of the country, either scarce or funnelled off to profit the privileged few, do not allow for maintenance of the equipment, for the regular supply of spare parts, or simply the decent maintenance of the workers. The paradox is that in this situation, some surprising forms of expertise appear. They are based on creativity, tinkering, recovery of parts and the deviated use of equipment.

This activity recalls the catachresis[23] of the researchers of the ECSC during the nineteen sixties[24]: in a deteriorated mode, the workers reveal themselves, by force of circumstances, absolutely ingenious. Without raising up the hasty analysis between the Third World Countries and the eastern countries, it is true that in the East the absence of maintenance of equipment, the difficulties of energy supply and the precarious material living conditions of the workers today creates functionings in deteriorated mode. These can lead to catastrophes; a recent example is the rupture of the pipe-lines in Siberia, due to the absence of maintenance of the installations. However, for a few catastrophic examples, some over and some no doubt yet to come, how many hidden catachreses have there been whose effects have been beneficial? We are discovering in the East, whether in social life or at work, a quantity of parallel activity traces that hallmark a latent ingenuity- even if up to now it has hardly been channelled. It is probably from this hidden wealth that we must start prevention initiatives- much more so than from a systematic exploitation of human errors whose origins are known from the start, and whose dangers are measured.

5. CONCLUSIONS

One year of functioning a research network does not allow us to draw conclusions without appeal, at best we can set up paths of reflection. It is with this in mind that the following remarks should be understood:

5.1. One can not imagine transfers of knowledge isolated from any context. But this latter is not always visible, nor clearly demarcated. One of the first interesting results in the confrontation of the eastern European/western European HCM network is to bring out that prevention must not at this time be based on the same methods. The analysis and classification of errors, their exploitation in view of technical or organizational modifications-or still for training purposes- is only thinkable in a system where error is viewed as *one of the factors*, often fortuitous, able to lead to a catastrophe. That is to say in a system where the premises regarding safety are well established and where the will to integrate this into the management is real- without which the

---

[21] De Keyser, V. & Nyssen, A.S. (1993). Op. Cit.

[22] Wisner, A. (1985). Quand voyagent les usines. Paris : Syros

---

[23] Catachresis characterizes the informal use of a tool, deviating from the goal for which it was designed, where the use is not in accordance with the work procedure. Unscrewing a screw with a coin is a catachresis. The catachresis has a double face--that of reliability and that of unreliability. Indeed, if in numerous cases the catachresis allows the worker to get out of a bad move, and to gain some time, in others, on the contrary, it leads to accidents

[24] Faverge, J.M. (1972). Op. cit.

denunciation of human error becomes an alibi and hallmarks the powerlessness to treat other factors.

5.2. If the capitalist transition is now neglecting to invest in safety on a day to day basis, to the benefit of a rapid accumulation necessary to its expansion, this critical phase will not be able to continue much longer without risk. Regulation loops and a control of economics will necessarily intervene. And this no doubt at the very moment when investments in safety, profitable in the middle term, will be supported. It is important to prepare this phase by placing emphasis on the integration of safety and economics. Methods of auto-evaluation of the state of safety in the firms, which are very simple, able to function without experts and *directed towards the installation of adequate structures* like the ones that exist in western Europe today[25], will be useful. The study of human errors, as possible precursors of accidents or of catastrophes can then find its place.

5.3. In the present phase there is certainly much more on which to capitalize from the workers' ingenuity in making the systems function in "deteriorated mode". Presently, it does not appear that this creativity can be positively channelled-as it was in western Europe in the quality circles. Flexible forms will have to be sought along with bottom-up reflection on safety, actualizing the workers' initiatives and human reliability- but equally sorting out the risky catachreses from those which pose no risk.

6. ACKNOWLEDGEMENTS

This study was carried out in the framework of the Prevention of Human Error in the high-risk industries network, of the program of Human Capital Mobility of the European Community. The participants to the Network are :
Prof. V. De Keyser : University of Liège (Belgium)
Prof. P. Millot : University of Valenciennes and Hainaut Cambresis (France)
Dr. T. van der Schaaf : Eindhoven University of Technology (The Netherlands)
Prof. M. Frese : Justus-Liebig Universität Giessen (Germany)
Prof. K. Duncan - Dr. J. Patrick : University of Wales Callege of Cardiff (United Kingdom)
Prof. G. Iosif : Romanian Academy (Romania)
Dr. M. Antalovits : Technical University of Budapest (Hungary)
Prof. A.B. Leonova - Prof. B. Velichkowki : Moskow State Univeristy (Russia).

---

[25] Like the ones that exist in western Europe today; cf. notably:
- CCE (1993). Sécurité et santé au travail à l'usage des PME. Manuel d'auto-audit. Luxembourg : Commission des Communautés Européennes, DG V.
- Reason, J., Free, R., Havard, S., Benson, M. & Van Vijver, P. (1994). Railway Accident Investigation Tool (RAIl) : A Step-by-Step Guide to New Users. Manchester : University of Manchester, Dpt of Psychology

PREVENTION OF ROUTINE ERRORS
THROUGH A COMPUTERISED ADAPTIVE ASSISTANCE SYSTEM

M. Masson

University of Liège, Belgium
Eindhoven University of Technology
Safety Management Group
Graduate School of Industrial Engineering and Management Science
P.O. Box 513, Pav. U-5, 5600 MB, The Netherlands

Astract: This paper starts by a presentation of the theoretical research context that has motivated the development of CESS, a prototype of an adaptive computerised assistance system aimed at preventing and correcting routine errors in tasks favouring routinisation.
It then summarises the system's structure and functions and the main results of an evaluation experiment carried out in laboratory. It ends up by a discussion of the system's potentialities and of the promises and limitations of adaptive systems regarding error prevention.

Keywords: human error, error modelling, prevention, adaptive systems.

1. HUMAN ERRORS AND ROUTINE ERRORS

In the areas of high-risk activities, such as aeronautics, thermoelectrical plants, steel industry, chemistry and petrochemistry, human errors can have tragic consequences. This means both the loss of human lives as well as material damages, environmental effects and economic losses. Such catastrophes as those of Chernobyl, Three Mile Island or Bhopal invite the scientific community and civil authorities to take a closer look at the nature of human error and to develop innovative error prevention and management measures.

Psychologists have indeed emphasised the necessity of maintaining human operators in complex systems because of their combined capacity to deal with unpredictable situations, to manage crises and conflicts, to compensate for the deficiencies of information systems and procedures, and to *acquire expertise* through interaction with the systems they control or supervise.

For a human operator, acquiring expertise involves not only the acquisition of knowledge, models, theories or methods concerning his technical and social environment, but also the development of repertories of interpretation and action schemes, know-how and automatisms that allow him to adapt himself in an economical way to work environments and to reduce their complexity. The acquisition of know-how is a highly adaptive mechanism. It provides the operator with cognitive and behavioural structures that guide interpretation, prepare action and serve as a support to anticipation and retrospection. There exists however one drawback to these advantages; to the extent that this mode of automatic functioning induces a risk of capture by habit, that is to say of *routine error*.

A routine error can appear harmless. They are commited in domestic as well as in work environments. The nature of the risks they bring about does not depend first and foremost on their nature. Indeed, they all present similarities on the psychological level and bring into play common mechanisms. What distinguishes them is the degree of tolerance of the environment in which they are made, which, depending on latent risks, can make of a routine error a truly harmless act or the source of a catastrophe.

2. A THEORY OF ROUTINE ERRORS

More than one century ago, William James (1890) pointed out that the successful practice of any type of activity results in the gradual devolution or delegation of control from a high level, closed loop and attention-driven control mode, to a low level, open loop and schema-driven processing style, characterising *routinised* or proceduralised activities. As illustrated by Reason in a video tape on human error (Reason and Embrey, 1980), devolution of control results in the setting up of semi-autonomous cognitive processors analogue to action demons in artificial intelligence. With practice, those processors progressively acquire a substantial autonomy in the release and control of activity. The more an operator gets skilled, the more his activity involves and relies on those ready-to-access and ready-to-use knowledge and control structures. Consequently, a substantial part of cognitive activity becomes governed by low level processors, which are hierarchically organised in long term memory. They operate mostly locally, in environments activating them through local calling conditions (Norman, 1981; Reason, 1987, 1990).

Conscious attention can control the selection of these low level processors, together with the order in which they are brought into play. High level control takes place in working memory and involves attentional checks upon progress of the current activity, that is upon the running of the lower order schemes that are involved. However, this control function is only intermittently - and sometimes inappropriately - exercised (Reason, 1987, 1990). The control of activity arises thus from the interaction of two control modes : the attentional or consciously directed mode, and the schematic or automatic mode (e.g. Norman and Shallice, 1980).

Some of the errors made by operators and more generally by human beings, are best understood as a property of that dual processing system - sequential and consciously driven versus parallel and schematic driven, and are thus both explainable and predictable. And among all human errors, routine errors are probably those that are the most predictable. The process of routinisation presents by the way the double advantage of increasing efficiency and autonomy by reducing mental workload (Bainbridge, 1989) and of freeing attentional resources from the situation at hand (Reason, 1987, 1990), both in its spatial and temporal components. Following Reason, there is however a *price to pay* for becoming skilled, as routines are sometimes automatically activated in an unsuitable context, in an inappropriate way or on an inappropriate object. And this because they were successfully, frequently or recently used in past circumstances, or because they are released by the environment through calling conditions. Erroneous activation of normally highly adapted and efficient activity segments results in what are usually called errors due to habit, routine errors or, more precisely, capture errors (Norman, 1981, 1990; Reason, 1990; Hollnagel, 1991).

3. NEED FOR ASSISTANCE AND METHODOLOGY OF DEVELOPMENT

Most of the support systems developed in the domain of man-machine interactions (e.g. Woods, Johannessen and Potter, 1990) are dedicated at supporting activities such as diagnosing, planning, estimation and regulation, that mainly resort to problem solving. But as pointed out by several authors in the fields of human factors engineering and applied cognitive psychology, in particular by Rasmussen (1987), a substantial part of operators' activities do not resort to problem solving but to know-how and skill. Operator problem solving oriented support systems are quite needed in those circumstances where problem solving is actually performed, that is essentially under degradated or incidental conditions, which create inhabitual, hard to manage situations. But they need to be associated with other types of assistance systems once the activity gets automated, which is the case when dealing with habitual tasks in familiar surroundings (Gersick, 1990), as can be found in aeronautics, power plants (Masson, Malaise, Housiaux and De Keyser, 1993) or train and car driving (Kruysse, 1992).

The methodology of development in this project is the folowing one.

Among human errors, routine errors are among the most frequent. They figure equally among the best explained and the most predictable. They are qualified, indeed, as systematic errors, because they are connected to the very characteristics of the cognitive system. Like any highly predictable phenomenon, routine errors can thus be modelled and then *simulated* by a computer. Such a simulation tool can then be integrated into an assistance system, aimed both at predicting their occurrence and to remedy them in a preventive or corrective manner.

4. CESS : A SYSTEM PROTOTYPE FOR ROUTINE ERROR PREVENTION

4.1. Object

Developing and testing an original support system dedicated at preventing and/or correcting routine errors has thus become the objective of this research. A software prototype called *Cognitive Execution*

Support System has been programed in lisp on a SUN Workstation and tested in the laboratory using a very simple task favouring routinisation (Masson, 1994).

4.2. Structure and functions

CESS works by sketching out anticipatively the way the user will act or react to a familiar task environment, by simulating in real time the basics of human routinised behaviour.

CESS actually anticipates the user behaviour on the basis of a combination of *similarity* and *frequency* criteria (Reason, 1987, 1990), in the way COSIMO simulates the cognitive process of knowledge extraction in operators (Cacciabue, Decortis, Drozdowicz, Masson and Nordvik, 1992).

In a routinised activity, the action a subject is most likely to carry out is determined both by the resemblance between the local situation and situations framed in memory (the similarity criteria), and by the frequency distribution of those situations in past circumstances (the frequency criteria). The system accounts for the first resemblance factor by computing and updating a similarity score between the situation faced by the user and similar exemplars held in memory. This similarity score is then combined to a frequency score, which indicates the number of times those exemplars were experienced in past occasions. The combination of the similarity and frequency scores gives each candidate situation its final activation level.

When activation reaches a given threshold, the candidate obtaining the highest activation score is selected, according to the "winner takes it all" principle, as the best candidate for estimating the subject's future action.

This candidate is then evaluated against the current situation. If it appears to be inappropriate or faulty, personalised warnings are sent by the preventive aid module, before the user has carried out any action on the system.

Errors that would still occur in spite of those preventive warnings are processed by the corrective aid module. This module gives the user the possibility to rectify his action before it would be released in the task environment. The corrective aid module thus operates like a filter, blocking selectively any unsuited action before it can bring about consequences.

In order to tune itself to the variability of the task and to what the subject learns, CESS is capable of enlarging its knowledge base by acquiring new situation exemplars, and of updating their frequency in the running of activity through an incremental learning mechanism.

5. EXPERIMENTAL EVALUATION IN LABORATORY

CESS was tested through the participation of 130 subjects (mostly psychology students) aged between 18 and 30, in a laboratory task favouring routinisation.

5.1. Experimental task and procedure

The task consists in transcoding of 240 series of 9 or 10 letters into corresponding series of digits, using an alpha-numerical code. The letter series are presented in sequence on the screen of a SUN 3.1 Workstation. Subjects have to enter the answers using the keybord. Pressing the "return" key is needed in order to validate the answers and display the next letter series on the screen. The speed of progression is thus under the subjects'control, and no emphasis is put on time. Subjects are indeed asked to pay more attention to accuracy than to speed, while performing the task within a "reasonable" time limit.

Routine errors are induced by controlling the frequency of occurrence and the similarity between a) a normal series and b) 12 modified series, differing from the former only by 1 or 2 digits :

- the normal series "A B C D E F G H I" was presented with a frequency of .95 (= 228 / 240);
- 10 one change series "A B C D E F G G H I" appeared with a frequency of about 0.04 (=10/240), and
- 2 two change series "A B C D F G G H I" were presented with a frequency of about 0.01 (=2/240).

The tasks begins by 40 presentations of the normal series. One change series are then inserted approximatively every 20 presentations. Double change series are presented only as the two last modified series. Ten experimental versions of the prototype differing either by the presence or absence of prevention and correction facilities or by characteristics of the interface, were tested. The main results of the experiment are summarised below.

5.2. Hypotheses

This experiment has four basic hypotheses.

1. In the absence of any preventive or corrective assistance, series featuring changes, which are considered to be propitious to routine errors, entail higher error rates than the normal series.

2. Preventive and/or corrective assistance provided by CESS entail a reduction of the error rates in these situations featuring changes.
3. The error rate of double change series is higher than the one of single change series.
4. Preventive and/or corrective assistance provided by CESS entail a higher error rate reduction in double than in single change situations.

5.3. Results

Results in general support these four hypotheses. They are reviewed hereafter in some details.

Presence and number of changes. From a comparison of the error rates in the three situations (without or with 1 or 2 changes), it appears that modified situations introduce a particular error risk, higher for situations with double than with single changes ($Pr > F = 0.0001$, df = 2).

Preventive assistance. Prevention decreases the use and need for correction in the course of activity, but is *not* sufficient to achieve a zero error risk level when used alone. Preventive assistance reduces in absolute terms the error rate in each situation type ($Pr > F = 0.0023$, df = 1) and this all the more as the situation is more propitious to error. This relation inverts when one considers this reduction in relative terms. However, as the interaction between preventive assistance and performance in each situation type is not significant ($Pr > F = 0.1782$, df = 2), these last two findings have to be considered only as tendencies.

Corrective assistance. Corrective assistance reduces in absolute terms the error rate in each type of situation ($Pr > F = 0.0001$, df = 1) and this all the more as the situation is more propitious to error. As for the preventive assistance, the relation inverts when considering this reduction in a relative manner. By contrast to preventive assistance, the interaction between corrective assistance and performance in each situation type is significant ($Pr > F = 0.0016$, df = 2). It thus confers to these results a more solid character. Corrective aid could also virtually have counteracted any risk of routine errors by forcing automatic correction. An option which was however not tested in the experiment.

Version of the assistance system. The version of the assistance system has no significant influence on the error rate in normal situations ($Pr > F = 0.2477$, df = 9) but has one in situations featuring single changes ($Pr > F = 0.0108$, df = 9) or double changes ($Pr > F = 0.0001$, df = 9). The modification of the significance threshold with the type of situations indicates that modified situations introduce a particular error risk, different (indeed higher) for double than for single changes, and that the different versions tested vary in their capacity to decrease the error rates in these modified situations.

Chronology of performance. Differences obtained in error rates when dividing the task into 24 series of 10 situations are very significant ($Pr > F = 0.0090$, df = 23). Differences between the rates of errors committed in the 12 situations featuring changes are also highly significant ($Pr > F = 0.0001$, df = 11). These differences can be interpreted by a difference in the nature of the control processes brought into play with the task chronology. From a psychological viewpoint, the beginning of the task corresponds to a learning phase, where full attention is allocated to the task. Then the beginning of the periodic presentation of single change situations corresponds to a phase of consolidation. These modified series entail an increase in the error rate, because they activate in the subjects' memory the answer unit that is normally used with the normal series. The consequence of this automatic triggering is that the the modified situation is not correctly processed and gives rise to the strong but locally wrong normal answer, provoking thus a *routine error*. In a subsequent phase, the single change series progressively acquire a status of exception. Subjects perform as if they were using the largest part of their attentional resources, liberated by routinisation, to the identification and processing of these modified series. Better mastered, these situations thus produce less errors, and even their detection process also seems to routinise itself. Finally, the two situations featuring a double change, presented at the end of the task, again entail an increase of the error rate.

Nature of errors. Beside the quantitative analysis of errors and the consideration of the performance chronology, the analysis of the nature of errors produces additional information, that supports the former interpretation in terms of error mechanisms. In single change situations, the success rate is approximately 90%, and about 70% of the errors have the form of the habitually correct but locally unsuitable answer corresponding to situations without change. The other forms of errors are a mix of the answers prescribed in normal or in single change situations. In double change situations, the success rate drops to 64 % at the first presentation, 55 % of the errors (i.e. 20 % of the answers) having the form of the answer prescribed in the single change situations and 22 % of the one prescribed in the normal situation. The success rate increases up to 74 % at the second presentation, with 42 % of errors corresponding to the single change situations, 12 % to the normal situation and other infrequent errors being a mix of the answers prescribed in the three situations. Double change situations thus entail a net diminution of the success rate comparing to the unchanged situations, followed by an improvement of performance at the second presentation, which testifies a better control in the perceptual and production stages of the answer. Double change situations also increase the variety of error forms. Consequently, if it turns to be rather easy to identify the situations propitious to routine errors, *the variety of answers highlights how difficult it is to predict precisely the form of errors*, despite the simplicity of the experimental task used.

5.4. Perspectives

The relevance of the task used during this first experiment must however be questioned, because of this extreme *simplicity* when compared to tasks human operators have to perform in real work environments. The potentialities of CESS have thus to be further investigated in a more valid task. This is by the way the object of a post-doc research project currently carried out in Eindhoven. This project is basically aimed at selecting a more complex application propitious to human errors, adapting and testing CESS with this new application, and evaluating the possibilities for using the system in industrial environments. The application chosen is the computerised fault diagnosis task initially developed by Rouse (1978)[1], and largely documented in the literature.

6. DISCUSSION AND CONCLUSION

Routine errors are among the most frequent human errors, both in everyday life and at the workplace. They turn also to be among the most resistent ones. They have indeed been qualified as "diabolic", because they seem to resist to any training, the person knowing in principle what to do.

Moreover, they are in their very principle the epiphenomenon of a highly adaptive mechanism of skill acquisition. The complexity of today's "high tech" industrial environments and the high number of variables to take into account can indeed sometimes exceed the limits of operators' cognitive resources. It is therefore desirable to favour the acquisition of interpretation and action schemes that contribute to reduce uncertainty and that provide, at low cost, actions that are *most of the time* appropriate. There is however a price to pay for this acquisition of know-how, to the extent that these interpretation and action schemes can be triggered in wrong situations, because of a similarity with the normal

---

[1] The fault diagnostic task of Rouse requires fault finding in graphically dispayed networks of interconnected and-gate components. The subject has to identify which component is faulty from outputs of the network and from a series of tests made either on connections or on components.

conditions of application and because of frequency and recency of prior use.

This project aims to demonstrate that there exists a reply to this paradox, not the only one possible but a way among others, that deserves to be further explored.

There are indeed several directions that can be followed to prevent or manage human error (e.g. Amalberti, 1992). The first one consists in improving man-system interfaces, by providing operators with more adapted information systems, both easier to interpret and more convenient to use. The development of expert systems supporting operators in areas where they are limited in resources, is the second one. Reason (1987) names this approach a "prothesis approach", to the extent that these systems' main function is to palliate a deficiency or a recognised limitation of the users, or to sustain or amplify some of their competences. The third direction is systems design, including techniques for risk assessment and their correlate, the modification of the nature of the system's units (components and subsystems) and of their relationship, so as to reach and to guarantee acceptable risk levels. The fourth one concerns the education and training of operators, with the objective to make them capable to adapt better and more rapidly to the work environments and tasks to which they are confronted. The development of training programs using high fidelity simulators is a good example of this tendency. The fifth one, which has been adopted in this research, is *adaptive assistance*.

Adaptive assistance consists of increasing the complexity and intelligence of assitance systems by giving them capacities to anticipate and to adapt themselves to the users actions and errors, conferring thus to them some *human character*. This solution rests indeed on an implicit idea of interactions between humans, according to which each actor builds and uses a model of his partners, so as to interpret and to anticipate their activity. It is then upon the existence and coherence of these models that depends the intelligibility, the predictability and finally, the success of the interactions. The objective is to supply assistance systems with these natural representation, interpretation and anticipation capabilities, so as to render them both more *foreseeable* and more *provident*.

To reach this objective constitutes a real challenge, as it appears extremely difficult to guarantee that these new capacities would yield efficient interactions in all possible figure cases. The passage from a conceptual perspective to practical applications induces indeed two main practical difficulties. The first one consists in managing the *unforeseen character of real situations*. All real situations cannot indeed be described in an extensive way. A classic solution to this problem consists in ensuring that the assistance system has an acceptable behaviour in extreme situations or that it could be disconnected easily and without introducing new danger. The second drawback comes from the *complexity of the human user* and its corollary, the difficulty to construct a valid model of his activity. Can such an assistance system have a model of the user that would be reliable enough to allow it to asses, in all circumstances, the quality of the user's actions and their adequacy to the context? And if this assessment turns out to be erroneous, would the system be capable of noticing it and of resetting its representation? These questions remain of course open questions. But they indicate that the presence of *procedures*, which limit the task variety by defining which actions are acceptable under which circumstances, is a favourable condition for adaptive systems. In particular, procedural work constitutes the best area of application for a tool like CESS, both for the reasons mentioned above and because it favours the routinisation of activity and consequently, is prone to routine errors.

There is however one question that still needs to be addressed: how would the user react facing a system with such human-like capabilities? Wouldn't there be a risk of rejection due to a feeling of loss of control? According to the experience acquired in this research, this risk seems to depend primarily upon the reliability and relevance of the system's interventions in the very first interactions with the user. It's a matter of *trust*, that does or does not establish itself in the user's mind. This is by the way a very general question to tackle when developing any adaptive assistance system for real work applications. A development that currently remains a long term goal of this research on automatic prevention of routine errors.

ACKNOWLEDGEMENTS

This research initiated at the University of Liège, Belgium, is continuing thanks to a grant from the Eindhoven University of Technology, The Netherlands. The presentation of this paper was supported by a CEC Human Capital and Mobility Network on Human Error Prevention.

REFERENCES

Amalberti, R. (1992). Safety in Process Control : an Operator-centred point of view. Reliability Engineering and System Safety. **38**, 313, 99-108.

Bainbridge, L. (1989). Development of Skill, Reduction of Workload. In: *Developing Skills with Information Technology* (L. Bainbridge and S. A. R. Quintanilla, Ed.), John Wiley and Sons, London.

Cacciabue, P.C., F. Decortis, B. Drozdowicz, M. Masson and J.-P. Nordvik (1992). COSIMO : a Cognitive Simulation Model of Human Decision Making and Behaviour in Accident Management of Complex Plants, *IEEE-SMC*, **22 (5)**, 1058-1074.

James, W. (1890). *The Principles of Psychology*, Henry Hott, New-York.

Gersick, J.G. (1990). Habitual routines in task-performing groups. *Organizational Behaviour and Human Decision Processes*, **47**, 65-97.

Hollnagel, E. (1991). The Phenotype of Erroneous Actions. In: *Human-Computer Interaction and Complex Systems* (G. R. S. Weir and J. L. Alty, Ed.). Academic Press, London.

Kruysse, H.W. (1992). How slips result in traffic conflicts and accidents. *Applied Cognitive Psychology*, **6**, 607-618.

Masson, M. (1994). Prévention Automatique des Erreurs de Routine. PhD thesis. 319 p. University of Liège, Liège.

Masson, M., N. Malaise, A. Housiaux and V. De Keyser (1993). Organisational Change and Human Expertise in Nuclear Power Plants : Some Implications for Training and Error Prevention. *Nuclear Engineering and Design*, **144**, 361-373.

Norman, D.A. (1981). Categorization of Action Slips. *Psychological Review*, **88**, 1-15.

Norman, D.A. (1990). *The Design of Everyday Things* (2nd ed.). Double Day Currency, New York.

Norman, D.A. and T. Shallice (1980*). Attention to action : Willed and Automatic Control of Behaviour*. CHIP 99. University of California San Diego, La Lolla.

Rasmussen, J. (1987). Cognitive Control and Human Error Mechanisms. In: *New Technology and Human Error* (J. Rasmussen, K. Duncan and J. Leplat, Ed.). John Wiley and Sons, London.

Rouse W.B. (1978). Human Problem Solving Performance in a Fault Diagnostic Task. *IEEE-SMC*, **8**, 258-271.

Reason, J.T. (1987). Generic Error Modelling System (GEMS) : A Cognitive Framework for Locating Common Human Error Forms. In: *New Technology and Human Error* (J. Rasmussen, K. Duncan and J. Leplat, Ed.). John Wiley and Sons, London.

Reason, J.T. (1990). *Human Error*. Cambridge University Press, Cambridge.

Reason, J.T. and D. Embrey (1980). *Human Error. A Fall from Grace* (BBC video tape, Reliability Associates, Ed.). Signal Vision, Knutsford, Cheshire.

Woods, D.D., L. Johannesen and S.S. Potter (1990). *Structured Bibliography on Human Interaction with Intelligent Systems*, CSEL Tech. Report 1990-009. The Ohio State University, Columbus.

HUMAN RECOVERY OF ERRORS IN MAN-MACHINE SYSTEMS

T.W. van der Schaaf

Safety Management Group
Graduate School of Industrial Engineering and Management Science
Eindhoven University of Technology
P.O. Box 513, Pav. U-8
5600 MB Eindhoven, The Netherlands

Abstract: This paper highlights the positive role that human operators often play in preventing small failures and errors from developing into an actual system breakdown. The resulting 'near misses' may provide an insight into a powerful alternative to human error prevention, namely: human recovery promotion. Theoretical approaches to modelling error recovery are discussed and translated into empirical research questions. These are partly answered by a number of pilot studies. The main conclusions are that error recovery is much more than simple luck or coincidence, that its root causes can be identified, and that these should have design implications for the technical and organisational context of the human operator's task.

Keywords: electric power systems, error correction, human reliability, medical applications, recovery, steel industry.

1. INTRODUCTION

The research project described in this paper focuses on the positive role that human operators often play in preventing an ongoing sequence of usually small failures and errors to develop into an actual total system breakdown or accident. This new concept of human recovery may provide designers, managers and researchers with a powerful alternative approach to the traditional one of human error prevention in process control, namely: human recovery promotion.

First, a simple incident causation model is presented in which the presence or absence of successful human recovery plays a decisive role in determining the effects of process deviations, technical failures and errors on the safety and reliability of man-machine systems. Also, the advantages of near miss reporting and analysis for human recovery research are briefly discussed.

Then, the process of human recovery is described to consist of three phases: detection of symptoms; localisation of their cause(s); and correction to return the system to its normal status.

The following section deals with the relationship of human error causes and the probability of recovery, and with the error detection lag. These theoretical predictions are mainly based on well known cognitive limitations and feedback-related aspects of the task situation.

Four ways of classifying (human) recovery in actual process control situations are proposed. The first classification deals with the type of preceding failure(s), for instance technical, organisational or human failure. Another way to look at human recovery is to distinguish the reaction after symptom detection: ignore the deviating status; repeat sequence of actions; or attempt fault localisation and correction. Thirdly and most importantly, the factors in the man-machine system that triggered or enabled recovery are categorized: technical factors related to process design (for instance to allow for reversibility), or interface design (e.g. to maximize observability of symptoms and effects); the organisational and management context (e.g. proper procedures, positive safety culture) and operator factors (e.g. accurate mental models). The fourth classification locates the phase in which a recovery factor contributes to the recovery process: detection, localisation or correction. These theoretical approaches are subsequently translated into the specific empirical research questions on which this project is focusing.

Finally the results of recent pilot studies in the energy production and steel industry, as well as those of medical errors in a surgical ward will be presented and their implications for designing recovery into man-machine systems will be discussed.

2. THEORETICAL APPROACHES

2.1 Incident causation model.

In Van der Schaaf (1992) a simple incident causation model is used (see fig. 1) to define accidents, near misses and their common root causes consisting of technical, organisational and human (operator) factors. When incident development cannot be stopped by the system's predetermined barriers and lines of defence, the only distinguishing factor between an accident and a near miss effect is the presence or absence of successful 'accidental' or unplanned recovery.

Although actual accidents also may contain attempts at recovery, it is obvious that near misses as defined above are the optimal source of data to study the phenomenon of recovery as the positive counterpart of failure.

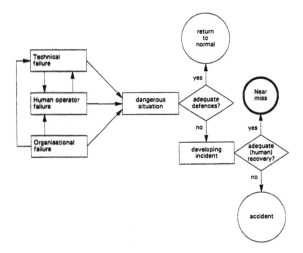

Fig. 1. The incident causation model.

These failure factors (or root causes) have so far been modelled successfully by the Eindhoven Classification Model (ECM) of system failure (Van der Schaaf 1991, 1992; Van Vuuren, 1995). In the pilot studies mentioned in section 4 of this paper, the ECM subcategories will not be used, only the main groups of Technical (T), Organisational and Management (O), and Human operator (H) failure factors.

2.2 Human recovery process phases.

The following definition of human recovery is proposed by Van der Schaaf (1988): "the (unique?) feature of the human system-component to detect, localize and correct earlier component failures. These component failures may be either his own previous errors (or those of colleagues) or failing technical components (hardware and software)". This definition implies the following phases in the recovery process:

* Detection: of deviations, symptoms, etc.
* Localisation: of their cause(s) (diagnosis in the strictest sense).
* Correction: of these deviations by timely, effective counter actions, after which these deviations are nullified and the system returns to a stable status again.

2.3 Dependency of recovery on preceding errors.

Embrey and Lucas (1998) discuss several factors affecting the probability of recovery from error, and the error detection lag. This relationship is highly relevant to understand the role of feedback in the recovery mechanism. Their main points may be summarized as follows:

*   Causes of skill-based slips and lapses are relatively unrelated to subsequent recovery factors; their human recovery probability is high, and the error detection lag will be small.
*   For rule-and knowledge-based mistakes the opposite holds: their recovery factors depend on the same preceding failure factors; probability of recovery is small, and the error detection lag large.

Main reasons for these predictions given by Embrey and Lucas (1988) include feedback related aspects and cognitive limitations: the awareness of an error possibility and the visibility of its effects is high for slips and lapses, but low for mistakes; cognitive limitations (e.g. confirmation-, fixation- and groupthink biases) would be small for slips and lapses, but large for mistakes. For the present paper the main implication is that the nature of the preceding human error(s) should be highly predictive of any subsequent recovery.

2.4 Classifications of human recovery aspects.

The preceding sections lead to the following four ways of classifying (human) recovery aspects: according to the preceding failure(s); according to the human operator's reaction after detecting an initial deviation or symptom; according to the type of recovery factor (or recovery root cause); and according to the phase in which this recovery factor makes its contribution.

2.4.1 Classification based on preceding failure.

Both the ECM (see section 2.1) and Embrey and Lucas (1988) provide the rationale for this taxonomy. Technical, organisational and human root causes of failures may be linked with their subsequent recoveries. Additional subcategories might include: recovery from one's own error, or from a colleague's (same or previous shift, when applicable); technical failure of equipment outside the central control room, of the interfaces within the CCR, of process control software, etc.

2.4.2 Classification according to operator reaction after symptom detection.

As noted by Reason (1990) in his GEMS model people seldom go through the entire analytic process of fault diagnosis when confronted with a deviation. This was confirmed by Brinkman (1990) who collected verbal protocols during a fault finding task. He observed the following three reactions after his subjects detected an error in their reasoning process:

*   Ignore the error and continue: rely on system redundancy and subsequent error recovery factors.
*   Simply repeat the most recent sequence of actions: try again, without any attempts at fault localisation.
*   Attempt fault localisation and optimize corrective actions: either by forward analysis (repeat the most recent action sequence and check every step) or backward analysis (trace back from symptom detection to previous actions, until the error is found).

By applying this classification, transitional probabilities between the recovery phases of section 2.2 might be established.

2.4.3 Classification according to type of recovery factor.

Such a classification should be the most important one for MMS-designers. The ECM for failure root causes could serve as a basis for recovery root causes too, with the following extensions:

*   Technical design of the process: aim at maximum *reversibility* of process reactions (Rasmussen, 1986) and 'linear interactions' plus 'loose coupling' (Perrow, 1984) of process components; these may be achieved by *structural* characteristics (e.g. buffers, parallel streams, equipment redundancy) and by *dynamic* characteristics (e.g. speed of process reactions, response delays).
*   Technical design of the man-machine interface: aim at maximum *observability* (Rasmussen, 1986) of deviations and their effects (e.g. transparency instead of alarm inflation).

*   Organisational and management factors: especially an updated, clearly formulated and well-accepted set of operating procedures, and a positive safety culture must be mentioned here (see also Van Vuuren, 1995).
*   Human operator factors: optimize the cognitive capabilities (e.g. accurate mental process model) of operators through selection and (simulator-) training, but also by supporting them with software tools to test hypotheses and avoid certain biases (see Masson, 1995).

3. EMPIRICAL RESEARCH QUESTIONS

Based on the proposals in section 2, the human recovery research project of the Eindhoven Safety Management Group is directed at the following empirical research questions:

1. Is recovery more than sheer luck or coincidence? If so, then recovery can be built into an MMS and managed!
2. Can recovery be classified with the same root causes as for failures? If so, what is the contribution of human recovery relative to technical and organisational failure barriers? How large is the contribution of human recovery in a variety of task situations and over a variety of system effects?
3. Are recovery factors identical to failure factors in a given MMS? If so, then preventing errors and promoting recovery would have to focus on the same MMS aspects.
4. In which phase(s) of the recovery process do recovery factors mostly contribute to system performance: symptom detection, fault localisation, or correction?

4. PILOT STUDIES

Pilot studies have recently been carried out in steel making, energy production, and surgery. A variety of system effects have been investigated: safety, reliability, and environmental effects of system breakdown.

4.1 Safety incidents in a steel plant.

In a Dutch steel plant Mulder (1994) identified failure and recovery factors in the same set of 25 safety related near misses. The results are given in fig. 2-4.

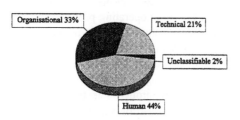

Fig. 2. Distribution of 154 failure factors in 25 safety incidents in a steel plant.

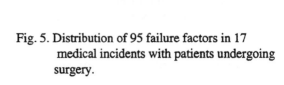

Fig. 5. Distribution of 95 failure factors in 17 medical incidents with patients undergoing surgery.

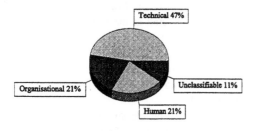

Fig. 3. Distribution of 34 recovery factors in 25 safety incidents in a steel plant.

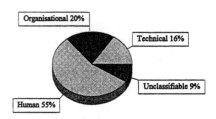

Fig. 6. Distribution of 22 recovery factors in 17 medical incidents with patients undergoing surgery.

4.3 Reliability and environmental incidents in an energy production plant.

In a small energy producing unit of a chemical plant Zuijderwijk (1995) classified failure and recovery root causes of 23 reliability and environmental near misses (fig. 7 and 8).

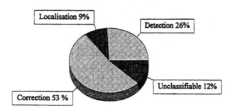

Fig. 4. Distribution of 34 recovery factors according to recovery phase in 25 safety incidents in a steel plant.

4.2 Medical safety incidents in a surgical ward.

In a large hospital in Eindhoven Van der Hoeff (1995) found the following failure and recovery factors in the same set of 17 medical near misses with patients undergoing surgery (fig. 5 and 6).

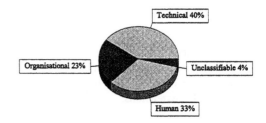

Fig. 7. Distribution of 86 failure factors in 23 reliability and environmental incidents in an energy production plant.

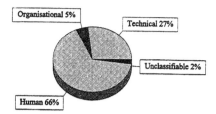

Fig. 8. Distribution of 56 recovery factors in 23 reliability and environmental incidents in an energy production plant.

4.4 General discussion of the pilot studies.

Figures 3, 6 and 8 show a range of 2 to 11 percent unclassifiable (that is: luck or coincidence) causes of recovery. This must be interpreted as a positive answer to question 1: around 90 percent or more of all recovery factors are clearly technical, organisational or human in nature and therefore researchable and eventually manageable.

The same figures show human recovery root causes contributing 21 to 66 percent. Comparison with the failure factors of figures 2, 5 and 7 shows this human recovery range to vary at least as much as the human failure range (e.g. 33 to 56 percent). The human component should therefore also be take seriously in terms of recovery possibilities (see question 2).

Zuijderwijk (1995) shows that the patterns of failure and recovery factors are clearly different. Rule- and skill-based factors dominate the operator failures, while human recovery also includes knowledge-based insights as very important. Similarly, 'material defects' are the most prominent technical failures, while 'design' covers all technical recovery factors (see question 3).

Finally, as an answer to question 4, figure 4 shows that hardly any recovery process goes through the more analytic localisation phase. Again, this could be interpreted as a confirmation of Reason's GEMS model, but there is also the possibility of an explanation in terms of time stress. If recovery is present only in the very last phase of the accident production chain of events (as was the case in most of the steel plant near misses) there may simply not be time enough for a time-consuming diagnostic effort; detection and correction 'just-in-time' may be all one can do in such cases.

5. IMPLICATIONS FOR MMS DESIGN

In spite of the immaturity of the proposed models and classifications, and of the small number of recovery incidents gathered so far, these ideas and results are intriguing enough to formulate the following tentative implications for designing a MMS:

* Consider recovery promotion as an alternative to failure prevention, especially when certain errors of failures are predictably unavoidable.
* Do not simply design out failure factors without considering the possible reduction of recovery factors: raising the level of automation in process control, or installing too many decision support tools for your operators may leave them helpless under certain situations.
* Try to support all recovery phases, primarily by means of an optimal man-machine interface: detection, localisation and correction (see 2.4.3).
* Invest in deep process knowledge of operators: reasoning beyond procedures appears to be essential for many recovery actions. Also consider error-management (Frese, ..): learning to learn from errors this way is perfectly in line with the concept of recovery promotion.

ACKNOWLEDGEMENT

The presentation of this paper was supported by a CEC Human Capital and Mobility Network on Human Error Prevention.

REFERENCES

Brinkman, J.A. (1990). *The analysis of fault diagnosis tasks: Do verbal reports speak for themselves?* Ph.D thesis, Eindhoven University of Technology.

Embrey, D.E. and D.A. Lucas (1988). The nature of recovery from error. In: *Human recovery: Proceedings of the COST A1 Seminar on Risk Analysis and Human Error* (Goossens, L.H.J., (ed.)), Delft University of Technology.

Frese, M. (1991). Error management or Error prevention: Two strategies to deal with errors in software design. In: *Human aspects in computing: Design and use of interactive systems and work with terminals* (Bullinger, H.J., (ed.)), Elsevier Science Publishers, Amsterdam.

Hoeff, N.W.S. van der (1995). *Risk management in a surgical ward (in Dutch).* M.Sc thesis, Eindhoven University of Technology.

Masson, M.A.R. (1995). Prevention of routine errors through a computerised adaptive assistance system. *This volume.*

Mulder, A.M. (1994). *Progress report 1993 of the SAFER project.* Report Hoogovens steel industry (in Dutch), IJmuiden.

Perrow, C. (1984). *Normal accidents: living with high-risk technologies.* Basic Books, New York.

Rasmussen, J. (1986). *Information processing and*

human-machine interaction. Elsevier Science Publishing Co., Inc., Amsterdam: North-Holland.

Reason, J.T. (1990). *Human Error.* Cambridge University Press, Cambridge.

Schaaf, T.W. van der (1988). Critical incidents and human recovery. In: *Human recovery: Proceedings of the COST A1 Seminar on Risk Analysis and Human Error* (Goossens, L.H.J., (ed.)), Delft University of Technology.

Schaaf, T.W. van der, D.A. Lucas and A.R. Hale (1991). *Near miss reporting as a safety tool.* Butterworth Heinemann, Oxford.

Schaaf, T.W. van der (1991). Development of a near miss management system at a chemical process plant. In: *Near miss reporting as a safety tool* (Schaaf, T.W. van der, D.A. Lucas and A.R. Hale (eds.)). Butterworth Heinemann, Oxford.

Schaaf, T.W. van der (1992). *Near miss reporting in the chemical process industry.* Ph.D Thesis, Eindhoven University of Technology.

Vuuren, W. van (1995). Modelling organisational factors of human reliability in complex man-machine systems. *This volume.*

Zuijderwijk, M. (1995). *Near miss reporting in reliability management.* M.Sc thesis, Eindhoven University of Technology.

HUMAN TUNNEL VISION AND WAITING FOR SYSTEM INFORMATION

Louis C. Boer

TNO Human Factors Research Institute
Soesterberg, The Netherlands

Abstract: Tunnel vision refers to the phenomenon of the human operator, supervising a complex system, absorbed so much in a part of the system that the overview is lost. For example, the pilot preoccupied with a technical malfunction and flying into the ground. High mental workload is commonly held responsible for tunnel vision. The present paper demonstrates otherwise. In a task of simulated process-control, most overview losses occurred during the low-workload part of fault management, where subjects waited frequently for diagnostic information from the system. The preliminary suggestion is that waiting for system information enhances operator passivity.

Keywords: Human error, human-centered design, fault diagnosis, man/machine systems, mental workload, personnel qualifications, supervision

1. INTRODUCTION

The goal of the Human Factors discipline is to improve design by applying knowledge about human operator characteristics such as errors. The present study aims at collecting knowledge about a particular human error called, after Moray (1981), *tunnel vision*.

Tunnel vision can be described as a narrowing of attention; as a concentration of mental capacity on a part of the system with, in consequence, a lost overview. According to Easterbrook (1959), narrowing of attention occurs as mental workload increases. For low or moderate levels of mental workload, performance improves as environmental stimuli not relevant for the task are excluded from consciousness. For high levels of mental workload, however, performance deteriorates as now also stimuli that are relevant for the task are excluded from consciousness.

Moray (1981) places narrowing of attention in the context of the human operator in process control. Typically, the human operator is responsible for several subsystems at once. The mental workload is fairly low, but fault conditions may arise, suddenly imposing a considerable mental workload. Tunnel vision—Moray stated—occurs when the operator concentrates on fault management activities and ignores the rest of the system meanwhile. How much deterioration of system performance this may produce depends on the joint probability that the system needs operator intervention urgently and the cost of delayed intervention.

Moray and Rotenberg (1989) used a task of simulated process-control to study tunnel vision. Subjects diagnosed and corrected faults occurring at random moments. Eye-movement recordings, used to trace the dynamics of the subjects' attention (mental capacity), revealed a narrowing of attention during fault management. Faults elsewhere were summarily noticed but they were not dealt with until the first fault was corrected. Moray and Rotenberg see this as a preference of humans to work in a serial fashion, doing one thing at the time. They did not explore whether their operators could learn to switch back and forth among simultaneous faults. It is possible that their operators had not enough mental capacity to switch among different faults.

Kerstholt, Passenier and Schuffel (1994) let subjects work in a situation in which a strategy of switching among different faults was advised. The subjects performed on a simulated ship bridge with three independent systems: navigation, platform, and cargo. Faults could arise in all systems, one at the time, two at the time, or three at the time. There was some degree of time pressure; the system "got lost" if the fault was not corrected within 120 s. A reset action could be taken to buy extra time. The study revealed a tendency *not* to use the reset action when it was needed most—three faults at once. Instead, subjects kept working on the first fault.

Mental workload is a potential cause to prefer a serial strategy. Switching among tasks demands more mental capacity than the execution of each task in isolation. One reason for the investment of extra mental capacity is prevention of *lapses* and *mistakes* that arise from confusion and cross-talk among tasks (Reason, 1990). The literature abounds with examples of mental *concurrence cost* (Allport *et al.*, 1995; Navon and Gopher, 1979; Wickens, 1990).

There is, however, another aspect to tunnel vision: attention captured by waiting for system information. This aspect is elucidated with some examples.

In 1972, the crew of EAL Flight 401 concentrated all mental capacity on a fault in the landing gear system. An unintended disengagement of the altitude-hold feature of the autopilot escaped attention; the ensuing gradual descent escaped attention; the low-altitude alarm escaped attention, a remark of the air-traffic controller failed to restore attention to the situation. The plane crashed 4 minutes after the disengagement of the autopilot, killing most people aboard. The National Safety Transport Board concludes "preoccupation with a technical malfunction". The concept of preoccupation entails connotations like "lost in thought" and "excessive concern" (Webster), pointing to mental capacity deployed unproductively and, as far as the descent was ignored, definitely counterproductively.

In 1993, one of my colleagues witnessed a less serious incident on the bridge of a large ferry. Kerstholt was there to collect data on the behaviour of crews. When the ship entered the harbour, the captain showed a concentration of mental capacity because of a technical malfunction: a failure to retract the stabilizers. (In shallow waters, stabilizers are retracted to avoid collision with the ground and damage to the hull. The captain's reaction was to reduce speed in order to gain time, to issue orders to the repair crew, and to prepare an emergency plan in case the fault would not be solved within 5 minutes. After these actions, the captain could only wait for the repair. The captain showed a definite pre-occupation with the progress of the repair crew. He was almost continuously questioning them through the intercom, annoying them rather than anything else.

Crawshaw *et al.* (1993) analyzed manoeuvres of ships in open waters. They observed "a disproportionate number of failures to monitor either the radar or the ARPA or the external visual scene" during overtaking manoeuvres (p. 7). That is, the officers of the watch concentrated their mental capacity on the position of the vessel they were overtaking. They did not use the time and the opportunity to mind other tasks.

A common element of these examples is *waiting for system information*; waiting for the findings of the flight technician, or the findings of the repair crew; and waiting until a ship is overtaken. Waiting time can, in principle, be used productively by paying attention to other tasks. The examples demonstrate that the human operator does not always use waiting times productively, but may concentrate on the source where the waited-for information will eventually come from. Moray and Rotenberg (1989) present evidence for this conclusion. The system their operators worked with was slow in accepting typed input: a single character needed 5 s; the complete command 15 s. They observed that the operator continued to attend the area of the display where the echo would appear, even during fault management.

Tunnel vision and overview loss may be the result of an unproductive concentration of mental capacity. Experiment 1 addressed the question whether waiting for information can create tunnel vision.

2. EXPERIMENT 1

Subjects were engaged in a task of simulated process control. The primary task was continuous supervision (at least once every 2½ min), a secondary task was occasional fault management (three problems during a 4-h work shift). The instruction was to always mind the primary task, irrespective of fault-management. The dependent variable was the number and the duration of supervision losses.

Fault management was a two-phased diagnosis, based on observation of the production system. The task was to identify which segment of a production line was faulty. Frequently, the subjects had to wait until the system reached a state in which information became available. During the first global phase, the subjects could do no more than to wait for system information. During the second detailed phase, the subjects played a more active role, because they used *segment disabling* to identify the segment that contained the fault.

2.1 Method

Subjects. Eighteen students participated as paid volunteers for an afternoon and the next morning. They were aged between 18 and 30 years.

Apparatus. There was a SUN workstation or a PC comparable to such a workstation. The windows and softkeys of the displays were operated with a mouse.

Tasks. The subjects earned "production units" by (a) supervising a temperature and (b) correcting faults. The two tasks were presented in separate windows, but never at the same time. The normal production rate was 0.36 units per minute. There was no production during fault management.

Supervision task. The task was to keep a temperature within the limits. A supervision window could be pulled down by clicking the softkey *overview*. The temperature path of the last 20 s was shown in the form of a graph (range 0-100), the setpoint (50) visible as a line in the middle. To restore the temperature, subjects clicked on "+" and "-" softkeys. Five seconds after the last click, the window disappeared. Subjects were advised to inspect the temperature at least once every 2 minutes. They were free to inspect more frequently.

The software added a systematic linear deviation to the temperature. One deviation brought the temperature to the limit (0 or 100) within approximately 2½ minutes; another brought the temperature to the limit within approximately 13½ minutes. If the subject did not restore the temperature, a crash followed with a direct production loss of 7.0 units and an indirect loss of 1.5 units because it took 5 minutes before the system resumed production. The total production loss of a crash—8.5 units—was equivalent to 30 minutes' continuous production.

Fault management task. The task was to solve faults. Faults were announced by an alarm. Clicking on the softkey *machines* brought down a window showing

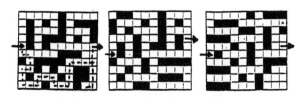

Figure 1. Machines and their production lines. Objects follow either the right wall (dotted line in Machine 1) or the left wall.

three empty machines, each containing two multi-segment production lines. One of the segments was faulty. To supervise the temperature meanwhile, subjects closed the machines window and opened the supervision window. To return to the fault management task, the subjects closed the supervision window and reopened the machines window.

The production lines of the machines were maze-likemulti-segment paths, illustrated in Figure 1. Objects travelled through the machines according to a "Production Line 1" algorithm or a "Production Line 2" algorithm. Passing a faulty segment damaged the algorithm whereupon the object began to travel randomly. A deviation from the normal production line was, therefore, a reliable sign of a faulty segment along the route.

The subjects started an object by clicking on a machine's entrance. The object appeared, and began to travel through the machine at a speed of 1 segment per s. In a first global diagnosis, the subjects had to find out which of the six production lines contained the faulty segment. The subjects first determined the algorithm guiding the object. There were two algorithms: "follow the left wall" and "follow the right wall", defining the two production lines. Then, they observed the behaviour of the object at crossings to see whether or not it wandered astray. On reaching the exit, the object disappeared and a new object started automatically at the entrance. Objects could be eliminated by clicking on them. Then, a new object could be started in one of the machines. Letting many objects run at once was not possible.

The average production line consisted of 35.8 segments (range 29 to 43). The average number of crossings per production line was 8.1 (range 7 to 10). The average time between successive crossings was 4.4 s (range 1 to 16). The maximum between-crossings interval over a given production line was always at least 8 s.

Once the subjects knew which production line was faulty (global diagnosis), they used the option of *disabling* for detailed diagnosis. After disabling, the segment could no longer influence or damage passing objects. Disabling occurred by clicking on a segment. The segment then became grey. The segment was restored by clicking on it again. The segment then became white again. One way to make a detailed diagnosis was to disable all segments of the first part of the faulty production line, and to observe whether or not objects still wandered astray. If the objects still wandered astray, the conclusion was that the fault was located in one of the segments of the second part of the production line, &c.

Once the subjects had identified the faulty segment, they clicked on it after activating a softkey *final repair.* This closed the machines window and restarted the normal production process.

Procedure. The subjects came for two sessions. The first session—3½ h—was for instruction and exercise. Both tasks were practised separately. Roughly two thirds of the time were devoted to fault management. At the end of the session, it was tested whether the subject understood the systematics of fault management. Only subjects passing the test participated in the second session.

Data were collected in the second session. Subjects were reminded about the payoffs. A financial bonus was announced for the best performing subject out of every four. Subjects were explicitly warned not to neglect the supervision during fault management. Then, the subjects worked for 4¼ h on the task. They did not know how much faults to expect. Faults were presented after 45, 120 and 190 minutes. Subjects were not allowed to wear watches.

Analysis. A distinction was made between periods with and periods without fault management. In addition, periods of 10 minutes directly after fault correction were analyzed separately as recovery periods. Within the fault-management periods, a distinction was made between global and detailed diagnosis (the first instance of segment disabling marked the transition between global and detailed fault management).

The dependent variable was supervision loss, defined as intersupervision intervals exceeding 2½ minutes.

2.2 Results

One of the subjects failed the test of the practice session, and did not participate in Session 2. Data of another subject got lost due to an administration error (lost file). The only thing known about this subject was a system crash during fault management. Data of the remaining 16 subjects were analyzed further.

The average duration of a supervision was 7.0 seconds, irrespective of whether supervision took place in periods with or without fault management.

Twenty three supervision losses were observed; 4 times in periods without fault management (always the same subject), 19 times during fault management (the same subject and 6 others), and never in the recovery periods 10 minutes after fault correction. The duration of the supervision loss was 3.10% of the time for periods without fault management, and 0.02% of the time for periods with fault management. Seven system crashes were observed, always during periods of fault management (6 subjects).

Table 1. Distribution of 23 supervision losses over the different periods of the process-control task

	no fault man'gt	fault man'gt	recovery periods
duration (min)	176	49	30
id., relative	69%	19%	12%
sup'v loss (chance distr)	16	4	3
id., actual	4	19	0
χ^2 component	8.9	48.7	2.7

A χ^2 was used to test whether supervision loss was more frequent during fault-management than at other times. The chance distribution of 23 supervision losses over the different periods was estimated, minding the duration of the periods. Then, the deviation between the chance distribution and the actual distribution was calculated as illustrated in Table 1. The χ^2 deviation was highly significant: $\chi^2=60$; $p<0.001$. The χ^2 was for 81% due to the fault-management periods: *Supervision loss was limited to fault-management periods almost exclusively.*

Further analysis of the fault-management periods revealed that most supervision losses occurred during global fault-management: 11 out of 19. Only 4 supervision losses occurred during fault-management. The χ^2 deviation was significant: $\chi^2 = 6.3$; $p < 0.05$. For this analysis, a third category was defined, namely, the transition between global and detailed diagnosis for supervision losses beginning in the global phase and continuing in the detailed phase. The number of supervision losses in the transition phase was as predicted by chance (see Table 2).

Table 2. Supervision loss (n=19) for the various phases of fault management

	global diagn.	trans-ition	detailed diagnosis
duration (min)	5.7	2.4	8.3
id., relative	35%	15%	50%
sup'v loss (chance distr)	6.7	2.9	9.4
id., actual	11	4	4
χ^2 component	2.7	0.5	3.1

2.3 Discussion

The present process-control task was able to evoke tunnel vision. Tunnel vision occurred contrary to instructions and despite the fact that the fault-management task contained waiting periods sufficiently long and sufficiently frequent to accommodate all activities required for adequate supervision.

An important observation was that supervision loss was most frequent in the first phase of fault management, a phase that required hardly anything else than waiting, observing, and interpreting.

Experiment 2 tested the mental workload in the different phases of fault management. One of the requirements of the first phase of fault management is to mentally trace the production lines. (The reader may have tried this with Figure 1). In the second phase, this has been done already. So, the mental workload in the first phase could have been higher than in the second phase, despite the fact that there was more overt activity in the second phase.

3. EXPERIMENT 2

Subjects did a task of continuous fault management (global, then detailed) under two different conditions. Under single-task condition, fault management was never interrupted. Under dual-task condition, fault management was ever interrupted by a secondary task every 20-60 seconds. The subjects were forced to pay attention to this task. If the secondary task would interfere more with global than with detailed fault management, the conclusion is that mental workload is highest during global fault management.

3.1 Method

Subjects. Twenty four new subjects participated as paid volunteers, the same way as in Experiment 1. They were aged between 18 and 30 years.

Single and dual task. The fault management task was made continuous by presenting the next fault as soon as the previous fault was corrected. The task was presented singly (no other task) and as a dual task, in combination with an extra task. For dual-task presentation, time was sliced in periods of 60 seconds. After 20 seconds, a time indicator appeared above the machines window, counting down 40 seconds. During the countdown, the subjects could open the supervision window. The information in the supervision window was the same as in Experiment 1, but system crashes were impossible, and no action was required. The supervision window disappeared after 12 s or earlier; then the subjects resumed the fault management task. The time the supervision window was active was 20% of the immediate preceding fault-management time. For example, the supervision window was active for 4 seconds if the subjects opened it as soon as the countdown indicator appeared (20% of the preceding 20 s fault management); the supervision window was active for 11.8 seconds if the subjects opened the supervision window 39 s after the countdown indicator appeared (20% of the preceding 59 s fault management). If the

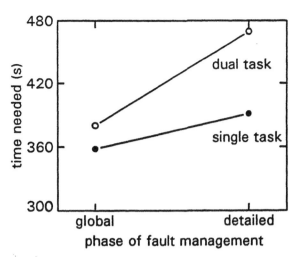

Figure 2. Time needed by the subjects for the two phases of fault management with task condition as the parameter.

countdown indicator reached zero, the machines window went dark. The machines window could be restored by opening the supervision window, waiting 12 s, and reopening the machines window.

Procedure. The subjects practised during Session 1 and came back the next day for data collection. The task was to solve faults efficiently. The number of faults to be solved was 18. Subjects worked in pairs, one resting, one working, and changing places after two or three faults. Single and dual task conditions switched if the subject returned after a resting break.

3.2 Result

The time needed for fault management was subjected to an analysis of variance with subjects as a random factor. Single vs. dual task was significant as a main effect [$F(1,23)=9.76$; $p<0.005$] and interacted with fault-management phase [$F(1,23)=5.10$; $p<0.04$]. Figure 2 shows an increment in the time needed for fault management for dual as opposed to single task condition. The time increment was 6% for global fault management and 20% for detailed fault management. Separate tests for the two phases revealed no effect for the global phase [$t(23)=1.31$] and a significant effect for the detailed phase [$t(23)=3,35$; $p<0.003$].

3.3 Discussion

Mental workload was certainly not higher in the first phase of fault management (global diagnosis). Rather the workload was higher in the second phase of fault management (detailed diagnosis). Adding the secondary task increased the time for global diagnosis with 6% (nonsignificant), but adding the secondary task increased the time for detailed diagnosis with a significant 20%. The conclusion is that global fault management left the subjects with sufficient spare mental capacity to pay attention to another task; by contrast, detailed fault management more or less depleted the subjects' mental capacity: When paying attention to another task, performance on the primary task suffered.

The concurrence costs of the present experiment must have been lower than those of Experiment 1, as the appearance of the supervision task was automated. The point of Experiment 2 is, however, that the extra "supervision task" added similar mental workloads to either phase of the fault management task. The conclusion of Experiment 2 thus remains: lower mental workload in the global phase of fault management.

4. GENERAL DISCUSSION

The experiments showed that the human error called tunnel vision can be evoked in a task of simulated process control. Roughly half of the subjects lost the supervision for some of the time during fault management. Moray and Rotenberg (1989) showed that subjects prefer to work in a serial fashion, sticking to fault management and not accepting other tasks. Kersholt, Passenier and Schuffel (1994) demonstrated that this occurs despite instructions and pay-offs. The present study makes the important point that tunnel vision occurs in the low-workload phase of fault-management where opportunities to attend to other tasks abounded.

Waiting for system information, logically an invitation to pay attention to other tasks, seemed to promote rather than to suppress tunnel vision: Supervision loss was especially likely if the operator waited frequently for system information. A possible explanation is operator passivity; the system presents its information in such a way that the human operator becomes a passive and, therefore, unproductive agent. A preliminary recommendation for design is to avoid operator passivity.

The present line of research may have consequences for personnel selection and assessment as well. If it is possible to assess the proneness for tunnel vision for the individual operator, then "weak" operators could get remedial training, and aspirant operators could be accepted or rejected on the basis of their test results. There is undoubtedly a market for such a test, for example, for pilots and crew aboard ships (crew resource management). The present performance test avoids a notorious weakness of personality tests, namely, the possibility of testees "faking good" or manifesting themselves along the lines of social desirability (e.g., see the recent Besco-

Kennedy debate, 1994). Use of the present task for personnel selection should, however, be preceded by sound psychometric study; for example, studies ascertaining whether the personality characteristic "proneness to tunnel vision" has stability over time, over situations, and so on.

5. REFERENCES

Allport, A., E.A. Styles, and S. Hsieh (in press). Shifting intentional set: Exploring the dynamic control of tasks. In *Attention and performance XV: Conscious and nonconscious information processing* (C. Umiltá and M. Moskovitch, (Ed.)). Academic press.

Besco, R.O. and R.S. Kennedy (1994). Pilot personality testing (letters to the editor). *Ergonomics in Design*, July, p. 36.

Crawshaw, C.M., A. Healey, G.R.J. Hockey and J.A.I. Lambert (1993). Task analysis and critical incidents. In: *The impact of new technology on the marine industries*. Proceedings, 13-15 September 1993.

Easterbrook, J.A. (1959). The effect of emotion on cue utilization and the organization of behaviour. *Psychological Review*, 66, 105-112.

Flavell, J.H. (1976). Metacognitive aspects of problem solving. In: *The nature of intelligence* (L.B. Resnick, (Ed.)), pp. 231-235. Erlbaum, Hillsdale.

Kersholt, J.H., Passenier, P.O. and Schuffel, H. (1994). *The effect of apriori probability and complexity on decision making in ship operation*, Report TNO-TM 94 C-22, TNO Human Factors Research Institute, Soesterberg NL.

Logan, G.D. (1979). On the use of a concurrent memory load to measure attention and automaticity. *Journal of Experimental Psychology: Human Perception and Performance*, 5, 189-207.

Moray, N. (1981). The role of attention in the detection of errors and the diagnosis of failures in man-machine systems. In: *Human detection and diagnosis of system failures* (J. Rasmussen and W.B. Rouse, (Eds.)). Plenum Press, New York.

Moray, N. and I. Rotenberg (1989). Fault management in process control: Eye movement and action. *Ergonomics*, 32, 1319-1342.

Navon, D. and D. Gopher (1979). *Psychological Review*, 86, 214-255.

Reason, J. (1990). *Human Error*. University Press, Cambridge UK.

Veenman, M.V.J. (1993). *Intellectual ability and metacognitive skill*. University of Amsterdam, Amsterdam.

Wickens, C.D. (1990). Processing resources and attention. In *Multiple Task Performance* (D. Damos, (Ed.)). Taylor & Francis, London.

A MULTIPARAMETER TACTILE DISPLAY SYSTEM FOR TELEOPERATION

Dimitrios A. Kontarinis and Robert D. Howe

Harvard University, Cambridge, MA 02138, USA

Abstract: Tactile feedback promises to improve the dexterity of telemanipulation by providing information about the contact between the remote robot's fingers and the grasped object. This paper reports the development of a system for relaying two tactile parameters, vibration and small-scale object shape. The appropriate sensors are mounted in the gripping surface of the remote slave robot's hand, while the tactile display devices are mounted at the operator's finger tips on the master manipulator. For vibrations, the sensor is a miniature accelerometer, and the display is a voice coil actuator. This system is capable of relaying small vibrations from 50 to 1000 Hz. For small-scale shape, the sensor is a 16x16 mm 64 element tactile array sensor which measures the pressure distribution across the contact area. The tactile shape display raises and lowers a grid of pins to recreate this pressure distribution on the operator's finger tips. Experimental tests with a force-reflecting teleoperated hand show that the system improves task performance and aids in object localization.

Keywords: Human-machine interface, telemanipulation, teleoperation, telerobotics.

1. INTRODUCTION

The sense of touch is a vital component of human manipulation skill, as illustrated by our loss of dexterity when wearing gloves. If telemanipulation is to approach human abilities in direct manipulation, it is essential to relay touch information from the remote manipulator to the operator. Our research is aimed at developing this tactile feedback capability. We are working to identify the mechanical parameters that must be sensed at the remote manipulator, to develop sensors to measure these parameters, and to build displays to convey them to a human operator. In this paper we describe a system for relaying two of the most important tactile parameters in dexterous manipulation, vibrations and small-scale object shape.

Vibrations are useful for identifying surface textures and for detecting slip between the gripper and grasped objects. Vibrations are also generated by a variety of transient events in manipulation tasks, such as the instant of contact between the finger tips and an object, and contact between a grasped object and a surface in the environment. Neurophysiological experiments show that humans make extensive use of vibratory information in controlling manipulation (Westling & Johansson 1987). A few previous studies have examined the display of vibrations in a task-related context.

Hawkes (1987) used an acceleration sensor to detect vibrations in the finger tips of a remote manipulator, but displayed the resulting signal to the teleoperator in audio form through a loudspeaker. Minsky et al. (1990) developed a system that can provide vibratory information in virtual environments. This joy-stick based device simulates the mechanical interactions, including vibrations, produced by stroking a stylus over various surface textures and features. Our system combines the functionalities of these previous systems by sensing the vibrations generated in telemanipulation and displaying them to the operator in their original form with high fidelity and good sensitivity.

Small-scale shape information is useful for object recognition, since geometric forms such as edges and corners are often distinguishing features. In control of manipulation, the contact location and the local geometry at the finger-object contact determine kinematic behavior, and controls whether rolling, pivoting, or sliding will occur in response to finger motions and task forces. Work on tactile sensing for autonomous robotics has provided many designs for array sensor devices; see Howe (1994) for a recent review. Several groups have reported work on tactile display of shape in teleoperation and virtual environments (e.g. Cohn, Lam, & Fearing, 1992; Hasser & Weisenberger, 1993). These devices use arrays of pins which are raised against the

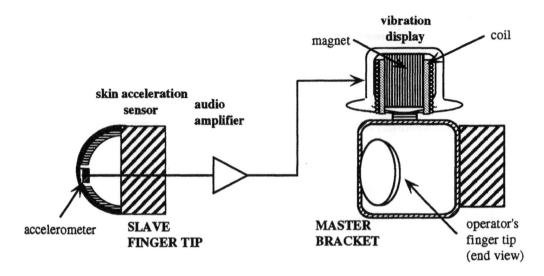

Figure 1. The vibration relay system.

operator's finger tips to approximate the desired shape.

In contrast to this previous work, our goal is to develop a tactile shape display for use in the grasping surface of a force-reflecting teleoperated master robot hand. This poses extremely difficult design challenges. First, the display must be small enough that it fits between the fingers when manipulating an object, and light enough to avoid loading the master and limiting force reflection range and responsiveness. In addition, because it is located at the point of contact between the manipulator and the operator's finger tip, it must be strong enough that the entire reflected force can be supported by the display while maintaining the desired shape. Finally, the display's spatial and temporal bandwidth should approach the capabilities of the human cutaneous sensing system. Our prototype display uses shape memory alloy actuators to control pin height against the operator's finger tip. Performance is adequate to convey basic geometric forms (Kontarinis & Howe, 1993). In the experiments described below, we show that these tactile sensing and display systems can assist an operator in executing telemanipulation tasks.

1.1 Human Tactile Sensing

Because tactile displays are designed to stimulate the human sensory system, we briefly review some pertinent results on the neurophysiology of human mechanoreceptors. There are four types of specialized nerve endings in the glaborous skin of the human hand which play important roles in manipulation tasks (Johansson & Vallbo, 1983). They may be categorized by two criteria: the size of their active areas and their response to static stimuli. Nerve endings with small receptive fields are called Type I units, while those with large fields Type II. Units that respond to static stimuli are denoted SA (for slowly adapting), while those with no static response are denoted FA or RA (for fast or rapidly adapting).

FAII units function as an extremely sensitive vibration sensor. They are located deep in the subcutaneous tissue and have large receptive fields, at least 20 mm in diameter. Often a single unit will respond to vibrations from a few dozen to a few thousand Hz applied anywhere on a finger or half of the palm. This suggests that FAII unit responses do not localize vibratory stimulus on the skin surface, so high frequency vibration information can be conveyed with a single vibration display for each finger tip.

Type I receptors (both SA and FA) are located close to the surface of the skin where the deformations and induced stresses are more pronounced. SA I mechanoreceptors are probably most important in small-scale shape perception, which suggests that a relatively low bandwidth shape display may suffice in many applications. The ability to separately perceive two pointed indenters on the finger tip requires that the points be separated by 1-2 mm, which suggests a specification for spatial resolution of shape display devices. The fourth type of receptors, the SAII units, responds to directional skin stretch across an extended area. This stimulus is best provided by conventional force reflection rather than a tactile display.

2. HIGH FREQUENCY VIBRATION SENSOR AND DISPLAY

Our prototype vibration sensing and display system consists of vibration sensors located in each slave robot finger tip, and a vibration display located on each master finger tip. From the human factors discussed above, a high frequency vibration relay system should convey mechanical vibrations in the range of 50-1000 Hz with variable amplitude and frequency. A number of different tactile sensors can detect task-related vibrations at the remote robot finger tips (Howe 1994). We use skin acceleration sensors, which consist of miniature instrumentation accelerometers

Figure 2. The shape display.

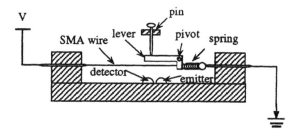

Figure 3. Design of one element actuator.

mounted on the inner surface of the rubber finger tip skin of the slave fingers, as shown in Figure 1 (Howe & Cutkosky, 1989). A layer of foam rubber beneath the skin provides passive compliance to improve grasp stability and isolate the sensor from vibrations in the robot mechanism. In the mounting configuration used here, the skin acceleration sensors have their greatest sensitivity to vibrations in the vertical direction; however, the compliant skin and foam readily couple vibrations in other directions to create an omnidirectional sensor. One advantage of the skin acceleration sensor is its excellent sensitivity to vibrations at the frequencies we are concerned with here.

Since the frequencies of interest are in the audio frequency range, miniature loudspeakers were easily modified to serve as vibration displays. These prototype devices consist of 0.2 watt loudspeakers mounted "upside down," with the outer cones and metal frames removed. The remaining structure containing the magnet, coil, and central diaphragm is then attached to the master manipulator near the operator's fingers (Figure 1). The base containing the permanent magnets is free to move in space. Passing current through the coil generates a force against the magnet, which accelerates the 35 gram base and produces an inertial reaction force against the manipulator. This inverted mounting results in a higher moving mass compared to the usual audio configuration, providing larger inertial forces. The range of motion is 3 mm, and the displays can, for example, produce up to 0.25 N peak force at 250 Hz. Vibrations are transmitted to the fingers of the human operator through aluminum bracket "handles" mounted at the ends of the master manipulator finger links (see section 4 below).

3. SHAPE SENSOR AND DISPLAY

Tactile shape information is important for both object recognition and control purposes. Manipulation of objects by multifingered hands can be described as a series of changing kinematic configurations involving the hand, the object, and the environment. The local geometry of the contact area determines these kinematic relationships. For example, when an object is grasped between the fingers for precision manipulation, the contact areas between the object and the finger tips form temporary passive joints. Depending on the shape of the contact area, these passive joints have a different number of degrees of freedom (DOF). This section describes a new system for conveying information in telemanipulation that allows the localization as well as the recognition of the kinematic type of contact.

The shape relay system consists of a tactile sensor and a tactile display. A tactile array sensor in a remote robot finger tip measures the distribution of pressure across the contact area. A dedicated computer system samples this pressure distribution and applies signal processing algorithms. The resulting signal is sent to the tactile display device mounted in the finger tip contact area of the master manipulator.

3.1 Tactile array sensor

The construction of the capacitive tactile array sensor used in this system is based on an earlier design by Fearing (Fearing, 1990). It is composed of two crossed layers of copper strips separated by thin strips of silicone rubber. As a force is applied to the surface above the point where two strips cross, the distance between the strips decreases, which increases the capacitance between the strips. By measuring the capacitance at each crossing point, the spatial distribution of pressure across the sensor can be determined.

We have devoted considerable effort to optimizing the manufacturing process for these devices. The sensor forms a thin, compliant layer which can be easily attached to a variety of slave finger tip shapes and sizes. By encapsulating the sensor in a layer of elastomer, the surface compliance of the contact area can be controlled. In this prototype system we used 8 strips at 2 mm spacing in each direction, providing 64 force sensitive elements. Special-purpose electronics scan the array to measure the capacitance at all elements in 5 msec. The noise level of each sensor element is less than 0.001 N, and the useful range extends to over 2 N at each element.

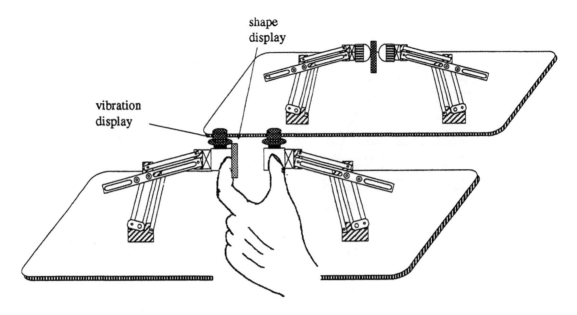

Figure 4. The teleoperated hand system.

3.2 Tactile Display Design

The shape display raises pins against the human finger tip skin to approximate the desired shape (Figure 2). We selected shape memory alloy (SMA) wires as actuators because of their very high power-to-weight, power-to-volume, and force-to-weight ratios. The mechanical design of one element of the shape display is shown in Figure 3. A length of SMA wire is attached to a rigid frame at one end and to a small lever at the other. A spring connected between the lever and the frame keeps the wire in tension and provides a restoring force. The SMA wires are actuated by heating with an electric current. The elevated temperature results in a material phase change which increases the tension and/or shortens the length between the ends of the wire. This causes the lever to pivot about a fixed shaft. The other end of the lever then forces up a pin which rests against the tip of the operator's finger. The displacements of the lever is measured by an optical emitter-receiver pair.

The levers provide a 3:1 reduction in force and amplification in displacement. The wires we used are 30 mm long and 0.075 mm in diameter. The shape display consists of four layers, each layer having six actuators. The center-to-center spacing of the tactors is approximately 2.1 mm.

3.3 System performance

SMA actuators are very difficult to control. In summary, the improvements we have achieved in the performance of the SMA actuators are:

    - Faster heating response by adding a feedforward derivative term;

- Faster cooling response by adding air-cooling;
- Proportional displacement by adding position sensing.

These improvements result in a bandwidth with a -3 dB point between 6 and 7 Hz. The response is close to linear. The slight asymmetry in the response is due to the slightly different time constants associated with heating and cooling of the wire. Furthermore, with our current control set up we are able to control accurate displacement of loaded tactors (see Kontarinis&Howe, 1994 for more details).

4. EXPERIMENTS

The experiments described below use a force-reflecting teleoperated hand system developed in our laboratory for the study of tactile sensing and display (Howe, 1992). This system trades a limitation on the number of degrees of freedom for a clean and simple mechanical design, which results in good control of fine forces and motions. Master and slave manipulators are identical two-fingered hands, with two degrees of freedom in each finger, so finger tip position or force can be controlled within the vertical plane (Figure 4). The design uses brushless DC torque motors in a direct-drive configuration, resulting in low moving mass and negligible friction, backlash, and torque ripple. Two axis strain gauge force sensors measure finger tip forces on both master and slave hands. The system is designed to execute tasks that humans usually accomplish with a precision pinch grasp between the thumb and index finger. For most tasks the operator's wrist rests on the table top and the operator makes contact with the master only at the tips of the thumb and index finger.

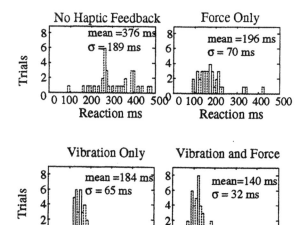

Figure 5. Histograms of the reaction times from the puncturing task

4.1 Puncturing Task

This experiment is designed to test the proposition that when vibrations indicate the progress of a manipulation task, vibrotactile display can improve performance by minimizing reaction times or force magnitudes. The task we selected is designed to emulate the key aspects of medical procedures such as biopsies and catheterizations, where a needle must penetrate a thin and relatively stiff layer of tissue but avoid damaging soft tissue underneath. A 0.05 mm thick plastic membrane (cellophane tape) is mounted on a rectangular frame 0.8 mm in height and approximately 10 mm x 30 mm across. Under the frame is a layer of soft latex rubber. This test structure is mounted on a force sensor that measures the applied force normal to the surface. Subjects were asked to pierce through the tape using a sharp needle held between the fingers of the slave manipulator without exerting excessive force that could damage the underlying rubber layer. They were also instructed that their performance would be penalized for applying excessive force.

Three subjects participated in these experiments. They executed the task with and without force feedback, and with and without vibration feedback. Visual feedback was always available. The order of feedback combinations was randomized to minimize sequence effects. In total, for each feedback case there were 38 sequences comprised of 152 trials. Subjects were permitted to practice the task until they became proficient, and each 10 trial session lasted for about 30 minutes.

The force on the apparatus was sampled at 10 kHz, then digitally filtered (first forward and then backward to eliminate phase delays) with a 250 Hz, ten-pole low pass filter, and finally decimated by a factor of ten. This force information was used to evaluate the subjects' reaction time to penetration of the membrane. The

piercing of the membrane was easily recorded as a transient in the force trace recorded from the force sensor under the apparatus. The time at which the subjects started retracting was defined as the first point of the trace (after piercing) that a line of slope of 2 N/s was tangent to the force and greater than all points of the force signal for the remainder of the trial. This establishes an objective measure of the time at which the force begins to significantly and monotonically decrease, independent of small-scale noise on the force signal.

Figure 5 presents reaction time results for the puncturing task experiment. In the force feedback only case the 15 ms delay introduced by the low pass filter has been subtracted from the reaction times. The width of the histogram in the no haptic feedback case in Figure 5 suggests that subjects had difficulty reliably detecting the puncture event without haptic feedback. The presence of either force or vibration feedback significantly decreased mean reaction times by approximately a half ($7.2 < t < 11$, $0.005 < p < 0.025$). This accords with previous reports that visual reaction times are considerably slower than tactile reaction times in manual tasks (Boff & Lincoln, 1988). The combination of both force and vibration feedback was similarly significant in further reducing reaction time by about 50 ms.

4.2 Feature localization in teleoperation experiments

This experiment was designed to test the ability of the shape relay system to convey significant tactile information in teleoperation. Sensors in the finger tips of the remote manipulator measured the shape encountered during task execution, and the shape display relayed this information to subjects operating the telemanipulation system. The phantom used in these experiments consisted of a cylindrical piece of hard rubber 4 mm in diameter embedded 5 mm beneath the surface of a foam block (Figure 6). A linear translation stage was used to change the vertical location of the phantom between trials to random locations within a 20 mm range. Subjects were asked to probe the phantom with the teleoperated hand until they located the tumor. The experimenter recorded the difference between the actual position of the tumor and the subject's reported location. For the initial planar experiments reported here, only one row of the tactile array and one row of the shape display were used to relay shape information. During the tumor localization task, full force reflection was available but no visual feedback was allowed. Subjects performed the task both with and without the shape feedback to determine the effectiveness of the shape relay system.

Figure 7 shows the results for a total of 60 trials by three subjects. Using tactile feedback, subjects were able to locate the tumor with an error of 1 mm or less

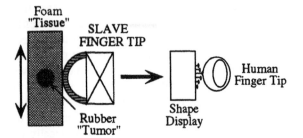

Figure 6. The feature localization experimental set up.

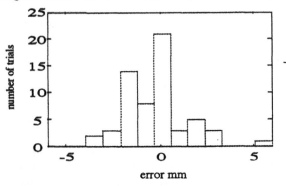

Figure 7. Results from the feature localization experiments.

in over 50% of the trials, and with an error of 3 mm or less 95% of the time. When the shape feedback was not available, the mean absolute error was over 13 mm. In the absence of tactile information subjects often guessed at the tumor location, or based their responses on spurious force signals due to interaction of the edge of the slave finger tip with the phantom. These results demonstrate that tactile feedback can provide information not available with conventional force feedback, and demonstrates the capability to sense and display significant shape features during telemanipulation.

5. CONCLUSIONS AND FUTURE WORK

We have constructed a prototype multiparameter system for tactile sensing and display in dextrous telemanipulation. Experimental results demonstrate that vibratory and small-scale shape information can play an important role in manipulation. Humans make use of this information, and relaying this information can improve performance in teleoperation. Further work will be directed at improving the design of the system components, particularly the shape display, which must meet extremely demanding design requirements. We are also working to delineate the appropriate roles of each type of tactile information, and to distinguish the task properties where they are most important.

6. REFERENCES

Boff, K. R., & Lincoln, J. E. (Eds.). (1988). *Engineering Data Compendium: Human Perception and Performance*. Ohio: H. G. Anderson Aerospace Medical Research Laboratory.

Cohn, M. B., Lam, M., & Fearing, R. S. (1992). Tactile feedback for teleoperation. H. Das (Ed.), *Proceedings of Telemanipulator Technology*, Proc. SPIE 1833, 240-254.

Fearing, R. S. (1990). Tactile sensing mechanisms. *Int. J. Robot. Res.*, 9(3), 3-23.

Hasser, C., & Weisenberger, J. M. (1993). Preliminary evaluation of a shape memory alloy tactile feedback display. H. Kazerooni et al. (Eds.), Proc. *Symp. Haptic Interfaces for Virtual Env. and Teleop. Syst., ASME Winter Annual Meeting*, New Orleans, DSC- 49, 73-80.

Hawkes, G. S. (1987). *Apparatus providing tactile feedback to operators of remotely controlled manipulators* (U.S. Patent No. 4,655,673), filed May 10,1983.

Howe, R. D. (1992). A force-reflecting teleoperated hand system for the study of tactile sensing in precision manipulation. *Proceedings of 1992 IEEE International Conference on Robotics and Automation*, Nice, France, 1321-1326.

Howe, R. D., & Cutkosky, M. R. (1989). Sensing skin acceleration for texture and slip perception. *Proceedings of 1989 IEEE International Conference on Robotics and Automation*, Scottsdale, AZ, 145-150.

Howe, R. D. (1994). Tactile sensing and control of robotic manipulation. *Journal of Advanced Robotics*, 8(3), 245-261.

Johansson, R. S., & Vallbo, Å. B. (1983). Tactile sensory coding in the glabrous skin of the human hand. *Trends in Neuroscience*, 6(1), 27-32.

Kontarinis, D. A., & Howe, R. D. (1993). Tactile Display of Contact Shape in Dextrous Manipulation. H. Kazerooni et al. (Eds.), Proc. *Symp. Haptic Interfaces for Virtual Env. and Teleop. Syst., ASME Winter Annual Meeting*, New Orleans, DSC- 49, 81-88.

Kontarinis, D. A., & Howe, R. D. (1994). Static display of shape. H. Das (Ed.), Proceedings of Telemanipulator Technology, Proc. SPIE 2351, 250-259.

Minsky, M., et al. (1990). Feeling and Seeing: Issues in force display. Proceedings of Symp. on Interactive 3D Graphics, 235-243.

Sheridan, T. B. (1992). *Telerobotics, Automation, and Human Supervisory Control*. Cambridge, MA: MIT Press.

Westling, G., & Johansson, R. S. (1987). Responses in glabrous skin mechano-receptors during precision grip in humans. *Experimental Brain Research*, 66, 128-140.

Issues and Design Concepts in Endoscopic Extenders

Ali Faraz, B.Sc., M.Sc.*, Shahram Payandeh, assistant professor*
and Alex G. Nagy, MD.**

*Experimental Robotics Laboratory (ERL), School of Engineering Science, Simon Fraser University,
Burnaby, British Columbia, CANADA V5A 1S6

**Department of Surgery, Head of Laproscopic Surgery, University of British Columbia, Vancouver,
British Columbia, CANADA

Abstract. Endoscopic surgery as a less invasive method of surgery, is more difficult to perform. The
associated problems are visual issues, hand/tool's movement, and force/tactile sensing. To overcome
some of these difficulties, in this paper new design concepts are reviewed and proposed such as
tools with flexible stem, suturing device, positioning stand, and robotic endeffectors for master-slave
systems.

Key Words. Master-slave systems, Endoscopy, Endoscopic tools/equipments, Flexible stem
graspers, Suturing/sewing devices.

1. Introduction

Endoscopic surgery has been used widely in the
past decade as an alternative to conventional open
surgery and it is also gaining more grounds for
other applications. This is due to many advantages such as:
- Shorter recovery time
- Decreased risk of infection
- Less pain/ trauma for the patient
- Reduction in hospital stay/cost

Generally, endoscopic surgery is a less invasive
method of surgery which is performed by long
surgical tools and endoscopes that are inserted
through small hole incisions for reaching the surgical site. The main draw-backs of the current designs are that they are not able to *extend* all of the
movements and sensory capabilities of the hand of
the surgeon to the surgical site. Inherently, this
method of surgery introduces many new problems
when compared to the conventional open surgery.
For example, some of these problems are:
- Indirect vision through endoscope and monitor,
that is usually two dimensional without depth perception.
- Limited degrees of freedom of tools movement
at the surgical site compared to open surgery.
- Very limited force/tactile sensing.

There have been many developments of new endoscopic extenders, vision systems, and a few
robotic assisting devices (Nagy,*et al.*,1994; Rininsland,1993; Mitchell,*et al.*,1993). However there
still remain a great deal of demand for research
and development to answer some of the basic needs of the surgeon, such as flexible stem

graspers, suturing devices, positioning stands, and
robotic systems. In this paper, the main attempt is to examine issues related to endoscopic
surgery and systematically specify the main problems. Then solutions and design concepts are proposed that facilitates the surgeon in performing
his tasks.

In the following sections of this paper, first the
issues and problems related to endoscopic surgery
are discussed, then some of the design variations
and concepts are provided as possible solutions for
them.

2. Endoscopic issues and problems

There are basically three categories of endoscopic
issues and problems: visual issues, hand/tool's
movement, and force/tactile sensing, that are described in the following sections:

2.1. *Visual Issues*

Endoscopes are basically a video camera where
the visual information is obtained through a long
tube (about 10 mm in diameter and 300 mm
in length). Two types of camera systems of direct and indirect designs can be used. In the
direct type the CCD array is located at the tip
of the tube and signals are transmitted through
the tube, while in indirect system, the image is
transmitted through the fiber optic or lenses to
the other end of endoscope where the CCD camera is located. Both systems provide a clear field
of view of $0-60°$, but still some problems remain

to be addressed such as:

- *Lack of stereoscopic view* (in the case of usual 2D vision systems): Even for simple positioning tasks with endoscopic tools, it takes almost twice the time to perform under direct monocular vision comparing to direct binocular vision, and it is even longer (almost 3 times) under endoscopic viewing condition (Tendick,*et al.*,1993).

- *Limited field of view* : Due to the size limitation of the monitor, as well as endoscopes field of view, the vision does not give the natural 120° field of view of human eyes. Therefore the vision is not perceived naturally and does not provide a natural control environment for the surgeon.

- *Limited resolution* : Visual resolution could be increased by decreasing field of view of endoscope as a trade off. But even in this case the final viewing resolution is determined by the resolution of monitor, which is much lower than resolution of human eyes viewing from a distance.

- *Limited contrast and color fidelity.*

There have been some technological advancement in the application of 3D vision systems in endoscopic surgery. 3D stereo endoscopes available on the market, from quite a few different manufacturers, have improved the depth perception and consequently performance (Mitchell,*et al.*,1993). On the other hand, there are some practical considerations that if taken into account can improve the performance greatly, such as:

- *Position of the monitor:* The distance of monitor from the surgeon should be arranged so that the angle of view of monitor is the same as the endoscopes field of view.

- *Position of the endoscope:* It is important to adjust the axial position of the endoscope for optimum resolution/magnification. On the other hand it is even more important to select the proper incision points for endoscope to give the natural viewing angle of the surgical site and surgical tools. If the angle between the endoscope and tools is more than 45°, then the performance of surgeon drops significantly (Tendick,*et al.*,1993). The best incision location for the endoscope can be in the region between the left and right surgical tools.

- *Angular orientation of the endoscope:* In order to have proper viewing orientation on the monitor, the endoscope should be rotated around its central axis, so that the general orientation of the vision on the monitor can be adjusted to be the same as the vertical orientation of surgeon.

2.2. *Movement of Hand/Tool*

In endoscopic surgery the main requirement is based on the performing the operation through small holes (of trocars on each incision point).

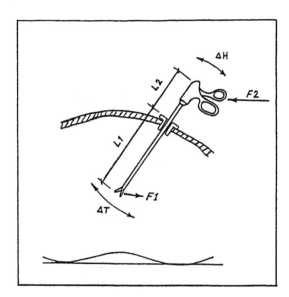

Fig. 1. Movements of the endoscopic tool/hand.

This introduces limitations to the required surgical movements as well as the available degrees of freedom. Basically the incision point and trocar act as a spherical joint on the abdominal wall that allows 3 rotational and one axial movement at the joint location. The inherent problems associated with this spherical configuration of movements are:

- *The sensitivity* of tool's movement (ΔT) with respect to the hand movement (ΔH) which is determined by the distance of incision point to the surgical site (or L1 the length of tool inside the body, Fig.1) over the outside length of the tool (L2).

$$\frac{\Delta T}{\Delta H} = \frac{L1}{L2} \qquad (1)$$

Since the total length of tool is fixed ($L1 + L2$), when the incision point is far from surgical sight the sensitivity of movement is high ($L1/L2 > 1$) and positioning control is more difficult to achieve. On the other hand when the surgical sight is close to the incision point ($L1/L2 < 1$) most of the tool is outside and sensitivity is low. But the positioning control is usually difficult in this case as well, since the surgeons hand is in a high awkward position that is tiring and difficult to control. The optimum case could be when L1 is almost equal to L2($L1/L2 = 1$), this can give equal hand/tool movement and provides more normal hand posture for the surgeon. Therefore whenever possible it is better to select the location of the incision points with respect to the surgical site in such a way that the distance is about half the tools length. However, due to other practical reasons this might not be possible all the time and create difficulties in hand control of the tool.

- *Reverse Motion:* The pivoting joint effect like

movement of the tool around the incision point gives the reverse effect at the handle. This means for example when tools tip should move to the right, the surgeon must move his hand to the left. It is a matter of long training and practice to get used to this unnatural way of position control. What makes it even more difficult is the fact that usually the vision systems provide images that are mirror image along the vertical axis of monitor (i.e. a tool on the left hand side on the monitor is actually in the right hand of the surgeon).

- *Fix Orientation:* For proper manipulation of tissue and suturing needle, 3 degrees of rotational movement is required at the surgical site. With current design of rigid stem tools, only one rotational movements around the axis of stem is possible. Specially in the case of complicated tasks the importance of tools orientation at the surgical site is prominent. For example, in suturing, the completion time with endoscopic tools is twice as much as with ordinary surgical hand tools that have the advantage of orientational flexibility (Tendick,*et al.*,1993).

2.3. *Force/tactile Sensing*

Force sensing at the tip of surgical tool is important for better and safer performance of tasks such as: cutting, suturing, moving, and testing tissues. Due to the endoscopic tools length, forces are transmitted very poorly to the hand. Also the lever effect of tool around the incision point changes the magnitude and direction of these forces $(F2/F1 = L1/L2)$Fig.1.

Also tactile sensing is an important sensory to control grasping force, evaluate surface texture, and to detect small movements such as artery pulse. In endoscopic tools all of these informations are lost and only the grasping force of the tool can be sensed to some extend by the surgeon. Of course even in this case its magnitude and stiffness is altered by all the intermediate mechanical linkages.

3. Solutions and Design Concepts

By improving surgical procedures, training, and more practice, it is possible to reduce completion time per task and increase level of skill. But there is a limit based on learning curve for each task, procedure, and set of tools being used. Any dramatical change usually comes by introduction of new tools or systems that in turn bring totally new procedures and new skills. In endoscopic surgery, there are many new tools and systems needed that can improve performance in many ways. Their success generally depends on their effectiveness, ease of use and natural interface with surgeons hands. In the rest of this paper some of

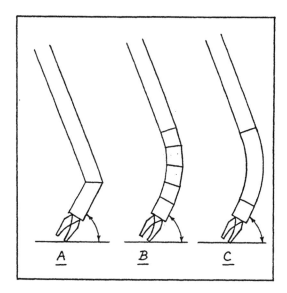

Fig. 2. Different designs of flexible stems

new design concepts are reviewed and proposed which are subject of further research and development. They span from basic mechanical components such as flexible stem to master-slave robotic systems.

3.1. *Flexible Stem*

As mentioned earlier, the present rigid stem (or cannula) endoscopic tools have 4 degrees of freedom and lack 2 rotational movements at the surgical site. The challenge and difficulty lies in creation of these rotational motion on a stem diameter of only about 10 mm which is deep inside the body. The design still should provide some room for the linkages and connectors to pass through the joint(s) to the other moving elements and sensors at the end of the stem. Basically there are three types of designs to provide the required angular movement: Single joint, Multi-joints, and Continuous stem, that are described in following sections.

a) *Single joint:* In this design all the rotational movement is performed by one pin joint that is actuated by linkage mechanism, or pneumatic actuators from outside (Fig.2a). The advantages of single joint mechanism are its relative simplicity, and low space requirements for its movement inside the body. The disadvantages are that it does not provide much room for other linkages to pass through, having a very sharp bending angle, and it only provides one degree of rotation.

b) *Multi-joints:* In this design the rotational movement is distributed among a number of joints with limited range (Fig.2b). These joints could be pin joints (with relative 1 DOF) or spherical joints (with 2 DOF). The control and actuation is usually performed by tendons on the periphery of the stem. In the case of spherical joints there are usually 4 wires while through their axial displacements and the amount of tension, it is possible to bend the stem to the desired angle and lock it in that orientation. The advantages of this design are in its wide rotational range, allowing large passage for other linkages to pass through the joint, and its capability to create gradual bend. The disadvantages are larger space requirements inside the body to bend to a desired orientation, and more complex control of wires axial movement/tension.

c) *Continuous Stem:* Due to recent Results in research and development Shape Memory Alloys(SMA), there is a new prospect of developing joint-less flexible stem (or multi-joint stem) that could be bent by heating the SMA part of the stem (Fig.2c). Although more research is needed to evaluate its potentials, but at this stage, it can be mentioned that its advantages and disadvantages are very similar to the previous multi-joint design. Except for the fact that main issues here are the control of SMA stem and its mechanical performance under dynamic conditions.

3.2. *Suturing Device*

One of the most difficult tasks in endoscopic surgery is suturing. It needs a lot of training and practice to perform the simplest sewing and knotting techniques inside the body. Basically it is performed by using two endoscopic graspers(which are called needle drivers). As mentioned earlier, problems of vision, tools movement, and force sensing in endoscopic surgery are most tangible and acute in suturing task due to the complex tools/needle movements. In this regard there have been some attempts to develop miniature suturing devices that facilitate sewing by eliminating manipulation of needle (Nagy,*et al.*,1994; Rininsland,1993; Neisius,*et al.*,1994, Melzer,*et al.*,1993). The main idea behind most of these designs (Fig.3a and 3b) are to transfer a needle between two jaws by a reciprocating motion and intermittent locking of the needle by one of the jaws each time (Melzer,*et al.*,1993). The needle has a central cross bore for the thread that is carried through tissues being sutured. The advantage of this design is its compactness and few moving parts.

There is another design under investigation and development that the needle is a circular arc shape(about 240°), that is moved in a circular path(Fig.4). The movement is provided by continuous motion of one finger, and the surgeon has total control of needle both in terms of position and direction of the needles movement.

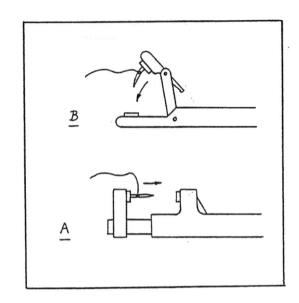

Fig. 3. Suturing devices with reciprocating motion design.

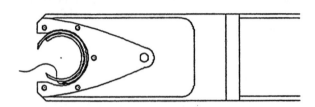

Fig. 4. The suturing device with circular motion design.

3.3. *Positioning Stand*

In the endoscopic surgery, positioning of the tool and keeping it fix in that position is a routine task that is carried out frequently. This is usually done by an assistant surgeon, that beside crowding the operating area, it is costly. An alternative could be the usage of a stand similar to the configuration of SCARA robot (Fig.5) with additional end-joints that can give the tools two rotational degrees of freedom about X and Y axis. Since the stand is naturally balanced, the surgeon can move the tools freely with minimal opposing frictional forces, and lock them at any position and orientation. Even in the unlock state of the stand, the surgeon does not have to carry the total weight of the tools. On the other hand, dexterity can be improved by locking the two joints at the base of the arm (i.e. joints A and B, Fig.5) when the tools are already inside the body through abdominal incision points. This provides a resting frame for the surgeon as well as a rigid base for the end-joints(i.e. joints C and D, Fig.5) to be moved in a much more controlled manner.

The positioning stand can be upgraded to a computer aided surgical system by adding measuring sensors to joints of the arms. Through the kinematic model of the mechanisms, one can monitor and approximate the position vector and the ori-

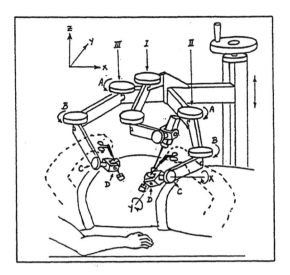

Fig. 5. The positioning stand with three arms.

entation vector of the end-point. Also by having a database of the abdominal cavity and internal organs, It is possible to provide graphical map of the body and the relative position of surgical tools with respect to them.

Specially this could be used for training purposes by developing interactive softwares tools to aid students with graphical, video, and text, that provide them step by step complementary information about each procedure.

By adding actuators (in a closed loop control) to the above passive system, it is possible to obtain full robotic capabilities for automatic positioning of tools.

This is possible without sacrificing other capabilities mentioned earlier. Here the surgeon is still able to move the arms manually with ease while performing the operation with tools attached to the arms. To achieve this, the arms not only must be light and frictionless, but also their engagement with motors should not interfere with manual handling of the arms by the surgeon.

3.4. Robotic Endeffectors and Telesurgery

This could be implemented in two stages:

a) *Hand-tool endeffectors:* Initially, it is possible to incorporate some of the more advanced mechanisms and sensors into endoscopic hand-tools to provide more dexterity for the surgeon. Such improvements can be:

- The implementation of flexible stem in the hand-tool that is actuated by the surgeon for obtaining the proper orientation at the surgical site. This extra actuation mechanism can be used in conjunction with the surgeon's own hand movements to accomplish a proper orientation. Though the surgical movements are totally generated by direct hand movement of the surgeon.

- The application of force sensors in the hand-tool to provide the surgeon with better force/tactile sensing.

b) *Master-slave endeffectors:* The main difficulty in endoscopic surgery is the usage of very long tools through fix small incision points. No matter how much the design of tool (both in terms of degrees of freedom and optimum interface with the surgeon's hand) is improved, still direct physical hand control of the tool is unnatural, remote, and physically demanding. Only with a lot of training and practice, it is possible for the surgeon to obtain a fraction of skill and dexterity level of open surgery. Therefore to obtain much higher dexterity, direct hand control of endoscopic tools can not be the solution. Further improvement lies in the development of robotic endeffectors (to replace the hand tool) which are indirectly controlled by the surgeon. This is actually a master-slave robotic system that surgical movements of the robotic endeffector inside the body is generated and controlled by hand movements of the surgeon on a telesurgical workstation.

The success of such a system not only depends on the general control characteristics of the master-slave system (such as accuracy, fast response, and force reflection), but also its ease of usage and being natural to control the slave. For example to control the endeffector by means of a "joystick" is not a natural interface for the surgeon, since all the endeffectors movements should be translated to movements of the joystick by logical step by step reasoning, instead of subconscious control. In order to achieve an easy to control master-slave system, which does not require substantial training, the mechanical movements (or DOF) of the endeffector should be similar to natural movements of human hand. In other words the master and slave should be kinematically similar. For example, the graspers jaws movement to be directly controlled by angular movement of the thumb with respect to other fingers, or the orientation of the grasper could be directly controlled by angular movements of the wrist.

Force reflection is an important sensing and safety issue that could be incorporated into the master-slave system. It gives force sensing to the surgeon during operation by friction control of the axes of master arm.

- *Slave Configuration:*

Based on previous discussion, there could be different design variations of master/slave configurations. The following is based on one such design that is subject to research and future development.

The slave endeffector (Fig.6) is mounted on the arm of a positioning stand similar to the one described in section 3.3. The positioning stand in this case is a passive mechanism that holds the endeffector in the proper position and orientation with respect to the incision point. The slave en-

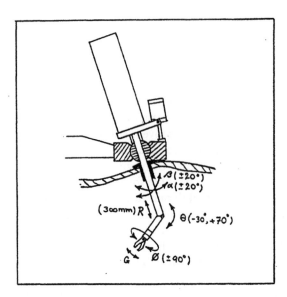

Fig. 6. The slave endeffector.

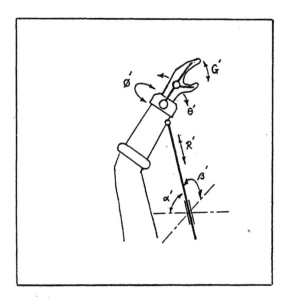

Fig. 7. The master arm controller.

deffector provides 3 positioning axes (R, α, β), 2 orientational axes (θ, ϕ), and one grasping action (G). This configuration provides a very compact robotic system which is essential for endoscopic surgery since in a normal operation at least 2 endeffectors are required in a limited space.

- *Master configuration:*
The master arm has a polar configuration (Fig.7) that hand's position is sensed by axes (R', α', β'), the wrist orientation by (θ', ϕ'), and the angle of thumb/finger for grasping by (G').

Since the master and slave are kinematically similar, not only it is easy to use, it is much less difficult to implement the system in terms of sensing, actuation, and control issues. Though the challenge lies in design and manufacturing of such miniatured endeffector with 5 degrees of freedom that is accurate and dynamically controllable by hand's movements.

4. Conclusion

Endoscopic surgery is prefered mostly over open surgery due to many advantages it has for the patient.Though special problems that surgeons are facing in endoscopic surgery , should be studied and analyzed in order to provide better visual information, easier movement of hand/tool, and better force/tactile feedback sensing. Surgical tools with flexible stem provides better access to the surgical site. The suturing device is a special tool for the difficult task of suturing and knotting that increases speed substantially. The positioning stand can be used for positioning tools, a resting frame for the surgeon, as well as a computer-based information system. Robotic technology can be used in endoscopic endeffectors and telesurgical manipulators. Master-slave systems of similar kinematics are suitable mostly, due to the ease of usage, and ease of implementation.

5. Acknowledgments:

The authors would like to thank the financial support of The Institute for Robotics and Intelligent Systems (IRIS), A Federal Networks of Centres of Excellence.

6. REFERENCES

Melzer, A. (1993). Intelligent Surgical Instrument System ISIS, *J.End.surg.*, pp.165-170.

Mitchell,T.N., J.Robertson, A.G.Nagy, A.Lomax (1993). Three-dimensional Endoscopic Imaging for Minimal Access Surgery. *J.R. Coll. Surg. Edinb.* ,Vol.38, pp. 285-292.

Nagy, A.G., S. Payandeh(1993). Endoscopic End-Effectors. *The national Design Engineering Conference, March 1994*, **ASME, 94-DE-5.**

Neisius, B., P.Dautzenberg, R.Trapp (1994). Robotic Manipulator for Endoscopic Handling of Surgical Effectors and Cameras. *First International Symposium on Medical Robotics and Computer Assisted Surgery (MRCAS)*, pp.1-7.

Rininsland, H.H. (1993). Basics of Robotics and Manipulators in Endoscopic Surgery *Endoscopic Surgery and Allied Technologies.* ,Vol.3, pp.154-159.

Tendick,F., R.W.Jennings, G.Thrap, L.Stark (1993). Sensing and Manipulation Problems in Endoscopic Surgery: Experiment, Analysis, and Observation. *Presence, Winter 1993, MIT,* **Vol.2, No.1,** pp. 66-81.

VIRTUAL TOOLS FOR INTERACTIVE TELEROBOTICS:
POTENTIAL FIELDS AND TERRACE FOLLOWING

D. J. Cannon and C. Graves

K. W. Lilly and C. S. Bonaventura

Department of Industrial Engineering
The Pennsylvania State University
University Park, PA 16802

Department of Mechanical Engineering
The Pennsylvania State University
University Park, PA 16802

Abstract: Virtual tools, placed using an instrumented glove, were used to point and give directives to a telerobot in an interwoven graphic/video environment. A local minimum and dynamic collision were avoided using a terrace potential method for modifying trajectories by interactively specifying a safe way-point. In a demonstration under development, an operator reaches into the scene of a hazardous tank mock-up and, with a cyberglove-controlled virtual pipe-cutting tool, directs a PUMA 560 robot to cut a piece of pipe "there" while avoiding "that" obstacle en route.

Keywords: Human-machine interface, potentials, telerobotics, trajectories, virtual reality.

1. INTRODUCTION

Dominant control paradigms in robotics today include telemanipulation and autonomy. In the past, these were often seen as mutually exclusive alternatives. Telemanipulation is a manual control paradigm that is often tedious, but it promotes direct human interaction which, in turn, supports work in an unstructured environment. The "autonomous" paradigm alternative is actually a misnomer. While it allows an operator free time for observation and planning in the execution phase, this so-called autonomy often requires extensive pretask human involvement to teach motion or custom scene interpretation. Both paradigms, telemanipulation and autonomy, may overuse valuable human resources where a more efficient human-machine relationship appears to be achievable.

The Virtual Environment-based Point-And-Direct (VEPAD) Program within the Industrial and Manufacturing Engineering Department of the Pennsylvania State University is developing the Interactive Specification (InterSpec) system as a generic facilitation technology for studying aspects of telerobotics related to the human-machine interface in applications such as hazardous waste remediation, space exploration, and flexible manufacturing. One component of this point-and-direct interactive specification research involves the development of new methods to modify trajectories proposed in simulation. To successfully designate a general task using natural gesture requests, the VEPAD/InterSpec operator uses an instrumented glove (Cyberglove) to manipulate graphic representations of robot tools flown in the workscene. Subsequent trajectory plans to perform such tasks, automatically generated as paths in a potential field comprised of repelling obstacles and attractive destinations, are, if necessary, altered by the human using a virtual way-point modification tool to incorporate geometric and dynamic improvements. By specifying a way-point, the operator, in effect, specifies a terrace potential other than zero as a path objective to follow around obstacles.

In a demonstration of the VEPAD/InsterSpec system now under development, an operator reaches into the scene of a hazardous waste storage tank mock-up, and, with a virtual pipecutting tool held in a Cyberglove representation of the hand, directs a PUMA 560 robot to cut a piece of pipe while avoiding a standpipe and barrel en route. Only the three key points of the trajectory are specified. A potential field-based trajectory is planned to complete this task. Instances of a local minimum and a dynamic collision are then overcome through interactive way-point modification. Specifically, a way-point is selected that is clear of the local minimum. A second way-point is defined, if required, at a location which ensures acceptable dynamic vibration and a safe approach to the final goal. Instead of seeking minimum potential, the autonomous navigation planner follows a terrace of constant potential at a field strength determined by the humanly specified way-point.

Through this natural task-oriented interaction between human and robot in nearly real time, a world model emerges with geometric and dynamic attributes without requiring a human to perform cumbersome telemanipulation. In principle, a convex aura of potential can be created about the obstacle concavity, and a precedent for safe dynamic approach to destinations of standpipe profile defined. Using this system, the operator may "look before leaping" (by virtually specifying points) and "try before buying" (by previewing and interacting as necessary with the automatically generated trajectory). A dynamically complex robot with multiple degrees of freedom is now directed to perform an unstructured task, yet no world model is initially required, and minimal supervisory control time is involved. A high level of confidence is thus insured before actually moving the robot in a complex and possibly dangerous environment.

2. BACKGROUND

This research shares features of telerobotic supervisory control and virtual environment technology with others now working in these fields. At the most basic level, the VEPAD/InterSpec concept is a robotics extension of a computer mouse used to manipulate icons on a display monitor. Apple Computer Company commercialized the original click-and-drag concept. At the MIT media lab, Bolt and colleagues (1984) extended click-and-drag to a wall-sized screen on which icons of naval vessels were moved in 2-D by pointing, while saying: "put that... there". The VEPAD program adopts the "put that there" phrase, to maintain continuity with the MIT work, while extending the concept to the manipulation of actual objects using physical robots which perform real world actions based on virtual environment input.

From a navigation standpoint, a JPL Mars Rover concept implemented by Wilcox (1987) used Computer Aided Remote Driving (CARD) in which a human operator moved a 3-D cursor in a Martian scene to design a safe vehicle path. Burtnyk (1991), of the National Research Council of Canada, also explored a supervisory control concept for exploiting autonomous robot functionality in an unstructured environ-

ment as an operator directed cursor position in conjunction with structured light machine vision.

Noyes and Sheridan (1984), Kim and Bajczy (1991), and Ince, Bryant and Brooks (1991) are among several research teams who have introduced superposition of preview displays on live video. Although the full cumbersome motion of the manipulator is still specified, these approaches, in addition to making the system controllable, result in time savings for the operator. Shared and traded control are being explored by Hayati and Venkataraman (1989) to restrain the robot from dangerous maneuvers and to provide guidance during telemanipulation. Included among others in similar lines of investigation are: Morgenthaler et al. (1990) and Backes et al. (1993). Sheridan spoke early, and often, of the need to explore the range of interactive telerobotics approaches (Sheridan, 1976, 1992).

Within virtual environment robotics, the objective has traditionally been either to program robots in the simulated world for parallel execution in the real world, or to train humans to perform telemanipulation better before assuming the master controller of a real robot. In collaborative development of virtual reality techniques for controlling robots in hazardous waste applications, Sandia National Laboratories is extending this concept and incorporating the principles of the VEPAD/InterSpec system to break away from pure simulation into an interwoven live/graphic environment. The Intelligent Mechanisms (IM) group at NASA Ames, also in collaboration, is expanding the area of graphic augmentation with video so that an operator graphically designates a course for vehicles while allowing autonavigation to complete the task should obstacles be encountered (Fong, Hine, and Sims, 1994).

Literature on path planning at minimum potential in a field of negatively charged obstacles, a negatively charged robot, and a positively charged goal includes Khatib (1986) and Krogh (1984). Lilly (1993) examined improved dynamic modelling for complex robotic systems interacting with general environments.

3. PROJECT DEVELOPMENT

The interwoven reality VEPAD/InterSpec concept bridges the telerobotics continuum between the extremes of telemanipulation and autonomy by including desirable features of each without incurring their individual drawbacks (Figure 1).

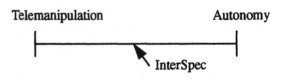

Fig. 1: InterSpec bridges the Telerobotics Continuum

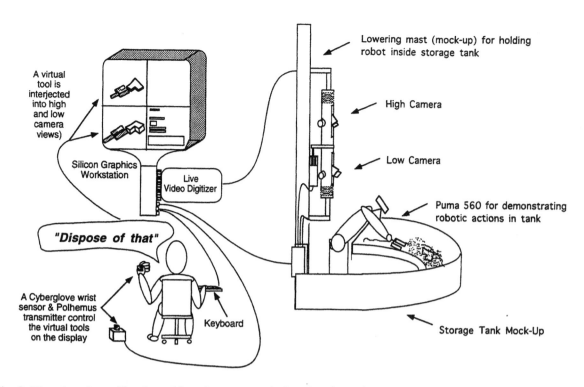

Fig. 2. Virtual tool specification with an instrumented glove in a hazardous tank mock-up

Allowing the operator to interactively control virtual tools at the object level in live video gives him/her the ability to specify a few key points in the live scene while leaving tool changes, trajectory planning, and end-effector actions such as grasping or cutting to the robot. The important interactive element of telemanipulation remains, but autonomous features are incorporated. This frees the human for task conception, scene interpretation, and other high level activities, and makes directing a robot possible for the non-expert. Preconfigured tasks can still be activated or whole paths manually defined, but exclusive reliance on either approach is eliminated. The human and robot do what each does best.

Cannon initially explored the interactive specification concept in the mid-1980's (Cannon, 1992). The resulting Stanford Point-And-Direct (PAD) mobile robot demonstrated that an operator could direct a robot to perform tasks in a natural and interactive way by pointing to objects and destinations while giving directives such as "put that there." To further this concept, interactive graphic representations of robot end-effectors were subsequently created at The Pennsylvania State University for hazardous waste tasks including excavation, cutting, radiation scanning, suctioning, and grappling. When selected from a toolbox, a virtual tool, associated with a specific task, rotates and translates in proper perspective anywhere in a three-dimensional graphic workspace. This virtual reality is interwoven with live video on a Silicon Graphics workstation (Figure 2) with depth correlation. The virtual tools are grasped, like any other real tool, using a graphic representation of an

instrumented glove worn by the operator. Each tool then flies simultaneously in two views that correspond to widely spaced camera viewing locations.

When an operator selects a virtual tool from the virtual toolbox and pushes it deep into the workspace, the virtual tool and graphic representation of the glove recede to appropriate size, just as a real tool would appear to shrink when travelling back into a real scene. A typical InterSpec camera targeting location is an item of cut pipe or debris in a hazardous waste enclosure. A simple mouse click on the same location in each view causes the cameras to quickly drive in pan and tilt to center upon the specified point and record its three coordinate position. A virtual tool is then selected from the toolbox and appears in the scene for movement to objects and destinations. A plane of depth-correlated intersection between physical and virtual reality is translucently draped in the live/graphic interwoven workspace at a distance corresponding to the triangulation depth of the object in the scene. When a virtual tool engulfs a video object by passing partially through this cut-plane, all protruding portions of the tool are re-rendered in phantom wire frame to indicate tool depth to the operator. This also makes hidden portions disappear as a real tool would disappear if passed behind an object. Wireframing is preserved as an enhancement to keep tools from being totally "lost" behind objects. At such points, an operator gives gestures to direct the robot to, for example, "put that there". The robot then plans and executes the task requested at specified positions and orientations defined in this manner.

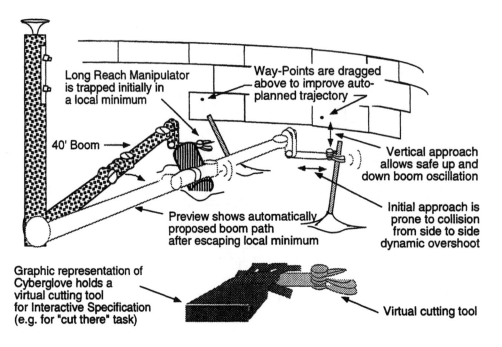

Fig. 3. A robot boom is directed using a virtual tool to cut a standpipe "there". A safe path must be planned.

4. CURRENT FOCUS

While most tasks are adequately specified by selecting tools and placing them relative to objects and destinations, there are often cases when human involvement must go beyond simple object specification. In hazardous waste storage tank remediation (Figure 3), for example, a 40 foot long-reach manipulator has significant dynamics (e.g. structural flexibility) with which to contend. A path planner may be unable to avoid all local minima and may be unable to plan safe and efficient paths without risk of collisions. Without turning to risky telemanipulation and sacrificing the benefits of automatic trajectory planning, a virtual path modification tool was implemented to guide trajectory development. A local minimum is avoided, for example, by reaching in with the graphic representation of the Cyberglove and moving a trajectory way-point to avoid the minimum. Similarly, a new way-point is added to ensure that the long boom oscillates vertically instead of horizontally during its approach to a 4-inch diameter pipe to be cut. Using this technique, the operator need not be trained in manually controlling multiple massive links in cluttered environments, yet (s)he can provide common sense input to assist the dynamics-based trajectory planner.

4.1 Potential Field Implications with Way-Point Modification

In general, a local minimum in trajectory planning occurs because of obstacle concavity or effective concavity in obstacle groupings. First, a single obstacle which is concave in shape results in a local minimum in the potential field when the planner takes the trajectory to minimum potential within the concavity itself. When the trajectory planner attempts to pass the robot through this configuration of obstacles on its way to an attractively charged destination point, it becomes mathematically "trapped". Movement in any direction increases the equivalent potential. The basic planning algorithm stalls at this stage, and an outside heuristic or other mechanism is needed to further guide the planning process.

Several levels of trajectory modification are possible in avoiding local minima and accommodating dynamic vulnerability. At the first level, an interactively specified way-point defines a terrace of mid-potential which keeps the trajectory from prematurely descending to a minimum potential level (from which it might not reemerge prior to acquiring the final destination). At the second level, world knowledge is acquired as a result of the human interaction, so a similar task involving similar circumstances will not again require human involvement.

As more difficult and complicated tasks are attempted, complimentary elaboration on the heuristics of interpreting human involvement may be added as desired. While any individual robotic task can be fully automated, there will remain a need for human involvement in the foreseeable future. As truly challenging tasks, such as those anticipated in hazardous waste remediation, high diversity flexible manufacturing, and space exploration are attempted, better and more natural ways for the human to interact with robotic systems must be developed.

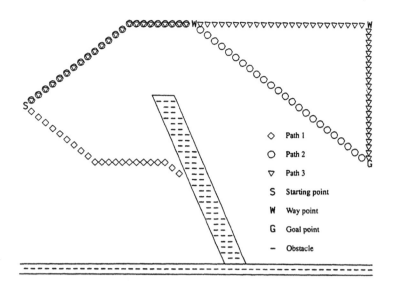

Fig. 4. Standard potential field approach with way-points as intermediate goal points

5. SYSTEM DESCRIPTION

The graphics techniques used to interactively modify way-points are similar to those used to manipulate other virtual tools in the InterSpec system. Indeed, a virtual way-point modification tool was developed along the same lines as the virtual cutting and other tools of the VEPAD program. Rendering of the virtual tool and PUMA 560 robot used in the current demonstration is done within a Deneb Telegrip visualization software package which incorporates dynamics modelling.

After the human interactively specifyies a task, the robotic system programs itself using the potential field method to move to tag points associated with end effector position and orientation (i.e. x, y, z, roll, pitch, yaw). At the current stage, the potential field trajectory is converted to Telegrip tag points by data entry. Once the tag points are placed in 3-D space (ultimately as an automated routine), the robot is programmed to move through or perform operations at the tag points. One of the inputs from the Cyberglove interface is the tag point location for tool operation. The other tag points are way-points en route derived from the potential field planner.

The position and orientation of the Polhemus sensor on the back of the Cyberglove worn by the operator is tracked, and position is displayed within the virtual scene. In this system, then, the operator reaches into the virtual scene with the graphic representation of the Cyberglove and clenches the way-point modification tool. (S)he then uses the tool to grab and drag individual way-points. When a way-point is moved, a new trajectory terrace potential replaces the default minimum potential criteria. In this approach, a new dynamic trajectory is then calcu-

lated, passed to Telegrip, and shown in simulation. An acceptable trajectory is executed if approved by the human operator after viewing the entire preview.

6. POTENTIAL TERRACING EXAMPLE

In a descriptive demonstration of the dynamics problem, a PUMA 560 robot arm carrying approximately 5 pounds in an outstretched side-to-side maneuver collided with a target object when the robot was programmed for high speed movement to a point approximately one inch shy of that target. A Silicon Graphics workstation with Deneb software and a PUMA emulation terminal utilizing the PUMA's native VAL II code were used to direct the robot in two test cases. The robot did not collide when using only its normal gripper. However, when the robot carried a test mass representing a heavy tool (such as an excavator or hydraulic shears), the robot moved with the added dynamics of the test mass and bumped into the obstacle. A longer more massive arm could be expected to deflect more seriously. A trajectory with safe way-points was then implemented to insure safe arrival along a different angle of approach.

One simple but computationally inefficient way to incorporate human interactive way-points into potential field trajectories is to make each way-point, in turn, an isolated goal point and successively recalculate the potential field (Figure 4). Trajectories are then linked with a way-point goal of one calculation as the initial point of the next. In the standpipe avoidance application, the initial entrapment in a local minimum (Path 1) is solved by inserting a way-point above the obstacle (Path 2). Two potential fields are then calculated, one where the minimum potential is at the way-point and the next, which starts at the way-point, recalculates with minimum potential at the final goal. When the concept is extended to two intermediate way-points, three potential field calculations are required (Path 3).

A second method requires only the original potential field calculation. Rather than seeking minimum potential, however, an intermediate potential equal to that of the way-point is followed until nearing the final goal (Figure 5). Several algorithms are possible for defining a threshold that is considered "near" to the goal. The present method follows the terrace of constant potential until the plane of the goal is reached and then applies the minimum potential criteria for the final approach. The plane of the goal, in this case, passes vertically through the goal point and contains the mast from which the arm is suspended.

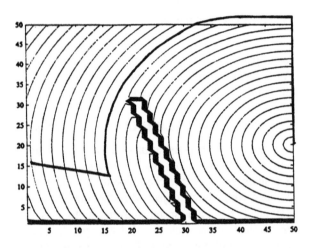

Fig. 5. Potential Terrace Following Avoids Obstacle

The net result of this interactive specification research is an elaboration on the potential field method (Khatib, 1986; Krogh, 1984), which has heretofore generally assumed a minimum potential criteria. Now a continuum of field potentials other than zero can serve as path generating criteria.

7. SUMMARY

First, virtual tools were developed to allow an operator to direct a robot in an unstructured environment. These were used to direct a robot to cut a standpipe with a gesture command to cut "there". Second, a terrace following approach to trajectory planning expands our repertoire of interwoven live/graphic tools for virtual environment-based point-and-direct robotics. This allows a human operator to interactively specify common sense information in a natural graphical way while preserving the robot's sophisticated autonomous navigation capabilities to a reasonable extent. Humans and machines do what each does best, and robustness is achieved by providing a means for human-machine interaction.

REFERENCES

Backes, P. G. Beahan, J. and B. Bon (1993). Interactive command building and sequencing for supervised autonomy. *IEEE Inter. Conf. on Rob. and Aut.*, **6**, pp. 795-801.

Bolt, R. A. (1984). *The human interface where people and computers meet.* Lifetime Learning, Belmont, CA.

Burtnyk, N. and J. Basran (1991). Supervisory control of telerobots in unstructured environments. *Inter. Conf. on Adv. Rob.*, Robotics Society of Japan.

Cannon, D. J. (1992). *Point-and-direct telerobotics: object level strategic supervisory control in unstructured interactive human-machine system environments.* Ph.D. dissertation, Stanford Univ.

Fong, T., Hine, B. and M. Sims (in press). The application of telepresence and virtual reality to subsea exploration. *Proc. of ROV'94: Second Workshop on Mobile Robots for Subsea Env.*

Hayati, S. and S. T. Venkataraman (1989). Design and implementation of a robot control system with traded and shared control capability. *Proc. of 1989 IEEE Inter. Conf. on Rob. and Aut.*, pp. 1310-1315.

Ince, I., Bryant, K. and T. Brooks (1991). Virtuality and reality: a video/graphics environment for teleoperation. *Proc. of 1991 IEEE Inter. Conf. on Sys., Man, and Cyb.*, **2**, pp. 1083-1089.

Khatib, O. (1986). Real-time obstacle avoidance for manipulators and mobile robots. *Int. J. of Rob. Res.*, **5**, 1:90-98.

Kim, W. S. and A. K. Bejczy (1991). Graphics displays for operator aid in telemanipulation. *Proc. of 1991 IEEE Inter. Conf. on Sys., Man, and Cyb.*, **2**, pp. 1059-1067.

Krogh, B. H. (1984). A generalized potential field approach to obstacle avoidance control. *SME Tech. Paper MS84-484.*

Lilly, K. W. (1993). *Efficient Dynamic Simulation of Robotic Mechanisms.* Kluwer Academic Publishers, Norwell, MA.

Morgenthaler, M. K., et al. (1990). A testbed for tele-autonomous operation of multi-armed robotic servicers in space. *SPIE Cooperative Intelligent Robotics in Space*, **1387**, pp. 82-95.

Noyes, M. and T. B. Sheridan (1984). A novel predictor for telemanipulation through a time delay. *Proc. of 20th Ann. Conf. on Man. Control*, NASA/Ames Research Center.

Sheridan, T. B. and G. Johannsen (1976). *Monitoring Behavior and Supervisory Control.* Plenum, NY.

Sheridan, T. B. (1992). *Telerobotics, automation, and human supervisory control.* MIT Press, Cambridge, MA.

Wilcox, B. H. and D. B. Gennery (1987). A mars rover for the 1990's. *J. of the British Interplanetary Society*, **40**, pp. 484-488.

A CONCEPT FOR SYMBOLIC INTERACTION WITH SEMI-AUTONOMOUS MOBILE SYSTEMS

M. Pauly* and K.-F. Kraiss*

**Aachen Technical University, Institute of Technical Computer Science, D-52074 Aachen, Germany,
E-mail: pauly@techinfo.rwth-aachen.de*

Abstract. When an operator has to control a remote mobile system he is faced with several problems: small visual field, difficult handling, non-humanlike communication, etc. This paper describes a new concept for a symbolic interaction with semi-autonomous mobile systems, using a symbolic world model and a humanlike communication-interface.

Key Words. Mobile robots; co-operative control; intelligent knowledge-based systems; man/machine interaction; remote control; system architectures; telecontrol; telerobotics

1. INTRODUCTION

Just a few years ago large and expensive remote controlled vehicles were mostly used in military area and space research. With cheaper and smaller sensor and computer technology, these systems receive more and more autonomy and occupy more parts of todays industrial work life. However, up to now the development in the field of human machine interaction has been slower than the progress in the hardware sector, especially in the area of service roboting. These service robots often need much more communication with the operator and the environment then robots used in industrial production.

To Improve the performance of service robots, the bottleneck in the man-machine communication must be reduced. On the one hand this can be done by training the operator on the system, on the other hand by adapting the system to the operator. Traditionally, the operator interacts with the mobile robot, like driving a car by using a steering wheel, sidestick, or pedals. However, in teleoperating systems there is usually no tactile feedback and the visual field is rather small. With these restrictions it is much more difficult to control a tele-robot than, e.g., driving a car (Browse and McDonald, 1992).

At the Institute of Technical Computer Science a semi-autonomous mobile system was developed for the purpose of monitoring, e.g., museums and office buildings at night. New concepts of computer-human interaction are tested using this vehicle as testing platform.

This paper focuses on the human-machine interaction part of this project. The communication between the operator and the system is based on a symbolic representation of the operating environment to make it more transparent and thus facilitate user interaction.

2. SEMI-AUTONOMOUS SYSTEMS

2.1. *Architecture*

Almost all 'autonomous systems' obtain their orders from a human operator. By this way these systems are strictly semi-autonomous. Today most control structures applied to semi-autonomous or autonomous systems possess complex vertical or horizontal control structures (Albus *et al.*, 1988). The disadvantage of these control structures is that the operator must usually have a complex internal model to give the vehicle effective commands.

A further disadvantage is that these systems need a lot of computer power to realize these complex control structures. All semi-autonomous and autonomous vehicles have their own local intelligence onboard, e.g., for path planning (Alexopoulos, 1992), (Barraquand *et al.*, 1992) and navigation. This local computer performance is restricted by the capacity of batteries, space, etc. (Zhao *et al.*, 1994).

At the Institute of Technical Computer Science a vehicle was developed having only the necessary intelligence onboard, in order to keep it small and lightweight. Due to this, a behaviour-oriented control is used onboard the vehicle (Gat *et al.*, 1994), (Dorigo and Schnepf, 1993). The remain-

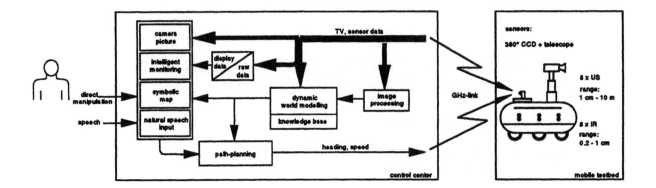

Fig. 1. system schematic

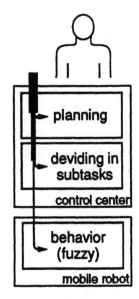

Fig. 2. software architectures

der of the software is located in the control center (see Fig. 1): dynamic world modelling, image processing, path-planning, knowledge base, connected wireless with the vehicle. As a result, this knowledge can be shared by several semi-autonomous agents working together.

The system architecture consists of three layers (see Fig. 2): a behaviour-oriented layer, a subtask layer and a planning layer. It is possible to interact with each of these three layers separately.

Both the subtask and the planning layer are located in the control center. In the top layer (the planning layer) complete mission and path planning is done. In a first step the operator must give the overall goal, like e.g. 'monitor the whole floor', or 'visit the next room'. The second step is done by the planning layer, composing a plan and offering this result to the human operator. The operator then confirms the result or modifies the plan via the symbolic human-computer interface.

In the second layer the plan is divided into tasks and subtasks.

Onboard the vehicle the behaviour-oriented layer is located. This layer consists of behaviour modules, e.g., for obstacle avoidance, wall following, etc. All behaviour modules are realized with fuzzy-logic (Enste, 1995), developed at the Institute of Technical Computer Science. The behaviour modules are selected in such a way that the vehicle acts like a human operator controlling the robot. Usually, the controlling commands are given from the top layer, but in special situations the operator can interact directly with the mobile platform.

2.2. Symbolic map

Symbolic interaction at the human-machine interface requires a corresponding underlying software. Hence, the kernel of the system is a symbolic map in combination with a dynamic world modelling.

World modelling starts with collecting sensor information. A vision system, ultrasound and infrared sensors onboard the vehicle deliver raw data about the environment. After onboard preprocessing of sensor data, the camera picture and preprocessed data are transmitted via HF-link to the control center, where the dynamic world modelling component translates the data from the ultrasound and infrared sensors as well as the information from the camera picture into a simple geometric representation of the environment.

The main information for the geometric map is generated from the onboard ultrasound sensors (Kuc and Barshan, 1992). Infrared and video sensors are only used for navigation near obstacles or for better orientation of the operator.

In a second step, the world modelling component

Table 1 command level

command hierarchy	setting a goal
high level command	take the second door left
intermediate command	60 m straight ahead then 90 degree turn to the left
low level command	manned control of speed and heading

transforms the geometric map into a non-metric symbolic map, using an expert system. This map is visualized in the control center.

2.3. *Human-machine interaction*

For the man-machine interaction several input- and output-devices are realized. To modify a plan or to drive the robot directly, speech input and 3-dimensional input-devices, like space-ball and space-mouse are used. The operator can interact with these IO-devices intuitivly in a humanlike way.

Using a spaceball or a space-mouse the supervisor can operate in two dimension, e.g. like setting new waypoints, changing the path, or moving the synthetic cameraview. Figure 3 shows the symbolic map with four waypoints. The operator can now change the path by moving the waypoints, inserting, deleting or accepting the plan.

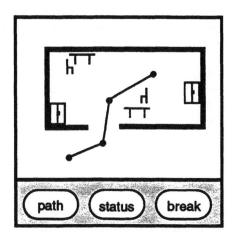

Fig. 3. symbolic map with 4 waypoints

Usually, all subtasks and the exploration of the environment are executed autonomously by the vehicle. However, in case of deadlocks or emergency situations, the remote human supervisor is prompted for further advice. The answer to a prompt is given in a humanlike intuitive way, e.g. by saying 'now go right'. The same philosophy is used to obtain goals or to set high level commands to the system, e.g. 'make a left turn and on your way avoid collision with obstacles'. Beside these high level commands the operator can also give adaptive commands at a lower level (see Tab. 1).

Such a system is easier to learn and easier to use for the operator, because he can use it in an intuitive way.

3. TESTBED

3.1. *Control center*

The display in the control center is divided into three parts: the camera picture, the instruments and a symbolic map. The operator controls the vehicle interactively using the symbolic map, where all important information from the sensors and the world model are visualized, and natural speech input. The navigation information for the vehicle is extracted automatically by the system out of the symbolic map and the dynamic world model.

Using the symbolic map the operator is supported by the system by a synthetic cameraview generated from the world model instead of the transmitted camera-picture. The advantage of this synthetic view is the possibility to change the virtual viewpoint of the camera. Fig. 4 shows the bird's-eye view.

Figures 5 and 6 show the top view and the driver's view. This synthetic view is generated by the symbolic map, using a Silicon Graphics Iris Indigo[2] Extreme UNIX-workstation. In emergency or special situations the operator can control the robot directly. In this case it is more reasonable to use spaceball or spacemouse than using the symbolic map.

3.2. *Mobile testbed*

The testbed of our system is a 30 x 40 cm large, 6-wheel mobile platform. The vehicle has three motors; two motors are used for the differential or skid-steer mechanism and the third motor is for shifting gears. Onboard the mobile robot a 486 PC controls the communication with the operator in the control center, manages the two onboard controllers - a Motorola 68000 controller and an 8051 controller - and preprocesses the sensor data.

Currently, two different modes for ultrasonic distance measurement are realized: One for short range measurement (1.5 cm to 50 cm), and one for long range (40 cm to 10 m). For the very

Fig. 4. bird's-eye view

short range up to 2 cm infrared sensors are used. A CCD-camera (see Fig. 7) transmits a picture of the front area to the control center, where the image processing is done by an Apple Quadra 950. The information linkage between the control center and the vehicle is realized by a 2.5 GHz data link.

4. EXPERIMENTS

Two separate experiments, one dealing with the interaction with the symbolic map, and one for direct navigation tasks using the virtual camera have been devised. The test environment was a simulated office area with three unknown rooms (see Fig. 5).

In the first experiment, the test-operator had to select, change and delete waypoints in the symbolic map. In a second experiment the operator had to control the mobile platform directly, using spaceball, spacemouse and keyboard.

First tests with this concept indicate, that beginners need only short time learning to work with the system (Kehr, 1994). They use the IO-devices intuitively to interact with the symbolic map, e.g. setting new waypoints.

The possibility to change their view makes it easier for the operator to navigate in the simulated bureau area as shown in the second experiment. Less collisions with obstacles and faster completion of the tasks are detected. For standard navigation purposes all operators preferred the bird's-eye view, only in near proximity of an obstacle they switch to the drivers view for fine tuning.

5. CONCLUSION

Teleoperated vehicles are mostly rather complex systems, which need a good cooperation between the operator and the system. To reach this aim a concept for symbolic interaction with a semi-autonomous mobile system to reduce the bottleneck in the human-machine communication was developed. First experiments demonstrate the usefulness of such a concept.

The system is a step forward in controlling semi-autonomous mobile robots applications, with potential for further advancements such as tactile feedback for direct controlling.

6. REFERENCES

Albus, J.S., R. Lumia and H. McCain (1988). Hierachicial control of intelligent machines applied to space station telerobots. *IEEE Transactions on Aerospace and Electronic Systems* **24**(5), 535–541.

Alexopoulos, Ch. (1992). Pathplanning for a mobile robot. *IEEE Transactions on Systems, Man, and Cybernetics* **22**(2), 327–338.

Barraquand, J., B. Langlois and J.-C. Latombe (1992). Numerical potential fields for robot path planning. *IEEE Transactions on Systems, Man, and Cybernetics* **22**(2), 224–241.

Browse, R.A. and M.L. McDonald (1992). Using tactile information in telerobotics. *IEEE Transactions on Systems, Man, and Cybernetics* **22**(5), 1205–1210.

Dorigo, M. and U. Schnepf (1993). Genetics-based machine learning and behavior-based robotics:

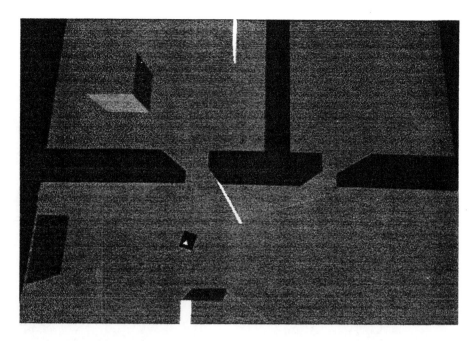

Fig. 5. top view

A new synthesis. *IEEE Transactions on Systems, Man, and Cybernetics* **23**(1), 141–154.

Enste, U. (1995). Autonome Navigation eines mobilen Roboters mit Hilfe von Fuzzy-Logic, Diploma Thesis, Institute of Technical Computer Science, Aachen Technical University.

Gat, E., R. Desai, R. Ivlev, J. Loch and D.P. Miller (1994). Behavior Control for Robotic Exploration of Planetary Surfaces *IEEE Transactions on Robotics and Automation* **10**(4), 490–503.

Kehr, M. (1994). Steuerung eines Telepraesenzfahrzeuges in einer teilweise unbekannten Umgebung: Umweltmodellierung und Entwicklung alternativer Navigationstechniken, Diploma Thesis, Institute of Technical Computer Science, Aachen Technical University.

Kuc, R. and B. Barshan (1992). Bat-like sonar for guiding mobile robots. *IEEE Control Systems* pp. 4–12.

Meng, M. and A.C. Kak (1993). Mobile robot navigation using neural networks and nonmetrical environment models. *IEEE Control Systems* pp. 30–39.

Zhao, Y., C.V. Ravishankar and S.L. BeMent (1994). Coping with limited on-board memory and communication bandwith in mobile-robot systems. *IEEE Transactions on Systems, Man, and Cybernetics* **24**(1), 58–72.

Fig. 6. driver view

Fig. 7. semi-autonomous mobile system

HUMAN-MACHINE INTERFACE FOR A MODEL-BASED EXPERIMENTAL TELEROBOTICS SYSTEM

T.T. BLACKMON* and L.W. STARK*

*Telerobotics and Neurology Units, University of California at Berkeley,
486 Minor Hall, Berkeley, CA, 94720-2020*

Abstract. A Human-Machine Interface (HMI) has been developed for a model-based experimental telerobotics (TR) system. Interaction between the human operator (HO) and the remote robots is facilitated through a shared model of the TeleRobotic Working Environment (TRWE). Using virtual environment technologies, the HO commands simulated robots, appropriately represented with high-fidelity computer graphics, to define a sequence of goal positions for the actual robots. Primary issues related to the HMI are the flexibility of the model in assisting the HO and the communication of information between the cognitive model of the HO and the varied forms of the computer model.

Key Words. Teleoperation, telerobotics, human-machine interface, human supervisory control, computer graphics, virtual reality.

1. INTRODUCTION

Teleoperation involves the extension of human sensing and motor capabilities to the control of remote machinery. Applications of teleoperation can be found in the areas of hazardous materials handling, space and underwater exploration and development, medical procedures such as endoscopy and surgical robotics, and others. The majority of research in teleoperation has been in the area of direct manual control with other investigations involving the utilization of human supervisory control for remote manipulation (Ferrell and Sheridan, 1967).

Performance difficulties arise in teleoperation in the areas of displays, controls and communications (Stark et al., 1987), primarily due to the 'remoteness' of the task. Difficulties in viewing 3d spatial information from a transmitted, 2d video image often lead to inadequate perception of depth and spatial layout of the remote environment. Also, limited communication bandwidth can cause excessive time delays which negatively affect the ability of the human operator (HO) to perform direct manual control, often resulting in the adoption of a 'move-and-wait' strategy (Ferrell, 1965).

Model-based approaches have previously been developed to help overcome some of these difficulties in teleoperation. Graphical simulators have become an essential part of HO training and experimental research platforms. Visual enhancements, based upon model knowledge of the remote site, have been used to augment displays for the HO

(Kim et al., 1987a). Also, local simulation models of the controlled plant have been used as a predictive display to help compensate for excessive time delay by superimposing the estimated effect of a commanded movement onto the video feedback, subsequently followed by the resultant delayed move (Noyes and Sheridan, 1984; Kim and Bejczy, 1993).

An alternative approach to direct manual control in teleoperation involves the use of supervisory control systems combining the higher level intelligence of the HO with the autonomous capabilities of the remote agents, referred to as telerobotics (TR) (Sheridan, 1989). With enabling technologies in the areas of advanced robotics, autonomous control, human-computer interfaces, and virtual environments, advantages of computer assistance for the supervisory control of remote machinery is ripe for exploration.

This paper describes a Human-Machine Interface (HMI) for an experimental TR system. The purpose of the experimental platform is to explore fundamental issues of model-based approaches in TR. Section 2 provides an overview of the experimental system, describing the modes of operation and the use of a model of the remote site to assist the HO in the control of remote robots. Section 3 details the HMI including the controls, displays, 3d graphical model and the TR configuration editor. Emphasis is given to the flexibility of the model for assisting the HO and the cognitive issues involved in the HO interacting with the varied forms of the computer model.

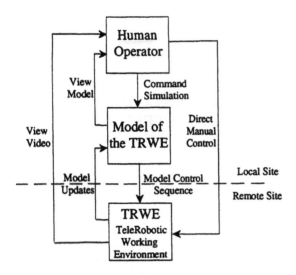

Fig. 1. Block diagram of the experimental telerobotics (TR) system utilizing a model of the Tele-Robotic Working Environment (TRWE) to assist with direct manual and human supervisory control modes.

2. MODEL-BASED EXPERIMENTAL TELEROBOTICS SYSTEM

The experimental TR system is composed of the HMI and the TeleRobotic Working Environment (TRWE), which includes the robots with controller circuitry and computational intelligence, video cameras, objects and the work space. Two robots - a Mitsubishi-RM 501 stationary, 5-dof manipulator arm (M-robot) and a mobile Armatron robot with 7-dof (A-robot) are controlled by a HO in a separate room of the laboratory. Three stationary video cameras provide visual feedback (a top view and two nearly orthogonal side views) of the remote site.

The emphasis of the system is the utilization of a model of the TRWE to help visualize the remote site and to efficiently utilize the levels of autonomous control capabilities of the remote robots (Fig. 1). The model of the TRWE consists of a priori knowledge of the content, geometry, kinematics, dynamics and other relational properties of the robots, work space and objects at the remote site. This model is dynamically updated by sensory information and position feedback from the remote site. The system supports two modes of operation: direct manual control and model-based supervisory control.

2.1. Direct manual control

In direct manual control mode, the HO commands the robots at the remote site using appropriate input devices and control strategies. Video images of the TRWE are transmitted to the local site and used as visual feedback for the HO. Such di-

rect manual control, with the HO closing the control loop, corresponds to more traditional forms of teleoperation.

Model-based approaches are incorporated into this control mode using feedback information from the remote site to display a graphical model of the TRWE as an additional viewing aid. Some advantageous uses of a supplemental graphical model include the incorporation of visual enhancements (Kim *et al.*, 1987*a*) for increased depth perception and the ability to arbitrarily orient the artificial camera viewpoint. Also, the command signal to the TRWE can be sent simultaneously to a local simulation model and used as a predictive display for compensation of time delay.

2.2. Model-based supervisory control

In the supervisory control mode, the HO interacts with the model of the TRWE for on-line planning of task segments for the robots at the remote site. The HO commands a simulation to define a model control sequence, or list of desired goal positions for a robot. Upon command, the model control sequence is transferred to the TRWE where the robots perform algorithmic path planning between subsequent desired positions and execute the planned trajectories autonomously under human supervision.

The model control sequence, which is defined using the on-line planning simulation, thus becomes the supervisory command language between the HO and the robots. If requested, the HO can also preview the planned robot trajectories for approval and possible modifications. During execution of a planned trajectory, the HO can issue emergency overrides if necessary, defaulting into direct manual control mode, or redefine the model control sequence.

Similar model-based approaches have been used for supervisory control in TR. A forward simulation and predictor display was extended to a planning strategy for a 'tele-autonomous' system with time and position clutches to enable exchange of control between the HO and the autonomous robot (Conway *et al.*, 1987). A real-time graphical simulation with kinesthetic interaction allowed a HO to 'teleprogram' a remote manipulator in the presence of a significant time delay (Funda and Paul, 1990). A computer graphic aid for TR has also been developed in which the HO specifies end points of robot goal positions using a computer mouse and can visualize the resultant algorithmic trajectories (Park and Sheridan, 1991; Hachenburg, 1994).

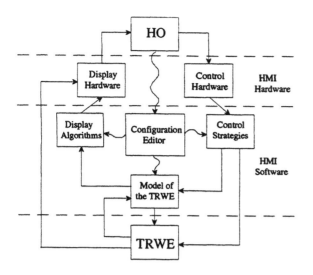

Fig. 2. Block diagram of the Human-Machine Interface (HMI) components relating the hardware and software for the controls and displays. Straight lines between blocks denote continuous links while wavy lines represent intermittent links.

3. HUMAN-MACHINE INTERFACE

The HMI for the experimental telerobotics system consists of the hardware and software used as the displays and controls (Fig.2). The HMI hardware consists of the input devices for commanding the robots and simulation models and the displays for visual feedback of the video from the remote site and the model of the TRWE. The HMI software consists of the model of the TRWE, the control strategy algorithms, the display mode algorithms, and the configuration editor software for setting system attributes and issuing symbolic commands.

3.1. *Displays*

Video feedback. A large screen television monitor (27" Sony Trinitron) displays the video feedback from the remote site and a 3-way selector switch is used by the HO to choose from the multiple camera views available (Fig. 3).

Graphical model. The HO also views a representation of the model of the TRWE using high-fidelity computer graphics (Fig. 4). The default display device is a standard high resolution (1280 x 1024 pixels) monoscopic display. The software was written in C programming language using Silicon Graphics GL (Graphics Library) and is currently running on a Silicon Graphics Indigo2 Extreme workstation.

Geometric models of the robots and work space objects are used to generate polygonal approximations which are rendered and animated using GL. The graphical model can be displayed using several drawing styles. The default drawing style is

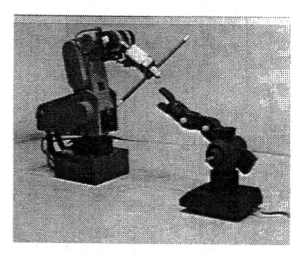

Fig. 3. Video image transmitted from the remote site for visual feedback showing the stationary M-robot (left) and the mobile A-robot (right).

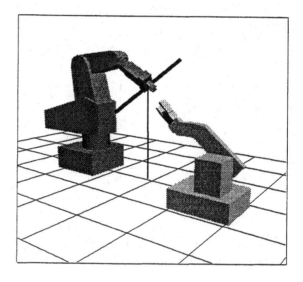

Fig. 4. Graphical display of the model of the TRWE. Note the benefit of the vertical reference line and horizontal grid as visual enhancements for depth perception.

solid model geometry with hidden surface removal and lighting calculations. Optionally, the HO can display selected parts of the graphical model using a wireframe drawing style with hidden line removal. In addition, a kinematic skeleton of a robot as a stick figure can also be displayed.

3.2. *Controls*

A pair of 2-dof joysticks are available as input devices to control the robots or simulation models. The joysticks provide analog output and are connected to the system computer using a 12-bit A/D converter (Metrabyte VMECAI-16). In the default configuration, the HO uses joint angle control (JAC) strategy with the joystick as a rate control device to command the individual joint angles of the robot manipulators.

3.3. *Extensions to displays and controls*

The HMI also provides optional extensions to the displays and controls which aim to improve the performance of the HO. Most extensions are built upon previous research and are beyond the scope of this paper. They will be only mentioned to provide a further description of the available options and capabilities of the HMI. The focus of this paper is to study issues concerning the cognitive processes for the HO as a supervisory controller for the model-based TR system.

Displays. Optionally, a set of StereoGraphics Crystal Eyes sequential field shutter glasses are available as a stereo viewing device. Also, the HO may select to use a head mounted display (HMD) with stereo viewing (Kim *et al.*, 1988) and an electro-magnetic head tracking device (Duffy, 1985) for control of viewing perspective. Previous studies have shown the robust benefits of stereoscopic HMDs for increased depth perception in teleoperation (Kim *et al.*, 1987c; Liu *et al.*, 1993).

Controls. A VPL DataGlove with a Polhemus 3Space tracker is also available as an alternative input device. Using an end-effector control (EEC) strategy, the HO specifies the position of the robot end-effector while inverse kinematic algorithms are used to solve for the corresponding joint angles in real time. Previous research has shown the use of the DataGlove as a legitimate input device for manipulator control in teleoperation (Halem, 1992). Both the DataGlove with 3Space tracker (in position control) and the joystick (in rate or position control) are available for EEC. Studies have also compared the advantages and trade-offs between rate control vs. position control for simulated 3-axis pick-and-place operations (Kim *et al.*, 1987b).

3.4. *Making the model fit the task*

Two issues should be paramount in the evaluation of a HMI for a model-based TR system. Can the HO exploit the flexibility of the model of the TRWE as both an additional visual aid and as a tool for outlining tasks for the robots to perform autonomously under human supervision? Can the HO learn through experience and via the graphical user interface (GUI) to understand and use the varied forms of the computer model to gain insight into the teleoperation task and implement effective control?

Model of the TRWE. The model of the TRWE, residing in computer memory as electronic bits of information and communicated to the HO using high-fidelity computer graphics, is partly controlled by the HO and partly by the remote site,

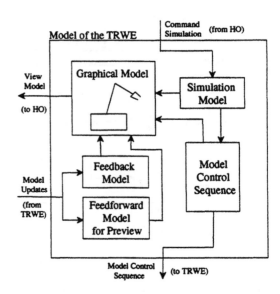

Fig. 5. Model of the TRWE is partially updated by the HO and partially updated by information from the remote site.

depending upon the current mode of operation (Fig. 5). The underlying knowledge base model of the TRWE provides a basis for the necessary forms of the model used for display and control.

Simulation models are used by the HO in supervisory control mode to define the model control sequences which outline the tasks for the robots. This simulation model can also be used for training of direct manual control or as a predictive display for time delay compensation. A feedback model is updated by sensed information from the remote site and used as an additional viewing aid to supplement the video view. Feedforward models are updated from the algorithmic path planning of the robots and used for supervisory preview. For visual presentation, a graphical model is used to communicate these forms of the model to the HO.

Content of the graphical model. A critical aspect of the HMI concerns the content of the graphical model with respect to the overall model of the TRWE. A combination of the feedback model, simulation models, model control sequences, and feedforward models must be appropriately selected to form the current graphical model which is displayed to the HO. Depending upon the mode of operation and system status, the HO will desire to view different combinations of these models to extract the information necessary.

Based upon the current mode of operation, some level of intelligence is instilled upon the configuration software for choosing the default attributes of the graphical model. The HO can also set the attributes of the graphical model to be dis-

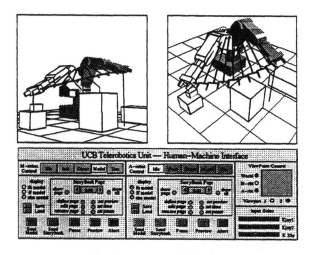

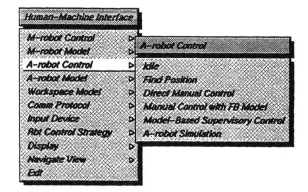

Fig. 7. Pop-up menu used as part of the configuration editor of the HMI.

Fig. 6. Graphical display for the HMI utilizing double viewports for the graphical model and a Graphical User Interface (GUI) input panel for quick access to system configuration and commands. The solid model of the M-robot represents the feedback position while the wireframe and skeleton models represent the desired goal position and corresponding feedforward trajectory preview.

played. However, caution must be used to ensure that a crowded or confused graphical model does not mislead the cognitive model of the HO. The variety of drawing styles available can be used as a means for distinguishing the different components of the graphical model. For example, a solid model of a robot can represent the feedback position while a wireframe model represents a desired goal position and skeleton models show a feedforward preview of the trajectory (Fig. 6).

Flexibility of the model. Another critical aspect of the HMI concerns how to effectively utilize the advantages provided by the flexibility of a computer model. For example, in defining a model control sequence, the HO can command the simulation unconstrained by the actual robot dynamics, if desired. The simulation model may be moved at faster speeds and the HO can try various robot positions without possible implications of damage from manual control error. In addition, the HO could use the simulation model with constrained dynamics for practice before execution of a difficult direct manual control operation.

The HO can also use the flexibility of the graphical model to arbitrarily position the viewing perspective, subsequently aiding the perception of the 3d spatial layout of the TRWE. Several methods are available to the HO to set the artificial camera viewpoint. The system also provides a number of pre-set viewpoints for the HO to select, such as front, side and top views of the work space or a specified robot. This allows the HO to more readily visualize the remote site, maintain spatial ori-

entation in the model, and possibly 'see' through objects obstructed in any particular camera view, or views, to uncover 'hidden' obstacles.

3.5. *Configuration Editor*

A configuration editor with a Graphical User Interface (GUI) coordinates the control modes and strategies, the attributes of the model of the TRWE, and the display mode algorithms. Thus, the configuration editor provides the cognitive link between the HO, the model of the TRWE, and configuration of the displays and controls of the HMI. The GUI consists of pop-up menus, a virtual input panel, and text screens for system messages and keyboard input.

Pop-up menus. Using a computer mouse to activate a pop-up menu, the HO can set the current control mode and model attributes for the robots and work space, establish communication connections, select the control hardware and strategies, and adjust the display hardware and viewpoint algorithms (Fig. 7). The use of a pop-up menu provides a user-friendly access to the variety of options available for the system configuration.

GUI input panel. An input panel with virtual push buttons, sliders and a 2d positioner allows the HO to more quickly set critical and frequently used aspects of the configuration editor. Subsets of the input panel function to switch control modes, issue system commands, define the model control sequence, set the graphical display content and viewpoint perspective, and adjust input control gains (Fig. 6). The input panel was created using the Forms Library, a GUI tool kit for Silicon Graphics workstations (Overmars, 1993). The GUI input panel also provides visual feedback of system status.

4. CONCLUSION

A HMI has been developed for a model-based experimental TR system. The emphasis of the system is the incorporation of a model of the remote site to assist the HO in specifying goals for the autonomous robots to perform under human supervision. Building upon technologies and previous research in the area of teleoperation, the HMI allows the HO an intelligent means for interacting with the controls and provides an effective visual display system. The salient issues of the HMI discussed in this paper relate to the cognitive processes of the HO as a supervisory controller for the model-based TR system.

Acknowledgements. The authors would like to acknowledge partial support from NASA Ames Research Center, through Cooperative Agreements: NCC-286 (Dr. Stephen Ellis, Technical Monitor) and NCC-757 (Dr. Robert Welch, Technical Monitor) and from Fujita Research Corporation (Drs. Ken Kawamura and Iris Yamashita, Technical Monitors). Thanks is also extended to close colleagues April Hachenberg, Yeuk Fai Ho and Yong Yu for assistance with the experimental set-up.

5. REFERENCES

Conway, L., R. Voltz and M. Walker (1987). Tele-autonomous systems: Methods and architectures for intermingling autonomous and telerobotic technology. In: *Proceedings of the 1987 IEEE International Conference on Robotics and Automation.* Raleigh, N.C.. pp. 1121–1130.

Duffy, M.K. (1985). A head monitor system using the search coil method. Master's thesis. University of California at Berkeley. Berkeley, CA.

Ferrell, W. R. (1965). Remote manipulation with transmission delay. *IEEE Transactions on Human Factors in Electronics* **HFE-6**, 24–32.

Ferrell, W. R. and T.B. Sheridan (1967). Supervisory control of remote manipulation. *IEEE Spectrum* **4**(10), 81–88.

Funda, J. and R. P. Paul (1990). Teleprogramming: Overcoming communication delays in remote manipulation. In: *Proceedings of the 1990 IEEE International Conference on Systems, Man and Cybernetics.* Los Angeles, CA.

Hachenburg, A.P. (1994). Simulations of a telerobotic testbed. Master's thesis. University of California at Berkeley. Berkeley, CA.

Halem, J. P. II (1992). Anthropomorphic teleoperation: controlling remote manipulators with the dataglove. In: *Proceedings of the Human Factors Society 36th Annual Meeting.* Santa Monica, CA.

Kim, W.S., A. Liu, K. Matsunage and L.W. Stark (1988). A helmet mounted display for telerobotics. In: *IEEE Computer Society COMPCON Spring '88.* San Francisco, CA.

Kim, W.S. and A.K. Bejczy (1993). Demonstration of a high-fidelity predictive/preview display technique for telerobotic servicing in space. *IEEE Trans. on Robotics and Automation* pp. 698–702.

Kim, W.S., F. Tendick and L.W. Stark (1987a). Visual enhancement in pick-and-place tasks: Human operators controlling a simulated cylindrical manipulator. *IEEE J. of Robotics and Automation* **RA-3**, 418–425.

Kim, W.S., F. Tendick, S.R. Ellis and L.W. Stark (1987b). A comparison of position and rate control of telemanipulations with consideration of manipulator system dynamics. *IEEE J. of Robotics and Automation* **RA-3**, 426–436.

Kim, W.S., S.R. Ellis, M.E. Tyler, B. Hannaford and L. Stark (1987c). Quantitative evaluation of perspective and stereoscopic displays in three-axis manual tracking tasks. *IEEE Trans. on Systems, Man and Cybernetics* **SMC-16**, 61–72.

Liu, A., G. Tharp, L. French, S. Lai and L.W. Stark (1993). Some of what one needs to know about using head-mounted displays to improve teleoperator performance. *IEEE Trans. on Robotics and Automation* pp. 638–648.

Noyes, M.V. and T.B. Sheridan (1984). A novel predictor for telemanipulation through a time delay. In: *Proceedings of the 20th Annual Conference on Manual Control.* NASA Ames Research Center, Moffet Field, CA.

Overmars, M.H. (1993). *Forms Library: A Graphical User Interface Toolkit for Silicon Graphics Workstations.* Department of Computer Science, Utrecht University.

Park, J.H. and T.B. Sheridan (1991). Supervisory teleoperation control using computer graphics. In: *Proceedings of the 1991 IEEE International Conference on Robotics and Automation.* Sacramento, CA. pp. 493–498.

Sheridan, T. B. (1989). Telerobotics. *Automatica* **25**(4), 487–507.

Stark, L., W. S. Kim, F. Tendick, B. Hannaford, S. Ellis, M. Denome, M. Duffy, T. Hayes, T. Jordan, M. Lawton, T. Mills, R. Peterson, K. Sanders, M. Tyler and S. Van Dyke (1987). Telerobotics: Display, control, and communication problems. *IEEE J. of Robotics and Automation* **RA-3**, 67–75.

INFORMATION EXCHANGE AND TEAM COORDINATION IN HUMAN SUPERVISORY CONTROL

Barrett S. Caldwell

Department of Industrial Engineering
University of Wisconsin-Madison
Madison, WI USA

Abstract: This paper addresses issues of "socio-technical" demands for effective team performance in supervisory control tasks. Team members must exchange data and system information, as well as interpersonal meanings and contexts, in order to develop shared and synchronized models of system behavior. The author's extensions of current human performance models, including SRK frameworks, incorporate multiple information exchange and task coordination performance issues required to support expert team performance and "optimal" information exchange in a dynamic task environment.

Keywords: Differential equations; Dynamic systems; Feedback control; Human supervisory control; Information flows; Knowledge-based control; Second-order systems; Sociotechnical systems design.

1. DEMANDS FOR TEAM-BASED SUPERVISORY CONTROL

Computer-mediated data exchange and information presentation technologies have come to substantially alter human-system interaction (HSI) in supervisory control task environments. Increasing complexity, demands for reliability, and consequences of failure in these task environments often requires the integration and coordination of multiple human supervisors to maintain effective system performance. Effective information access and exchange is required to permit supervisors to share information in a timely manner, and perform based on synchronized understandings of system activity (and, when necessary, interventions by one or more supervisors).

A socio-technical approach to team-based supervisory control must recognize the domains of task integration and information exchange required for team-based performance, and the social contexts and meanings developed through the common experiences and shared situation awareness or shared system models that teams create over time and repeated interaction.

2. DOMAINS, STYLES, AND UNITS OF PERFORMANCE

Researchers addressing human performance in complex systems have recognized that optimal task performance requires a combination of skills operating at multiple levels. Human performance frameworks which address these multiple *domains* of performance include knowledge-skills-abilities (KSA) performance frameworks (Swezey and Salas, 1992), and particularly skills-rules-knowledge (SRK) approaches to human performance and error (Rasmussen, 1983; Reason, 1990; Sheridan, 1992). These domains range from rapid execution of well-learned open-loop motor programs (Laszlo and Bairstow, 1985; Schmidt, 1982), through recall and execution of lists of procedures, and development of situational awareness, problem solving strategies, and other aspects of mental models (Moray, 1990; Payne, 1991; Rouse, Cannon-Bowers, and Salas, 1992). However, performance domains only represents one dimension of information exchange and task coordination in team performance.

2.1 Performance domains and KRSA

The author's elaboration of SRK and KSA performance frameworks has led to the development of a knowledge-rules-skills-abilities (KRSA) performance matrix (Caldwell, 1994a) (see Figure 1). The rows of the KRSA matrix are similar to those of SRK. Rasmussen's primary development of SRK suggests that expertise causes the operator to transfer skills across domains from knowledge-based to skill-based performance (Rasmussen, 1983). KRSA maintains, however, that development of expert *styles* of performance can be acquired independently within each *domain* of performance.

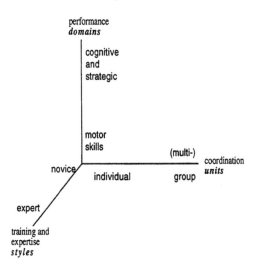

Figure 1. Conceptual framework of dimensions of performance: styles, domains, and units.

A second modification distinguishes performance **skills** from performance **abilities**, or physiological capacity limitations. This distinction may be implicit in SRK (J. Rasmussen; K. Vicente, personal communication, September 19, 1994), but KRSA makes the distinction explicit. One benefit of this approach is the recognition of differing training strategies for enhancing performance at those levels. (For example, training in general abilities, such as weightlifting, will enhance this performance domain without providing directed benefits for specific skilled tasks, such as kicking or throwing drills in sports.) In summary, the KRSA performance domains are:

• **Knowledge:** strategic situation awareness, problem solving, and development of mental models;

• **Rules:** Effective memory recall of task sequences and ordered procedural lists;

• **Skills:** Task-directed physical (sensorimotor) performance criteria and competence;

• **Abilities:** Overall physical (physiological and sensorimotor) performance limits and capacity

2.2 Coordination styles and units

The KRSA matrix also includes multiple columns based on the *unit* of integration required for information access and exchange and coordinated task performance. While SRK usually addresses individual performance, KRSA focuses on three units of integration: the individual, the small group or team, and the organizational unit composed of multiple groups.

Expert *styles* of performance, regardless of *domain* or *unit* of coordination, require the development of two competencies resulting from training and experience. The first is the ability to access and communicate large amounts of context-relevant information quickly and reliably, at the appropriate performance domain. The second, and related, competency is the capacity to develop a set of "implicit" cues to further increase effective information exchange (compare Rasmussen's (1983) signals, signs, and symbols).

At the unit of the group, these competencies can be described as fluid, dynamic allocation and coordination of roles and functions (McGrath, 1990; Sundstrom, DeMeuse, and Futrell, 1990), or shared mental models (Majalian, Kleinman, and Serfaty, 1992; Thordsen and Klein, 1989). The establishment of jargon and group-specific communication patterns, based on shared experience, are evidence of this implicit performance style. Shared experience and implicit communication usage allows groups more effective coordination and balance between rapid, open-loop performance and robust, error-correcting feedback performance.

3. CONTEXTUAL ISSUES IN INFORMATION ACCESS AND EXCHANGE

A persistent problem in statistical approaches to information theory and cybernetic feedback control is that these approaches do not fully consider context and meaning in the process of information exchange (Shannon and Weaver, 1949; Wiener, 1961). These contextual influences on human information access and communication performance suggest that the criteria for "optimal" information exchange from a context-based performance perspective differ from those of a statistical information theory perspective. Statistically, as message redundancy and message length increases, performance (as defined by reduction of message uncertainty) should increase. "Optimal" information flow in HSI or team communication requires balancing both volume and level of detail of information corresponding to time available.

Effective individual performance in HSI is also strongly related to context factors and the strength of the individual's model of the system being controlled (Brehmer, 1992; Moray, 1990; Moray, Dessouky, Kijowski, and Adapathya, 1991). Navigation through information systems often suffers from a combination a combination of "too much" (excessive volume of responses to a query) and "too little" (inadequate cues for determining relevance) information (Eberts, 1994; Smith, Shute, Galdes, and Chignell, 1989). Situational variables such as message urgency and volume of message content have a major influence on technology acceptance in organizational settings (Caldwell, 1993; Caldwell and Uang, 1994).

Prior work in both social and technical systems dynamics provides support for achieving balance in information exchange. Effective interpersonal contact in social interactions occurs when exchange of personal information is dynamically regulated within a tolerance range based on the context of the social setting (Altman and Taylor, 1973). Statistical process run control focuses on maintaining critical system variables (in this case, information) within an appropriate tolerance to optimize resulting system performance (Box, Hunter, and Hunter, 1978). Research in parallel and distributed computer processing has also addressed the issue of balancing costs and benefits of information exchange in various situations (Billard and Pasquale, 1993; Pasquale, 1992).

Thus, context-based optimization criteria for effective information exchange in both HSI and team performance can be framed in terms of the following questions:

1) How much information can be processed?

2) At what level of abstraction (jargon or domain expertise) can it be understood?

3) How much time is available for information exchange and resulting performance?

Human information processing limits provide upper bounds to acceptable information exchange rates and volumes. In addition, situational and contextual factors (including performance styles of the operator) may further affect design goals and tolerance ranges for appropriate information exchange. Recognizing context in dynamic information exchange has substantial potential benefits in improving performance. A recognition of the time required for accessing and processing information (compared to time available for performance), as well as the level of expertise of the supervisory operator, permits "tailored" information exchange that can be better

integrated into efficient task performance. These relationships are illustrated in Figure 2.

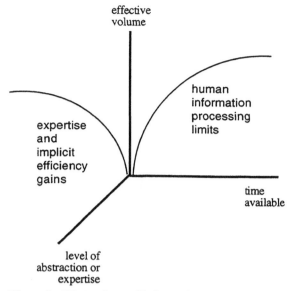

Figure 2. Dimensions of information exchange and contextual influences on human performance.

This strategy is spontaneously adopted in interpersonal communication and team performance (McGrath, 1990). Additional work is required to enable intelligent interfaces to accomplish similar tasks. However, in both cases, even an initially incorrect evaluation of context factors has significant time and performance advantages. The operator is more likely to have usable information quickly, and will have to conduct fewer additional requests for the correct information.

3.1 Applications to HSI database query systems

Intelligent advisors and information support systems for supervisory control HSI can particularly benefit from a context-based approach to query responses. A general-use system (a public information database, such as a library catalog, journal reference database, or citation index) can simplify the query process by asking "probe" questions of the user regarding expertise, time available, and volume of information requested. System responses would be varied based on these user context variables. If the user finds the result more or less than desired, an incrementing function for either volume or expertise would allow a more evolutionary search strategy for the user.

Limited-use HSI systems (such as control rooms) have an additional advantage in that the expertise (and preferred style of information presentation) of each user can be determined and "trained" into the system with frequent user experience. Critical-event systems (such as emergency procedures) would also include estimates available of time available for task performance. A combination of these functions would permit directed, tailored information to each user to improve performance and enhance understanding of system behavior.

4. INFORMATION EXCHANGE AND SECOND-ORDER FEEDBACK SYSTEMS

This paper has thus far highlighted information flow as a critical element of coordinated task performance in a feedback-based human performance context. This focus on information flow and feedback systems suggests applications of tools familiar in studies of other electrical and mechanical feedback control systems.

Second-order differential models have been effectively applied to human manual control (Proctor and VanZandt, 1994; Rouse, 1980; Sheridan and Ferrell, 1974), as well as servomechanisms and guidance control systems. However, applications of these second-order models in cognitive performance, particularly team-based performance, have not been as widely attempted. This author has successfully demonstrated the application of the following equation:

$$a'' + 2c \cdot a' + b^2 \cdot a = 0 \qquad (1)$$

to describe acceptance of message transmission time delay in organizational information technology use (Caldwell, 1993; Caldwell, 1994c). This approach considers the coefficients $\underline{b}$ and $\underline{c}$ as benefit and cost parameters of information gain vs. cost of delayed information. Studies of information flow and synchronization of distributed computer processors has utilized the same conceptual approach (Billard and Pasquale, 1993; Pasquale, 1992).

The successful application of equation (1) to interpersonal information exchange is due in part to two differences in model development. Most studies of human manual control attempt to elaborate frequency-domain (s-plane) transfer functions of motor behavior, including specific transfer functions for particular performance components (such as perception, afferent and efferent nerve signals, and inertial elements of physical structures of the human body). These component transfer functions would be prohibitively difficult to elaborate in cognitive performance, because of the lack of agreement on how such performance is organized and completed.

The studies of information exchange and communications performance have avoided these problems by assuming a "black box" conception of human behavior (not an unreasonable assumption for cognitive and interpersonal behavior). Thus, the studies do not attempt to determine component transfer functions. Additionally, the author's research emphasizes time-domain, rather than frequency-domain, examination of human performance in complex systems. Time-domain models more explicitly demonstrate the influence of situational and temporal factors on information access and exchange.

4.1 Interpretation of information exchange dynamics variables

The interpretation of concepts such as inertia, coupling, and other analogous system variables implied in equation (1) has been recognized as an issue requiring further study (Caldwell, 1994c; Panel on Human Performance Modeling, 1990). Models of user acceptance of transmission delay using equation (1) distinguished user expectations of synchronous (damping ratio $\zeta > 0$) vs. asynchronous (damping ratio $\zeta < 0$) communications behavior (Caldwell, 1993; Caldwell, 1994b). Drawing analogies from other mechanical systems examples of negative damping, one can interpret the coefficient $\underline{b}$ as indicating *time sensitivity*. Although negative damping is a condition of system instability in mechanical systems, negative time sensitivity of information in communications suggests greater HSI system performance which is more robust to unplanned delays in information transmission.

4.2 Specifications of supervisory control task demands as forcing functions

How supervisors make the transition to active system intervention and control, particularly after long periods of passive monitoring activity, is a problem of continuing interest (Ryan, Hill, Overlin, and Kaplan, 1994). Further investigation of a second-order model of task demands and supervisor control behavior can assist in meeting this challenge. Rapid shifts in task demands and required information flow suggest use of a time-domain description of input forcing functions. The differential equation with forcing function F(t):

$$a'' + 2c \cdot a' + b^2 \cdot a = F(t) \qquad (2)$$

cannot be solved in general. Thus, efforts must proceed with caution, with special attention to reasonable choices of F(t). It is useful to consider a **task demand function** of the form:

$$F(t) = A_1/t_r \cdot f_1(t_0) - A_2/t_f \cdot f_2(t - t_e) \qquad (3)$$

116

for a task beginning at time t_o and ending at time t_e. Coefficients $A_{1,2}$ are the magnitude of task demand shifts at the beginning and end of the task, measured either in bits or natural task demand units (such as number of subtasks or communication acts required to complete a project). Both t_r and t_f are defined as traditional rise and fall times (from 10% to 90% of full demand shift scale). These coefficients allow an examination of both magnitude of task demand transitions, and the effects of transitions that happen at rates faster than the sampling capabilities of the HSI ($t_{r,f} \rightarrow 0$, or an "instantaneous" transition).

If f_1 and f_2 in equation (3) are defined to be "tractable" transition functions (such as ramps, parabolas, unbounded ($f(t) = e^{kt}$) or asymptotic ($f(t) = 1 - e^{-kt}$) exponentials, or sinusoids), equation (2) can be solved over a wide range of types of workload transitions. Solutions can expressly identify the nature of human supervisory response as a system response to the forcing function $F(t)$. Experimental studies can examine the effects of the form of $f(t)$, the magnitude A of the workload transition, fatigue or panic effects associated with the task duration $t_e - t_o$ or system change rates $t_{r,f}$. Future research in this area should attempt to elaborate these effects.

5. CONCLUSIONS

Human supervisory performance in complex systems requires further integration of systems engineering and feedback control models effectively applied in electrical, mechanical, and human motor performance systems. Information flow and exchange in supervisory performance does suggest use of second-order time-domain feedback control models. Interpersonal communications acceptance and task performance are significantly affected by situational and contextual forcing functions. Careful definition, interpretation, and application of these task demand transition functions may provide substantial insights to permit improved design and robust performance of complex human supervisory control systems.

ACKNOWLEDGEMENTS

This work was partially supported by a National Science Foundation Grant (IRI-9320719).

REFERENCES

Altman, I. and Taylor, D. A. (1973). *Social Penetration: The Development of Interpersonal Relationships*. New York: Holt, Rinehart and Winston.

Billard, E. A. and Pasquale, J. C. (1993). Effects of Periodic Communication on Distributed Decision-Making. In *IEEE International Conference on Systems, Man, and Cybernetics*. New York: Institute for Electrical and Electronics Engineers (IEEE).

Box, G. E. P., Hunter, W. G., and Hunter, J. S. (1978). *Statistics for Experimenters: An Introduction to Design, Data Analysis, and Model Building*. New York: John Wiley and Sons.

Brehmer, B. (1992). Dynamic decision making: Human control of complex systems. *Acta Psychologica* **81**, 211-241.

Caldwell, B. S. (1993). Situational and information constraints affecting communications with 1-1000 second transmission delays. In G. Salvendy and M. J. Smith (Eds.), *Human-Computer Interaction: Applications and Case Studies--Proceedings of the Fifth International Conference on Human-Computer Interaction, vol. 1* (pp. 167-170). Amsterdam: Elsevier.

Caldwell, B. S. (1994a). Coordination and Synchronization of Skilled Performance in Groups Conducting Space-Based Tasks. In *Proceedings of the 1994 Human Interaction with Complex Systems Symposium*, (pp. 126-132). Greensboro, NC: North Carolina A&T State University.

Caldwell, B. S. (1994b). Developing Robust Mathematical Models of Information Technology Use in Organizations. In *Human Factors in Organizational Design and Management – IV: Proceedings of the Fourth International Symposium on Human Factors in Organizational Design and Management held in Stockholm, Sweden, May 29-June 2, 1994* (pp. 526-530). Amsterdam: North-Holland.

Caldwell, B. S. (1994c). Robust Interpersonal Communication in Complex Systems: Dynamics of Socio-Technical Components of Information Flow. In *Proceedings of the 1994 Human Interaction with Complex Systems Symposium*, (pp. 122-125). Greensboro, NC: North Carolina A&T State University.

Caldwell, B. S. and Uang, S.-T. (1994). Response Surface Analysis of Effects of Situation Constraints on Communications Media Choice in Organizations. In *Proceedings of the Human Factors and Ergonomics Society 38th Annual Meeting*, (pp. 744-748). Santa Monica, CA: Human Factors and Ergonomics Society.

Eberts, R. E. (1994). *User Interface Design.* Englewood Cliffs, NJ: Prentice Hall.

Laszlo, J. and Bairstow, P. (1985). *Perceptual-motor behaviour: Developmental assessment and therapy.* New York: Praeger.

Majalian, K. A., Kleinman, D. L., and Serfaty, D. (1992). The Effects of Team Size on Team Coordination. In *IEEE International Conference on Systems, Man, and Cybernetics* (pp. 880-886). New York: IEEE.

McGrath, J. E. (1990). Time Matters in Groups. In J. Galegher, R. E. Kraut, and C. Egido (Eds.), *Intellectual Teamwork: Social and Technological Foundations of Cooperative Work.* Hillsdale, NJ: Lawrence Erlbaum Associates.

Moray, N. (1990). Mental Models of Complex Systems. In N. Moray, W. R. Ferrell, and W. B. Rouse (Eds.), *Robotics, Control and Society: Essays in honor of Thomas B. Sheridan* (pp. 133-149). London: Taylor & Francis.

Moray, N., Dessouky, M. I., Kijowski, B. A., and Adapathya (1991). Strategic Behavior, Workload, and Performance in Task Scheduling. *Human Factors* **33**, 607-630.

Panel on Human Performance Modeling (1990). *Quantitative Modeling of Human Performance in Complex, Dynamic Systems.* Washington, DC: National Academy Press.

Pasquale, J. (1992). Decentralized control in large distributed systems: Coordination theory and collaboration technology at UCSD. In *Coordination Theory and Collaboration technology (CTCT) Workshop* (pp. 85-93). Washington, DC: National Science Foundation.

Payne, S. J. (1991). A descriptive study of mental models. *Behaviour and Information Technology* **10**, 3-21.

Proctor, R. W. and VanZandt, T. (1994). *Human Factors in Complex Systems.* Boston: Allyn and Bacon.

Rasmussen, J. (1983). Skills, Rules, and knowledge; Signals, signs, and symbols, and other distinctions in human performance models. *IEEE Transactions on Systems, Man, and Cybernetics* **13**, 257-266.

Reason, J. (1990). *Human Error.* Cambridge, UK: Cambridge University Press.

Rouse, W. B. (1980). *Systems Engineering Models of Human-Machine Interaction.* New York: North Holland.

Rouse, W. B., Cannon-Bowers, J. A., and Salas, E. (1992). The Role of Mental Models in Team Performance in Complex Systems. *IEEE Transactions on Systems, Man, and Cybernetics* **22**, 1296-1308.

Ryan, T. G., Hill, S. G., Overlin, T. K., and Kaplan, B. L. (1994). *Work Underload and Workload Transition as Factors in Advanced Transportation Systems* (EGG-HFSA-11483). Idaho National Engineering Laboratory / EG&G Idaho.

Schmidt, R. (1982). Generalized motor programs and schemas for movement. In J. Kelso (Ed.), *Human Motor Behavior: An introduction.* Hillsdale, NJ: Lawrence Erlbaum Associates.

Shannon, C. E. and Weaver, W. (1949). *The Mathematical Theory of Communication.* Urbana, IL: The University of Illinois Press.

Sheridan, T. B. (1992). *Telerobotics, Automation, and Human Supervisory Control.* Cambridge, MA: MIT Press.

Sheridan, T. B. and Ferrell, W. R. (1974). *Man-Machine Systems: Information, Control, and Decision Models of Human Performance.* Cambridge, MA: MIT Press.

Smith, P. J., Shute, S. J., Galdes, D., and Chignell, M. H. (1989). Knowledge-Based Search Tactics for an Intelligent Intermediary System. *ACM Transactions on Information Systems* **7**, 246-270.

Sundstrom, E., DeMeuse, K. P., and Futrell, D. (1990). Work Teams: Applications and Effectiveness. *American Psychologist* **45**, 120-133.

Swezey, R. W. and Salas, E. (Eds.) (1992). *Teams: Their Training and Performance.* Norwood, NJ: ABLEX.

Thordsen, M. L. and Klein, G. A. (1989). Cognitive Processes of the Team Mind. In *IEEE International Conference on Systems, Man, and Cybernetics (Cambridge, MA)*, (pp. 46-49). New York: IEEE.

Wiener, N. (1961). *Cybernetics* (2nd ed.). Cambridge, MA: MIT Press.

DESIGNING HUMAN-MACHINE SYSTEMS TO MATCH THE USER'S NEEDS

Paul Fuchs-Frohnhofen, Ernst A. Hartmann, D. Brandt

Department of Informatics in Mechanical Engineering (HDZ/IMA),
University of Technology (RWTH), Aachen, Germany

Cognitive compatibility of human-machine interfaces implies that the structure of the interface should match the cognitive structures of the users. A theory of mental models (structures of analogue knowledge) is described. A methodology incorporating this theory has been used to analyse the user's mental models in work settings. It has been embedded into a research-based set of criteria for system design to match the user's needs. Furthermore, it has been employed to generate variants of human-machine interfaces with direct user participation. As an example, a new CNC system was developed.

Keywords: Human-centred design, Human-machine interface, Systems design

1. INTRODUCTION

It is well known that human-machine interfaces, as they appear today, have been designed by (software) engineers based on what they know about how people work with these systems. During recent years, some areas of psychological research have contributed to a better understanding of what goes on in the minds of workers end engineers working with computerized systems. This methodology should be combined with direct user participation and observe criteria worked out in the work sciences.

An important aspect of human-centered system design is cognitive compatibility, which means that the structure of the human-machine interface of the computer (or some other machine) should match the cognitive structures of the users.

In the following, a theoretical approach will be presented to describe various types of mental models. It will be shown how this theory can be applied to analyse the user's mental models in order to achieve cognitive compatibility. These mental models can provide designers with a conceptual framework to generate prototypes of human-machine interfaces to be tested and elaborated in a cooperative process involving the future users.

As an example, CNC machine tools today are mostly controlled by putting abstract codes into a computer keyboard. The process as understood by the users, however, is a quasi-continuous process of a tool cutting into material. Thus the mental model of the CNC worker does not correspond to the design of the CNC human-machine interface. It factually mirrors the mental model of the software engineer who designed it. Hence, it seems necessary to design CNC interfaces corresponding closely to the mental models of experienced workers using the system.

2. CRITERIA FOR TECHNOLOGY DESIGN

The research project reported here was to design production systems which are cognitively compatible with the mental models of the users (i.e. the workers). For this aim, existing knowledge from work science and the results from different research projects of the HDZ/IMA have been used. The criteria thus developed are suitable for skilled workers individually and in group work settings, i.e. in controlling machine tools and manufacturing systems. (Fuchs-Frohnhofen, 1994)

These criteria have been integrated in a comprehensive checklist to ascertain that the system designed is user-friendly in its widest sense. The system is, thus, considered as a whole including all its elements- i.e. interface, intrinsic structure, cognitive compatibility, work organization and working environment etc.. The list of aspects following here are a condensed version of this checklist of criteria.

The workers should be able to: learn about and adapt easily to the system and its different uses, experience directly and influence manually the manufacturing process, maintain and repair the system, make use of the system's self-explanation features and documentation, make use of data bases and decentralized information networks (but with protection of personal data), simulate manufacturing processes before starting the `real` process, develop communication and cooperation in groups and networks.

These criteria may be used as a framework in designing new tecnological systems. In addition to using these criteria, the users (i.e. the workers themselves) need to be integrated into the design process from the very beginning. Thus, their`mental models`are likely to be taken into account in the design as discussed in the following paragraph.

3. A THEORY OF MENTAL MODELS

Human knowledge may be divided into two domains of cognitive structures: intrinsic and extrinsic representations (Palmer, 1987). Research on intrinsic – or analogue – knowledge focussed on two major theoretical concepts: mental images and mental models. Mental models seem to be the more general concept, integrating mental images: Following Johnson-Laird (1983), images are those aspects of mental models that rise into consciousness. Evidence concerning the psychological reality of mental images and mental models may be found -among other references- in Gentner and Stevens (1983)..

Inspired by Johnson-Laird (1980, 1983), Hartmann and Eberleh (1991) proposed a taxonomy of mental models. Six basic types of physical mental models can be distinguished by their way of representing spatial and temporal structures as shown in figure 1. Physical mental models deal with the perceptible world, wherereas conceptual models represent abstract concepts. The columns of the matrix combine types of mental models according to how they deal with time: There may be no temporal variation at all *(static models)*, a temporal sequence of static structures *(discrete models)* or a completely continuous process, comparable to a movie, which may be in real time or not *(continuous models)*.

The upper line of the matrix contains mental models which do not represent the world in a metric fashion (e.g. without regard of actual size and location); their spatial structure is *topological*. Three types of models can be distinguished: *Relational* models represent properties of objects and relations between objects. A good example is an organigram of a company, which only depicts the functional relations between departments but not the physical location of these departments on the company site. In *causal* models, these objects and relations can be represented at several discrete points of time. *Temporal* models allow the same in continuous time.

Models in the lower line of the matrix do contain *metric* spatial information like – as an example – a mental image of a location visited on holiday. As causal and temporal models, *sequential* and *kinetic* models add the possibility of representing discrete and continuous time, respectively.

A general taxonomy of mental models is discussed in Hartmann (1994).

In the following paragraph, a methodology will be presented to analyse mental models and to use this taxonomy for the generation of prototypes of human-machine interfaces.

4. A METHODOLOGY FOR THE DESIGN OF COGNITIVELY COMPATIBLE HUMAN-MACHINE INTERFACES

Carroll *et al.* (1991) describe a circular paradigm of human-machine interface design (Figure 2). It starts with an analysis of existing technology. Every artifact (technical system, tool, piece of art ...) can be interpreted as an operationalisation of (implicit) hypotheses about its users. The human-machine interface of a technical system, for example, is designed according to the designer´s mental model of how the system will be used in practice. This includes assumptions about the users, their tasks and their way of working and thinking.

These assumptions can be inferred from an analysis of this technology in situations of use, and the design rationale can be reconstructed. The design rationale describes how the system specifically fulfills the requirements of the application domain. Using concepts from psychology, especially German work psychology (Hacker, 1986; Greif, 1991), a vision of desired future work in this application area is generated. From this a scenario-based design representation is derived. A scenario is a description of a future system in situations of use. It explains how the specific features of the system will support users in doing their tasks. The system – or a prototype of it – being implemented, studies in work settings will allow to evaluate the design scenario.

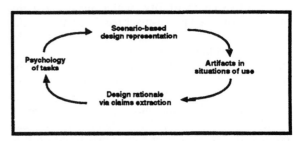

Figure 2: The task-artifact cycle (Carroll *et al.*, 1991)

This task-artifact cycle is illustrated in figure 3. It is applied to the design process of a manufactoring system. The four phases of the circular process are outlined.

Mental model structure spatial \ temporal	static	discrete	continuous
topological	relational	causal	temporal
metric	spatial	sequential	kinetic

Figure 1: Taxonomy of physical mental models

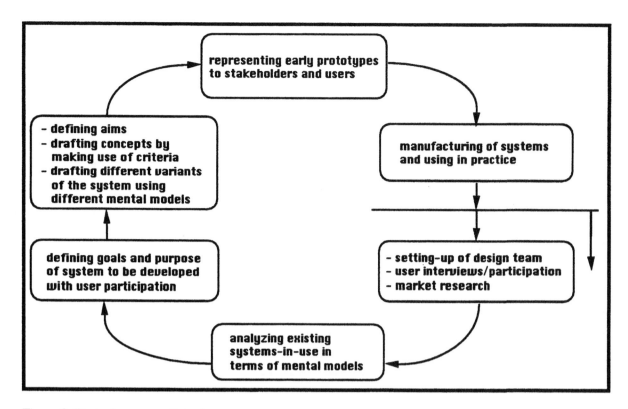

Figure 3: Designing a manufacturing system

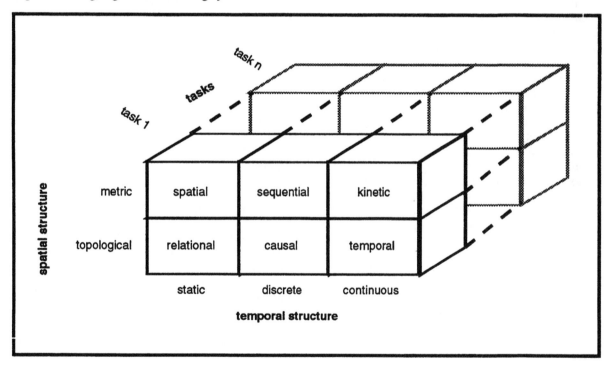

Figure 4: The analysis space

5. THE COMPLEMENTARITY OF ANALYSIS AND DESIGN

This task-artifact cycle shows two different uses of the mental model concept: analysis and design. It means, the taxonomy of mental models can be used in two ways within this circular process: Firstly, when extracting the implicit hypotheses about the users from studies of technologies in use, the taxonomy may serve as an analytic grid to diagnose the user's mental models by evaluating these hypotheses. Secondly, in drafting the scenarios, the taxonomy suggests a variety of human-machine interfaces for the system modules which are to support different tasks according to different mental models.

Figure 4 shows the analysis space for physical mental models. Two of its dimensions are made up

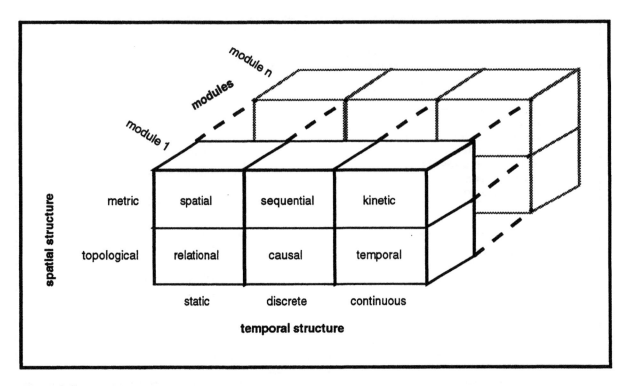

Figure 5: The design space

by the dimensions of the taxonomy of physical mental models: spatial and temporal structure. The third dimension depicts the tasks performed by the system's users. The user's mental models can be analysed following this rationale: The (implicit) hypotheses about the cognitive structures of the users as operationalized in the human-machine interface may be more or less correct. If the structure represented in the interface fits the user's mental model of this task well, he will feel supported by the technical system. If it does not fit, usability problems will arise. Observing and interviewing users at their workplace will provide the data to estimate the degree of compatibility between mental models and interface structures. If the interface structure is known, the presumptive type of the user's mental model can be inferred from the specific kind of compatibility or incompatibility observed.

As a counterpart of the analysis space, the design space combines mental models with modules of the technical system (figure 5). First a desirable task structure has been defined based on psychological concepts (see above). Subsequently variants of human-machine systems are constructed for each of those tasks – according to the taxonomy of mental models – as a part of a scenario-based design representation. This representation is evaluated and reformulated in cooperation with future users. For some tasks, the type of mental model, or the most probable types, may be known from an analysis like the one described above, so that the range of variants to be designed may be narrow. This information may also be lacking, because this task is specific for the new technology to be designed or it was not contained in the task structure before its redesign. In this case, conceptions for human-machine interfaces should be set up for all types of mental models

except those obviously irrelevant for this specific task.

In this construction process, some aspects should be kept in mind: (1) Mental models are task-specific. Therefore, the same user is likely to employ different models for different tasks. (2) For these same task, different users may think in terms of different models. (3) Even the same person may use different models for the same task depending on his actual condition (degree of skill, fatigue ...). To allow for these differences, human-machine interfaces should be adaptable to different mental models.

6. EXPERIENCES IN PRACTICE

In a research project funded by the German Federal Minister of Research and Technology in the research programme "Work & Technology", this approach was applied to the design of a computerized numerical control (CNC) system for lathes which is especially suited for skilled workers (Sell and Henning, 1993). The software was developed by the company R. & S. Keller (Wuppertal, Germany), which cooperated with the HDZ/KDI in this research project. Special effort was made to take into account the demands of experienced elderly workers. In this project, engineers, practicioners in vocational education, social scientists and psychologists cooperated with users of machine tools.

The so-called WOP-systems (Workshop-Oriented Programming), a special category of CNC-systems, were the starting point of the project. They show the following features: (1) The description of the geometry of the workpiece to be made is separated from the technological specifications (sequence of cutting operations like rough cutting, threading, fine cutting etc. and the corresponding specifications of tools, feedrates, spindlespeeds etc.). (2) The interaction style at the human-machine interface does not correspond to the ISO programming standards for CNC machine tools. It is rather a kind of graphic-interactive dialogue.

Because of this graphic-interactive style of human-machine interaction, WOP-systems were considered to be especially suited for shopfloor users not accustomed to computer aided work. In this research project, this technology was analyzed in two settings: (1) Experienced users of these systems were interviewed at their working place: how they felt supported or hindered by these systems in their work, and how they would like future systems to be designed. (2) Workers experienced in conventional turning were observed how they learned to use a WOP-system. The difficulties they encountered and the mistakes they made were analyzed.

As mentioned above, the use of a WOP-system implies the separation of geometric and technological input. The geometry module has to be used right at the start unless a geometry has been imported from CAD. Interpreted in terms of the taxonomy of mental models (figure 1), this means that the user is supposed to make up a static spatial model of the part, because the manufacturing process has to be neglected during geometry specification. When using the technology module afterwards, the user has to think of the part in the way of a sequential model. The research reported here showed that this is not the way experienced workers think about turning. Rather, they tend to consider the manufacturing process right from the start, so that the separation of geometry and technology modules does not fit their style of thinking and working.

Hence, the company Keller decided to develop a new CNC control system. In their software prototype, the human-machine interface allowed sequential or kinetic models to be used right from the start (Figure 6). This prototype offers three modes of operation: (1) In the "direct manual control" mode the user moves a virtual tool in the computer simulation with an electronic handwheel. These cuts are recorded by the system and the internal model of the part is changed accordingly (kinetic model).(2) In the "positioning by the handwheel" mode the handwheel is used to move a cursor on the screen connected with the tip of the tool's edge by a "rubber band".When the cursor has been positioned at the desired end point of the tool path, pressing the "return"-key will cause the tool to follow the rubber band to this point. This tool path will be automatically stored by the system as a program sentence according to the ISO programming standard (sequential model).(3) Additionally, geometrically complex elements may be described in a graphic-interactive way as in the geometry module of a WOP-system (spatial model). Unlike WOP-systems, however, this method may be restricted to those elements of the workpiece where it is really necessary; there is no need of defining the whole geometry of the workpiece. These contours can be used as "templates" for manual turning. Alternatively, tool paths may be generated automatically.

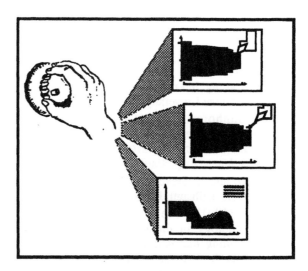

Figure 6: The CNC$_{plus}$ control system, modes of operation

7. EVALUATION OF SYSTEM DESIGNED

The prototype of the system thus designed was presented to and tested by users. They generally appreciated the straightforward way of working with this system. It depended on their personal style of work and their previous experiences, whether they preferred the direct manual control or the positioning-by-the-handwheel mode.

Encouraged by this positive feedback, a prototype machine was built and tested. Now, additionally to the virtual tools on the screen, the real tools of the machine can be used for manual turning and "implicit programming".

The overall evaluation of the "design philosophy" of the machine was very positive. Several improvements were suggested. As an example, the electronic handwheels were considered to prevent the "feel of the cut" known from mechanical handwheels. In the final machine, special handwheels are available optionally as a tactile display that feed the cutting force back to the user. Thus the concept of cognitive compatibility- embedded into a larger framework of user-oriented criteria- has proved its value in practice.

REFERENCES

Carroll, J. M., W. A. Kellogg and M.B. Rosson (1991). The Task-Artifact Cycle. In: *Designing Interaction* (J. M. Carroll, (Ed.)). Cambridge University Press, Cambridge.

Fuchs-Frohnhofen, P. (1994). *Zur facharbeiterorientierten Gestaltung der Mensch-Maschine-Schnittstelle bei CNC-Drehmaschinen.* Augustinus, Aachen. [German]

Gentner, D. and A. Stevens, (Eds.). (1983). *Mental Models.* Erlbaum, Hillsdale, N.J.

Greif, S. (1991). The role of German work psychology in the design of artifacts. In: *Designing Interaction.* (J. M. Carroll, (Ed.)). Cambridge University Press, Cambridge.

Hacker, W. (1986). *Arbeitspsychologie.* Huber, Bern. [German]

Hartmann, E. A. and E. Eberleh (1991). Inkompatibilitäten zwischen mentalen und rechnerinternen Modellen im rechnerunterstützten Konstruktionsprozeß. In: *Software-Ergonomie '91 – Benutzerorientierte Software-Entwicklung.* (D. Ackermann and E. Ulich (Eds.)). Teubner, Stuttgart. [German]

Hartmann, E.A. (1995). *Eine Methodik zur Gestaltung kognitiv kompatibler Mensch-Maschine Schnittstellen.* Augustinus, Aachen. [German].

Johnson-Laird, P. N. (1980). Mental models in cognitive science. *Cognitive Science,* **4**, 71-115.

Johnson-Laird, P. N. (1983). *Mental Models.* Cambridge University Press, Cambridge.

Palmer, S. E. (1978). Fundamental aspects of cognitive representation. In: *Cognition and Categorization.* (E. Rosh and B. B. Lloyd (Eds.)). Erlbaum, Hillsdale, N.J..

Sell, R. und K. Henning, (Eds.). (1993). *Lernen und Fertigen.* Augustinus, Aachen. [German]

ENTERPRISE SUPPORT SYSTEMS:

HUMAN INTERACTION WITH COMPLEX ORGANIZATIONAL SYSTEMS

William B. Rouse

Search Technology

4898 South Old Peachtree Road

Atlanta, Georgia 30071 USA

Abstract: Probably the most ubiquitous class of human-machine system involves managers, designers, and other personnel operating and maintaining complex organizational systems. These people perform tasks such as situation assessment, planning and commitment, and execution and monitoring at strategic, tactical, and operational levels of the enterprise. To assist people in performing these types of tasks, this paper outlines the structure and components of an Enterprise Support System.

Keywords: Man/machine systems, business process engineering, decision support systems, database systems, knowledge-based systems, training.

1. INTRODUCTION

When thinking about human-machine systems, we have traditionally tended to focus on aircraft pilots, automobile drivers, and process control operators. More recently, we have begun to pay significant attention to manufacturing and shop floor personnel. People in such systems are very deserving of our attention.

There is another, much larger class of personnel that receives very little attention from the human-machine systems community. This class includes millions of managers, designers, and other people who operate complex enterprises. This ubiquitous set of people are very much in need of support in terms of interface design, aiding, and training.

This paper is concerned with understanding the tasks of these people and developing means to support performance of these tasks. In particular, the emphasis is on computer-based Enterprise Support Systems (ESS) that enhance people's abilities and help them to overcome their limitations in processes of performing tasks in complex organizational systems.

2. TASKS OF ENTERPRISES

To consider what information and related types of support people in enterprises need, we should consider people's information requirements in the context of their tasks. By focusing on tasks, we can then determine what information is needed, when it is needed, and how it can best be provided.

Representation of the tasks of people in enterprises involves several issues. First, of course, it involves defining the scope of these tasks. In this paper, the focus is on how organizations deal with change. Hence, the concern is with tasks that relate to anticipating impending needs to change, recognizing the emergence of needs to change, and responding to these needs.

Beyond identifying tasks, the context within which these tasks occur must be defined. This involves specifying the functional areas and/or hierarchical levels where tasks are performed. Further, since the effects of information on task performance are of interest, it is important to consider how the nature of questions asked and information needed vary across areas and levels.

In a very broad sense, there are three types of tasks that enterprises perform in the process of anticipating, recognizing, and dealing with change: 1) situation assessment, 2) planning and commitment, and 3) execution and monitoring.

2.1 Situation Assessment

Situation assessment involves answering questions such as: What is happening? What is likely to happen? Answering these questions requires both seeking information and formulating explanations. Information seeking includes identification of alternative information sources, as well as evaluation and selection among these sources. Explanation includes generation, evaluation, and selection among alternative explanations.

While this definition of situation assessment correctly describes the task, it does not portray how this task is typically performed. Situation assessment is seldom as analytical and systematic as this definition might lead one to think. In particular, the generation, evaluation, and selection activities that are noted as elements of information seeking and explanation are usually performed in parallel and often without a conscious sense of, for example, generating alternatives.

2.2 Planning and Commitment

Assessment of current and likely future situations is often a precursor to attempting to take advantage of situations, or perhaps change situations. This goal leads to the task of planning and commitment. This task involves generation, evaluation, and selection among alternative courses of action, as well as commitment of resources to the course(s) of action selected.

A particularly challenging aspect of planning is the generation of alternatives. It is common for enterprises to view this activity too narrowly, in part at least, due to a tendency to narrowly define the world. Consequently, they do not consider potentially attractive courses of action.

2.3 Execution and Monitoring

Planning and commitment should lead to execution and monitoring. This task includes implementation of plans, observation of consequences, evaluation of deviations from expectations, and selection between acceptance and rejection of these deviations. One of the greatest difficulties associated with this task is lack of execution.

2.4 Levels of Performance

The above three tasks are performed at three levels of an enterprise -- strategic, tactical, operational.

At the strategic level, the focus is on the enterprise's mission, goals, strategies, and strategic plans. Consideration of these constructs occurs within the context of trends and events that are affecting, or could affect, the enterprise. These trends and events include at least market and technology considerations, and often economic, political, and social issues.

The tactical level focuses on the objectives specified in strategic plans and is concerned with tactics and tactical plans necessary for achieving these objectives. These objectives might include sales targets, financial goals, technology capabilities, and so on. Tactics and tactical plans are considered in the context of opportunities available and, unfortunately, an abundance of diversions.

The operational level is concerned with schedules, budgets, and project/production plans. At this level, considerable attention is paid to demands and problems. Demands typically focus on results and resources. Problems are often related to a lack of expected results and difficulties with resources.

2.5 Key Questions

The three types of tasks and three levels at which these tasks are performed can be combined to define the key questions faced by enterprises. The result is shown in Figure 1.

The 27 questions shown in this figure represent key concerns that all enterprises must -- or at least should -- address. They range from broad, externally-oriented concerns to narrow, internal issues. Typically, most of an enterprise's resources are devoted to dealing with the issues in the lower right cells of this matrix. In contrast, most of an enterprise's major mistakes occur when dealing with the issues in the upper left cells of the matrix.

3. ENTERPRISE SUPPORT SYSTEMS

The purpose of an Enterprise Support System (ESS) is to assist in asking and answering the types of questions shown in Figure 1. In this section, an ESS architecture, necessary information infrastructure, and component tools are discussed.

3.1 ESS Architecture

To elaborate the ESS concept, we need to consider the relationships among the constructs underlying all of the tasks in Figure 1. To do this, one first lists all the constructs. For example, the constructs for situation assessment at the strategic level include consideration of external trends and events, strategic assumptions, and market opportunities and threats. Next, potential relationships between pairs of constructs are considered and linkages noted. Finally, constructs and linkages are rearranged to create a visually comprehensible diagram.

The result of performing this type of analysis is the overall ESS architecture shown in Figure 2. Note that elements of this diagram may be linked without explicit lines being shown. For example, all elements include inputs labeled "results," but lines connecting the outputs of all the rightmost elements

LEVELS/TASKS	SITUATION ASSESSMENT	PLANNING & COMMITMENT	EXECUTION & MONITORING
STRATEGIC	What external trends and events are affecting me? Are my strategic assumptions still valid? Are there new market opportunities or threats?	What are my overall strengths and weaknesses? How might I pursue market opportunities and avoid threats? Are expected consequences believable and desirable?	Are strategies being appropriately implemented? Is implementation producing expected results? Is rethinking needed?
TACTICAL	What specific opportunities and diversions are affecting me? Are my tactical intentions still valid? What are the implications for ongoing efforts?	What are my opportunity-specific resources and constraints? How might I pursue specific opportunties and avoid diversions? Are expected consequences defendable and affordable?	Are tactics being appropriately executed? Is execution producing acceptable results? Is redesign needed?
OPERATIONAL	What problems and demands are affecting me? How are these problems and demands affecting my results? Are my expectations still reasonable?	What are my task-specific resources and constraints? How might I solve problems and satisfy demands? Are expected consequences justifiable and acceptable?	Are tasks being appropriately performed? Is performance producing necessary results? Are problems evident?

Fig. 1. Key Questions for Enterprises

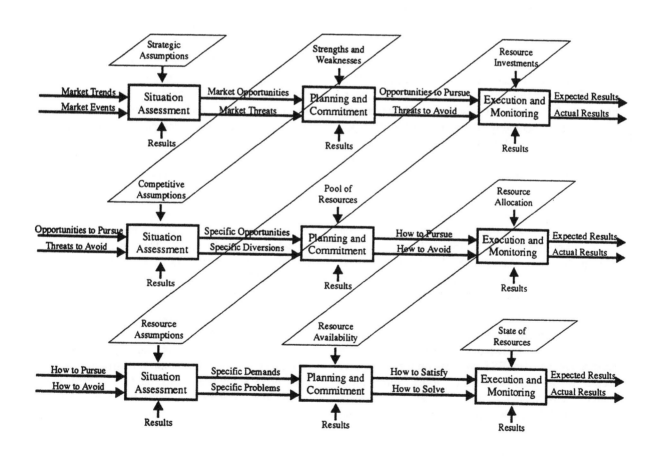

Fig. 2 Overall Enterprise Support System Architecture

in this figure are not shown to avoid a very cluttered diagram.

Consider each of the elements of Figure 2, starting with the upper left. Market trends and events are interpreted in the context of strategic assumptions to yield a situation assessment expressed in terms of market opportunities and threats. Opportunities and threats are then translated into plans for pursuing a subset of these opportunities and avoiding threats based, in part, on the enterprise's strengths and weaknesses. Once appropriate resource investments have been initiated, these plans are then executed and results monitored.

Strategic plans for pursuit of opportunities and avoidance of threats are assessed at a tactical level, premised in part on competitive assumptions, to identify specific opportunities and diversions. Based on the expected pool of resources, plans are developed for pursuing specific opportunities, as well as avoiding specific diversions. Resources are then allocated which enables execution of these plans and monitoring their results.

At the operational level, the plans for how to pursue opportunities and avoid diversions are assessed relative to resources assumptions. Plans are then revised and refined to reflect the availability of resources. Plans for satisfaction of demands and solution of problems are created and executed, with results that are affected by the actual state of the resources being utilized.

The support process outlined in Figure 2 may seem overly formalized, and perhaps very rigid. This would be true if this process was used in the linear, hierarchical manner just described. However, the wide variety of people typically involved in the types of activities depicted in Figure 2, as well as the asynchronous nature of their involvement, dictates that people be able to access and utilize an ESS at any point in the process shown.

Thus, the prescription implied by this figure should be viewed as a nominal process at most. This process provides a path to follow if there are no other preferences. An ESS should provide guidance on what is likely to be the best thing to do, but also support users in whatever they choose to do.

Another way to view the process depicted in Figure 2 is as a schematic for integration of a set of tools, databases, and knowledge bases. From this perspective, an ideal ESS provides uniform and easy access to a wide variety of tools, information sources, and related utilities that enable seamless utilization of these resources to address the tasks and questions in Figure 1.

3.2 Information Infrastructure

This ideal has yet to be realized. It is, therefore, useful to consider the ingredients that are likely to be essential to providing the types of support

envisioned for this ideal ESS. First of all, a key element of the ultimate ESS is the information technology infrastructure necessary to seamless integration and utilization. This information technology includes both computer and communications technologies.

At the moment, there are a wealth of these technologies. However, it is difficult to get them all to play together. Recently, because of a one-month working vacation, I needed to add modem, fax, and network capabilities to my high-end laptop. Now, with a 40+ character access code I can communicate throughout the U.S. I cannot yet fax worldwide because that requires another string of digits for which there is no room in the setup software. Clearly, seamless integration is yet to come.

Nevertheless, the infrastructure necessary for the ideal ESS is emerging. It may take a few of years, but it will not take a decade. It is quite reasonable to assume that the necessary underpinnings for the ultimate ESS will soon be here.

However, such accomplishments will not be sufficient. One also has to be concerned with the applications software that will use this infrastructure. Seamless integration of these applications requires that one move beyond syntactic compatibility and focus on semantic and pragmatic compatibility. The various applications of interest have to deal with compatible meanings and practical value added for users.

To illustrate, an ESS is very likely to include word processing, spreadsheet, and database capabilities. These capabilities are likely to be linked to systems for order entry and processing, as well as accounting. Semantic integration involves assuring that all constructs, e.g., words and objects, mean the same thing throughout these applications. Pragmatic integration is concerned with the value added for the enterprise.

Semantic and pragmatic integration are difficult to achieve they depend upon a common information model across applications. This, in turn, requires a common representation of problem domains. Such agreement is very difficult to achieve, at least at this stage of the industry's maturity.

A key to gaining competitive advantage in this process is the ability to lead in setting industry standards. Microsoft is rapidly gaining this advantage for the information infrastructure level of Enterprise Support Systems, based to a great extent on the work of various industry pioneers. However, at the semantic and pragmatic levels, many issues remain open.

3.3 Component Tools

Beyond infrastructure, there are tools that access and utilize infrastructure. It is useful to discuss several of these component tools. Some of these tools are obvious and readily available. The more obvious include packages and systems for word processing, report writing, database access and utilization, spreadsheet modeling, computer-aided design, project management, materials resource planning, process control, order entry and processing, accounting, and forecasting.

These types of applications software provide most of the capabilities necessary for the functionality depicted in the lower right elements of Figure 2. While the levels of integration of these capabilities remain fairly rudimentary, many of the elements are well defined and evolving toward standardization. Consequently, it is unlikely that new word processing or spreadsheets packages, for instance, will appear and gain significant market share.

Moving from lower right toward the upper left of Figure 2, the state of affairs is much more open. This is due to the nature of the functionality depicted. While the lower right is concerned with schedules, milestones, and budgets, as well as comparing actual results to planned results, the upper left is more concerned with what should be pursued and how to pursue it.

Methods and tools that support these needs are far less numerous than those just mentioned, and far from being standardized. We have been involved in developing three tools that fit into this category. These tools are called "advisors" in that they provide nominal methodologies -- in the sense explained earlier -- and include online aiding and tutoring in the use of the tools. While these tools can be used separately, they also include rudimentary abilities to share information at both semantic and pragmatic levels. These tools also rely on several of the lower level tools for word processing, spreadsheet, and project management capabilities.

The product planning advisor is concerned with representing the interests of all of the stakeholders in the product chain (Rouse, 1991, 1994). This chain might include, for example, designer, manufacturer, distributor, seller, customer, user, and maintainer. A representation of the needs and wants of all of the stakeholders is used to evaluate the relative utility of alternative market offerings, including those of competitors.

The business planning advisor is focused on the market characteristics of the business proposition (Rouse, 1992, 1994). Using the outputs of the product planning advisor, or equivalent, the business planning advisor deals with issues such as market segmentation, size, and share; marketing, distribution, and sales channels; nature and extent of competition; and competitive strategies and tactics. The result is a set of action plans for pursuing specific opportunities and dealing with competitive forces.

The advisor for dealing with organizational changes emerged from problems associated with implementing new strategies. Planning and implementing major changes can result in substantial internal conflicts. Good examples include defense conversion and shifts to more environmentally-friendly manufacturing. Typically, these conflicts are expressed in terms of differences regarding strategies. However, the primary sources of such conflicts are usually below the surface and involve differing belief systems about customers, innovation, technology, performance and so on. This advisor supports the diagnosis of such differences and development of plans for remediating these differences and building a consensus (Rouse, 1993, 1994).

The three advisors just described are exemplars of the types of support for activities toward the upper left of Figure 2. There are a variety of other tools that support these elements of an ESS. The range of offerings is very diverse and, by no means, standardized in terms of having a common representation. While, for example, the three advisors have a common conceptual basis, their abilities to communicate with each other are rudimentary. Fortunately, trends in software technology are accelerating such communication capabilities. However, it will be quite some time before this range of offerings shares underlying problem representations.

3.4 What Users Want

It is useful to discuss the types of support that users want as they pursue the activities depicted in Figure 2, particularly the activities toward the upper left of this figure. These observations are based on having worked with roughly 100 enterprises and over 2,000 participants in assessment and planning activities associated with change.

It is important to note that many of these activities are performed in groups. While individuals may prepare background materials and initial analyses, most decisions are made in group settings. In such settings, groups often want a structured process that provides a nominal path for proceeding. This desire is quite similar to that noted for the overall ESS process in Figure 2. Succinctly, people want a clear and straightforward process that can guide their discussions, with a clear mandate to depart from this process whenever they choose.

A second desire expressed by groups involves capturing the information compiled, decisions made, and the linkages between these inputs and outputs. People want such an audit trail so that they can justify decisions and sometimes reconstruct decision processes. Further, since the group may not all be together at the same time, they want group members to be able to asynchronously access the audit trail in order to understand what went on since they were last involved.

A third desire is for facilitation of the group's process. While human facilitation is often a key element in such group settings, computer-based tools can also provide elements of facilitation. For example, large screen displays linked to computer-based tools can present and manage the nominal process, provide prompts in terms of questions, and give advice based on what has transpired thus far. The neutrality of this type of facilitation -- the computer has no explicit stake in the proceedings -- is often quite compelling. It provides a clear means of short circuiting tangents and getting back on track.

A fourth desire that people have of such tools is for the tools to tell them something that they did not know. While structured processes, audit trails, and facilitation are greatly valued, they may involve inputting much information to the computer. This process may result in considerable sharing of information among group members. However, this information is inherently such that at least one of them already knew it. Why should they invest the effort necessary to provide the tool all this information?

The answer has to be that the tool is able to digest this information, see patterns or trends, and then provide advice or guidance that the group perceives they would not have thought of without the tool. This can be accomplished in a variety of ways. For example, the product planning advisor searches through multi-dimensional market and product models to provide advice on how to best improve products. The business planning advisor employs a rule-based expert system to suggest how to best improve the market and technology potential of business plans. The advisor that deals with conflicts underlying organizational changes looks for patterns among needs and beliefs underlying conflicts and prompts focusing on key differences that are affecting reaching consensus.

These types of advice and guidance have to be such that they are not viewed as magic. Thus, the tools have to able to explain the sources of their suggestions. People have to be able to explore the basis of tools' outputs until they can realign their intuitions with these results. Otherwise, they are not likely to accept the advice or guidance.

4. CONCLUSIONS

This paper has focused on the nature of human interaction with complex organizational systems, particularly as these systems attempt to anticipate, recognize, and respond to change. The tasks involved in processes of change were described and the nature of a computer-based system for supporting these processes was discussed. The information infrastructure necessary for such a support system was also considered, as were component tools.

Evaluation of this type of a system is very difficult. To date, the methods and tools described in this paper have been, and continue to be, employed by roughly 100 enterprises. A variety of success stories has emerged. However, much more importantly, this usage has resulted in a plethora of comments and suggestions for how an ESS in general, and methods and tools in particular, could be even more supportive. The result has been an evolutionary ESS concept that is being driven by its stakeholders. This is likely to be the most crucial factor in success of the ESS concept.

REFERENCES

Rouse, W. B. (1991). Design for success. Wiley, New York.

Rouse, W. B (1992). Strategies for innovation. Wiley, New York.

Rouse, W. B. (1993). Catalysts for change. Wiley, New York.

Rouse, W. B. (1994). Best laid plans. Prentice Hall, Englewood Cliffs, NJ

Information Technology in support of Living Oriented Society

Hiroshi TAMURA

KYOTO INSTITUTE of TECHNOLOGY
Department of Information Technology
Matsugasaki, Sakyoku, Kyoto, 606 JAPAN
fax: +8175-701-7211, email: tamura@hisol.dj.kit.ac.jp

abstract: The need of establishing human information technology (HIT), which is aiming to enrich living of the people is proposed. Human information technology should enhance various process of human communications, which are not limited to message transmission. HIT should further promote every member of a family to take broader roles of collaboration, support effective energy saving and resource recycling, and enhance interactions among individuals.

keywords: human communication, information age, energy saving, resource recycling, comfort, network karaoke, image performance

1. Computer Information Technology

Information technology so far might be better named as computer information technology. It started from processing the industrial and business data, and has contributed in improvement of efficiency, in cost reduction, etc. The computer information technology has been the key technology to advance the industrial societies. In the early days of computer information technology, the key problems were the way to reform the business so that they could utilize the computer facility of data processing. Most typically in the countries like Japan, some companies stopped using kanji (Chinese characters) to specify their products and name of workers and began using katakana which was only operable Japanese letters on computers in those days. Computer seemed so effective and useful even worth to abandon their own literacy. This was the ways advanced business men preferred to reform their company.

As the computer improved its capacity, it became capable of processing Kanji. The direction of computer information technology was then reversed. They began adapting the technology to the information systems consistent to the conventional culture. Information technology essentially should be to enhance communication and understanding among people, by using computers and other technical systems.

Same reading (katakana) but different letter (kanji)
カノウ (katakana)
加納, 叶, 嘉納, 狩野 (kanji)
ヒロシ (katakana)
博, 弘, 洋, 浩, 宏, 寛, 裕, 浩志, 弘司, 博司 (kanji)
Many Japanese names sound equal but written differently. When a name is written by the reading (katakana), postmen were confused in delivery, and nurse could not realize who the patiant is. More over sentenses in katakana was difficult to read.

Fig. 1 Katakana reading and kanji names

According to the classical concept, advancement of the industry is achieved by mass production and automation. They tried to introduce new or high technologies in order to advance the degree of automation and manpower saving. The office automation was introduced to reduce man work engaged in processing routine business data. Encouraged by the preliminary success of data processing, business men began to realize a new possibility of using data processing. Business automation did not reduce the data processing work, but continuously created the work of data processing. Computer market was continuously expanded, not because they succeeded in manpower saving by office automation, but they succeeded in creating new markets and new type of jobs.

People have learned that the mass production is less flexible to the change of market needs and that in order to provide innovative products to the market, human commitment is essential in cooperation with automation. The key for the modern industry is to organizing the human communication with sufficient knowledge to the recent technology among designer, planner, worker, service men and the users. Introduction of scientific measuring systems in factories has enable the workers objectively estimate the cause and effect of various phenomena, and let them understand the ways to prevent errors or defects, and let them improve their process or products.

Introduction of high technology and new machine as well as improvement of the product require creative ideas and problems finding among various sort of people related. Automation does not remove human work, but it changes the quality of work. People have to invent new jobs within or in relation to the new and old work environment.

High technology and new technology have to be applied not to remove human work, but to create new work chance. By making the agricultural machines smaller and easy to operate, old people with no technical back ground could continue working in the field, which is of benefit both to the farmer and for the society.

The POS systems are one of the most heavy duty information systems. The way of using the systems are different by the manager. Most typically the system is used for automatic shop keeping, that is to order the commodity equivalent to the amount of sales automatically. The other way of using the POS systems is to use it to decide the amount of order. So the POS is used as a decision support tool, and the shop keeper might decide what should be supplied. A technical system can be used either for automatic shop keeping or human centered. In the human centered shops, the manager has to analyze the reason for the sales change, and related changes in shop environment. The manager will learn from the daily success and failure.

In the automatic shop keeping, the programmer will be satisfied if the cost for the shop manager could be saved by the intelligent programming. Automatic shop keeping is possible or effective when environments around shops are homogeneous and no competitive shops are located around there.

2. Information Age

Links among a family are changing in the information society. Formerly children learned how to cope with nature in a family. They learned how to cut wood to make fire, or how to protect housing against rain and snow. Now they learn how to cope with economy. They learn the value and the functions of money, and how it changes by the economy. They learn to establish an economy unit in the society. In the near future they have to learn how to cope with the flood of data. The role of the family was to provide physical and economical shelters for their children. In the information age it is to provide shelters against the flood of data and to bring up the problem finding and problem communicating competence to their children. In the former times it was valuable to have a copy of data, but now it is rather easy to make a copy. It is nonsense for an individual or for a business organization to keep large storage of data. Of the essential significance is to select necessary data and throw away less significant data. Selection of necessary data out of flood of less significant can be done only by having proper understanding to the problems. Roles of HIT are to support the family to bring up and to enhance the problem finding capability of the young, the adults as well as the aged.

Education in advanced industrial countries are still based on the memory training. They are trained how to read and write in the primary school, trained to remember the conjugate of irregular verbs, trained how to solve second order equations in the high school, and again in the computer course at university, learn correct use of command languages. They are continuously trained to solve problems, as quickly as possible without making the mistake. But now a days computer can solve many problems, if properly defined. And robots can make camera, or radio and many more thing better than the man. It is of worthless for the man to compete with computers or robots, in the tasks which machines can do. The task of the human worker in the highly robotized factory has been to find a new problems in the factory, in the product and in scheduling. The problems have to be discussed within the shop floor and between different work sections. The workers have to be trained to communicate with various levels.

Communication in limited sense was the transmission of message. Various technical tools have been introduced to support clear and exact message transmission, such as microphone and speaker, slide projector, OHP and copy machine. Communication, however is not a mere transfer of message. Communication is the process of mutual understanding and reaching consensus. To promote mutual understanding, people has to be aware of the barriers of communication. There might be various barriers of communication, distance, time, language, knowledge bariiers. Communication may comprise such processes as finding the opponent of talk, specifying the problems, catching the chance to talk, preparing proper expression, follow up commitments, etc. [Tamura, 1990]. In evaluation of media communication not only the message transfer, but also the whole processes of communication should be examined whether they are properly supported.

Narrow sense
    Communication = message transmission
Broader sense
    communication = mutual understanding.
      communication technology =
    to find out the barriers of communication
        and to remove them

Processes of communication
      search for whom to contact
        refine problems
         catch chance
       proper expression
      transfer or redirect
        follow up

Fig.2 Concept of communication

3. Everything, Everywhere

Computer work station has been proud of being occupating the major area of desk top in the office and other work places. In the near future, the size of computer will be as small as pocketable or wrest mountable and associated display will be mountable on head like eye glass. With such predictions in mind, we have to think of the information technology tightly involved in the daily life of the people.

Small computer is not a big data center nor all mighty in handling multimedia. They are transferring media data to the center near by. They should be small but reliable agent to access centralized or distributed data base. As mentioned previously, communication is not mere transfer of message. It start from finding the person to communicate. Since people are moving more frequently and larger distance, and engaged in diversified tasks, the service to find proper person and provide appropriate media of communication become significant.

Small computer is beautiful because it could be installed in every piece of articles in the kitchen, living room, bath room, etc. It will be installed in walls, celling and floor as well as movile phone, and AV machines. They should be linked together by the wireless, and could be accessed every body at home and out side. In order to assure free and flexible access, member of family including pets and plants, as well as various sensors and transmitters installed in machine and walls shall have addresses.

Machine like kitchen oven, washing machine, VCR are normally monitored by specific person today at home. When a programmed task is complete, they make some message sound to acknowledge the completion to the people near by. It is not necessary to make message sound to be heard by everybody. In house locator will sense the location of the particular member of family and transfer the message sound to the person. If the locator detects the person is out or not in accessible locations, the message should be transferred to alternative person near by. Of course if necessary, the system will try to reach other responsible agents.

The address systems today are not convenient, when people begin to use several access number. Some people already have several addresses, which are fixed to location (home, office, car,etc) or limited function (phone, facsimile, telex, video phone). It is inconvenient, because people have to remember many addresses to access same person. It is more convenient if a person could be accessed by a same address. The sender will forward person access code and certain function code that will specify his selection of which communication media(phone, facsimile, telex, video phone).

The recipient terminal will analyze the function code and transfer the call to the terminal available to the recipient, depending on the location and situations. The access control communication and media communication will be clearly differentiated. Even in case of normal telephone, it takes considerable time to access the person to whom to call. Normally the phone call comes most frequently to the person who are out of office or too busy to answer the call. When a recipient is busy and unable to answer the call, the call might be transferred to fax or voice recorded terminals. If the recipient is attending to a conference, and not be able to answer a phone call, access control agent will leave message where the fax or voice recordings are stored.

Sensors, transmitters and movile networking will be especially useful to help people with special needs. They are often not capable of sending commands punctually. For such people network agent will analyze the behaviors of the person, and try to understand his intention. The system will try to manage the situation automatically based on the understanding.

4. Comfort of Coordination

Now a day, a society might be regarded poor, not because of absolute lack of goods but by the lack of flexible delivery systems. People like to live on their own preference, and they have to find ways of coexisting with the people of conflicting demands and preferences. At home, people of three generations are often living together. The aged and the adult, or the children like and need to have different foods. Or the meal time of is different by person in a family and different meal is served upon each demands or preference. Recently it is not by no means strange that each individual on the dinner table is eating different food. The preference changes irregularly depending on the previous menu and physical conditions. It is not bad to have different food in different time independently, but it is also enjoyable to have food jointly together. People like to have such diversity of living independent and interacting together.

To keep such mutually independent but coordinated living, no individual should be constrained in the house or kitchen keeping. The house hold or cooking machines have to be conveniently operatable to many of the family members including aged and children. It is not appropriate to design these machines assuming the principal users to be the young ladies. It is also not appropriate to to think the main objectives of machine use is to minimize the house hold labor. The machines have to be designed to increase the freedom of use and non-use, freedom to adapt various members of the family to create new possibilities. It

is most desirable to create a new chance of communication and collaboration by introduction of a new machine. Word processors created chance to write sentences for many people, why not kitchen processor create a new food communication. Kitchen processor is not the tool to automate food processing, and to minimize the cost. In stead it is the tool to reform the family life colorful and rich. It should be pleasant enough like game computers or karaoke machines. Because it is fashionable and pleasant school children like to cook, husband proud to cook for their family.

5. Resource Recycling

Wealth of a nation used to be measured by the amount of production, or by the number of owners of such goods like car, telephone and TV, or by the amount of energy and resource consumption. It is not the amount of goods, but the coordinated use of the goods which is essential to define wealth of the nation in the information age.

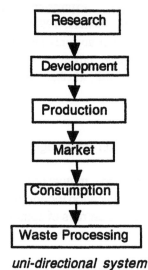

uni-directional system

Fig.3 Resource flow in the industrial society

Human information technology (HIT) shall provide effective tool for building ecologically sustainable societies. Information technology should be used more effectively to prevent over consumption of energy and resources. In the industrial ages a lot of effort have been done to make manufacturing process effective and less resource consuming. But the delivery as well as consumption process was by no means scientific. Information technology has to be based on the living, in stead of production. So far research and development were more interested in innovation of production process. For the effective and comfortable use of resource and energy, studies on the living of the people are most essential. Fig.3 shows the flow new technology in the industrial societies. The scientific findings and technical inventions gave influences to the living in one direction, from up to down. The science and technology in the living oriented societies have to be situated near to the living, and have to consider the recycling process as well as the production, as shown in Fig. 4.

In collection and classification of used goods, higher information technology is required than in the production process. The sales market of the used goods needs more information technology than the sales market of the new product, since the used goods are different one to another in every detail, and often require some repairing. The users of used goods might have much diversified requirements, and the users have to check the match to his requirements. The collected used goods have to be classified into reusable and not reusable. The used goods have to be transferred to the delivery market. From the not reusable goods , reusable components should be decomposed to be used as parts. The decomposed goods shall again be separated by different materials, like paper, plastic, iron, copper, aluminum, gases, etc.

For the collection, decomposition and separation robots and information technology should be introduced, in the same way in the production.

Efficiency of the resource recycling is essentially determined by the purity of the recycled materials. Toimprove the quality of resources, citizen's consciousness and their participation is most important. This is something related to the worker's participation in manufacturing.

Actual recycling process is by no means effective nor elegant, so the some people do not like to do, if permitted.

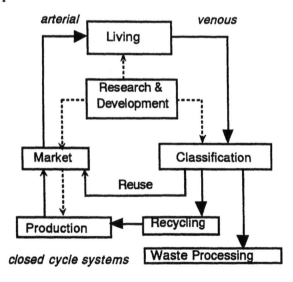

Fig. 4 Resource flow in the living oriented society

Works not beloved by people should be automated. For collecting used paper for recycling, the fax paper should be careful separated in order to preserve quality. At present, people separate it visually. Such use of manpower is by no means effective. It is most desirable to invent a physical sensor to detect fax paper from normal. Even if the sensors are not invented, necessary robot has to be developed. One possible option is to print invisible bar code on each sheet of paper, and the robot will identify the code and separate it. For such purpose automatic machine will most properly used. Developing a recycle robot is tough challenge, since the processing cost should be much lower than in the POS systems or in game computers. But information technology have to challenge continually to the field where social demands exist.

HIT will contribute in developing the clean and automatic recycling process, so that people feel like to participate in the recycling activities spontaneously on the one hand, and in making it enjoyable by reforming the activities more fashionable, or by including some game features and fun.

6. Enjoyable Interaction

In the early days of computer information technology people had to adapt to the technology. Thanks to the advances in computers and digital networks, human information technology tightly connected to the living of people are now feasible to grow. Human information technology is not a mere tool of manufacturing, nor business management. it is now an essential part of our living environment. Information technology to be implanted in the social environment should be lovable, enjoyable and well understandable to everybody, while it is safe, reliable and resource saving.

Japanese gardens are beautiful, they look natural. But it is an artificial nature built in a small part of nature. The artificial nature is beyond nature. They provide ever lasting relax and relief. Actually they are always changing by season and by growth of plants and grass, and by rain and snow. Also mind of the people changes. Nevertheless the nature beyond nature are beloved beyond the changes. People now expect for the human information technology such interfaces like Japanese garden.

As the computer information technology advanced, people are required to remember more and more commands, to monitor flood of the data carefully, to operate key or handles in regular order, and to watch carefully to detect errors. In spite of advances in computer information technology, people have to read more and more papers. In spite of advanced network technology, people have to move here and there more quickly. Now people are no longer interested in high speed and high precision computations. People are more interested in more relaxed and peaceful living environment. They are attracted by the computer use with wonder and surprise. People who are not currently the user of computers, will become the use only when the information systems could provide such environment, where people share the experience of tears and laugh with and through the environment.

Among various computer technology screen savers are the one providing rest and relief. Some of them make use of full computer power to colorful artistic motion. In place of gift articles exchanged by friends and family, computer gifts will be exchanged among people in future. Computer gifts are like a flower in the room, they will decorate the computer screen. When computer works are not busy they come out and present beautiful screen performance and melody. If the user interact with the gift program, it learns the preference of the users and adapt its performance to the preference.

Each country and people love each culture. One country may prefer top hero, and top sales. Other country may enjoy a culture they could participate together. In Japan, there have been a music movement which encourage every one to sing. It started from NodoJiman, MittsunoUta, Utagoe and to Karaoke. Popular singers will present the time catching ways of singing to a song. The player who are by no means trained in singing, challenges to sing the song he loves. The audience send the applause to his/her participation and contribution to make homely environment. A lot of electronics have been introduced to enhance Karaoke performance. They normally use laser disc for prompting words of song on the display together with the back music. In the most recent Karaoke, called network Karaoke, they use ISDN digital service to send MIDI code and words from computer located hundreds of kilometers far. At present ISDN service are available in most part of Japan, but the actual user is still not much. Under such circumstance, the network Karaoke is the top user of the ISDN service, which was never expected by the network investors.

Players like network Karaoke, because the new titles of songs are included in the list quicker than the disc Karaoke. The time for publishing the optical disc is too long to compete with network Karaoke. In case of network Karaoke, as soon as the code of a song is registered in the central computer, the song can be played everybody everywhere. At the same time Karaoke studio need not have to stock big file of optical discs.

Finally future of the Karaoke has to be predicted. Player want to pretend himself virtual environment. Player might want to sing it at the Waikiki shore, or in Pat O'Brien house in New Orleans, or in the front of Heian Shrine in Kyoto. Multi media technology will enable him sing in front of the back ground image reproduced from optical video.

Reference

H.Tamura(1990) Invitation to the Human Interface, JI-TV Eng.Japan, vol. 44, pp. 961-966

H. Tamura (1991) Human Interface in Manufacturing, Human Interface, vol. 7, pp. 639-644, .

J. Mey, H.Tamura (1992) Barriers of Communication in a Computer Age, AI & Society, vol.6, pp.62-77

H.Tamura (1994) Human Information Technology, Systen, Control and Information, vol.38, pp.245-251

AUTOMATED CLASSIFICATION OF PILOT ERRORS IN FLIGHT MANAGEMENT OPERATIONS

S. ROMAHN* and D. SCHÄFER **

*Aachen Technical University, Institute of Technical Computer Science, D-52074 Aachen, Germany.
E-mail: romahn@rwth-aachen.de

**German Aerospace Research Establishment (DLR), Institute of Flight Guidance,
D-38108 Braunschweig, Germany. E-mail: dirk.schaefer@dlr.de

Abstract. This paper introduces SmartTranscript, a tool for automated analysis of pilot action protocols recorded while operating the Flight Management System (FMS) of modern commercial aircrafts. Since errors are valuable indicators of pilots' mental workload and therefore of the user interface quality, SmartTranscript checks pilot action protocols for occurred user errors. The conceptual model of operator's tasks, proposed by Rouse and Rouse (1983), together with its human error categories, was found suitable to classify pilot errors detected in experiments. In addition to traditional approaches, SmartTranscript is a means to evaluate pilot-FMS interfaces objectively and enables comparison of interface alternatives.

Key Words. Human error; error analysis; classification; aircraft operations; flight control; protocols; tasks; human-machine interface; man/machine interaction

1. INTRODUCTION

In the last decades many cockpit features have been automated in order to avoid flight accidents and incidents caused by so-called pilot errors. Improved but more complex technical systems on-board created a working environment in which pilots have to manage a flight rather than to fly an airplane. Recent research (Billings, 1991) indicates that further automation leads to a dead end. Situation awareness and pilot workload suffer from the strategy which in the past has been trying to remove pilots from the control process by delegating tasks to automated systems (Sarter and Woods, 1991). Future developments require a workshare between operator and machine according to their particular abilities in order to optimize human-machine cooperation (human centered automation). Because pilot workload is closely related to errors in the operation of cockpit equipment (Kantowitz and Casper, 1988) this paper introduces a tool, SmartTranscript (Schäfer, 1994), that automatically searches for pilot errors in interaction protocols. Errors found during FMS operation are classified to specific categories, introduced by Rouse and Rouse (1983).

2. BACKGROUND

2.1. The Flight Management System

A key player in cockpit automation is the Flight Management System (FMS) which supports pilots in planning and controlling a flight. Precise route definition and fuel calculation shall enhance flight performance according to economical and safety purposes. Different degrees of automation can be selected which range from flying manually to completely handing over control to the FMS.

Experiments have been conducted with modern civil aircraft pilots investigating interaction with the FMS, e.g. Sarter and Woods (1992). These indicate that FMS usage is uncritical in standard situations like en-route flight. In more complex situations however, in which the FMS could provide an important benefit by reducing pilots' workload, pilots tend to ignore the FMS and perform manoeuvres either by autopilot or by hand. According to pilots' statements, main reasons for this are insufficient expectation conformity, lacking system architecture overview and improper mode awareness. As the basic FMS-features are mostly unchanged since the 70's and great progress has since been made in both computer hard- and software design and in the knowledge about human operators, improvements on the FMS layout could be achieved that surely contribute to safer and more efficient aviation.

2.2. Experimental system evaluation

There are, in general, two ways to evaluate new or existing systems, either analytically or experimentally. Analytical methods help to derive forecasts about system performance and reliability from a set of parameters describing the system. Models, like GOMS, e.g. Irving *et al.* (1994), are used to represent the cognitive relationships in human operation. Since a seizable relation between system parameters and quality is required, analytical evaluation easily becomes very complex. Experimental tools have therefore been designed to observe operators' behaviour and interpret it with regard to HMI quality. Errors in this context provide a rich source of information as they are directly related to distinct steps in the problem solving process.

Traditionally, experimental approaches to analyse pilot-FMS interaction were taken either by interviewing pilots or by interpreting situations in which FMS usage was unsuccessful or insufficient. More reliable and objective data can be obtained by recording pilots' actions while operating the FMS. The resulting protocol has then to be interpreted, considering knowledge about the system and the situation. An advantage of this approach is the possibility to automate protocol interpretation. As, however, information is obtained on a relatively low level, other ways of situation interpretation such as pilot questioning should be considered as additional sources of information.

Among a variety of methods to record operator-machine interaction such as video-recording or think-aloud-protocols the keystroke protocol was found most suitable. Pilot interactions are recorded in terms of at what time which key has been pressed. Additional parameters determining the actual system state are also recorded to support off-line interpretation. Information complexity is comparatively low, so both protocol recording and system interpretation can be automated (Smith *et al.*, 1991). Interpretation of the interaction protocols can be performed with respect to plan recognition (i.e. what was the pilot's objective while performing a set of actions) and error identification and classification (i.e. which actions were non-optimal and what kind of error occurred).

2.3. *Error classification*

The objective is to examine user interface quality with respect to performance and reliability of the human-machine interaction. Because errors are linked to the mental process the pilot is occupied with, a classification scheme is needed that supports evaluation of operators mental workload state which is responsible for the occurrence of a particular error. Among others the error classification scheme proposed by Rouse and Rouse (1983) as well as the underlying conceptual model of operator's tasks, depicted in figure 1, was found particularly suitable to meet these demands.

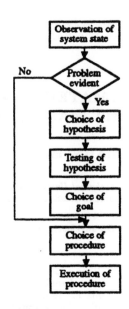

Fig. 1. Conceptual model of operator's tasks

In the course of each of these six steps errors may occur, but by regarding keystroke protocols, understanding can hardly be gained about what happened in the steps Observation of system state, Choice of hypothesis and Testing of hypothesis. Therefore only the steps Choice of goal, Choice of procedure and Execution of procedure are relevant for interpretation of interaction protocols.

3. SYSTEM DESCRIPTION

This section describes the system components as shown in figure 2. The system consists of a software simulation of the Control and Display Unit (CDU) through which the FMS is programmed, a session recording facility, a knowledge-based tool for session protocol analysis (SmartTranscript), and possible applications of the outcome.

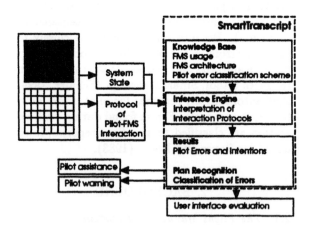

Fig. 2. Architecture of SmartTranscript within simulation environment

3.1. CDU simulation and session recording

In order to separate FMS analysis from costly flight simulator sessions and to facilitate pilot-FMS interaction recording a software simulation of the CDU was designed (Romahn, 1991). Emphasis was laid on realism of user interface behaviour rather than FMS functionality which is only implemented as far as necessary for correct data output. The CDU simulation was implemented on a Macintosh PC using the HyperCard development kit. All CDU elements and the CDU itself are prototyped in their original size. The software simulation of the CDU as used in the experiment described later in this paper is shown in figure 3.

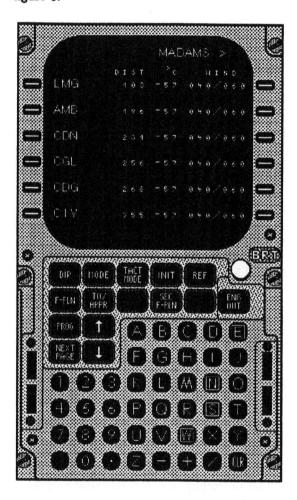

Fig. 3. Software simulation of the CDU (reduced)

Among the various FMS-CDU versions that only differ in minor hardware or software details the Airbus A310 FMS was chosen for the simulation. With minor adjustments to SmartTranscript, however, any other simulation or real CDU may be used.

An interaction protocol can be recorded on-line while the pilot uses the FMS in a simulated flight mission. This protocol contains both the pilot-FMS interactions (keys pressed) and the system state in terms of a parameter set (actual page, scratchpad content, etc.). Table 1 shows a sequence of a sample keystroke protocol (simplified).

Table 1   Sample protocol

Line	Time	Key	Page	Scratchpad
...				
82	3034	4	Init A	
83	3244	1	Init A	4
84	3523	0	Init A	41
85	3867	LK 5	Init A	410
86	4133	CLR	Init A	Format Error
...				

As the protocol contains further data that is not displayed in figure 1, the system state at the time of each interaction is fully determined and thus off-line processing of the protocol is possible. The recorded protocol is stored in a database.

3.2. SmartTranscript methodology

SmartTranscript has been designed to execute protocol processing after reading the files from the database. For off-line protocol processing several requirements have been posed:

- Processing must be provided by an 'open system' to support future FMS-functionality enlargement.
- Processing must be performed in an easy-to-comprehend way.
- The system must comment its conclusions.

To meet these requirements, a knowledge-based system has been found most suitable. The knowledge required by a software system in general consists of actual data (facts), knowledge about the system under investigation (domain knowledge) and knowledge about how processing should be conducted (inference knowledge). As traditional software systems do not separate domain knowledge and inference knowledge procedures become very inflexible and difficult to comprehend. Especially opening the system to new requirements becomes a difficult task. Therefore, the requirements discussed above could hardly be met by traditional software systems. Knowledge based systems clearly separate domain knowledge (which is stored in the knowledge base) and inference (which is provided by the inference engine). Additional or altered domain knowledge does therefore not effect the inference structure.

For these reasons SmartTranscript has been implemented as a knowledge based system. The domain knowledge is stored in rules, facts are represented by objects. Pilot actions and system state parameters are read from the session protocol via database bridges. For each set of actual data describing a particular interaction and the actual system state an instance of an object class called 'action' is generated that stores the data in its properties. To provide results, further objects are generated in the course of processing.

3.3. *Knowledge base*

The domain knowledge of SmartTranscript contains information about FMS usage, FMS architecture and the pilot error classification scheme. Knowledge about FMS usage and FMS architecture was transferred from operating manuals to a set of rules. With these rules the following questions can be answered: What can be done using the FMS? How has it to be done? How is the user interface controlled? Which errors may occur?

For an appropriate application of the Rouse and Rouse error classification scheme, an environment-specific definition of goals and procedures is required. In the context of FMS-usage it has been found suitable to define a goal as "a particular modification of the information the FMS possesses about the flight mission". To obtain a goal, a procedure must be applied which itself consists of particular actions. The following example illustrates the definitions used:

Goal Entering Cruise Flight Level
Procedure
- Opening Init A-page by function-key (if necessary)
- Entering Flight Level 410 by typing F,L,4,1,0
- Assigning FL410 to the parameter CRZ FL by pressing a corresponding linekey

Minor modifications to the specific categories of the classification scheme help meeting the particular environment conditions of FMS-usage. The error classification scheme used by SmartTranscript is depicted in table 2.

Using the above definition, errors can now be linked to one of the error classes. Examples are:

Choice of goal (cat. 4) It was chosen to enter a parameter the FMS does not accept at the current system state.
Choice of procedure (cat. 5)
An incorrect procedure was chosen to enter a parameter, e.g. the input format was incorrect.
Execution of procedure (cat. 6) The procedure was executed incorrectly, e.g. by unintentionally double-clicking a key.

Table 2   Specific error categories

Category	Description
4c	non-productive goal
4d	goal not chosen
5a	incomplete procedure
5b	incorrect procedure
5c	unnecessary procedure
5d	procedure not chosen
6	error during execution

3.4. *Inference engine*

SmartTranscript interprets interaction protocols in a two-phase process: In a first phase, each interaction is interpreted per se. Knowledge about the FMS-architecture helps to understand which particular action the keystroke effected. Interactions are then classified to support further interpretation. Formal errors such as invalid keystrokes etc. can already be identified at this level.

In a second phase, coherent sets of interactions are identified and isolated. These sets are related to procedures. Pilots' intentions (goals) can be identified from these sets of interactions. After the goal has been identified, an optimized procedure for this particular goal is extracted from a database for comparison. Errors can now be identified and classified according to table 2.

3.5. *Pilot errors and intentions*

SmartTranscript provides its users with the results-files described below. According to users' preferences, each file provided by SmartTranscript is available in text format and/or database format. Further processing, e.g. for statistical purposes is thus supported.

Pilot actions The result of the interpretation of every single pilot action per se.
Pilot navigation The order of FMS-pages accessed during mission completion.
Pilots intentions Pilot's intentions (goals) and the way in which he or she worked towards a specific goal. Comparisons to an optimum for reaching each particular goal are given.
Pilot errors Occurrence of errors of different categories is listed in this file.

Examples are given in chapter system demonstration.

3.6. Applications

The information provided by SmartTranscript may serve several tasks. Firstly, frequency of errors of different categories helps to evaluate the HMI quality with respect to particular design criteria such as hardware design, system architecture etc. Future FMS developments may be tested by a SmartTranscript version adapted to the new environment, thus allowing specific comparison between rivalling layouts in an early design phase. Secondly, user assistance may be provided after the user's intention has been identified (Romahn and Kaster (1993)). An enhanced FMS version may thus provide the pilot with useful information about a particular situation. Thirdly, after identification of both intention and errors a warning may be initiated, explaining to the pilot why SmartTranscript suspects occurrence of an error and what could be done instead. Both pilot assistance and warning require on-line processing which is not yet implemented by the current version.

4. SYSTEM DEMONSTRATION

4.1. Test scenario

To demonstrate SmartTransscript's ability to isolate pilot errors from keystroke protocols and classify found errors to the Rouse and Rouse categories an experiment was conducted. In this experiment 11 pilots were asked to perform a standard pre-flight check using the CDU software simulation. All pilots were experienced in flying a glass cockpit aircraft and were familiar with FMS functionality. During test sessions pilot keystrokes were recorded as well as the situation-determining parameters. Table 3 shows the data to be entered into the FMS during the pre-flight check.

Table 3 Pre-flight check data

Performance factor
Airport pair and route selection
Aircraft position and IRS alignment
Cost index
Cruise flight level
Airpressure at destination airport
Fuel weights
Standard instrumental departure
Standard arrival routine

4.2. Experimental results

After the experiment protocols were analysed by SmartTranscript. In this section some of the results are presented to demonstrate possible ways of further statistical data processing, error interpretation and task analysis.

Figure 4 depicts necessary page-transitions for task completion. As an example for a statistical analysis of SmartTranscript's results table 4 shows the mean page transitions per pilot. The last two rows of the table compare the average number of pilot's page openings to the optimum number corresponding to figure 4.

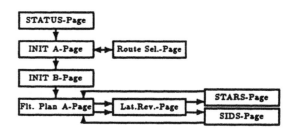

Fig. 4. Optimum FMS page transitions for task completion

Table 4 Mean FMS page transitions

to page	Status	Init A	Init B	RteSel	F-Pln	LatRev	SIDs	STARs	others
from page									
Status	1.00								
Init A		1.27	1.18		0.36				0.09
Init B	0.36				0.91				
RteSel	1.18								
F-Pln	0.27					1.81		0.36	0.72
LatRev							1.09	0.82	
SIDs					1.09				
STARs					1.09				
others	0.09				0.09				
opened mean	2.90	1.27	1.18		3.54	1.81	1.09	1.18	0.81
optimum	2.00	1.00	1.00		3.00	2.00	1.00	1.00	

Pilot errors during the pre-flight check detected by SmartTranscript are collected in table 5. Mean incorrect actions per pilot are plotted, ordered by both the pages on which they occurred and by the categories they were classified to.

There are striking numbers of pilot errors on the Init pages (A+B) and the Flight Plan page (F-Pln). Errors on the Init pages mostly belong to categories 5a (incomplete procedure) and 6 (error during execution of a procedure). For the Init pages are where most of the numerical data is entered, there is obviously a handicap that keeps pilots from correctly typing in data. A cause may be found in the need to enter information in a fixed format. The Flight Plan page's main objective is to provide pilots with flight progress information and gain access to the Lateral Revision page

(LatRev) that allows a more precise view on flight sections. The small size of the CDU screen forces the pilot to scroll the flight plan. Scrolling the flight plan in the wrong direction is the detected cat. 6 error, obviously a discrepancy between the expected scroll direction and the key lables.

Table 5   Mean incorrect pilot actions

errors	Cat. 4c	Cat. 4d	Cat. 5a	Cat. 5b	Cat. 5c	Cat. 5d	Cat. 6
page Status		0.27	0.36	0.36	0.18	1.27	0.45
Init A	2.00	0.73	15.55	1.28	0.09	1.36	10.27
Init B	2.82	0.27	8.82	0.36	0.09	1.36	1.91
RteSel					0.09	0.27	
F-Pln	1.45		0.55	-0.18	0.09	0.18	4.73
LatRev							
SIDs		0.09		0.27	0.18		
STARs		0.18	0.18	0.09			
others	0.18		0.09			0.09	

A comparison with human expert protocol analysis shows that error identification by SmartTranscript causes very little false alarms. The reason is that normative behaviour can be described very precisely and any deviation from correct procedures may be seen as an error. First experiences with experiments show that, although a little number of false alarm usually goes along with a little number of hits, SmartTranscript however detected most of the errors occurred in the recorded protocols.

Results of SmartTranscript's task recognition facility are presented by the following excerpt of the output. Raised tasks are printed with the result achieved, the effort made and the related position in the protocol.

FINAL APPROACH FUEL was provided and set to 2.0 . 4 actions were performed. This took 26.25 sec. (lines concerned: 81 to 84).

ROUTE RESERVE was proposed to 3 but rejected by the system. 2 actions were performed, 1 of which was incorrect. This took 15.48 sec, 2.55 sec for incorrect actions. (lines concerned: 85 to 86).

ROUTE RESERVE was corrected and set to 3.0 . 4 actions were performed. This took 9.98 sec. (lines concerned: 87 to 90).

STAR was inserted after having selected runway 07 and approach TOF07. 3 actions were performed. This took 72.53 sec. (lines concerned: 94 to 96).

5. CONCLUSION

This paper introduced SmartTranscript, a knowledge-based tool for automated analysis of keystroke protocols. Thanks to the high degree of determination of the FMS' user interface procedures, rules could be found that describe pilot-FMS interaction very precisely. This description includes both normative and erroneous behaviour. Therefore, pilots' subgoals when performing a given task can be extracted by SmartTranscript as well as errors on the latter three levels of the Rouse and Rouse conceptual model of user's tasks.

6. REFERENCES

Billings, C.E. (1991). Human-centered automation: A concept and guidelines. Technical memorandum. NASA Ames Research Center. Moffet Field.

Irving, S., P. Polson and J.E. Irving (1994). A goms analysis of the advanced automated cockpit. In: *Human Factors in Computing Systems (CHI'94)*. Boston. pp. 344–350.

Kantowitz, B.H. and P.A. Casper (1988). Human workload in aviation. In: *Human Factors in Aviation* (E.L. Wiener and D.C. Nagel, Eds.). pp. 157–187. Academic Press, Inc.. San Diego.

Romahn, S. (1991). *Gestaltung und Bewertung von Cockpitinstrumenten durch Rapid Prototyping*. Forschungsinstitut für Anthropotechnik. Wachtberg-Werthhoven.

Romahn, S. and A. Kaster (1993). Creating intelligent user interfaces using prototyping and knowledge based support technologies. In: *International Workshop on Intelligent User Interfaces*. Orlando. pp. 259–262.

Rouse, W.B. and S.H. Rouse (1983). Analysis and classification of human error. *IEEE Transactions on Systems, Man and Cybernetics*, 13, No. 4, 539–549.

Sarter, N.D. and D.D. Woods (1991). *Pilot Interaction with Cockpit Automation I: Operational Experiences with the Flight Management System*. Department of Industrial and Systems Engeneering. The Ohio State University.

Sarter, N.D. and D.D. Woods (1992). *Pilot Interaction with Cockpit Automation II: An Experimental Study of Pilots' Model and Awareness of the Flight Management System*. Department of Industrial and Systems Engeneering. The Ohio State University.

Schäfer, D. (1994). Entwicklung eines Expertensystems zur wissensbasierten Gestaltung von Cockpitinstrumenten. Technical report no. 94/5. Institute of Technical Computer Science. Aachen.

Smith, J.B., D.K. Smith and E. Kupstas (1991). *Automated Protocol Analysis: Tools and Methodology*. University of North Carolina at Chapel Hill.

EMPIRICAL MODELLING OF PILOT'S VISUAL BEHAVIOUR AT LOW ALTITUDES

A. Schulte* and R. Onken*

*Universität der Bundeswehr München, Institut für Systemdynamik und Flugmechanik, Werner-Heisenberg-Weg, 85577 Neubiberg, Germany. E-mail: ld1baxxl@rz.unibw-muenchen.de

Abstract. The paper presents an experimental approach to establish empirical models of pilot's visual behaviour in certain low-level flight situations. The experimental design mainly includes the presentation of typical flight mission segments to pilots as video replay and the measurement of their eye movements in the visual out-of-the-cockpit scene. The experiments were carried out with pilots who had low-level flight experience. The pilot's subjective evaluations concerning the realism of the flight replay and their task assignment are positive throughout. Objective evaluations lead to the elicitation of dependencies between low-level flight situations and the pilot's visual perception strategies.

Key Words. Aerospace engineering; Artificial intelligence; Behavioural science; Fuzzy modelling; Intelligent knowledge-based systems; Knowledge acquisition; Self-organizing systems; Simulators; Vision.

1. INTRODUCTION

In low-level flight pilot's heavily rely upon visual information from the out-of-the-cockpit visual scene. Altering flight situations characterized by task and scene contents require different strategies of visual perception in order to resolve the task. These perceptual skills can be aquired in flight training. (Richards and Dismukes, 1982) For some reasons it might be desirable to shift parts of training to flight simulators. In order to reinforce correct behaviour patterns with respect to transfer of training the simulator visual system should allow the utilization of those visual strategies pilot's using in real flight (Beckett, 1992). Today flight simulator visual system design is based upon investigations concerning psychophysical properties of the human visual system (Padmos and Milders, 1992), flight performance measurements (Lintern *et al.*, 1989; Wolpert, 1988; e.g.), and pilot's ratings (Kleiss, 1990; e.g.). Not only Padmos and Milders (1992) pointed out the research needed for a better understanding of task specific perceptual behaviour.

This paper presents a perspective of how to incorporate eye movement measurements as an indicator of mental perceptual processes in the design of human operator behavioural models and the design of human/machine interaction on the basis of the operator model. Fields of application might be visual system design as pointed out above and the use in expert systems for pilot assistance (Onken, 1994).

2. EXPERIMENTAL DESIGN

2.1. Approach

Pilot's visual routines depend on the flight situation. Schulte and Onken (1994) investigated examples in the domain of simulator training. So, the characteristics of visual scanning will change under different situational conditions. In this context the pilot's input situation (see Fig. 1) is described by a set of features. The features of the situation are the visual stimuli from the cockpit outside view and the displays, mechanical cues from the aircraft movement and the control actions, the pilot's task assignment, and his physical and mental resources conditions. The pilot is assigned to the flight tasks by a preflight briefing and inflight crew communication. He reacts upon

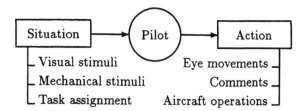

Fig. 1. Pilot as input/output system

the current situation with certain output actions such as eye movements, verbal speech output, and aircraft control operations (Fig. 1).

The aim of the investigation is to follow an experimental approach in order to establish an empirical model of pilot's visual behaviour and to prove respective hypotheses. Therefore, an experimental design had to be chosen in order to observe pilot's visual activity performing low-level flight tasks in a combat aircraft. The most straightforward way of carrying out these observations would be real flight. The experimental steps could be:

1. Briefing: A preflight briefing contains information about the general mission plan and the navigation tasks. The pilot plans the visual navigation and builds up a mental representation of navigational landmarks in order to comply with the mission goals.
2. Flight mission: After the preflight briefing the test subjects are confronted with the situational stimulus material from the out-of-the-cockpit view in low-level flight. During the experiment the test subject's eye movements are measured.
3. Debriefing: The test subject has the opportunity to give comments and make subjective ratings.

When considering this approach in more detail, the difficulties become apparent to perform eye movement measurements under reproducible experimental conditions in a combat aircraft regardless of other principal operational restrictions of low-level flights. Therefore, the experiment was designed to be conducted in a laboratory environment.

2.2. *Stimulus material*

The stimulus material is a three-camera video recording of the out-of-the-cockpit visual scene of one reference mission over the South of Germany at altitudes of approximately 250 feet ($< 100m$) above ground at 420 knots ($\simeq 220m/s$) speed in the low-level segments. The mission contains the segments Transit, Rapid Descent, Terrain Following, Visual Fix, Terrain Masking, and Target Run-in. The stimulus material is presented as video replay. Additional stimuli are crew communication and topographic maps for planning and performance of visual navigation.

2.3. *Apparatus*

In order to realize the presentation of typical flight mission segments to pilots and to conduct the measurement of their eye movements a flight replay system (see Fig. 2) was developed in the research flight simulator of the *Universität der Bundeswehr München*. The video replay is materialized on a collimated visual display. The video scene provides a field-of-view width of $135° \times 35°$. Additionally, the aircraft head-up display (HUD) and the front cockpit instrument panel (Head-

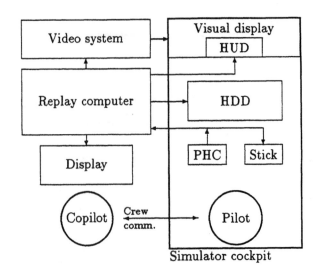

Fig. 2. Flight replay system (schematic)

down display, HDD) were reproduced by means of computer graphics based on phototextures. Mechanical stimuli are provided by the replay of the control stick displacements. Pilot's actions concerning navigation tasks are supported by a pilot's handcontroller (PHC). The copilot uses an additional display in order to cope with the flight situation for crew communication.

2.4. *Subjects*

The experiments were carried out with 15 fighter-bomber pilots (mean age = 33 years) from the German Air Force and the Royal Air Force as test subjects with low-level flight experience on combat aircrafts (mean hours total flying time = 1305).

2.5. *Eye movement measurement*

The eye movements of the test subjects were measured with the NAC EMR-600 eye mark recording equipment during the replay. The sensors for corneal reflection measurement (Young and Sheena, 1975) are mounted in a head unit worn by the test subject. The system provides a video recording of the visual scene viewed at by a head mounted NTSC camera. The eye marks of both eyes are superimposed on the video image and also available as digital data. In order to minimize errors due to parallax and vergence eye movement while accommodating different viewing distances the optical calibration is done with the virtual image of the calibration frame, provided by the collimation optics, to achieve identical viewing distances while calibration and flight replay measurement.

Further analysis of the eye movement data is done in off-line procedures with the aim to extract features such as local distribution of fixations, fixation duration and frequency, type of fixated scene elements, etc. In order to extract the desired measures from the EMR-600 data stream basically three processing steps have to be conducted:

1. Errors in measurement have to be reduced.
2. The eye marks must be transformed in a head motion independent description.
3. The fixated scene elements must be derived from the eye mark's position in the scene.

The following subsections deal with some special considerations of the postprocessing of eye movement data. They might be skipped by readers with less interest in technical details without further loss of understanding.

2.5.1. *Error reduction.*

Due to movements of the head unit relative to the test subject's head and other influences the eye mark position tends to drift away from the point of true fixation. The frequency of this process is far below the eye movement frequency. Other error effects are due to eye positions far from centerline. This deviation can be measured under the following conditions: (a) The test subject fixates points on a test grid, (b) the test subject comments on certain values displayed in the HUD (e.g. 'Speed is OK.': Fixation of speed readout), (c) the test subject comments on very special scene features (e.g. 'Powerline pylon in sight.': Fixation of pylon). In order to estimate the error in between the samples a backpropagation neural net (Rumelhart and McClelland, 1987) was trained in the following configuration: five input units (time, left/right eye position x/y); one hidden layer, ten units; two output units (left/right eye error). For each test subject and period not longer than 15 minutes two nets were trained, one for the errors along the x-axis (left/right eye) and the other for the errors along the y-axis. 10% data were used as test data in order to control the number of learn data presentations. The trained nets Δ_{BPN}, mapping the characteristics of error development, were used to compute a correction p_{cor} for the raw eye movement data p_{raw} according to equation 1.

$$p_{cor}(t) = p_{raw}(t) + \Delta_{BPN}(t, p_{raw}) \qquad (1)$$

The following results were obtained: (a) The ambiguity of measurement due to the distance between the two eye marks (mean distance in raw data = 23 pixels) was reduced significantly (mean distance in corrected data = 7 pixels) for fixations out-of-the-cockpit, (b) the correction of the y-error makes sense to the observer.

2.5.2. *Head motion eliminaton.*

The corrected point-of-regard positions are represented in pixel-coordinates of the head mounted EMR scene camera. Further processing requires the transfor-

mation into flight replay coordinates. Therefore, the head movement was tracked manually in the scene camera video by measuring the positions of three HUD instruments. The positions of these instruments in flight replay coordinates are known. Thus, a linear transformation of eye movement data is possible and allowed for small angular head movements. The treatment of the equations used is beyond the scope of the paper.

2.5.3. *Fixation determination.*

Without knowledge about individual test subject's eye movements the scene contents of the flight replay video recording was extracted manually. Point features and line features including edges of area features were taken into consideration as well as HUD instruments. In order to decide which visual object is fixated, the eye movement data are segmented into fixations, smooth pursuit, and saccades. An evaluation is done for each object which is visible during the whole fixation or pursuit movement between two saccades. Therefore, a set of rules is applied. The rules are based upon the consideration of the mean distance between the point-of-regard and the evaluated object, the variation of that distance during the fixation, and the direction and length of the point-of-regard's movement with respect to the object's movement during the fixation. The object with the highest score is supposed to be fixated.

3. DISCUSSION

3.1. *Subjective ratings*

Besides the objective data recording the pilots had the opportunity to give subjective ratings concerning the realism of the flight replay, their task assignment, and the acceptance of the replay system. Fig. 3 visualizes the almost positive pilot's ratings. The dots mark mean values of ratings concerning the listed topics.

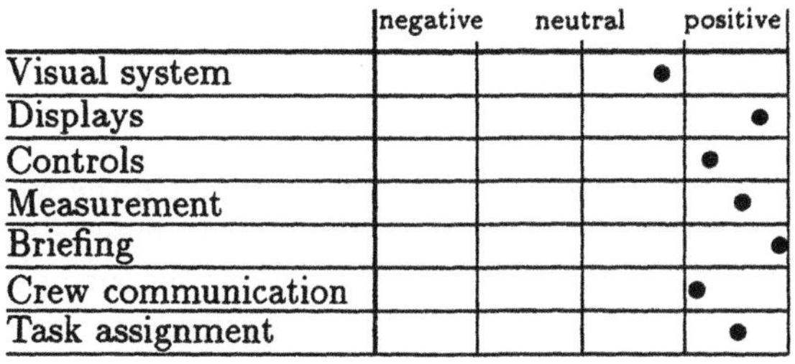

	negative	neutral		positive
Visual system			●	
Displays				●
Controls			●	
Measurement			●	
Briefing				●
Crew communication			●	
Task assignment				●

Fig. 3. Subjective ratings

3.2. *Objective evaluations*

The evaluation of the objective experimental data leads to the elicitation of distinct dependencies between the various situations during low-level flight and the corresponding pilot's actions of visual information perception.

The evaluation of the objective data is done with a special focus on the pilot's visual behaviour in two typical flight tasks corresponding with levels of human performance (Rasmussen, 1983).

1. The pilot's strategies in visual navigation for position update: rule-based behaviour.
2. The pilot's head-up scanning: skill-based behaviour.

3.2.1. *Navigation task.*

Concerning the navigation task, it was found that a set of rules exists leading pilots to choose certain visual features as navigation aid. During the experiments each test subject had to perform so called *Visual Fixes* (each subject four times). The Visual Fix is a specific technique for updating the aircraft's inertial navigation system. Therefore, the pilot selects a suitable landmark (fixpoint) near the planned track from a topographical map in the preflight briefing. Thereby, he elaborates a strategy for visual identification of the landmark. During the flight (flight replay) the pilot applies the strategy in order to find the selected fixpoint and to mark it with a HUD symbol controlled by the PHC.

The experiments show that test subjects tend to select almost the same fixpoints out of a great multitude of alternatives independently. Tab. 1 lists the experimental findings for each Visual Fix leg. The number of alternatives was estimated from the length of each leg (1/km) which is rather the minimum estimation. For modelling the

Table 1 Distribution of selected landmarks for Visual Fix

No.	Alternatives	Selections	Distribution
1	80	7	5, 3, 2, 1 (3×)
2	67	4	8, 4, 2, 1
3	30	9	4, 3, 2, 1 (6×)
4	26	5	10, 2, 1 (3×)

process of pilots selecting fixpoints from the map Visual Fix No. 2 was chosen. The purpose of the model is to predict the pilots' selection out of an evaluation of a digitized map of the area along the track. Fig. 4 shows the basic structure of the model. Based upon digitized feature maps and terrain elevation data the hypotheses generator constructs hypotheses for landmarks to be used as fixpoints or lead-in features (features that help the pilot to unambiguously identify the fixpoint) along the track in a given search area. The knowledge is represented in a rule base: Candidates for fixpoints are (a) point features, (b) bends of line

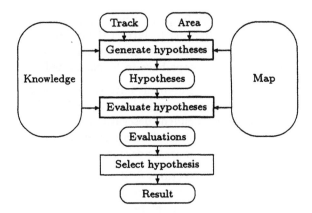

Fig. 4. Structure of the Visual Fix predictor

features, (c) corners of area features, (d) intersections of line features, and (e) intersections of line features with area features. Each of these hypotheses is evaluated with respect to the knowledge base. The hypothesis with the highest evaluation score is selected. The knowledge for the hypotheses evaluation is represented as an evaluation diagram as successfully used by Prévôt and Onken (1993) for planning tasks in instrument flight. Figures 5 and 6 show the evaluation diagrams for

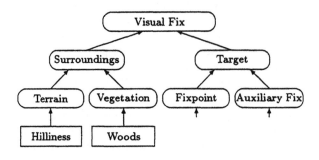

Fig. 5. Evaluation diagram for Visual Fix planning

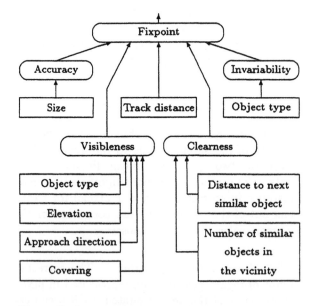

Fig. 6. Evaluation diagram for fixpoint

146

the Visual Fix planning. The evaluation of each node is calculated as function of the subnodes' evaluations taken out of the interval [0, 1]. Each path of the diagram can be traced down to a terminating node (represented by rectangles). The values of the terminating nodes are determined by functions of the map features with respect to the fixpoint hypotheses. These functions are membership functions according to fuzzy set theory (Zadeh, 1965) in most cases. Tab. 2 shows the results of a digital simulation of the model for Visual Fix No. 2 (126 fixpoint hypotheses, up to 5 auxiliary fix hypotheses per fixpoint hypothesis, mean score = 0.47). With Tab. 2 and Fig. 7 it is obvious that

Table 2   Simulation results compared with experimental findings for Visual Fix

No.	Fixpoint	Pilots	Model
1	Street/railway intersection	8	0.94
2	Bridge over motorway	4	0.86
3	Factory near village	2	0.94
4	Ferry station in river bend	1	0.32

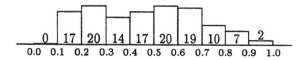

0	17	20	14	17	20	19	10	7	2

0.0 0.1 0.2 0.3 0.4 0.5 0.6 0.7 0.8 0.9 1.0

Fig. 7. Distribution of scores

the fixpoints No. 1-3 (pilots' most frequent selections) achieved the highest evaluations by the digital simulation of the model. Fixpoint No. 4 was debriefed by the particular pilot as having been a faulty selection.

3.2.2. *Head-up scanning.*   Scanning the visual scene for aircraft stabilization tasks is found to be a stereotype visual routine. It seems to be almost independent from situational features except the display symbology. For modelling aspects of skill-based scanning process Kohonen's neural self-organizing maps (SOM) (Kohonen, 1987) are chosen. The algorithm consists of the following steps:

1. initialize weights,
2. randomly select sample from input space,
3. select best matching weight vector,
4. apply activation function on similarity matching winner and neighbours,
5. modify activated weights with current input vector,
6. decrease activation function and neighbourhood size and continue at 2.

The input vectors were sampled from the experimental eye movement recording data. It consists

of the components: (a) point-of-regard x position (metric), (b) point-of-regard y position (metric), (c) fixation duration (metric), (d) fixated viewing area (non-metric: HUD, left/right top (LT/RT), left/center/right bottom (LB/CB/RB), left/right out (L/R), HDD), and (e) task (non-metric: Terrain Following, Visual Fix). 473 samples were taken. In order to process non-metric data steps 3 and 5 of the Kohonen algorithm had to be modified. Figures 8-11 show the results of digital simulation (Input vector presentations total number = 5000; Processing units = 50×50; Activation function: $\alpha(t) = 0.3 \cdot e^{-0.0001 \cdot t}$; Neighbourhood shrinking function: $N_c(t) = 75 \cdot e^{-0.001 \cdot t}$; Lateral interaction: triangle function; Neighbourhood geometry: circle).

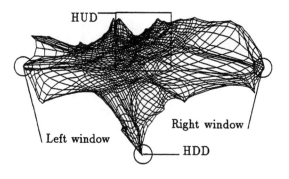

Fig. 8. SOM of visual field (Visual Fix)

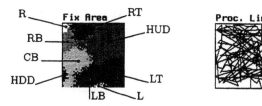

Fig. 9. SOM component map fixation area; Process line. (Visual Fix)

Fig. 8 visualizes the visual field in the central window as used for the Visual Fix navigation task. Fixations in the left/right window and HDD are reduced to one point each. Within these areas no eye movement measurement was conducted. The SOM weight vectors show a fairly regular density over the central visual field. In Fig. 9 this finding is represented by the nearly equal sized areas in the component map of the fixation area. So, during Visual Fix no special area is preferred for fixations. The process line (see Fig. 9) is built by lines connecting the SOM's best representatives for successive input vectors. For the Visual Fix task no preferred process line segments could be detected.

Figures 10 and 11 show the corresponding diagrams for the Terrain Following task. The usage

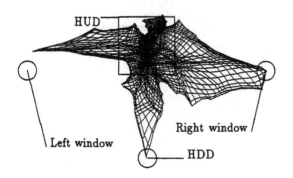

Fig. 10. SOM of visual field (Terrain Following)

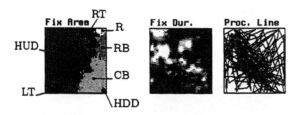

Fig. 11. SOM component map fixation area; Fixation duration (light grey: longer, dark grey shorter); Process line. (Terrain Following)

of the visual field is dominated by a strong concentration on the HUD field-of-view and the horizon. Fixations in the HUD are longer than those outside. So, the pilot concentrates mostly in the direction straight ahead. The process line (Fig. 11) shows frequent jumps in the HUD (Radar altimeter $\leftrightarrow$ Aircraft symbol $\leftrightarrow$ Heading readout). So, a stereotype scanning behaviour could be detected.

The incorporation of more situational features in the input vectors in order to describe the visual processes more detailed is planned for future works.

4. CONCLUSIONS

The results suggest that the visual behaviour of pilots differs between different low-level flight tasks. Certain characteristics of the behaviour could be incorporated in suitable operater models. Differences between the pilot's behaviour in a navigation task and a stabilization task are discussed. The pilot's selection of landmarks for navigation is described in detail. The use of such models is suggested to be in a straightforward way as predictor of visual fixations for flight simulator visual system design. A more sophisticated application could be the use in expert systems for the recognition of pilot's intents based upon inflight eye movement measurement. A considerable amount of research work is needed in the application domains.

5. REFERENCES

Beckett, Peter (1992). Effective cueing during approach and touchdown - Comparison with flight. In: *Piloted Simulation Effectiveness*. AGARD Conference Proceedings 513.

Kleiss, James A. (1990). Terrain visual cue analysis for simulating low-level flight: a multidimensional scaling approach. In: *IMAGE V Conference*. Phoenix, Arizona.

Kohonen, Teuvo (1987). *Self-organization and Associative Memory*. Springer Series in Information Sciences 8. Springer. Berlin Heidelberg.

Lintern, Gavan, Daniel J. Sheppard, Donna L. Parker, Karen E. Yates and Margaret D. Nolan (1989). Simulator Design and Instructional Features for Air-to-Ground Attack: A Transfer Study. *Human Factors* 31(1), 87–99.

Onken, Reiner (1994). Basic Requirements Concerning Man-Machine Interactions in Combat Aircraft. In: *Workshop on Human Factors/Future Combat Aircraft*. Ottobrunn, Germany.

Padmos, Pieter and Maarten V. Milders (1992). Quality Criteria for Simulator Images: A Literature Review. *Human Factors* 34(6), 727–748.

Prévôt, Thomas and Reiner Onken (1993). Onboard Interactive Flight Planning and Decision Making with the Cockpit Assistant System CASSY. In: *HMI-AI-AS*.

Rasmussen, Jens (1983). Skills, Rules, and Knowledge; Signals, Signs, and Symbols, and other Distinctions in Human Performance Models. *IEEE Transactions on Systems, Man, and Cybernetics* SMC-13(3), 257–266.

Richards, Whitman and Key Dismukes (1982). Vision Research for Flight Simulation. Technical Report TR-82-6. AFHRL. Williams AFB, AZ.

Rumelhart, David E. and James L. McClelland (1987). *Parallel Distributed Processing. Explorations in the Microstructure of Cognition*. MIT Press.

Schulte, Axel and Reiner Onken (1994). Knowledge engineering in the domain of visual behaviour of pilots. In: *Eye Movement Research: Mechanisms, Processes and Applications* (John M. Findlay, Robert W. Kentridge and Robin Walker, Eds.). Elsevier. North Holland.

Wolpert, Lawrence (1988). The active control of altitude over differing texture. In: *Proceedings of the Human Factors Society 32nd Annual Meeting*.

Young, Laurence R. and David Sheena (1975). Survey of eye movement recording methods. *Behavior Research Methods & Instrumentation* Vol. 7(5), 397–429.

Zadeh, L.A. (1965). Fuzzy sets. *Information and Control*.

AUTONOMY, AUTHORITY, AND OBSERVABILITY: PROPERTIES OF ADVANCED AUTOMATION AND THEIR IMPACT ON HUMAN-MACHINE COORDINATION

Nadine B. Sarter and David D. Woods

Cognitive Systems Engineering Laboratory
The Ohio State University

Abstract: In a variety of domains, problems with human-machine coordination are a matter of considerable concern. These problems are the result of a mismatch between human information processing strategies and the communication skills of automated systems. Results of recent empirical research on pilot interaction with advanced cockpit automation on the Airbus A-320 illustrate how problems with man-machine coordination are changing in nature in response to the ongoing evolution of modern systems from passive tools to active yet 'silent' agents. The highly autonomous and powerful nature of modern technology in combination with low system observability fails to support operators' expectation-driven approach to monitoring.

Keywords: Automation, Human error, Human-centered design, Monitoring, Supervisory Control

1. INTRODUCTION

In a variety of domains, the introduction of advanced automation technology has led to unexpected problems with overall system performance. Many of these problems are the result of breakdowns in the communication and coordination between man and machine. One of the major difficulties is the failure of operators to track, anticipate, and understand the status and behavior of the automation -- a lack of situation, or more specifically, mode awareness which can lead to mode errors and 'automation surprises' (Sarter and Woods, 1994, in press). A lack of mode awareness can be viewed as the consequence of a mismatch between required and available feedback on system behavior, between operators' monitoring strategies and the communication skills of the automation. Operators are known to monitor automation based on a mental model of its functional structure. This model provides the basis for forming expectations about system behavior which, in turn, guide operators' attention allocation across information-rich system displays. This expectation-driven monitoring approach is likely to be affected by recent trends in automation design. Advanced automated systems are increasingly autonomous in the sense that they are capable of carrying out long sequences of actions without requiring operator input

once they have been programmed and activated. They are also powerful as they are capable under certain circumstances to override or modify operator input. These capabilities have changed the role of modern systems from a passive tool to an active agent in the overall system. As a consequence, effective communication and coordination between humans and advanced automated systems has become more critical. To support the increased need for coordination, systems would need to become more transparent. Feedback design would need to be improved to support the human operator in keeping track of and in anticipating the status and behavior of his machine counterpart. A look at modern systems suggests, however, that the development of effective feedback has not kept pace with the demand for it. Few changes to feedback design have been made, and it is not clear whether those changes result in improved, stagnant, or even decreased system observability, thus further increasing the gap between required and available feedback.

One area where the trend towards 'strong and silent' machine agents is ongoing is the aviation domain. Empirical research on pilot interaction with one of the most advanced automated aircraft currently in operation was carried out to study the impact of the above described trends in automation design. The results of this research indicate that 'automation surprises' and mode errors -- problems that had already been identified on earlier generation 'glass cockpit' aircraft -- still exist on highly advanced flight decks. However, their nature and circumstances have changed as a consequence of new system properties and capabilities.

2. PILOT INTERACTION WITH HIGHLY ADVANCED COCKPIT AUTOMATION

One of the areas where automation technology is rapidly evolving is the aviation domain. Even with early generation flight deck automation, pilots experience problems with tracking and anticipating system status and behavior (see Wiener, 1989; Eldredge, Dodd, and Mangold, 1991; Sarter and Woods, 1992, 1994). Their lack of system or mode awareness results in 'automation surprises' and mode errors where the pilot takes an action that is not appropriate to achieve his goals given the actual (not the perceived) status of the system. Mode errors and 'automation surprises' are the result of a mismatch between the communication skills and requirements of human and machine. Changes in system design can therefore be expected to have an impact on the nature and severity of difficulties.

To study the effects of recent trends in automation design -- higher system autonomy and authority without the required parallel improvement in system observability --, a line of research on pilot-automation coordination on the most advanced automated aircraft currently in operation, the Airbus A-320, was carried out. This research involved observations of transition training and line operations of the airplane. A survey of A-320 pilots was conducted to gather information on pilots' experiences with advanced cockpit automation and to analyze reasons and circumstances of reported difficulties. Based on the results from these activities, an experimental simulation study of human-automation communication and coordination was carried out. Eighteen experienced A-320 pilots flew a 90-minute scenario on a full-mission simulator. The scenario for this study was designed to probe pilot-automation coordination in situations where a mismatch between pilots' expectations of system behavior and actual system behavior was likely to occur. The scenario events and tasks were instantiations of situations where the automation either carried out an unexpected action or failed to do what was expected by the pilot. Given the expectation-driven nature of system monitoring, these mismatches were expected to result in 'automation surprises' and mode errors.

Behavioral and verbal data were collected on pilots' mode and overall system awareness. In addition, their ability and strategies to cope with 'automation surprises' and to prevent or recover from mode errors were studied.

Converging evidence from the above research activities indicates that 'automation surprises' and mode errors still occur in the context of highly independent and powerful automation. Compared to results from studies of pilot interaction with earlier generation cockpit automation (Sarter and Woods, 1992, 1994), there seems to be an increase in the occurrence of a particular kind of mode error. Mode errors on less advanced automated flight decks tend to be errors of commission where a pilot action is the necessary precursor for a problem to evolve. Such errors still occur on highly advanced aircraft. In addition, however, errors of omission seem to occur more frequently. Here, a problem develops because of the pilot's failure to act, or more precisely, to detect and intervene with undesired system behavior that was not directly commanded by him.

As indicated by this research and recent incidents and accidents (e.g., Lenorovitz, 1990; Sparaco, 1994), mode errors of omission are on the increase as a consequence of increasingly independent and powerful systems that fail to keep their users informed about their status and behavior.

3. CONTRIBUTING FACTORS TO BREAKDOWNS IN HUMAN COORDINATION WITH ADVANCED AUTOMATION

In the following paragraphs, system properties will be discussed that contribute to the observed breakdowns in the coordination between humans and modern technology. These properties are related to the high level of system autonomy, authority, and complexity. As discussed in the introduction, none of these properties is problematic per se but can result in problems when combined with low observability because the need for coordination is increased while, at the same time, the opportunity for it is being reduced (see figure 1).

The monitoring of an automated system requires an adequate mental model of the functional structure of the system. A mental model allows for the formation of expectations of system responses to input from various sources. In that sense, the model is a necessary prerequisite for the timely allocation of attention to critical information, and it guides the exploration of subsets of the usually large amount of available data (Carroll and Olson, 1988).

There are several factors associated with modern automation design that are not fully compatible with this expectation-driven approach to system monitoring. In the following paragraphs, some of these properties such as a high degree of coupling, system reactivity to input from several different sources, and inconsistencies in system design will be discussed.

3.1 Mismatches Between Expected and Actual System Behavior Related To

...*Input from Multiple Sources.* One problem with an expectation-driven monitoring approach in the context of advanced automation technology is the

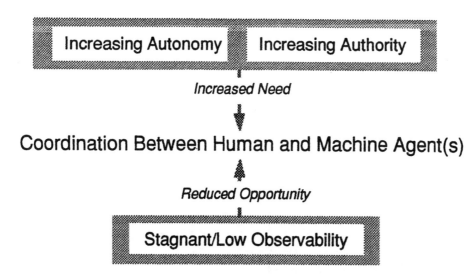

Fig. 1. Opposing Trends in Automation Design and Their Impact on Human-Machine Coordination.

associated requirement to know about all input to the system to be able to anticipate and monitor for transitions in system behavior. Most advanced automated systems can change their status and behavior in response to input from a variety of sources. Multiple users can enter data and issue commands to the system. Sensor input indicating that a predetermined target is being reached or deviated from or that a predefined envelope is being violated can also result in transitions in system behavior. Such sensor input or input by other users is not necessarily accessible by the system user (see Segal, 1990). As a consequence, he may not expect the consequences of the unobserved system input, and he is likely to miss its effect(s) unless the feedback provided by the system is salient enough to attract his attention to ongoing or imminent transitions.

Examples of changes in system behavior on the flight deck of advanced automated aircraft that occur without immediately preceding operator input are the transition to a different automation mode upon reaching a pre-entered target altitude or the activation of the so-called Alpha-Floor protection mode once an exceedingly high angle of attack is detected by the sensors and communicated to the automation.

...Related To System Coupling. The high degree of coupling of many advanced automated systems provides another opportunity for breakdowns in mode awareness. Coupling refers to the fact that operator input to these systems can have more than the expected and therefore monitored consequences. It can also have side effects or 'effects at a distance' which may not be expected nor desired by the operator (Woods, 1988).

Examples of such coupling effects on the flight deck of advanced automated aircraft are the loss of pilot-entered altitude or speed restrictions after changing a pre-selected runway or approach, or the coupling of lateral and vertical navigation modes where the selection of a new desired lateral mode automatically results in the transition to a new vertical mode independent of the pilot's intentions.

Such unintended side effects of operator input are particularly troublesome as they are difficult to detect. The operator who anticipates and looks for only part of the set of consequences of his input may be satisfied to see his expectations fulfilled. This partial overlap of expectations and actual system status is likely to prevent further search for changes.

...Related To Inconsistent System Design. Another contributor to breakdowns in mode awareness are inconsistencies in the design of a system. Operators form expectations of system behavior based on their mental model of the system's functional structure. Such a mental model can be thought of as a representation or collection of consistent rules underlying system behavior. If, under certain circumstances, a system acts in a way that is not consistent with the user's model, an 'automation surprise' is the result, and it can be very difficult or impossible for the operator to explain what happened and how the automation behavior will continue to develop. The same problem can occur if the required input for a desired outcome is not consistent across tasks or circumstances.

The following figure illustrates how the above discussed factors -- system coupling, multiple inputs to systems, and inconsistent system design -- can contribute to unexpected and possibly undesired system behavior. It is these factors that played a major role in the occurrence of observed difficulties with pilot-automation interaction in the above study.

4. CONCLUSION

Problems with human-automation interaction are the result of a mismatch between the abilities and limitations of both human and machine (Billings, 1991). As technology keeps evolving and changes in nature, the opportunity for new kinds of errors is sometimes created. One recent example in the aviation domain is the incompatibility of the current trend towards highly autonomous and complex yet 'silent' cockpit automation with pilots expectation-

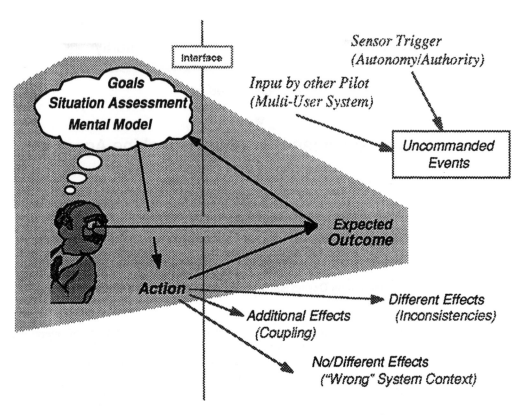

Fig. 2. Challenges to Expectation-Driven Monitoring of Advanced Automated Systems.

driven monitoring strategy. This example illustrates that the term 'automation' does not refer to a group of homogenous system. The impact of different trends in automation properties and design on human-machine interaction needs to be analyzed to be able to develop specific countermeasures.

5. ACKNOWLEDGMENT

The preparation of this document was supported, in part, by NASA-Ames Research Center under Cooperative Agreement NCC 2-592. The authors would like to thank the technical monitor of this work, Dr. Everett Palmer.

6. REFERENCES

Billings, C.E. (1991). *Human-centered Aircraft Automation Philosophy* (Technical Memorandum 103885). NASA-Ames Research Center, Moffett Field, CA.

Carroll, J.M. and J.R. Olson (1988). Mental models in human-computer interaction. In: *Handbook of Human-Computer Interaction* (M. Helander, Ed.), pp. 45-65. Elsevier Science Publishers, New York, NY.

Eldredge, D., R.S. Dodd and S.J. Mangold (1991). *A Review and Discussion of Flight Management System Incidents Reported to the Aviation Safety Reporting System*. Battelle Report, prepared for the Department of Transportation, Volpe National Transportation Systems Center, Cambridge, MA.

Lenorovitz, J.M. (1990). Indian A320 crash probe data show crew improperly configured the aircraft. *Aviation Week & Space Technology*, **132** (6/25/90), 84-85.

Sarter, N.B. and D.D. Woods (1992). Pilot Interaction with cockpit automation: Operational experiences with the Flight Management System. *The International Journal of Aviation Psychology*, **2** (4), 303-321.

Sarter, N.B. and D.D. Woods (1994). Pilot interaction with cockpit automation II: An experimental study of pilots' model and awareness of the Flight Management System. *The International Journal of Aviation Psychology*, **4** (1), 1-28.

Sarter, N.B. and D.D. Woods (in press). "How in the world did we ever get into that mode?" Mode Error and Awareness in Supervisory Control. *Human Factors*.

Segal, L.D. (1990). Effects of Aircraft Cockpit Design on Crew Communication. In: Contemporary Ergonomics 1990 (E.J. Lovesey, Ed.). Taylor and Francis, U.K.

Sparaco, P. (1994). Human factors cited in French A320 crash. *Aviation Week and Space Technology*, (1/3/94), 30.

Wiener, E.L. (1989). *Human Factors of Advanced Technology ("Glass Cockpit") Transport Aircraft* (NASA Contractor Report No. 177528). NASA-Ames Research Center, Moffett Field, CA.

EXPERIMENTAL STUDY OF VERTICAL FLIGHT PATH MODE AWARENESS

Eric N. Johnson & Amy R. Pritchett

Graduate Research Assistants, MIT Aeronautical Systems Laboratory

Abstract: An experimental simulator study was run to test pilot detection of an error in autopilot mode selection. Active airline air crew were asked to fly landing approaches by commanding the *Flight Path Angle* mode while monitoring the approach with both a Head Up Display and Head Down Displays. During one approach, the *Vertical Speed* mode was intentionally triggered by an experimenter instead, causing a high rate of descent below the intended glide path. Of the 12 pilots, 10 did not act to decrease the high descent rate prior to significant glide path deviation.

Keywords: Automatic Operation, Automation, Human Factors, Human Supervisory Control, Man/Machine Interaction, Aircraft Simulators

1. INTRODUCTION

Loss of pilot awareness about the commanded autopilot descent modes is a speculated or reported cause in several recent incidents involving Airbus A320 aircraft. With the new autopilot systems of this aircraft, the pilot can command several vertical flight path modes, including a specified Flight Path Angle or a specified Vertical Speed. These two modes share the same selector knob and display, have only a simple push-button toggle to switch between them, and have similar mode indicators. Therefore, the presentation and selection of these modes combined with the potentially severe consequences of an error generate several serious questions about the supervisory control task required of the pilot by these new systems.

1.1 Motivation

On January 20, 1992, an A320 aircraft crashed during a non-precision approach into Strasbourg airport, killing 86 passengers. The descent rate of the aircraft has been estimated to be 3300 fpm, resulting in impact with high terrain. This differs dramatically from the descent detailed in the approach plate; the aircraft should have followed a gradual 'step-down' approach, which can also be approximated by a 3.3 degree descent.

It is speculated the flight crew inadvertently placed the aircraft into the wrong descent mode and did not recognize the problem during the following 47 seconds up to impact. This problem with mode awareness may have been influenced by the command pilot's likely primary flight reference, a Heads Up Display (HUD) with no mode annunciation, and by the lack of a Ground Proximity Warning System (GPWS).

Two other incidents involved the same confusion about descent mode during approach. Fortunately, the errors were recognized in time to prevent ground impact and were later reported by the pilots of the aircraft.

1.2 Background

In modern air carrier aircraft, the aircraft trajectory can be controlled in three ways: manual control by the pilot, automatic flight guidance over a specified route by a Flight Management System, and selected autopilot control over specified aircraft states. This selected autopilot control is commanded by setting specified 'Modes'; for the vertical flight path these modes include Altitude Hold, Altitude Select, Vertical Speed, Approach, and Missed Approach. In some of the new air transport aircraft, a new mode, Flight Path Angle, is also available.

The primary display and selectors of autopilot modes and their target values are presented on the Flight Control Unit (FCU). This instrument is located in the center console on the glare-shield, within reach of both the captain and the flight officer. The four rotary knobs can be used to select the desired state values of Speed, Heading or Track, Altitude, and Vertical Speed or Flight Path Angle. To select autopilot control over a particular state, the pilot sets the desired state value and then pulls on the knob to command the mode governing that state. An additional recessed push-button on the FCU toggles between the modes Vertical Speed or Flight Path Angle.

With the original A320 FCU, a selected vertical speed or flight path angle would both be shown as two digits, with a plus or minus sign indicating a climb or descent respectively. For example, descents at 3200 fpm or at 3.2 degrees would be depicted as '-32'. A retrofit modifies the display to show four digits in the case of a commanded vertical speed, and to put a decimal between the two digits shown with a Flight Path Angle.

When a mode is selected on the FCU, its target value is shown in its respective window. In addition, a text annunciation "VS" is shown on the FCU in white if Vertical Speed is commanded. If Flight Path Angle is selected, an "FPA" is shown, also in white, in the same area. The current autopilot modes are also annunciated at the top of each pilot's Primary Flight Display (PFD).

The pilot in the Strasbourg accident had a Heads Up Display (HUD) available for use as his primary flight reference. Mounted on top of the instrument panel, this display presents essential state information to the pilot while also allowing the pilot to see, through its glass surface, the forward out-the-window view. However, the HUD does not display any of the autopilot mode annunciations.

The pilots' vertical flight path mode awareness may be determined by several different factors. First, the pilot is responsible for determining each autopilot mode through FCU selectors. Next, the selected mode can be displayed through explicit autopilot mode annunciation, and by displays which illustrate the current and target values of the aircraft states. Finally, the attention the pilot can dedicate to supervising the autopilot can affect his/her awareness of the commanded autopilot modes. The Vertical Speed and Flight Path Angle modes are often commanded during Terminal Area operations and on Final Approach. These phases of flight have a high pilot workload and require frequent changes in the aircraft guidance commands. On a non-precision approach, the pilot must also be concerned with obtaining visual contact with the runway, requiring significant 'Heads Up' time. Therefore, on Final Approach the pilot's available attention to supervisory roles may be limited.

1.3 Objectives

To examine the issues of vertical flight path mode awareness, the objectives of this study were:

1) To examine the length of time required by pilots to recognize both
- a problem with the aircraft state (excessive vertical speed)
- the cause of the problem (incorrect autopilot mode), and

2) To determine the primary and secondary display cues the pilots used for vertical flight path mode awareness.

In order to accomplish these objectives, a part-task simulation was developed on the MIT Advanced Cockpit Simulator. This simulation was used to examine an inadvertent selection of Vertical Speed when a desired Flight Path Angle was intended, roughly modeling the situation speculated in the Strasbourg accident.

2. EXPERIMENT DESIGN

Using the MIT Advanced Cockpit Simulator (ACS), active airline pilots flew a series of non-precision approaches with reference to a Heads Up Display (HUD), a Flight Control Unit (FCU) and other instruments. During subject briefing, the pilots were told the use of the Flight Path Angle autopilot mode was being tested to allow for lower minimums in non-precision approaches. They were asked to command the flight path angle, supervise the approach and complete the approach manually when visual contact was made with the runway. During the fourth approach, an experimenter, acting as Pilot Not Flying, commanded Vertical Speed mode instead, and the pilots' recognition time of the excessive vertical rate and its cause was recorded. During a debriefing, the pilots were also asked for subjective comments about autopilot modes and their display.

This section first details the setup of the MIT ACS and its displays. Then the procedures followed during the experiment are described.

2.1 Simulator Setup

The MIT ACS is based upon two Silicon Graphics workstations providing both the graphics emulating the cockpit displays, and the computation to simulate the aircraft dynamics and to drive the ancillary controls. The aircraft has the level of performance approximating a Boeing 737. A side-stick was provided for manual control.

A Head Up Display (HUD), based on the Flight Dynamics HUD, was situated in front of the pilot. Although it differs somewhat from the A320 HUD, it presents similiar information. A radar altimeter indication was shown at heights below 500 feet above ground and no ground proximity aural alerts were given, imitating the Air Inter A320's lack of a Ground Proximity Warning System (GPWS). The HUD's in both the 3.2 degree and 3200 fpm descents are shown in Figure 1. For the 3.2 degree descent, the flight path indicator is on the 3.2 degree position on the pitch angle scale. For the 3200 fpm descent, the flight path angle is much steeper and the numerical value of -3200 appears on the lower right.

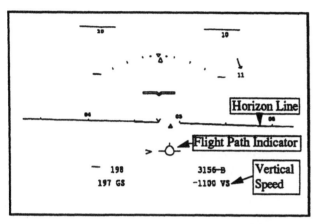

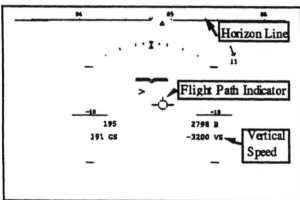

Figure 1. Heads Up Display in 3.2° and 3200 fpm Descent Modes

An out-the-window view was presented behind the HUD symbology. Until the aircraft was below 1500 feet above ground, the flight was in Instrument Meteorological Conditions (IMC). Then, a basic terrain map became visible. During the nominal approaches, an airport was shown on the ground.

All of the 'Heads Down' displays were shown on a second graphics screen, placed to the front and right of the pilot. These displays included: a representation of the Flight Control Unit (FCU), a generic Primary Flight Display (PFD) modeled after those in the Boeing 747-400 and Airbus A320, and miscellaneous indicators for gear and flaps.

The FCU displayed all annunciations and selected values in the same manner as the A320, but none of the selectors or dials could be used. In order to change the modes and values commanded by the FCU, the pilots were told to call out the settings they wanted selected and an experimenter, acting as Pilot Not Flying, entered them via a keyboard. The FCU is

shown in Figure 2 for descents of -3.2° and 3200 feet/min.

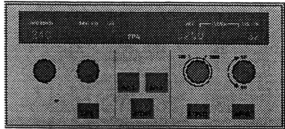

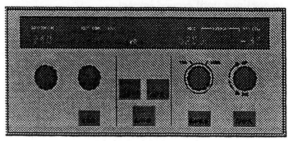

Figure 2. Flight Control Unit (FCU), Set for ' Descents of 3.2° and 3200 fpm

2.2 Experiment Procedure

The 12 subjects were current airline pilots. They were briefed before the flights on the displays and controls of the simulator. Especial care was taken in explaining the HUD, including the speed, pitch, altitude and vertical speed indicators, and in describing the modes available on the A320 and their differences from the subject pilot's normal aircraft.

The subjects were then told that the objectives of the study were to: Test use of Heads Up Display (HUD) and new autopilot systems for non-precision approaches in low visibility. They were told to execute a sequence of five final approaches in low visibility conditions by commanding the autopilot to follow a localizer and a specified flight path angle. After making visual contact with the runway they were to take manual control of the aircraft. During the briefing they were told that the point at which they took manual control was the metric of interest.

At the start of each approach, the aircraft was two miles outside of the Final Approach Fix (FAF). Therefore, the start of the approach involved a high workload, during which the pilot had pre-select the required flight path angle, trigger the autopilot mode at the FAF, and start commanding appropriate landing speeds, flap positions and gear extension. Once pilots were established on the approach and in landing configuration, pilot workload drops and subjects had ample time to monitor the approach and search for the runway through the HUD.

The experiment objectives given to pilots were deliberately misleading. By having an emphasis on visual contact with the runway, it was hoped their attention would be on the HUD and 'out-the-window', rather than fixating on the FCU and other 'heads down' displays.

Instead of flying the five approaches for which they were initially briefed, the pilots flew only four. During the fourth approach, the erroneous Vertical Speed mode was triggered by an experimenter, acting as Pilot Not Flying.

Once the pilot recognized the severe descent of the vertical speed mode approach and took any action to change it, the time of recognition was recorded. If the pilot took no action the descent was allowed to continue until ground impact.

After the last approach, pilots were asked several questions about their recognition, or lack of recognition, of the extreme descent and its cause. Other data recorded included the aircraft state data throughout the run, and a summary of the pilot's background information.

3. RESULTS

3.1 Subjects

Twelve air carrier pilots participated in this study. Ten flew larger transport aircraft such as the MD-80 or Boeing 767; the remaining two flew commuter aircraft with glass cockpits. Seven were captains and five were first officers, with an average of 9800 flight hours (the least experienced had 2200 hours, the most had 18,500 flight hours). None of the pilots had flight experience in the Airbus A320 or had flown HUD equipped aircraft.

3.2 Recognition of Severe Descent and Incorrect Descent Mode

One pilot of the twelve immediately noticed the extreme pitch down of the aircraft on the HUD. One pilot safely took manual control after the aircraft had descended approximately 1500 feet. Six pilots took control after a descent of approximately 2800 feet, when the aircraft was approximately 500 feet above ground level and the descent of 3200 fpm was well established, which would have resulted in ground impact in the Strasbourg accident. Four of the pilots did not take action before the aircraft impacted the ground as it was set in the simulation. These results are shown pictorially in Figure 3.

The depicted altitudes indicate only the points at which the pilots were aware that a serious problem existed with the aircraft configuration and took action. The pilots' actions in all but one case were to take manual control of the aircraft; the subject who immediately recognized the incorrect mode needed only to request that the mode be corrected. Questioning revealed that pilots who took manual control were reacting to the severe descent, the cause of which they had not been able to identify. The pilots intended to stabilize the aircraft manually and then attempt to ascertain the cause of this severe descent.

All of the pilots, even those who did not take any action, were confused and concerned by the increasing descent rate and its accompanying speed build up when in the unanticipated Vertical Speed mode. Some pilots attempted to reduce these particular aircraft states directly by requesting speed brakes and/or a somewhat shallower flight path angle.

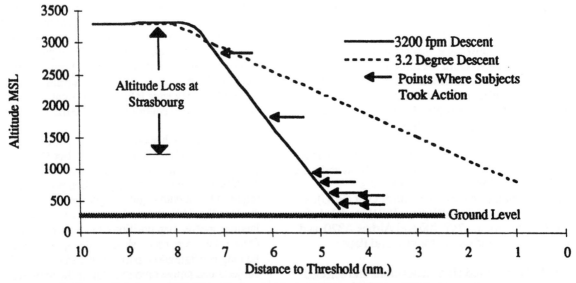

* **Four Subjects Did Not Take Action Before Impact With Terrain**

Figure 3. Altitudes At Which Pilots Took Action

3.3 Variations in Recognition Based on Pilot Characteristics

The altitude at which pilots took action has been examined for differences between pilots with different characteristics. No significant differences can be found between pilots with high and low levels of experience (as indicated by their flight hours), nor can differences be found between pilots of different ages.

Five of the twelve pilots received their initial flight training in the military. Only two of these five took action before the aircraft reached the ground, and at very low altitudes. Of the seven pilots whose initial training was not military, two pilots recognized the error at safe altitudes, four took action at low altitudes and only one did not take action. The number samples is too small to provide any statistical significance.

Six pilots identified themselves as captains; four pilots identified themselves as flight officers. Of the four flight officers, only one pilot took action (at a 'safe' altitude). Of the six captains, only one pilot did not take action. Again, there isn't enough data to provide statistical significance.

3.4 Primary and Secondary Cues of Extreme Descent

By noting what the pilots said during their simulation runs and the following debriefings, the dominant and supporting cues to the pilot of the extreme descent were identified. These did not appear to vary between the pilots who did or did not take action, nor did they vary between pilots with any other identified characteristic.

During the simulation runs, all of the pilots commented on the high descent rate. During the debriefing, five of the eight pilots who took action cited it as the strongest cue to take action. Another four pilots also cited it as a secondary cue. The indication of the high descent rate was described by six of the pilots as exceeding a certain threshold and by the other three as not conforming to a formula or cross-check they are accustomed to performing on approach (e.g. "Vertical speed should be about five times the ground speed.") One of the pilots indicated concern over the descent but then attributed it to the flight path angle mode he believed was commanded at the time.

Another cue that was commented on by most of the pilots, once the descent rate was established, was the speed buildup that occurred as the autopilot attempted to maintain the high rate of descent. In the debriefing, one pilot who took manual control cited this as the first cue of an abnormal situation; another two pilots also cited it as a supporting cue.

Once the 3200 fpm vertical speed is established, the aircraft descends at a flight path angle of approximately eight degrees. This steep descent prevents the aircraft from maintaining the commanded approach speed. With flaps fully extended, this caused the HUD's aircraft attitude indicator to drop below the flight path vector. This appearance of the HUD was cited by two pilots as the compelling cue for them to take action. Another four pilots cited the extreme nose down attitude as a supporting cue in the debriefing.

The appearance of the radar altimeter on the HUD at 500 feet AGL was cited during the debriefing by only one pilot as a supporting cue. However, seven of the pilots took action soon after this indication appeared on the HUD, suggesting that this indication was more compelling than the pilots' after-the-fact responses indicate.

Another supporting cue, mentioned by two pilots, was a cross-check of the altitude with the distance to the threshold as shown by the Distance Measuring Equipment (DME). These cross-checks were quite simple, usually a comparison between the fraction of altitude descended to the fraction of the distance covered.

The mode annunciators on the PFD or FCU were noticed by only two pilots, and only after they had taken action. Pilot comments indicate they did not feel the annunciators were compelling because of their 'heads-down' location and similar appearances for different modes.

3.5 Pilot Subjective Opinions on Mode Presentation

During the debriefing, all the pilots felt the presentation of Flight Path Angle and Vertical Speed modes could be improved. In free responses, six pilots stated that the mode annunciations should be made more distinct and identifiable. Three pilots stated the selector for these two modes should be physically separated. One suggestion was to use a different color to highlight the 'non-normal' mode, although no opinion was given about what should be the 'normal' mode.

Six pilots suggested mode annunciation or graphical cues on the HUD, although three also expressed concerns about cluttering the HUD and information overload. Two pilots suggested aural alerts for 'stupid' mode selections. One pilot suggested changes in the procedures used for selecting modes, such as calling out the mode and commanded state value, with a response from the pilot-not-flying.

3.6 Simulation Fidelity

During the debriefing, pilots were asked questions about the simulation fidelity. Eight of the pilots felt they understood the HUD and 10 of the pilots felt they understood the FCU. 10 of the pilots felt the workload in the simulation was realistic; one differing pilot felt the workload was too low, the other differing pilot felt the workload was too high. Overall, the pilots felt the simulation was realistic.

The simulation differed from normal approaches in that the subject-pilot was expected to recognize the error without relying on the experimenter-copilot. Recognition of the extreme descent and its cause may happen more quickly in a two-crew cockpit.

Finally, the pilots used in this study were new to HUD's, the mode Flight Path Angle and the A320 Mode Control Panel. Being relatively untrained in the use of Flight Path Angle mode, their recognition may have been slower than that expected from flight crew trained on these systems.

4. CONCLUSIONS

Several valuable conclusions about issues with vertical flight path mode awareness can be made. However, this experiment was not intended to examine any one aircraft's exact displays or systems; any such studies would require further simulation with exact displays and pilots trained on these the specific systems.

1) Most pilots showed a lack of awareness of the commanded descent mode and were confused by the resulting aircraft states.

All but one of the subjects allowed the aircraft to deviate significantly from the intended glide path, with ten pilots allowing the aircraft to reach altitudes where ground impact either happened or would be difficult to avoid. This indicates that pilots had a serious lack of autopilot mode and aircraft state awareness when given the displays used in the study. All of the pilots were concerned and confused by the vertical speed, pitch attitude and speed buildup that ensued from the descent, but many were reluctant to act because of confusion or a belief that these extreme states were required to maintain the expected flight path angle.

2) Pilots evaluated the condition of the aircraft by supervising aircraft states.

When flying the aircraft, the pilots monitored the aircraft states on their customary primary flight displays, rather than monitoring the commanded modes on the Mode Control Panel. This was shown by the pilots' comments during the simulations, when all of them mentioned the numerical value of the vertical speed, and most mentioned airspeed and/or altitude checks. These states were evaluated in two ways: as comparisons to allowable thresholds, such as "We should not be descending this fast"; and simple memorized rule manipulations, such as "Vertical speed should be five times the ground speed".

3) The display cues cited by the pilots and the instruments in their scan suggest study of some changes in mode presentation and pilot training.

To monitor autopilot conformance, pilots must compare between mode annunciations, commanded values selected on the Mode Control Panel, and the aircraft states shown on their Primary Flight Displays and HUD. This requires the pilot to reference several displays and compare between displays in different formats on different screens, sometimes referencing states that are not distinctly quantified (such as Flight Path Angle).

Several simple display improvements were suggested by the pilots, including: physical separation of the descent mode selectors and mode annunciations, more identifiable mode annunciations than just two or three letter identifiers, and the use of different colors for different modes.

More elaborate display improvements also warrant investigation. For autopilot modes involving altitude and speed, the commanded state values are shown graphically on the same displays as the actual aircraft states. This method of presentation reinforces the pilot's awareness of the commanded modes and their target states, and allows for easy supervision of autopilot conformance. These types of displays could also be shown for all target aircraft states, and could be included in a HUD type display without adding text annunciators.

Improvements in training and procedures were also suggested by some of the pilots. For example, pilots did not have a standard protocol for cross-checking these values with those selected on the Mode Control Panel. Also, none of the pilots cross-checked the mode that was selected by experimenter who was acting as co-pilot; such a procedure would help in quick detection of erroneous mode selection.

Finally, Flight Path Angle is a new autopilot mode that pilots are not accustomed to control manually. Only in the most recent glass cockpit aircraft is any indication of the flight path vector shown, and then only in reference to a pitch ladder on the direction indicator. This autopilot mode and any other mode referencing unfamiliar states requires additional pilot training so that they can quickly predict the underlying dynamics of the aircraft condition they are commanding, and thus monitor its conformance.

FIXIT: A CASE-BASED ARCHITECTURE FOR COMPUTATIONALLY ENCODING FAULT MANAGEMENT EXPERIENCE

Andrew J. Weiner and Christine M. Mitchell

Center for Human-Machine Systems Research
School of Industrial and Systems Engineering
Georgia Institute of Technology
Atlanta, Georgia 30332-0205

Abstract: An important requirement in the development of increasingly automated systems is a better understanding of how operators manage faults in complex systems. A system which computationally encodes operational expertise about anomalies and faults, and which potentially applies this knowledge to solve future problems or to aid to operators, is an important interim step. To support this goal, an architecture for understanding fault diagnosis in supervisory control systems is presented. The Fault Information Extraction and Investigation Tool (FIXIT) architecture uses recognition-primed decision making to describe how experts make decisions in real-world settings. Case-based reasoning is used to encode expert knowledge and experiences, with additional structure for fault management drawn from Rasmussen's strategies for state identification and diagnosis.

Keywords: Fault detection, fault diagnosis, fault identification, complex systems, human supervisory control.

1. INTRODUCTION

Fault management (i.e. detection, diagnosis, identification, and remediation) is an important aspect of human supervisory control. Sophisticated control systems are beginning to automate, or partially automate, this function (e.g., Sheridan, 1992). Prerequisite to computer-based fault management, however, is a detailed understanding of how humans carry out this function and an associated computational representation.

This research explores the development of a computational model of fault management expertise. Such a model would be useful both as an aid to operators who are faced with increasingly complex systems and as a step towards increasingly automated systems.

2. BACKGROUND

This research integrates ideas from recognition-primed decision making (Klein, 1989), case-based reasoning (Kolodner, 1993), and Rasmussen's (1981,

1984) representations and strategies for how experienced operators carry out fault management in complex systems. Figure 1 shows the theoretical underpinnings of the FIXIT architecture.

2.1 Klein's Recognition-Primed Decision Making

Klein's (1989, 1993) recognition-primed decision making offers a theory for how decision makers solve problems in real-world settings. Klein's theory states that decision making in such settings is based on the

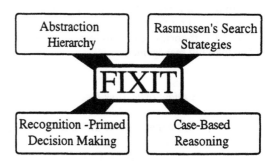

Fig. 1. The FIXIT architecture

recollection of previous experiences, which are then modified to meet the needs of the current situation. Experienced decision makers do not rely on formal models of decision making, but rather on their previous experience. They use their expertise to adapt previous experience to fit the current situation, and make decisions accordingly.

Previous fault management experiences can provide expectations for understanding and acting in similar situations (e.g., Kolodner, 1993). Rasmussen suggests that the heuristics and short cuts used by experienced operators may be "recognition-primed" (Rasmussen, 1993). This research makes a fundamental assumption that recognition-based decision making is both a behavioral characteristic of experienced operators and diagnosticians in familiar circumstances and an effective decision making strategy to underpin a normative model of fault management. Such an assumption is reasonable, for such behavior has been observed of experienced operators (Rasmussen, 1984; Sheridan, 1992).

2.2 Case-Based Reasoning

Complementing the cognitive theory of recognition primed decision making, case-based reasoning (CBR) offers a cognitive *model* describing how people use and reason from past experiences and a *technology* for finding and presenting such experiences (Kolodner, 1993; Domeshek and Kolodner, 1991). CBR uses information about previous events to guide its reasoning. This information is organized as a library of cases, or more specifically, as a 'case base'. Cases, which represent specific knowledge tied to specific situations, represent knowledge at an operational level; that is, they make explicit how a task was carried out, how a piece of knowledge was applied, and/or what strategies were used to accomplish a goal (Kolodner, 1993).

A case consists of two parts: the case body and the case indexes. The case body contains information about experienced situations in the form of text and pictures. The case indexes are combinations of descriptors which distinguish a case from other cases and uniquely identify it (Kolodner, 1993).

Both recognition-primed decision making and case-based reasoning are somewhat domain independent descriptions. Applications include design, computer programming, and command and control. Additional structure tailored for fault management is drawn from Rasmussen's theories and models of operators undertaking fault management.

2.3 Rasmussen's Diagnostic Task Description

Rasmussen (1984) suggests that the diagnostic task in supervisory control consists of two parts: selection of a search strategy for fault diagnosis; and selection of an appropriate level of description with which to represent the problem. Rasmussen (1981,

1984) proposes a set of fault identification and diagnostic search strategies from which experienced operators choose. Characterizing various useful problem representations, Rasmussen offers an *abstraction hierarchy* together with a whole part decomposition as the way in which operators characterize engineering systems (Rasmussen, 1981, 1984).

Search strategies. Rasmussen identifies two main search strategies: topographic and symptomatic. In topographic search, a template representing the normal or planned operation of the system is compared to the current system state. A fault in the system shows up as a mismatch and is identified by its location in the template. Symptomatic search starts with the observed symptoms of a fault, representing the abnormal state of the system. This set of symptoms is used as a template to search through a library of abnormal system conditions and retrieve a set of potentially useful cases (Rasmussen, 1984). Rasmussen identifies symptomatic search strategies as those most often used by experienced troubleshooters. Symptomatic search is very efficient from the point of view of information economy, and a precise definition of a fault can frequently be obtained from a single decision (Rasmussen, 1981).

Problem Representation. Rasmussen's abstraction hierarchy (AH) is a multi-level system description. Higher levels of abstraction represent the system in terms of purposes or functions, while lower levels represent the system in terms of physical implementation. Rasmussen augments this means-end dimension of system description with a part-whole decomposition of the system. The part-whole dimension is conceptually orthogonal to the means-end dimension represented by the abstraction hierarchy. Topological connections between system components are also represented. These connections reflect functional, physical, or spatial relationships between system objects (Rasmussen, 1985; Bisantz and Vicente, 1994). A system representation which contains the abstraction hierarchy, the part-whole decomposition, and topological connections is called the "means-end/part-whole space", or, occasionally, the "abstraction/decomposition space."

3. FIXIT

The Fault Information Extraction and Investigation Tool (FIXIT) is a computational representation of operator fault management experience. FIXIT has three components: the system model, the diagnostic reasoner, and the library of fault management cases.

3.1 System Model

FIXIT's system model is a multi-layered representation of the system, based on Rasmussen's general problem representation. This representation provides information on the physical form, functions, and goals of the various system components, and is intended to provide a global view of the system (Rasmussen, 1984). The structure permits the comparison of physically distinct engineering systems under a single representation, and allows fault management experience learned on one system to be applied to other systems. Fig. 2 is a small portion of a representation of a NASA satellite in means-end/part-whole space showing means-end links.

There are five levels to the abstraction hierarchy used in the system model: physical form, physical function, generalized function, abstract purpose, and functional purpose. At the level of *physical form*, the appearance, condition and location of each component and sensor are described. *Physical function* describes the states of system components. Because only individual sensors have measurable states in the system, the descriptions of physical function are at the sensor level of decomposition. The *generalized function* level describes the electrical, thermal, mass, and information flows between system components. Generalized function is represented in the means-end/part-whole space at the "block-diagram" level of description, a level which is appropriate for directing the reasoner to the appropriate location in the system model (Rasmussen, 1984). *Abstract function* describes the functions performed by the individual subsystems, and is described primarily at the subsystem level of decomposition. *Functional purpose* corresponds to overall system goals, and is therefore described at the system level of decomposition.

The part-whole dimension also consists of five levels of description: sensor, component, subsystem, system, and system class. The objects at the *sensor* level are components of the system which receive and respond to signals or stimuli, such as volt meters or thermistors. The objects at the *component* level of

decomposition are the individual physical entities which make up the system, such as gyroscopes, batteries, valves, pumps, and heaters. At the next level, these components are aggregated into meaningful *subsystems*, such as power, thermal control, or attitude control. The *system* level describes the entire system as a single whole, while *system class* refers to the group of engineering systems of which the specific system is a part.

3.2 Diagnostic Reasoner

The diagnostic reasoner contains representations of Rasmussen's symptomatic search strategies, as well as mechanisms to compare the current system state (represented as a current-state template) to cases in the fault library. The reasoner is responsible for both the construction of templates and the retrieval of matching cases from the fault library.

In FIXIT, as in all supervisory control systems, the diagnostic task is a search to identify and locate a change from normal or planned system operation (Rasmussen, 1981). The structure of this search consists of a sequence of search routines, which are used in an attempt to sequentially limit the current field of attention of the search. These routines are used to identify the appropriate subsystem, component, or state involved in the fault (Rasmussen, 1984). The reasoner receives information about the state of the system in the form of error messages. Fig. 3 shows an error message generated by a NASA satellite when a sensor detects a low current condition in a battery.

The symptomatic search strategies in the reasoner are used to relate the error message to a particular location in the system model's means-end/part-whole space. The results of each search give the reasoner a description of the current state of the system, which

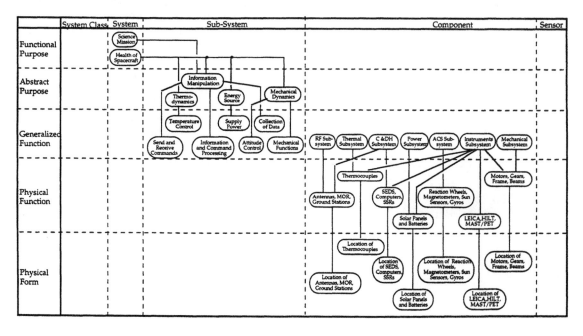

Fig. 2. A portion of the means-end/part-whole space for a NASA satellite.

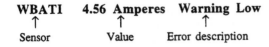

WBATI   4.56 Amperes   Warning Low

Sensor   Value   Error description

Fig. 3. A typical error message.

is then used to fill in slots of a current-state template. Fig. 4 shows the slot-and-filler structure of a current-state template.

The diagnostic reasoner searches the fault library to find matches to the current-state template. As the search progresses, the reasoner calls on matching procedures to assess the degree of match between the current-state template and the cases that are encountered. Retrieval algorithms return a list of partially-matching cases, each of which contains information which is potentially useful to an operator for resolution of the fault. Ranking procedures analyze the set of cases to determine, based on the cases' indices, which cases have the most potential to be useful. That set is retrieved by the reasoner. If no cases are retrieved, the reasoner restarts the process and uses a different search strategy to relate the fault to a location in the means-end/part-whole space. The result of this search is used to fill in the slots of a new template, which is, in turn, used to search the fault library. This process continues until a set of sufficiently useful cases is found and retrieved.

The diagnostic reasoner uses three types of symptomatic search strategy: pattern recognition, heuristic search, and hypothesis and test.

Pattern Recognition. Pattern recognition involves matching low level fault information, in the form of sensor and error message data, directly to the indexes of cases in the fault library. The diagnostic reasoner fills in the slots of the current-state template with received sensor information, without reference to the system model. Experts use pattern recognition to efficiently identify familiar system states and

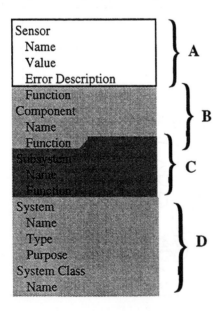

Fig. 4. A current-state template.

disturbances directly (Rasmussen, 1981, 1984).

The goal of the pattern recognition strategy is to retrieve any cases from the fault library which feature the error message just received by FIXIT. The name of the sensor, its value, and any additional descriptions of the error contained in the error message are used to fill in the slots of the current-state template. Part A of Fig. 4 shows the slots in the current-state template which are filled in by pattern recognition. If FIXIT receives the error message shown in Fig. 3, these slots are filled in as shown in Fig. 5.

Heuristic Search. If the diagnostic reasoner is unsuccessful in using the pattern recognition strategy it invokes the heuristic search strategy. Heuristic search differs from pattern recognition in its use of 'tactical' search rules. These rules are used relate to the error messages received by FIXIT to other locations in the means-end/part-whole space. Heuristic search uses two types of tactical search rules to relate objects in the system model. The first rule relates the received error messages to other parts of the means-end/part-whole space by navigating the functional links between components. The second type of tactical rule relates the error message to other parts of the system model by navigating the part-whole links between the sensors identified in the error messages and other components. The process the reasoner uses to relate sensor information to other system components amounts to a 'search' of the system model. The reasoner uses the information found in the system model to fill in slots of a current-state template. Rasmussen broadly refers to this type of search as "decision table search" (Rasmussen, 1981, 1984).

The use of heuristic search results in a broadening of the reasoner's search criteria. Rather than search for cases which match specific error messages, the diagnostic reasoner looks for cases which involve the same component (or type of component) as the one in which the failure occurred. The template slots filled in for this 'comparable component' search are shown in part B of Fig. 3. If FIXIT is using heuristic search to try and resolve the low battery fault from Fig. 3, a search of the system model tells the reasoner that 'WBATI' identifies the battery current sensor. The reasoner then searches the fault library for cases which contain battery current problems. The slots filled in for this comparable component search are shown in Fig. 6a.

If the comparable component search is unsuccessful the reasoner uses the heuristic search strategy to broaden the search again. The reasoner now focuses on the function of the failed component and the

Sensor
    Name: WBATI
    Value: 4.56 Amperes
    Error Description: Warning Low

Fig. 5. Template slots filled by pattern recognition.

a) **Sensor**
   Function: Electrical Current Sensor
   Component
   Name: Battery
   Function: Power storage

b) **Component**
   Function: Battery
   Subsystem
   Name: Power
   Function: Generate and Supply power to system

Fig. 6. Template slots filled by heuristic search.

subsystem to which the failed component belongs. This information is used to fill a different set of slots in the current-state template, shown in Fig. 4 Part C. If FIXIT is still resolving the low battery current problem, the reasoner now looks for cases which describe Power subsystem problems, concentrating on those cases which mention components which perform the same function as the battery. The relevant portions of the template this search are shown if Fig. 6b.

Hypothesis and Test. The strategy of hypothesis and test differs from heuristic search in the type of tactical search rules it uses. In human troubleshooting, hypotheses result from uncertain topographic search or fuzzy recognitions. FIXIT extends Rasmussen's (1981, 1984) hypothesis and test strategy by placing greater emphasis on the use of topological links between components to navigate the system model. In FIXIT, "hypotheses" are generated from searches of the system model. These searches relate the error message to other parts of the means-end/part-whole space. A "hypothesis" is tested by using the results of the search of the system model to fill in the slots of a current-state template.

The hypothesis and test strategy uses two types of tactical search rules to navigate the topological links in the system model. Topographic search rules control the use of the map of the system, and are used to navigate the spatial relationships between system components. This type of search is used primarily for thermal problems. Evaluative (i.e. causal) search uses both forward and backward propagation from a failed component to navigate block diagrams and schematics in the system model. Evaluative search is used for electrical and power problems.

If the diagnostic reasoner is implementing a strategy of hypothesis and test to resolve the battery current problem from Fig. 3., evaluative search is used to back-propagate the electrical fault along a causal chain The diagnostic reasoner references subsystem schematics in the system model to find a component which may caused the battery current fault (i.e., components which are upstream in a causal chain). The solar panels supply power to the batteries, and may have induced the low current state in the battery, so the reasoner conducts a search for cases featuring solar panel faults. The relevant portion of the template for this search is shown in Figure 7a. If the first hypothesis fails to find any matching cases, hypothesis & test is used to generate a second hypothesis. Evaluative search is used again, but FIXIT now attempts to forward-propagate the fault along a causal chain to find components which may have been affected by the failure of the current component (i.e. components which are downstream in a causal chain). The batteries supply power to the power bus, and FIXIT looks for cases which contain information about power bus problems (see Fig. 7b).

3.3 Library of Fault Management Cases

Guided by recognition-primed decision making theory, FIXIT encodes substantial parts of its knowledge as cases – records describing the occurrence of previous fault management situations. These cases reside in the library of fault management cases (or fault library). The fault library has two parts: a *case library,* which serves as a repository for fault management cases, and a set of *access procedures.* Together, the library and its access procedures make cases accessible when retrieval cues are provided to it from the reasoner. The body of a FIXIT case contains a description of a previous fault, information about how an operator resolved that fault, and data which describe the overall state of the system at the time the fault appeared. Table 1 shows the type of information contained in a case.

The first type of fault management information stored in a case is a description of the fault. A detailed description of the fault is provided, which a controller can use to verify or rule out the identity of a fault. A high-level conceptual description of the fault, called an 'explanation,' is also provided. The case contains a set of suggested responses to the fault, as well as a list of personnel to contact for additional information. System schematics, engineering drawings, and any accompanying documentation are also provided. The case contains a list of related faults to guide the operator to additional sources of information. The case also contains a flag

Table 1. Types of fault management information recorded in a FIXIT case

Description of the Fault	Previous Occurrence Information
Detailed description and explanation	Time and date of event
Potential operational impacts	State information
Suggested Response	Previous operational impacts
Accompanying documentation	Previous resolution of the anomaly
List of related anomalies	Documents used to investigate the fault:
Status	Source of information and date recorded

a) **Component**
    Name: Solar panel
    Function: Supply power to battery
  Subsystem
    Name: Power
    Function: Supply power to system

b) **Component**
    Name: Power bus
    Function: Distribute power to system
  Subsystem
    Name: Power
    Function: Supply power to system

Fig. 7. Template slots filled by hypothesis and test.

showing the status of any investigation of this fault.

The second type of fault management information stored in a case provides the operator with information about previous occurrences of the fault. The time and date of the fault occurrence are given. Information about the state of the system at the time of the fault (e.g., attitude information) is provided. Any documents which were used to investigate the fault are provided. These might include video display print-outs, engineering schematics, or x-t plots of sensor data. The source of the information is provided, as well as information relating to the experience of the author (to give the operator a way of gauging the reliability of the information contained in the case). The date the information was recorded is also provided, to guard the operator against using old or outdated information.

A case is indexed by the features of the fault. These features describe how the fault affected the operation of the system, including the affected system components and their functions. The set of indices which make up a case's index are identical to the slots of the current-state template shown in Fig. 4. A case may have several indexes associated with it, each representing a description of a different type of situation in which it might be useful.

Cases which share common symptoms are grouped together. The case library is thus organized into a 'fault hierarchy.' The use of this structure reduces the time required to search the case library, because only a small subset of cases needs to be considered during retrieval (Kolodner, 1993). Fig. 8 shows the structure of a portion of a FIXIT case library for a NASA satellite ground control application.

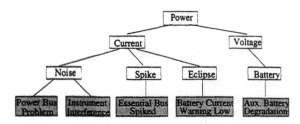

Fig. 8. Structure of the fault library

4. CONCLUSION

This research proposes an architecture for the computational encoding of fault management experience. The architecture rests on a combination of recognition-primed decision making, case-based reasoning, Rasmussen's symptomatic search strategies, and the abstraction hierarchy representation of systems. This architecture is suitable for a wide range of engineering systems. In particular, FIXIT will be demonstrated in proof-of-concept form for testing and evaluation for the domain of satellite ground control.

5. ACKNOWLEDGMENTS

This research was supported in part by NASA Goddard Space Flight Center grant NAG 5-2227 (A. William Stoffel, Technical Monitor).

REFERENCES

Bisantz, A. M., and K. J. Vicente (1994). Making the abstraction hierarchy concrete. *International Journal of Human-Computer Studies*, **40**, 83-117.

Domeshek, E., and J. L. Kolodner (1991). Toward a Case-Based Aid for Conceptual Design. Georgia Institute of Technology College of Computing. Atlanta, Georgia.

Klein, G. A. (1989). Recognition-Primed Decisions. *Advances in Man-Machine Systems Research*, **5**, 47-92.

Klein, G. A. (1993). A Recognition-Primed Decision (RPD) Model of Rapid Decision Making. In: *Decision Making in Action: Models and Methods* (G. A. Klein, J. Orasanu, R. Calderwood, and C. E. Zsambok, (Ed.)), 138-147. Ablex Publishing Corporation, Norwood, NJ.

Kolodner, J. L. (1993). *Case-Based Reasoning.* Morgan Kaufman, San Mateo, CA.

Rasmussen, J. (1981). Models of Mental Strategies in Process Plant Diagnosis. In: *Human Detection and Diagnosis of System Failures*, (J. Rasmussen and W. B. Rouse, (Ed.)), 241-258. Plenum Press, New York.

Rasmussen, J. (1984). Strategies for State Identification and Diagnosis in Supervisory Control Tasks, and Design of Computer-Based Support Systems. *Advances in Man-Machine Systems Research*, **1**, 139-193.

Rasmussen, J. (1993). Deciding and Doing: Decision Making in Natural Contexts. In: *Decision Making in Action: Models and Methods*, (G. A. Klein, J. Orasanu, R. Calderwood, & C. E. Zsambok, (Ed.)), 158-171. Ablex Publishing Corporation, Norwood, NJ.

Sheridan, T. B. (1992). *Telerobotics, Automation, and Human Supervisory Control.* MIT Press, Cambridge, MA.

HUMAN ERROR AND THE HOLON COGNITIVE ARCHITECTURE

Michael J. Young

Armstrong Laboratory
2698 G Street
Wright-Patterson AFB, OH 45433
(513) 255-8229
myoung@alhrg.wpafb.af.mil

Abstract: A new implementation of the schema concept is described and its role in generating human error is discussed. This new operationalization of the schema concept, called a thread, is an abstraction of neuron ensemble activity in the brain. A thread is both an ongoing, dynamic, process and a static "path of lowered resistance". The dynamic aspect of a thread gives rise to conceptually-driven processes, while the static component influences data-driven processes. The role of threads in generating conceptual and perceptual biases and creating prototypes, exemplars, and typicality effects is discussed.

Keywords: Human-centered design, human error, human perception, cognitive science, knowledge representation, modelling errors, neural dynamics, symbols.

1. INTRODUCTION

In most contemporary models of human error, perceptual and cognitive biases and overlearning of control routines are major factors in producing errors or mistakes. Reason (1990), for example, has proposed a Generic Error-Modeling System (GEMS) that consists of three classes of error each associated with a class of operator behavior. Skill-based control behaviors (Rasmussen, 1986) are stored sensorimotor routines that are triggered via an intention and then proceed automatically. Error slips occur when an overlearned routine is activated due to a lack of attention. In rule-based control behaviors (Rasmussen, 1986) the operator is aware there is a problem and is using rules of the form "If some condition exists then follow this course of action" to correct the error state. Rule-based mistakes typically occur when the wrong rule is applied. This happens when there is inappropriate matching of situational clues to the if clause of the rule due to a perceptual bias; that is, the operator sees the situational clues of well known rules and misses other clues that would have, if they had been seen, activated an alternative, but less well known rule. In knowledge-based behaviors (Rasmussen, 1986) the operator is aware

there is a problem and further aware that he or she does not know the appropriate rule to follow. Here the operator must first gather information to determine the current state, and then formulate a plan of action to correct the error state. Knowledge-based mistakes occur when the operator formulates an inappropriate plan due to either a perceptual bias leading to the picking-up of the wrong situational clues or a cognitive bias leading to the formulation of a well-learned plan of action when a novel approach would have been more appropriate.

Fundamental to most contemporary models of human error is the schema concept (Barlett, 1932). A schema (Mandler, 1985) is a temporally or spatially organized structure consisting of a set of variables or slots that are filled or instantiated by values. Schema are formed on the basis of proximities experienced in space or time. Schema are usually implemented as frames (Minsky, 1975), which are data structures consisting of variable slots that can be filled with specific kinds of information. Frames/schema become active when a sufficient number of variable slots are filled. Schema are used in models of human error to represent behavior routines that can be activated, objects that can be perceived, and plans that can be instantiated.

In recent years, however, schema-based models of behavior, perception and memory have often failed to explain or account for several types of experimental findings (c.f. Alba & Hasher, 1983, for a review), resulting in a near abandonment of this powerful concept. Perhaps the problems with schema-based models lie in the way the concept is operationalized. The schema concept proposed by Barlett (1932) was not a data structure, but a *process*. For Barlett, schema were "...an active organization of past reactions, or past experiences" (pg. 201) and "active developing patterns" (ibid.). In this paper a new operationalization of the schema concept is proposed as part of a new cognitive architecture, the holon cognitive architecture (HCA) (Young, 1993). We will call this new operationalization of the schema concept a *thread* to distinguish it from the existing definitions of schema as fixed data structures.

Our overall research goal is to develop Human Performance Process (HPP) models sufficiently sophisticated in their emulation of cognitive behavior to support operability investigations of alternative designs for complex real-time management information systems typically found in command and control and process control applications (c.f., Young, 1993 and Duetsch, et al. 1993, for additional information on HPP models and operability analysis). Our work in developing HPP models sophisticated enough to be used to make design decisions for management information systems has required us to propose a new cognitive architecture that more readily supports research on multi-tasking behavior and human error.

2. THE HOLON GENERAL SYSTEMS THEORY

Koestler (1967) proposed that the world consists of interacting hierarchies of entities and offered a general systems theory approach for describing this hierarchical world. The basic unit in Koestler's framework is the holon. Individual holons are simultaneously parts and wholes. Holons behave as if they were quasi-autonomous wholes, yet they are constituents parts of a hierarchical organization. Holons, at any given level, are both subordinate components of holons higher in the hierarchy and, simultaneously, the inclusive whole of holons lower in the hierarchy. They are similar to sub-assemblies which, in turn, are made-up of other sub-assemblies. Holons are arranged in hierarchy, called a holarchy.

Individual holons behave as if they are autonomous wholes. For example, each holon displays its own timing basis for patterns of activity. In addition, each holon has its own internal representation (working memory) of the environment. The "environment" for a holon may be the external world, the internal system, or both. A holon's

behavior is governed by fixed rules (collectively called the holon's canon), which are executed through flexible strategies. Holons are thus analogous to subroutines that consist of fixed code, but are passed variable parameters while executing. This combination of fixed canon and situation specific parameters allows a holon to generate a wide range of behaviors. Koestler's general systems theory provides the start of a framework to support development of a new cognitive architecture (c.f., Young, 1993, for additional information on the holon cognitive architecture).

3. THREADS

The HCA has been extended to incorporate the premise that the basic computational element of the brain is a neuron ensemble: a highly interconnected group of neurons that collectively function as a circuit (Damasio, 1989; Edelman, 1989; von der Malsburg, 1987). Neuron ensembles within the brain are always in a state of activation. They transmit information when their level of activation changes. A neuron ensemble's level of activation can be defined as a baseline pattern of firing along its axonal synapses. The baseline pattern of firing has both spatial and temporal components. The spatial component is defined by the subset of neurons within the ensemble that is firing. It determines where information is passed. The temporal component is defined by the frequency of the firing of individual neurons, summed across a given time-frame. It determines what information is passed. At any given moment in time a neuron ensemble's firing pattern can be considered as a baseline rate of activity, with respect to previous and subsequent moments. Information flows through a set of ensembles as *changes* in the baseline firing patterns of individual neuron ensembles.

von der Malsburg (1987) has proposed that neuron ensembles can serve as the basic building blocks of symbols. In his approach, each neuron ensemble codes a specific type of information depending upon where in the data stream it resides (e.g., neuron ensembles in the visual areas code form, motion and color information). Neuron ensembles can "combine together" by synchronizing their firing (von der Malsburg, 1987, Singer, 1993, Gray, 1992) to create symbols of varying sizes and complexities. Building symbols in this fashion has powerful implications for models of information-processing. If neuron ensembles are bound together through synchronization of firing then many such groupings can exist simultaneously as long as their activity is desynchronized relative to other groupings. This creates a plausible computational strategy for parallel processing of information and provides a means for the brain to keep separate objects separate (e.g., as in a visual scene) (von der Malsburg, 1987). In addition, separate ensembles, representing different component features, can be incorporated into different objects at different times. This eliminates the requirement to have a token (or a neuron) for every conceivable object. Objects are composed as required out of the available building blocks of neuron ensembles.

The HCA models neuron ensembles abstractly as holons. Furthermore, the HCA models the continuous neuron ensemble activity *discretely* as state changes and the synchronization of firing of ensembles as message passing. In the HCA, a neuron ensemble's continuum of activation (the "baseline" firing rates) is modeled discretely as a set of states. These states collectively define the functional purpose of the ensemble by defining the patterns or features the ensemble can recognize and respond to. Changes in baseline activity of a neuron ensemble are modeled through message passing (i.e., an ensemble, modeled as a holon, sends out a message when its state changes). The distributed type of representation through which disparate neuron ensembles combine to create symbols by synchronizing their firing is modeled through a new type of data structure, called a *thread*.

Threads are the HCA's engrams or memory traces. Threads have both an active and static component. An active thread corresponds to a set of neuron ensembles changing their spatial/temporal firing patterns and becoming temporally linked. Active threads can be variable in "size", depending upon the number of holons (neuron ensembles) that are temporally linked, and variable in duration, depending upon how long the holons stay linked.

Linkage of holons/ensembles is achieved by synchronizing firing patterns brought about through reentrant signaling (Edelman, 1989). Neuron ensembles have reentrant connections allowing them to feedback information (coded as a changes in firing patterns) to the ensembles which are feeding them information. When the input to an (downstream) ensemble is modified, it provides feedback to (upstream) ensembles with which it is connected. The reentrant signal produces synchronization by pushing active ensembles (e.g., those detecting a feature) across the action potential threshold simultaneously. Synchronization of the firing thus brings a referent into "focus" through a co-joining of its features.

The static components of threads are called thread segments. They correspond to paths of "lowered resistance" among ensembles; paths which develop as the result of co-activation. Several authors have proposed (Edelman, 1989; Hebb, 1946; von der Malsburg 1987) that the simultaneous activation of neuron ensembles leads to a strengthening of connections among the active ensembles. These paths develop and are modified dynamically; that is, the more experience you have associating certain features together, the easier it is to perceive the

attributes that define the referent, no matter if the referent is an object, situation, or plan of action.

The HCA models this process abstractly through the behavior of holons. Holons function as feature detectors, detecting specific patterns in the data stream dependent upon where they are in the brain. As such, they define the categories the mind can perceive and conceive. When a holon detects a pattern, it assumes a specific state and sends out a message to that effect. The set of states that a holon can detect are modeled as a *thread system*. A thread system defines the functional purpose of the holon by delimiting the patterns of stimuli a holon can recognize and respond to.

To help illustrate these ideas we will briefly consider the formation of a symbol, in this example a dog thread. Figure 1 is a notional depiction of six holons that code, colors, mammals, sounds, forms, habitats, and autobiographical information. In a real HPP model, each holon would receive input from many other holons not shown (i.e., each holon would sit atop a holarchy branch). Further, each holon can assume a limited number of states, representing the detection of specific sounds, colors, forms, etc. The set of states a holon can assume is its thread system. A holon assumes a state when its firing reflects a specific spatial/temporal pattern of firing, which is modeled in the HCA through message passing.

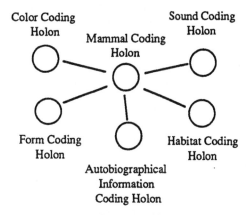

Figure 1. Depiction of Six Holons.

Figure 2 is a notional depiction of a dog thread within the mammal recognition holon. The dog thread consists of five segments: color, form, habit, sound and autobiographical information. Segments are formed through the co-activation of holons. They represent features that have frequently been detected together. The default values of the dog thread segments are shown in bold. Neurophysiologically, they represent the paths of spreading association that have the least "resistance". Symbolically, they define the prototypical dog. In Figure 2, the prototypical dog is a black shepherd named Shelby who had a specific bark and was usually seen in a specific house. The mammal coding holon, in addition, can recognize dogs other than Shelby. The set of dogs that can activate this thread are defined by the acceptable values of the thread segments, shown in brackets in Figure 2. Activation of the thread with other than the default values instantiates an exemplar, a subordinate instance of the prototype (Barsalou, 1982). This occurs when the holons comprising the thread signal a pattern other than the default values. Threads are a neurophysiological plausible alternate to the schema as a frame concept. Rather than being a static data structure, threads provide a dynamic mechanism for creating symbols of varying sizes, complexities, and temporal durations.

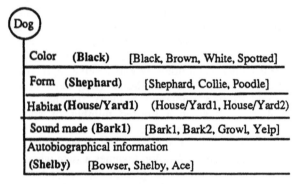

Figure 2 A Dog Thread.

An actual thread (within someone's mind) would have considerably more thread segments and acceptable values than shown here. The number of potential associations concerning dogs, for example, is probably very large. Further, there is natural variance among individuals in what segments (features) are associated with a dog thread, and in the acceptable values the segments take. Variation in thread segments and values arise from experience. Among individuals, there are both common and unique experiences concerning dogs. Common experiences produce conceptual similarities about dogs, while unique experiences give rise to distinct associations about dogs. As a result of these common and unique experiences, most individuals agree upon what is a dog, though individuals differ as to what is a prototypical dog; some think it is a collie, while others think it is a poodle.

Variation in thread segments among individuals when mapped experimentally give rise to typicality effects. Typicality effects are a statistical correlation found among individuals as to what features define a category. Typicality effects arise from commonalties in experience. Even though different individuals encounter different dogs, most dogs still share similar attributes. These shared attributes give rise to shared associations, which are measured as typicality effects. The result is that some features are judged more typical or representative of the category than others when measured by judgments of

a large number of subjects. The holon/thread concept provides a plausible model for the development of category structure, prototypes, exemplars and typicality effects.

4. FUNCTIONAL MODEL OF ATTENTION

The HCA does not require a central executive program to control behavior. Rather, the HCA employs a functional (Neumann, 1987) model of attention as a control mechanism. From a functional perspective, attention is not a unique sub-system with inherent capacity limitations. Rather, attention is a generic term used to describe several kinds of processes and sub-systems operating to limit the behavioral chaos that would ensue if an organism attempted to simultaneously perform several actions (Neumann, 1987). In the HCA, attentional processes synchronize the asynchronous activity of the holons, allowing features to form into objects.

Attentional processes (and limits) result from feed forward and reentrant signaling among manifold holons. Attentional limits reflect the time it takes for holons to decouple and recouple; to shift attention. This time factor is influenced by both static and active thread components. The static component is the "resistance" associated with a thread. A "strong" thread requires less of a signal to trigger its formation. The active component of a thread is product of reentrant signaling. Reentrant signaling reinforces threads already formed, creating a form of "dynamic resistance" that must be overcome for a state change to occur.

Static and dynamic thread factors interact to produce relatively stable states of mind. They provide a form of inertia that must be overcome for "attention" to shift. In addition, these factors ensure that when holons do change from one state to another, they will change states in mass. When one holon changes state its messages spread *association* through the holarchy, producing a rippling effect among holons. It is important to note that holons change states (i.e., change their spatial/temporal pattern of firing), not degree of activation. Spreading association thus differs from spreading activation in that during spreading association state changes either occur or do not occur, whereas in spreading activation the level of activation of a node changes. In systems employing a functional model of attention behavior *emerges* from the architecture as a temporal linking and enhancement of neural activity located in distinct parts of the brain (Damasio, 1989).

5. TOP-DOWN AND BOTTOM-UP DATA PROCESSING

Human information-processing is normally a combination of both data- and conceptually- driven processing. Data-driven (or bottom-up) processing involves a series of processing steps for which the output of one step serves as the input for the next. During data-driven processing, information is transformed from small perceptual units into larger aggregate chunks. In contrast, conceptually-driven (or top-down) processing uses situational context and general world knowledge to guide the processing and interpretation of information. The perceiver employs conceptual structures (e.g., schema or threads) to filter incoming information. During conceptually-driven processing, contextual factors are used to make appropriate selections and interpretations from the data stream.

Sensory processing during its initial stages is an example of a data-driven process. During sensory processing, information is transduced by holons residing at the lowest level of those holarchy branches which process sensory information, resulting in the formation of threads. As threads within the sensory-processing holons form, they trigger in a feed forward manner the formation of other threads in holons higher in the holarchy. This "flow of information" is a temporal constellation of holons. Individual holons assume specific states representing the detection of specific patterns in the incoming data stream. Detecting a pattern causes a holon to send a message to other holons higher in its chain, which in turn, may cause them to assume new states.

Perception is an ongoing process through which incoming information is visually parsed into objects by the co-joining of features detected by separate holons. Objects are co-joined as threads. As the perceptual scene changes, holons decouple causing existing threads to dissipate, which is followed by the formation of new threads representing the new objects. It is important to stress that holons are always in some state. The perceptual issue becomes, therefore, a question of whether the information flowing through the system is sufficient to cause holons to change states, to trigger the dissipation of one thread and the formation of another.

Individual holons are continually "summing-up" arriving signals. If the summation is "not equal" to the holon's current state (i.e., if the signals coming from diverse holons result in a different rate of firing), the holon changes its state, it changes its spatial and temporal pattern of firing both to those holons further along the processing chain and, through reentrant signaling, to those holons lower in its processing chain. These changes in signaling, in turn, may or may not cause changes in other holons, depending upon how the receiving holon's sums "add up". As described in the section on attention, both static and dynamic thread components influence this ongoing summation process.

The static component of a thread influences this process by being a path of lowered resistance, making it easier for information cascading through

the holarchy to flow in some directions rather than others. Threads, as paths of lowered resistance, make it easier to associate particular pieces of information together. They act as a *perceptual or cognitive biases*, diverting the flow of information into existing channels, making it easier to chunk specific data, associate select features, and build up a particular referent.

The dynamic thread component (i.e., reentrant signaling) influences the selection of information by providing a synchronized reentrant boost to the those holons comprising the existing "state". Reentrant signaling thus gives rise to *conceptually-driven* processing. Active threads have a processing advantage over non-active threads. They are part of the *context*, the active set of associations influencing the processing and interpretation of information. In addition, active threads "lower the threshold" for some select associations due to their patterns of connectivity. The holons comprising an active thread are synchronized in their firing and *signaling other holons*. These other holons, correspondingly, need less of a new signal coming from sensory data to change states and become part of a larger association. Therefore, active threads confer a conceptual advantage to ideas associated with the current context or state. As information begins cascading through the holarchy these two factors, static and dynamic "thread resistance", interact to determine which holons change states and form new threads. Their interaction determines what is perceived or what is "attended to".

6. HUMAN ERROR AND THREADS

Skill-based control behaviors (Rasmussen, 1986) are stored sensorimotor routines or behaviors that are triggered through an intention and then proceed automatically. Error slips in this class of behavior occur when the wrong routine is activated through a lack of "attention" (Reason, 1990). The HCA accepts Damasio's (1989) proposal that the brain/mind is composed of multiple domains, which are separate brain areas where specific types of information are processed. Some domains support cognitive functioning (e.g., planning, thinking, problem solving, etc.) others support perceptual processing, still others process motor movements. Activation of a sensorimotor control routine is the result of spreading association emanating in a perceptual domain triggering thread formation in a motor domain. Skill-based slips occur when the spreading association triggers a default routine (i.e., it follows the path of least resistance) instead of the appropriate routine. Normally, the triggering of the routine (i.e., the formation of a thread) would spread association to a cognitive domain which would "compare" the action to the intended action. However, in this case the spreading association is insufficient to trigger activity in the cognitive domain. The spreading association fails to overcome the existing holon/thread activity in the cognitive domain. The action is thus "unattended to", and the slip goes unnoticed.

In rule-based control behaviors (Rasmussen, 1986) the operator is aware there is a problem and is using rules to correct the error state. Rule-based level mistakes occur when the wrong rule is applied due to inappropriate matching of situation clues to the if clause of the rule as the result of a perceptual bias (i.e., the operator "sees" the situational clues of a well known rules and misses other clues that would have activated an alternative rule). Rules are implemented in the HCA as threads. The conditional clause of a rule corresponds to a thread proper; the component conditions of the rule are thread segments. The action part of the rule corresponds to the thread becoming active and having asociation spread (as a changes in the spatial/temporal patterns of firing of the holons comprising the tread) to other holons, thus triggering the formation of new threads (which directly or indirectly cause actions to occur). Mistakes occur when either a thread which is similar to the actual condition but has a lower activation threshold (because it has been seen more often) becomes active and triggers an inappropriate response, or when the thread corresponding to the actual perceptual situation becomes active but the spreading association triggers a sensorimotor routine with a lower threshold than the appropriate one. Both cases result in mistakes occurring.

In knowledge-based behaviors the operator is aware there is a problem and that he or she does not know an appropriate rule to follow. Here, the operator first must develop a plan to gather information on the current state and then formulate a plan of action to correct the error state. Knowledge-based level mistakes occur when the operator formulates an inappropriate plan due to either a perceptual bias leading to the processing of the wrong situational clues or to a cognitive bias leading to the formulation of a well-learned plan of action when a novel approach would have been more appropriate. Threads influence the perceptual process by being paths of lower resistance. They operate as biases "allowing" some information to be more readily perceived than other information. Threads also influence the planning processes. They again operate as biases, making it more likely that some courses of action will be followed than others.

7. Conclusion

The modelling of neural ensemble activity via threads permits a process-based implementation of schema. The dual nature of threads as both static and dynamic structures explains how a single system of neuron ensembles can lead to both data-driven and conceptually-driven error.

8. REFERENCES

Alba, J. W. and Hasher, L. (1983). Is memory schematic? *Psychological Bulletin*, 93, 2, 203-231.

Barsalou, L. W. (1982). Context-independent and context-dependent information in concepts. *Memory and Cognition*, 10, 82-93.

Bartlett, F. C. (1932). *Remembering*. Cambridge: Cambridge University Press.

Deutsch, S. E., Adams, M. J., Abrett, G. A., Crammer, N., L. & Feehrer, C. E. (1993). *Operator Model Architecture: Software Functional Specification*. (AL/HR-TP-1993-0027) Wright-Patterson AFB, OH: Armstrong Laboratory, Logistics Research Division.

Damasio, A. R. (1989). Time-locked multiregional retroactivation: A systems-level proposal for neural substrates of recall and recognition. *Cognition*, 33, 25-62.

Edelman, G.M. (1989). *The Remembered Present: A Biological Theory of Consciousness*. New York: Basic Books.

Gray, C. M. (1992). Rhythmic activity in neuronal systems: Insights into Integrative Function. In L. Nadel and D. Stein (Eds.) *Santa Fe Institute Studies in the Sciences of Complexity, Lecture Volume 4*. New York, NY: Addison-Wesley

Hebb, D. O. (1949). *The Organization of Behavior: A Neuropsychological Theory*. New York: Wiley.

Koestler, A. (1967). *The Ghost in the Machine*. London, GB: Hutchinson.

Mandler, G. (1985). *Cognitive Psychology: An Essay in Cognitive Science*. Hillsdale, NJ: Lawrence Earlbaum Associates.

Minsky, M. (1975). A framework for representing knowledge. In P. H. Winston (Ed.) *The Psychology of Computer Vision*. New York, NY: McGraw-Hill.

Neumann, O. (1987). Beyond capacity: A functional view of attention. In H. Heurer & A. Sandres (Eds.), *Perspectives on Perception and Action*. Hillsdale, NJ: Lawrence Erlbaum Associates.

Rasmussen, J. (1986). *Information Processing and Human-Machine Interaction*. Amsterdam: North Holland.

Reason, J. (1990). *Human Error*. Cambridge: Cambridge University Press.

Singer, W. (1993). Synchronization of cortical activity and it putative role in information processing and learning. *Annu. Rev. Physiol.*, 55:349-374.

von der Malsburg, C. (1987). Synaptic Plasticity as Basis of Brain Organization. In (Eds.) J. P. Changeux and M. Konishi *The Neural and Molecular Bases of Learning*. New York, NY: Wiley.

Young, M. J. (1993). Human Performance Models as Semi Autonomous Agents. *4th Annual Conference on AI, Simulation, and Planning in High Autonomy Systems*. Los Alamitos, CA: IEEE Computer Society Press.

RISK PROBABILITY ESTIMATION IN SYSTEMS USING DISTRIBUTED CONTROL

Edward J. Lanzilotta

Human-Machine Systems Laboratory
Massachusetts Institute of Technology
Cambridge, MA, USA

Abstract: The safety state model has been developed as a mechanism for estimating dynamic risk probability in surface transportation systems. Given the identification of a particular consequence and the conditions which can lead to that consequence, the model is used to observe the system in operation, and subsequently to provide an estimate of the risk probability of that consequence as a function of system state. In this paper, the method has been generalized for any type of system which utilizes a distributed control paradigm. Issues pertaining to practical model size are discussed. A worked example is included to illustrate the method.

Keywords: Distributed control, Human reliability, Man/machine interaction, Markov models, Risk, Safety Analysis

1 INTRODUCTION AND BACKGROUND

With any human-machine system, safety is a key concern. In industrial settings, the primary concern is the personal safety of the trained people that interact with the machinery, either by directly operating it or by merely being in the vicinity of operation. The safety issues in passenger transportation systems extend to the passengers (consumers) that use the services provided—the carrier has made an implicit agreement to deliver the passengers to their destination without harm. In the case of operation of consumer products, the "operator" is the consumer, typically with a low level of training—this places an additional constraint on safety considerations.

The dictionary definition of safety refers to freedom from risk of human injury or death [Webster's, 1994]. Risk includes the notion of both probability and outcome of some event occurrence [Lowrance, 1976; Rescher, 1983; Gratt, 1987]. Typically, an unexpected event which may lead to human injury or death is known as an accident. In most cases, an accident represents a demarcation point in time—the events occurring before the accident contribute to the probability of accident occurrence, while the events occurring after the accident are in the domain of outcome analysis. Risk probability is difficult to estimate, largely because most accident events are extremely rare. In addition, accidents are often due to compounding of intermediate failures, the causality of which are often quite difficult to determine. In principle, it is clear that the risk probability is not constant over time, but instead varies as the system state changes.

Many human-machine systems are distributed control systems. A distributed control system can be characterized as having multiple control elements, each of which has some domain of control with respect to the overall plant. These control elements may be either human operators or automatic controllers. The human operators are almost always obvious, while the automatic control elements are often "invisible" (e.g., consider the entire class of embedded controllers, which includes examples like automotive fuel injection, programmable home thermostats, intelligent bank cards). Many systems utilizing distributed control have profound safety implications—nuclear power plants, manufacturing plants, medical care, and transportation systems are some examples .

Human-machine interaction, by nature, raises the potential for safety issues. First of all, safety has been defined as relative to the risk of human injury— if there is no human-machine interaction, then by definition there is no safety problem. Second, human behavior is generally not deterministic, and the inclusion of a human control element in a system represents a significant opportunity for error, which can lead to a higher level of risk.

Fitts (1951) published a comparison of the performance of humans and automation for certain classes of tasks. His list clearly identifies classes of tasks which are well suited for human operators, and those which are not. In many cases, the use of automatic control would eliminate the potential for human error leading to safety problems. Yet, for a variety of non-technical reasons, human controllers are often employed in systems where automation would be more sensible. Part of the problem lies with the difficulty of qualifying more sophisticated automatic control elements with regard to overall system safety. However, safety qualification of human-operated systems is at least as difficult, especially when the human operator must interface with increasingly complex systems.

In order to address these issues, a method is proposed to estimate the risk probability of the overall human-machine system. Using this method as a system performance measure, the control architecture of the system can be "tuned" (i.e., redesigned) to best utilize the relative strengths of human operators and automatic control elements.

The safety state model has been developed as a method for estimating risk probability, as a function of system state. It is based on the discrete Markov process model, as presented by Howard (1971), and was first developed for the purpose of estimating operator performance, with respect to safety, in surface transportation systems [Lanzilotta, 1995a; Lanzilotta, 1995b]. Inspiration for the safety state model was drawn from techniques commonly used in quantitative system reliability and safety analysis. In particular, the methods of fault tree analysis and event tree analysis provide useful background [Swain and Guttmann, 1983; Lewis, 1987; McCormick, 1981; Gertman and Blackman, 1994]. Both methods utilize a tree structure to organize the events which can lead to failures in complex systems. The event tree method is considered "forward-looking" in that the analysis commences with a precipitating event and explores the possible consequences in light of subsequent decisions and actions. By comparison, the fault tree method is "backward-looking"—the analysis starts with a failure event and works backward to evaluate the possible combination of events that might have led to that failure. Both techniques can be used quantitatively or qualitatively. While these are powerful techniques with wide application, an acknowledged weakness in both is difficulty in accounting for the time relation of events.

Markov process models have been employed in the related area of system reliability. Reliability engineering is effectively a super-set of safety engineering—reliability techniques are used to estimate the failure or unavailability of a machine system, while safety techniques are concerned with those failures which pose risk to humans. Babcock

(1986) demonstrates the use of Markov process models to simplify the quantitative reliability analysis in fault tolerant systems. In this work, he contrasts the Markov process modeling technique to both fault tree methods and mean-time-to-failure methods. In addition, several methods for reducing the size of the Markov network are demonstrated, to ease the loads of data storage and computation. Lewis (1987) discusses the use of Markov process modeling to evaluate system reliability. He takes the significant step of using Markov states to represent the system states (in contrast, Babcock uses Markov states to represent events). This concept is used in the safety state model.

In this paper, safety state model analysis is described for human-machine systems using distributed control. A summary of the method is discussed, and a simple example is presented. In addition, issues pertaining to the practical sizing of the model are discussed.

2 DESCRIPTION OF THE SAFETY STATE MODEL

The fundamental goal of safety state model analysis is to provide a method for estimating dynamic risk probability. This goal is based on the notion that risk probability is not constant, and the level of risk can be profoundly affected by decisions and actions of one or more human operators. Dynamic risk probability can be used to measure and compare operator performance with respect to safety.

The risk probability is considered as a function of the system state, which varies in time. In general, a state variable is any time-changing property of a system. State variables can be continuous or discrete. A collection of state variables may be used to describe the system state, with the specific selection of the variables used being dependent on the objectives of the analysis. For the safety state model, the state variables are defined as binary conditions which can contribute to a safety-critical event (i.e., an accident). A set of these conditions can be concatenated into a series of binary digits, which represent a number. As a result, a single discrete integer value can be used to represent each combination of the specified state variables. This number is termed the "safety state."

The safety state model is a probabilistic model of the behavior of a human-machine system. Each safety state is represented by a state in a discrete Markov process model. One additional state is included, which represents the state of the system after an accident event has occurred. This state is termed the "failure state," and is a trapping state—once the model process enters this state, it never exits.

The states in the model are interconnected via transition paths. These transition paths represent events that cause a change in safety state. Each transition path has an associated probability. At a fixed time interval, the state of the model process is updated, resulting in either a transition to a new state or holding the current state.

An example of a safety state network is shown in figure 1. Assume that each binary digit (1 or 0) in the safety state number of each node represents the presence or absence of some condition (e.g., obstruction in path, driver awake, brakes functional, etc.). In each node, the decimal representation of the safety state number is shown, as well as the binary representation (which identifies the combination of conditions). This network is based on a safety state composed of three binary conditions, leading to eight safety states (states 0 through 7) and one failure state (state 8). The arrows show the transitions between states, which are bidirectional. (The diagram does not explicitly show the holding transitions, for clarity. It is assumed that all states in the network have a non-zero holding transition probability.) Because state 8 is a trapping state, there are no transition paths out from it.

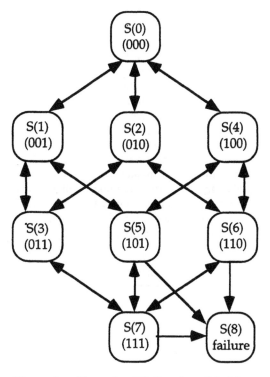

Figure 1 — Example of Safety State Model

A fundamental difference between the safety state model and other quantitative safety methods lies in the representation of events. In both fault tree and event tree methods, events are represented as nodes in the tree structure. By contrast, the safety state model uses nodes to represent states of the system, while the transition paths represent the events corresponding to a change in state. This is significant, as any given state can be reached from a number of other states, via the occurrence of different event chains. The most significant implication is that the calculated risk probability considers all possible causality chains, even those which might not be obvious to the analyst.

Analysis of the model allows determination of the relationship between system state and risk probability. The collection of state transition probabilities is organized in matrix form. The discrete form of Markov model is used, with state

transitions occurring at fixed time intervals. Assuming that the elements of this matrix are known, Markov process analysis is used to predict the mean time to the failure state (MTTF). The MTTF is converted into an equivalent risk probability. Since the MTTF is different for each state in the network, this procedure results in a set of risk probability values, as a function of safety state.

Once this function has been calculated, it is stored as a look-up table. To use this function, an operational system is observed, and the conditions that comprise the safety state are measured. The result is a safety state trajectory, a function of time. This trajectory can be directly transformed into a risk probability trajectory, which is used for measurement and comparison of operator performance.

To obtain the probabilities required to populate the state transition matrix, an operational system is observed and the safety state trajectory is recorded. Statistical analysis is performed on the safety state trajectory to obtain the state transition probabilities. The mean and distribution of state occupancy times are used to compute the holding probabilities for each state, while the distribution of state transitions is used to compute the transition probabilities from each state

In summary, safety state model analysis has four distinct phases: risk identification, system observation, model calibration, risk estimation. A flowchart of these phases is shown in figure 2.

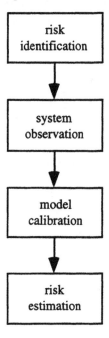

Figure 2 — Flowchart of Safety State Analysis

The risk identification phase is used to define the risk event of interest and identify the potential contributing factors. The risk event is defined as a single type of event which might result in personal injury to a human interacting with the system. The risk identification phase is the primary point of interaction on part of the analyst; the "output" of this phase is a specified risk event, with a set of potential contributing conditions.

Decisions made in the risk identification phase determine the measurements made in the system observation phase, during which an operational system is observed and data is collected. The resultant set of safety state trajectories is used to obtain an output set of statistical summary data.

The model calibration phase is the point where the model probabilities are calculated, and the risk probability function is determined. Using the output from the system observation phase, the state transition matrix is compiled. The matrix is then used to generate the risk probability transform function, which represents the output of the model calibration phase.

In the risk estimation phase, the system is once again observed, this time converting the newly-obtained system state trajectory into a risk probability trajectory via the risk probability transform function.

Safety state model analysis offers two distinct types of measures. The risk probability transform function (which is the output of the model calibration phase) serves to identify operational states which have a much higher level of risk than others. This information is of use to the system designer or analyst, as it provides a mechanism for identifying design weaknesses. In addition, evaluation of risk probability trajectories for individual operators provides a measure of human decision performance with respect to safety.

The most significant strength of the method, as compared to other methods of quantitative safety analysis, is the generality of the model structure. Its generality lies in the fact that there are no assumptions about the topology of the Markov process model—any state can transition to any other state. Therefore, when using the model to predict the MTTF, any and all possible causality paths are automatically included, without explicit consideration by the analyst. Thus, this method simplifies the task of the analyst to specifying a particular risk event and the conditions that might be contributory to that event. The generality of the method also enables straightforward automation of the computational phases.

The most significant weakness is the complexity of the model. The Markov process model of a complex system is not conducive to an intuitive understanding, in part because of the sheer number of states in the system. The resultant state transition matrix is necessarily large and unwieldy to inspect. However, this effect is mitigated through automation. As a result, the analyst is not required to have any direct contact with the internals of the model; once the risk event and contributory conditions have been specified, the next output of interest is the equivalent risk probability function.

A related weakness is the implicit requirement for an operational system. This requirement occurs in the system observation phase—it is difficult to observe and measure a system if it is not yet implemented. Although estimation of this data can be made in the absence of an operational system, any system of an interesting level of complexity will require a daunting amount of manual parameter estimation.

The method described above was initially conceived for measurement of human operator performance in surface transportation networks. However, the approach may be applied to virtually any human-machine system using a distributed control paradigm. The challenge for the analyst lies in defining an appropriate risk event and selecting system conditions which might lead to that risk event.

3 AN EXAMPLE

To demonstrate the use of the safety state model for risk probability analysis in systems with distributed control, the following simple example is presented. In most modern automobiles, fuel supply for combustion is metered under microprocessor control. If the control system has a failure, it can result in power surging at idle, which can lead to a scenario known as unintended acceleration. The driver-vehicle system represents a distributed control system—the driver makes control decisions with regard to gear selection and brake actuation, while the fuel control system represents a separate control loop. In order to provide clarity of the analytical method, the scenario is intentionally simplified.

state num	cond 3: apply brake	cond 2: wrong gear	condition 1: idle surge	failure potential	remarks
0	false	false	false		"correct" driver action
1	false	false	true		"correct" act with failure
2	false	true	false	√	rarer driver error
3	false	true	true	√	driver error with failure
4	true	false	false		"correct" driver action
5	true	false	true		"correct" act with failure
6	true	true	false	√	common driver error
7	true	true	true	√	driver error with failure
8					risk event has occurred

Table 1 — List of Safety States

The first phase of the analysis is risk identification. The risk event of interest is a collision between the vehicle and an obstruction. (A simplifying assumption is that there is always an obstacle present in the "wrong" direction of travel.) The following three contributing conditions are identified: 1) the engine idles higher than normal, due to a control malfunction, 2) the driver selects the wrong transmission gear (e.g., reverse instead of drive), and 3) the driver actuates the brakes. This risk event and set of conditions gives us nine different safety states in the model, as listed in table 1. (Note that the binary sense of the conditions does not matter in the analysis—it would be equally valid to specify the third condition as "driver does not actuate brakes." The only requirement is that the interpretation of the safety states is consistent with the specification.) The resolution of the time clock (and thus the discrete time interval) is one-tenth of a second.

state number	mean hold time	condition 3 changes	condition 2 changes	condition 1 changes	risk event occurs
0	1800	78%	21%	1%	0%
1	50	56%	4%	40%	0%
2	20	36%	4%	19%	41%
3	10	22%	2%	3%	73%
4	150	64%	18%	18%	0%
5	50	1%	26%	73%	0%
6	100	8%	67%	19%	6%
7	50	4%	16%	22%	58%

Table 2 — Results of Data Collection Phase

Considering the individual states, states 0 and 4 represent the "normal operation" states — there is no failure of the automatic control element, and the driver has not erroneously selected the wrong gear. The driver is merely using the brake properly, as driving conditions warrant. States 5 and 6 represent single point failures. In state 5, the automatic control fails, but the driver has taken precautionary action to compensate for this possibility. In state 6, the error is on part of the driver, but "correct procedure" (having the brake applied when selecting gears) is a precaution against this error. The remaining states between 0 and 7, inclusive, represent situations where compound errors have occurred. State 8 represents the state of the system after the risk event has occurred. This is a trapping state—once it is entered, it can never be left. This notion corresponds to the physical inability to "un-do" an accident. Once state 8 has been entered, the individual states of the conditions are of no consequence.

$$P = \begin{matrix} 0.9994 & 0.0000 & 0.0001 & 0 & 0.0004 & 0 & 0 & 0 & 0 \\ 0.0080 & 0.9800 & 0 & 0.0008 & 0 & 0.0112 & 0 & 0 & 0 \\ 0.0020 & 0 & 0.9500 & 0.0095 & 0 & 0 & 0.0180 & 0 & 0.0205 \\ 0 & 0.0020 & 0.0030 & 0.9000 & 0 & 0 & 0 & 0.0220 & 0.0730 \\ 0.0045 & 0 & 0 & 0 & 0.9930 & 0.0013 & 0.0013 & 0 & 0 \\ 0 & 0.0002 & 0 & 0 & 0.0146 & 0.9800 & 0 & 0.0052 & 0 \\ 0 & 0 & 0.0008 & 0 & 0.0067 & 0 & 0.9900 & 0.0019 & 0.0006 \\ 0 & 0 & 0 & 0.0008 & 0 & 0.0032 & 0.0044 & 0.9800 & 0.0116 \\ 0 & 0 & 0 & 0 & 0 & 0 & 0 & 0 & 1.0000 \end{matrix}$$

Table 3 — State Transition Matrix

The second phase is system observation. Assume that data has been collected on a simulation system, and the results listed in table 2 have been obtained. The mean holding time represents the average number of time intervals that an individual state was held. The remaining numbers show the distribution of state transition from each state. For example, from state 5, the system holds this state for 5 seconds (on the average), 1% of the state transitions resulted from

a change of condition 3 (resulting in transition to state 1), while 26% of the state transitions resulted from a change in condition 2 (resulting in transition to state 3), and so on. In the cases where the transition is a result of the occurrence of the risk event, the state transitions to state 8.

The model calibration phase converts the collected data (table 2) to an array of risk probabilities (table 5). In the process, the state transition matrix (table 3) is calculated, as is the array of MTTF values (table 4). Tables 2 through 5 are all in terms of a single time interval, specified at one-tenth of a second.

state number	MTTF
0	9340.3
1	7863.1
2	2972.8
3	854.1
4	8740.1
5	7216.8
6	6712.8
7	2716.3

Table 4 — MTTF Array

The MTTF array represents the average number of time intervals that will elapse in the future (i.e., from this point in time) until the system reaches the risk event, given the current state and no other knowledge about the past. The equivalent risk probability states that, if it were possible to have the risk event occur in the next time interval, what is the likelihood that the risk event would actually happen. This quantity provides a direct means of comparing risk levels as a function of system state. [Note that state 8 is not included in either table 4 (or table 5), as these tables reflect the time to get to state 8 (the trapping state) from the other states.]

state number	equiv risk probability
0	0.0001
1	0.0001
2	0.0003
3	0.0012
4	0.0001
5	0.0001
6	0.0001
7	0.0004

Table 5 — Equivalent Risk Probability Array

Finally, in the fourth phase (risk estimation), the system is again observed in operation. This time, the safety state trajectory can be converted, in real time, into a risk probability trajectory (that is, the varying risk probability, expressed as a function of time) via the look-up table given in table 5.

4 ISSUES OF MODEL SIZE

One of the disadvantages of using a Markov process model is that the number of states tends to grow very quickly with increasing number of variables, and the number of elements in the state transition matrix grows as the square of the number of states. In the case of the safety state model, this growth occurs exponentially—the number of states is (approximately) 2^n, and the number of elements in the state transition matrix is (approximately) 2^{2n}, where n is the number of conditions (binary variables). In any computer-based system implementation, there are two fundamental constraints: processor speed and memory size. Both of these constraints will have implications on the maximum practical size of the safety state model.

During the model calibration phase, the state transition matrix is converted to an equivalent risk probability function (as a function of safety state) via matrix inversion. This matrix inversion is the single most costly computation required in the method. The number of multiplications required to perform a general matrix inversion is roughly $O(N^3)$, and $N \approx 2^n$. Therefore, a 15 condition system would require roughly $3*10^{13}$ multiplication operations, which would require about 300,000 seconds (about 3.5 days) using a dedicated computer with 100 Mflop performance. Each additional condition raises the required processing time by a factor of 8.

In terms of data storage, the largest load is placed by the model calibration phase for storage of the state transition matrix (which contains approximately 2^{2n} elements). Assuming that high-precision floating point numbers ("double") are used (each 8 bytes), this means that a 15 condition model will require about 8 Gbyte of disk memory. Each additional condition raises the memory requirement by a factor of 4.

In summary, under the constraints imposed by current conventional computer technologies, safety state model analysis is limited to roughly 15 contributing conditions. Larger models could be accommodated using more exotic technologies, such as super computers or parallel-processing machines.

5 SUMMARY

The focus of this paper has been to discuss the issue of safety analysis with respect to systems incorporating distributed control. Distinction is made between risk probability and risk event outcome. The risk event (or accident) marks the time relationship between risk probability and outcome. A discussion of distributed control, and implications regarding human controllers, is included.

A summary description of safety state model analysis is provided. Safety state model analysis is decomposed into four phases of analysis, each of which is described qualitatively. A brief discussion of strengths and weaknesses of this approach is included. Applicability to general distributed control systems is discussed.

To illustrate the method, a simple example is included. This example explores the interaction between a human control element and an automatic control element in a simple system. The example serves to illustrate the mechanics of safety state model analysis and its general applicability to distributed control systems.

A brief discussion of practical implications on model size is included. The constraints are based on available memory size (disk space) and processor speed. Based on computer technology which is currently available, a nominal maximum model size of 15 conditions was determined. Implications of increasing the model size are included.

REFERENCES

Fitts, P.M. (1951). *Human engineering for an effective air navigation and traffic control system*. National Research Council, Washington, D.C.

Gertman, David I., and Blackman, Harold S. (1994). *Human Reliability & Safety Analysis Data Handbook*, John Wiley and Sons.

Gratt, L.B. (1987). "Risk analysis or risk assessment: a proposal for consistent definitions," in *Uncertainty in Risk Assessment, Risk Management, and Decision Making*, Plenum Press, NY.

Howard, Ronald A. (1971). *Dynamic Probabilistic Systems, Volume I: Markov Models*. John Wiley and Sons.

Lanzilotta, E. (1995a). *Analysis of Driver Safety Performance Using Safety State Model*. TRB Technical Paper 951071, Transportation Research Board.

Lanzilotta, E. (1995b). *Using the Safety State Model to Measure Driver Performance*. SAE Technical Paper 950968 (also, included in SP-1088).

Lewis, Elmer E. (1987). *Introduction to Reliability Engineering*. John Wiley and Sons.

Lowrance, William W. (1976). *Of Acceptable Risk*, William Kaufmann, Inc., Los Altos, CA.

McCormick, Norman J. (1981). Reliability and Risk Analysis. Academic Press.

Rescher, N. (1983). *Risk: A Philosophical Introduction to the Theory of Risk Evaluation and Management*, University Press of America.

Swain, A.D., and Guttmann, H.E. (1983). *Handbook on Human Reliability Analysis with Emphasis on Nuclear Power Plant Applications*. U.S. Nuclear Regulatory Committee.

Webster's New Universal Unabridged Dictionary (1994).

FUZZY DOCS: FAILURE DETECTION
AND LOCATION IN LARGE SYSTEMS

JIE REN* and THOMAS B. SHERIDAN*

*Massachusetts Institute of Technology, Cambridge, MA 02139, USA

Abstract. Failure diagnosis is the task of isolating and identifying the causes of malfunction of a plant from observable symptoms. An aggregated model-based diagnostic system will saturate subsystems downstreams and therefore make it difficult to isolate the cause of the failure in an abnormal condition. This paper extends a methodology which disaggregates a large system into several subsystems that are modeled and driven by the corresponding actual measurements. Any significant deviation between measurements and simulation indicates not only failure but also the location. The dynamic simulation enables the detection of failures during transient operation as well as the detection of emerging failures before any process signals reach their alarm levels. Fuzzy rules, which reflect human expertise for each subsystem, are used further to expose details of the malfunction and suggest ways of returning the plant to normal.

Key Words. Failure Detection; Complex Systems; Fuzzy Systems; Power Plant

1. Introduction

Failure detection and location is one of the most important functions of human supervision. The increasing complexity of systems in such technological fields as power generation, process engineering or transportation requires sophisticated decision aids for human operators to detect and locate system malfunctions. The goal of these aids is to help human operators to diagnose abnormal operating conditions, identify the source of failure and put the plant back to a normal condition. Many computerized on-line failure diagnosis methods have been developed in the past three decades, such as fault/event tree methods, disturbance analysis methods and voting techniques. These various approaches, which can be mainly classified as model–based or rule–based approaches, have been explored by Isermann (1984), Dhillon (1988), Himmelblau (1978), Rasmussen (1985) and Kramer (1988). A model-based approach will incorporate a structured plant model based on fundamental engineering principle, while a rule–based approach uses examples or heuristics to represent possible plant behavior.

In the model-based approach, Lind (1981) proposed to use a model of the system's normal operation mode to diagnose failures by comparing the system measurements with their corresponding model values. In this method, a single (aggregated) model is compared with the physical system. This method is advantageous because it can detect failures before they reach threshold levels. But when an aggregated model is used, any failure in an upstream subsystem may cause saturations of downstream subsystems and consequently trigger all the alarms like lights in a Christmas tree. The source of failure would be difficult, if not impossible, to discover.

To obtain the capability of locating failure, Sheridan and Hassan (1981) suggested breaking the aggregated model into submodels, each of which is forced by the appropriate measured plant variables. Tsach (1982) further developed the methodology and used the ideas of *power flow* balance to detect failures. The outputs of these submodels are compared with their corresponding plant outputs. A deviation of the outputs of any submodel indicates not only the failure but the location.

After possible failures have been narrowed to a subsystem, they are verified by more detailed logic or a rule-based model. Since there are uncertainties involved in the system, such as time-varying parameters and modeling errors, and a human expert is able to check the failure and determine the possible causes, a fuzzy rule base may be used to interpret the discrepancies between the system and the model. This helps operators cope with abnormal situations.

The extended methodology has been demonstrated with a fossil power plant.

2. Disaggregated On-line Comparative Simulation

This work is done on a methodology called *DOCS* – acronym for Disaggregated On-line Comparative Simulation – which breaks an aggregated model into submodels (Tsach, 1982). The inputs of each submodel are taken from the sensors (observations) and the outputs are compared with the corresponding measurements. Any deviation of a submodel indicates that this subsystem has potential failures. The dynamic simulation enables the detection of a failure during the transient operation as well as the detection of an emerging failure before any process signals reach their alarm levels. Because the simulation model is disaggregated into various submodels that correspond to different components of the plant, the true source of a failure can be easily located. Conversely, upsets in non-failure components which follow from abnormal conditions at the location of the failure are not falsely diagnosed as component failures of their own - as could happen in the case of conventional alarm schemes.

DOCS can therefore be applied to systems which are continuously monitored by human operators or computers. By continuously comparing the system outputs with their corresponding model variables, failures can be detected sooner than they would have been by observing system outputs crossing their thresholds, as illustrated in Figure 1. Here the top rectangles represent a real plant that is disaggregated into two subsystems, the middle ones are the corresponding simulations, and at the bottom are the comparators that show the discrepancies. Subsystem A in Figure 1 has two outputs: x and y. Both of them are assumed measurable. When a failure occurs in subsystem A, the comparator for x will detect a difference from A_m. Similarly, a difference from subsystem B_m indicates clearly a failure in subsystem B, assuming B_m is a good model of B.

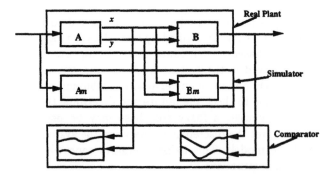

Fig. 1. Disaggregated On-line Comparative Simulation

3. Human Expertise and Fuzzy Rule Base

DOCS is designed to take tedious monitoring work from the human operator and help him detect and locate failures faster and more efficiently.

A human operator decides whether or not a system has failed based on:

1. The size of the discrepancy between the system and model output;
2. The length of time during which such discrepancy exceeds an allowable threshold.

In practice, under abnormal conditions, an operator has to cope with difficulties of understanding what is happening with various alarms in the control room. In a large system, there are uncertainties involved: system parameters are time-varying and models are not accurate. The discrepancies in *DOCS* are not steady and crisp. Fuzziness is generic to the model inaccuracy and system noise.

A human operator, however, is able to make decisions based on these *fuzzy* observations. It is desirable to have this kind of expertise stored in a rule-based form. A fuzzy rule base is a good candidate, since it can accept natural language from a human expert.

<u>Fuzzification of DOCS Variables</u>   DOCS variables can first be converted into fuzzy variables as follows:

In the given time duration T, a DOCS variable (display) is sampled at time intervals ΔT which is at least three times the associated system time constant to represent accurately the system dynamics. The magnitude discrepancy of the display window is divided into n zones from top to bottom. Summing the display points which fall in each zone, one can obtain the statistical discrepancy distribution of a process variable. The percentages of the points falling into the n zones constitute a fuzzy set $\tilde{A}$ which represents the human operator's observation. The fuzzy set is written symbolically as:

$$\tilde{A} = \sum_{i=1}^{i=n} \mu_i(i)/i \qquad (1)$$

where $i = 1, \cdots n$ is the universe of fuzzy set $\tilde{A}$ whose the membership is defined as $\tilde{A}$:

$$\mu_i(x) = \frac{sampling\ points\ in\ the\ ith\ zone}{Total\ Sampling\ Points} \quad (2)$$

The *standard* fuzzy sets defined in natural language are determined by experiment. For an example, assume the universe of discourse is $n = 5$ (*very low, low, medium, high, very high*). The statistical results are :

$$\tilde{A}_{low} = \{1.0, 0.9, 0.6, 0.1, 0.0\}$$
$$\tilde{A}_{verylow} = \{1.0, 0.3, 0.1, 0.0, 0.0\}$$
$$\tilde{A}_{medium} = \{0.2, 0.6, 1.0, 0.6, 0.2\}$$
$$\tilde{A}_{high} = \{0.0, 0.2, 0.8, 1.0, 0.6\}$$
$$\tilde{A}_{veryhigh} = \{0.0, 0.1, 0.6, 0.9, 1.0\}$$

After establishing the *standard* fuzzy set, one can translate the discrepancy observation fuzzy set $\tilde{A}$ back into natural language using the Euclidean distance.

$$D_1(\tilde{A}, \tilde{A}_{verylow}) = \{\sum_{i=1}^{i=5}[\tilde{\mu}_A(i) - \mu_{verylow}(i)]^2\}^{1/2}$$

$$D_1(\tilde{A}, \tilde{A}_{low}) = \{\sum_{i=1}^{i=5}[\tilde{\mu}_A(i) - \mu_{low}(i)]^2\}^{1/2}$$

$$D_1(\tilde{A}, \tilde{A}_{medium}) = \{\sum_{i=1}^{i=5}[\tilde{\mu}_A(i) - \mu_{medium}(i)]^2\}^{1/2}$$

$$D_1(\tilde{A}, \tilde{A}_{high}) = \{\sum_{i=1}^{i=5}[\tilde{\mu}_A(i) - \mu_{high}(i)]^2\}^{1/2}$$

$$D_1(\tilde{A}, \tilde{A}_{veryhigh}) = \{\sum_{i=1}^{i=5}[\tilde{\mu}_A(i) - \mu_{veryhigh}(i)]^2\}^{1/2}$$

The natural language corresponding to the smallest distance D_i is the best fit for the observation fuzzy set.

Weighted Sensitivity

After all DOCS variables in a subsystem are translated into fuzzy variables, human expertise regarding this subsystem's working condition is explained in the form of sensitivity factors, such as *this variable is very sensitive to failure* and *that variable is less sensitive to failure*, etc. These factors concerning each process DOCS variable are represented by another set of fuzzy variables.

Two fuzzy sets are associated with each variable. One is the observation discrepancy $\tilde{A}_i$ and the other is a sensitivity vector $\tilde{W}_i$, $i = 1, 2, 3$ (suppose a subsystem has three *DOCS* variables). Figure 2 shows this subsystem, for which the node represents the total sensitivity (possibility of failure for this subsystem), to which each descendant of this node contributes.

The generalized normal weighted mean at any node (measuring joint) contains the information

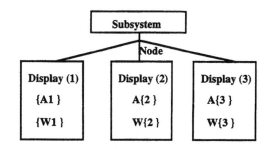

Fig. 2. Fuzzy Information of One Subsystem

of the total subsystem's behavior.

$$\{\bar{W}_0\} = \frac{\sum_{i=1}^{i=N}\{\bar{W}_i\} * \{\tilde{A}_i\}}{\sum_{i=1}^{i=N}\{\bar{W}_i\}} \quad (3)$$

where : $\{\bar{W}_0\}$ is the weighted mean (fuzzy set) representing the conclusion of the subsystem's performance; $\{\tilde{A}_i\}$ is the fuzzy set of the i-th process variable observation; $\{\bar{W}_i\}$ is the fuzzy set of the sensitivity for the i-th display variable.

To determine a weighted mean, one could choose as simple-minded a method as ranking each of the natural language expressions and then averaging these expressions. Such a scheme would result in:

$$\frac{(low + low)}{2} = low \quad (4)$$

$$\frac{(high + high)}{2} = high \quad (5)$$

$$\frac{(low + medium + high)}{3} = medium \quad (6)$$

When such a method is used, it fails to model the subtleties of the fuzzy expressions and allows for the possibility of failure. A suitable operation must be found for the formula as follows:

$$\bar{W}_0 = \frac{\tilde{A}_1 \cdot \bar{W}_{low} + \tilde{A}_2 \cdot \bar{W}_{high} + \tilde{A}_3 \cdot \bar{W}_{verylow}}{\bar{W}_{low} + \bar{W}_{high} + \bar{W}_{verylow}} \quad (7)$$

One should therefore not only extend the arithmetic operations defined on the real number domain to the operators defined on the fuzzy sets, but also use membership convex assumptions and Zadeh's extension principle (Zadeh, 1984), because of fuzzy multiplications and divisions.

Including Non-*DOCS* Variables There may be some other measurements which one is not able to model in the *DOCS* form, such as vibrations of bearings or the temperatures of lubrication oils.

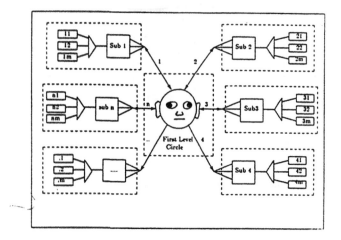

Fig. 3. Fuzzy Switch

One would prefer to include these signals into our failure detection system. For practical purpose, these observations are fuzzified using a piecewise linear function.

Fuzzy Switch Each subsystem has a node associated a failure possibility $\{\bar{W}_0\}_i$. If the i-th subsystem shows high possibility of failure, then the computer will switch its attention to that subsystem, and consequently look for an expert data base to assist the operator in fault diagnosis. Since the switch is based on the fuzzy variables from the nodes of subsystems, this is called a *fuzzy switch*.

Logic Inside Subsystems

The logic to further identify the location and the cause of a failure inside a subsystem is a failure-detection oriented expert system. This expert system is employed so that when the i-th subsystem shows a high possibility of failure, the intelligent interface tries to find the primary reason for the unexpected situation on the basis of knowledge in the expert base and knowledge of the state of the process variables at the time. The results of the analysis are shown to the operator. The computer can also include another expert base which gives the operator the corresponding safety actions to be carried out.

The expert base consists of conditional statements which represent all the causal relationships between failures and process variables. It is necessary to formulate general rules where the causal connections between failures and corresponding deviations in process variables are specified. This anticipation is made by a safety specialist on the basis of engineering analysis. This experienced part of the expert base will increase considerably

the forecasting power of the system and make results more reliable, especially if there are inconsistencies in the knowledge of the record of former incidents.

In the expert base, the relationship between failure and process variables is given as a set of fuzzy conditional statements :

$$
\begin{array}{llllll}
\text{If} & \tilde{A}_{1,1} & \ldots & \text{and} & \tilde{A}_{1,n}, & \text{then} & B_1 \\
\text{If} & \tilde{A}_{2,1} & \ldots & \text{and} & \tilde{A}_{2,n}, & \text{then} & \tilde{B}_2 \\
\cdot & & \cdot & & & & \\
\cdot & & \cdot & & & & \\
\text{If} & \tilde{A}_{m,1} & \ldots & \text{and} & \tilde{A}_{m,n}, & \text{then} & \tilde{B}_m
\end{array}
$$

The diagnostic system is decomposed with the aid of fuzzy switches and diagnostic algorithms. The aim of a fuzzy switch is to choose a certain specialized diagnostic algorithm on the bases of limited input data. After this, the diagnostic algorithm usually asks for some more input data to decide what the failure really is.

4. Procedure of DOCS

The procedure for failure detection and location by means of DOCS is summarized as follows:

1. The whole system is disaggregated into n subsystems, each of which has a node n_i and a diagnostic algorithm.
2. At every node, *DOCS* variables and some other important process variables of the corresponding subsystem are taken into account by a weighted mean of failure possibility for this particular subsystem. The computer will keep checking in the first level circle. If every thing is all right, no fuzzy switch will occur.
3. If one node has a high possibility of failure (decide by α-*level-set*, where α is selected

180

according the importance of the subsystem), the fuzzy switch leads the computer to this particular subsystem - the second level of checking.

4. In this subsystem, the corresponding expert algorithm will be active. From the knowledge base, the computer will help operators find the failure element and suggest what appropriate safety action can be taken.

The weight factors of the sensitive fuzzy sets can be changed by the operator under certain circumstances, and the expert base can be expanded to give increasing reliability as time goes on and the operators get more experience.

5. Conclusion

This paper describes a disaggregated on-line comparative simulation method for failure diagnosis. By disaggregating a large system into subsystems and by using fuzzy variables and fuzzy rules, one can use both engineering mathematical model and engineers' expertise to detect failures, locate their causes and suggest possible actions.

6. REFERENCES

Isermann, R. (1984). Process fault detection based on modeling and estimation method – a survey. In: *Automatica*, 20, p387.

Dhillon, B.S. (1998). Chemical system reliability: a review. In: *IEEE Trans. on Reliability*, 37 (2), p199.

Hassan, M. R. (1981). *Use of Disaggregated Model as an Aid to the Human Operator in Failure Diagnosis in Complex Automated Systems*. S.M. Thesis, Mechanical Engineering Dept. Massachusetts Institute of Technology.

Himmelblau, D.M. (1978). *Fault Detection and Diagnosis in Chemical Petrochemical Process*. Elsevier, Amsterdam.

Lind, M. (1981). The use of flow models for automated plant diagnosis. *Human Detection and Diagnosis if System Failures*. Rasmussen, J. and Rouse, W. B. (Ed.), Plenum Press, N.Y.

Kramer, M.K, and Finch, F.E. (1988). Development and classification of expert systems for chemical process fault diagnosis. In: *Robotics & Computer–Integrated Manufacturing*, Vol. 4, No. 3/4, p437-446.

Negoita, C.V. (1985). *Expert System and Fuzzy Systems*. Meulo Park Calif. Benjan/Cummings Pub. Co.

Rasmussen, J. (1985). The Role of hierachical knowledge representaion in decisionmaking and system management. In: *IEEE Trans. Systems, Man and Cybernetics*, SMC-15, p234.

Tsach, U. (1982). *Failure Detection and Location Method (FDLM)*. Ph.D. Thesis, Mechanical Engineering Dept. Massachusetts Institute of Technology.

Zadeh, L. (1984). Fuzzy probabilities. In: *Information Processing and Management*, 20, p363-372.

A DESIGN FOR A MAN MACHINE INTERFACE
FOR THE TELEMAN DEXTEROUS GRIPPER

André C. van der Ham, Ger Honderd, Wim Jongkind

*Control Laboratory, Department of Electrical Engineering, Delft University of Technology,
PO Box 5031, 2600 GA Delft, The Netherlands.
e-mail: a.c.vanderham@et.tudelft.nl*

Abstract: This paper describes the design considerations and implementation of a Man Machine Interface to a dexterous gripper and telemanipulator arm with a total of 17 degrees of freedom or more. The goal of the research is to eventually arrive at an interface which will give the operator a feel of telepresence. The mechanical design and control algorithms are explained. Results of preliminary testing of the MMI are presented.

Keywords: Man/Machine Interfaces, Ergonomics, Force Control, Position Control, Adaptive Control, Predictive Control, Redundant Manipulators.

1. INTRODUCTION

In 1989 the European Community started project Teleman, which is aimed at the development of tele-manipulators for hazardous environments. A group at the Control Laboratory of the Delft University of Technology is working on a dexterous gripper and the control system for the manipulator together with the Université Libre de Bruxelles and the company Vermaat Technics b.v. which is specialized in robots for the nuclear industry (van der Ham, *et al.*, 1993). The gripper has now become available for testing and the autonomous and manual control systems are now being implemented. This paper only deals with the manual control system for the gripper.

The entire gripper and arm construction may have as many as 18 degrees of freedom (DOF). In order to interface with the human operator, care should be taken in mapping these DOF, as not to overload the operator. In this paper, first the structure which is to be controlled is explained. The following part will point out the specifications that have been derived. Then the design of the man machine interface hardware is shown after which the control software architecture is explained. The results so far and

conclusions are presented at the end.

2. THE TELEMANIPULATOR

The Telemanipulator consists of the gripper, the arm and perhaps a mobile platform.

In nuclear industry, telemanipulators are mostly long reach devices with a snake-alike kinematical structure in order to move through the complex of tubing and obstructing walls. In a new Teleman project LACWAP, a self-expanding telemanipulator is being developed which can reach through 20 story structures and clamp itself to the structure for stability. At the end, a 6 DOF arm will connect to the gripper.

The gripper, as shown in figure 1, has 11 DOF. The three fingers each have 3 DOF of which the upper joints are alined perpendicular to the links and the bottom joint is aligned long the link to allow the finger to rotate at the base. The rotation at the base allows the gripper to change its posture to arrive at spherical and cylindrical grips. The gripper can also grasp pipes from the inside. The palm of the gripper has two degrees of freedom. It can move forward and rotate around its axis in order to do machining tasks.

A tool adapter in the palm facilitates the use of several specialized tools.

Sensorization is a problem in the nuclear industry. Therefore sensors have been used which can withstand radiation for a considerable amount of time. LVDTs for position, and force sensors based on the same principle as the LVDT have been mounted on the fingers. The finger base orientation is measured using resolvers.

The upper two joints of the fingers are actuated hydraulically and the base is actuated by a DC motor. The palm forward motion is controlled hydraulically and the rotation is powered by a pneumatic system.

3. THE SPECIFICATIONS

The specifications of the telemanipulator arm are not known exactly, for it will depend on the task at hand which arm will be used. The research assumes 6 DOF in cartesian space at the end effector. The gripper specifications are shown in table 1.

Table 1. Gripper characteristics

Fingers open/close	2 seconds
Finger base joint	0.2 revolutions/second
Link lengths (base-tip)	63 mm, 88mm, 92 mm
Max. angle (base-tip)	360°, (-45°;+45°), 45°
Grasp force	50 N maximum
Palm length	0-20 mm
Palm rotation	0.5 revolutions/second
Palm maximum torque	200 Nm

The goal for the Man Machine System (MMS) is to provide the operator with a sense of telepresence. Telepresence implies force, velocity and position feedback and control between the operator and the telemanipulator itself. Thus, for most operations, this will mean that the specifications for the interface to the operator will be the same as those for the MMI.

Fig 1. The dexterous gripper.

4. THE DESIGN

When the dynamics of the MMS is of the order of human limb dynamics, position control is necessary. When the dynamics of the MMS is much slower, then rate control is applicable. Experiments by Kim (1992) and Hannaford (1991) support this observation. Considering the specifications given in table 1, position control was considered for implementation. Rate control could become practical when the configuration of the gripper or MMI disallows further position control. Therefore rate control will be added as an enhancement to the system.

For the telemanipulator arm it is obvious that, in order to be able to operate over the full range of the manipulator arm, rate control will be necessary in order to move the system manually. The dynamics of the manipulator arm may be quick enough to necessitate position control, especially in case of fine motion control. Thus, the manipulator arm interface to the operator is now also implemented with position control. Rate control can easily be added later on.

The mapping of the 17 or more degrees of freedom to the operator interface must be performed carefully, for a bad design will increase the workload on the operator, perhaps even rendering it useless. Considering the tasks that have to be performed, the following MMI architecture has been defined:

4.1 Configuration

In order to make manipulation with the system feel as natural as possible, a mechanical glove interface has been chosen. The mapping of the gripper joints to the human finger joints directly can only be done for the top 2 joints of the gripper. Unfortunately, the human hand does not allow rotation of the base of the finger over the full range of 360°. However, in the process of grasping, one usually sets the base orientation of the fingers before hand; provided that there is enough knowledge about the object to be grasped. Therefore the base rotation is set using a potentiometer on a control panel. Accurate position control of the base rotation has proven to be problematic because of the slowness of the base joints. Probably, a spring-loaded potentiometer using rate control or active feedback of the position would make controllability better.

The palm is currently controlled using a 2 dimensional joystick, which is often used for throttle and aileron on model aeroplanes. The aileron control, which is spring loaded, is used for control of the rotation speed of the palm. The in/out motion of the palm is controlled by the throttle.

The configuration of the glove is planned to make use of the thumb, the index and middle finger placed in a more or less spherical fashion to imitate the gripper finger configuration (i.e. 3 fingers at 120° angle around the palm). Of the fingers the 2nd and 3rd joints will be used, as the 1st and 2nd joints are not inde-

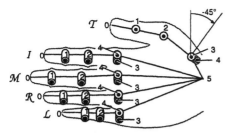

Fig 2. The joints within the human hand.

pendent with respect to actuation (see figure 2). For the thumb, a choice for the 1st and 2nd or 2nd and 3rd joints needs to be made. In order to limit the chance of collision, joints 1 and 2 have been chosen.

4.2 Actuation

Actuation of the glove needs to be done hydraulically to be able to feedback the high forces during manipulation. Tendon driven systems are usually incapable of exerting high forces and have problems of dry friction and elasticity, which make the control of such a system difficult. A robust/tough system is also necessary when used in practice. Hydraulic actuators are commonly used in practical applications where a high force vs. weight/size ratio is needed.

The problem of actuation lies in the placement of the actuators. The point of rotation must be exactly at the joint of the finger to avoid damage to the fingers themselves. Joint 3 is in the hand, thus a joint for the glove can not be placed directly near the joint of the finger. In order to rotate about the joint within the hand, the rotation could be enforced by a rotation from outside as shown in figure 3a. Although it is possible to fabricate such a curved cylinder, to reduce the cost and complexity a straight cylinder is chosen (see figure 3b). From the transfer of force point of view solution figure 3a does have its merits, as the other scheme may cause a bit of shearing and tearing due to the indirect manner in which the force is applied. The straight cylinder relies on the joint of the finger as a structural element for the movement while the other scheme could work well without the finger. The actuator will be connected to a frame that is fastened to the hand using straps. A gel patch will be used as padding.

The glove will also be adaptable to serve a large variety of hands. The length of the links of the frame and the center of the joint will be made adaptable. This adaptability does give problems for calibration of the system, which may be solved by choosing selected settings for the adaptable frame.

4.3 Control system

The control system takes care of the coupling of the gripper and arm to the glove and arm construction. The architecture of the control system is depicted in figure 4.

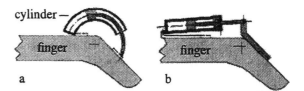

Fig 3. Hydraulic actuation schemes for the dexterous mechanical glove: a) a curved cylinder, b) a regular, straight cylinder.

In the beginning, the joints of the glove and of the gripper will be mapped directly to each other. Thus a 1 dimensional bilateral control scheme may be used for each pair of joints. The base joints, as explained, will be controlled from the control panel which also controls the palm.

The bilateral position force controller is divided into two parts each on either side of the communication channel. The algorithms are explained in §5.

5. RESULTS

5.1 Bilateral position/force control

Bilateral position/force control is a hot topic in teleoperation. In the past there have been great stability problems with force reflection in case of time delays of about 1 second. Anderson and Spong (1989) first described a way to deal with time delay. The technique used is similar to the idea behind robust control (an example of telemanipulation using this technique is given by Kazerooni, et al. (1993)). The basis lies in the small gain theorem, where if the open-loop gain of a system is smaller than one for all frequencies, it will be strictly stable. Thus, for good tracking behaviour the design rule becomes: by taking care that each side of the communication network is strictly passive or strictly damped (no overshoot) and the total DC open-loop gain is smaller than or equal to 1, the bilateral control scheme is stable.

A number of control schemes have been designed

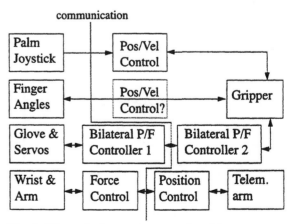

Fig 4. Bilateral teleoperator control architecture.

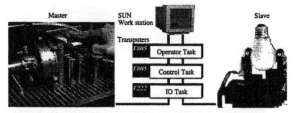

Fig 5. A 1 degree of freedom master slave system.

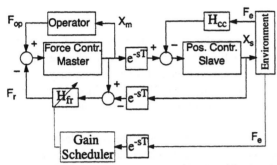

Fig 6. Bilateral position/force control with gain scheduling. H_{fr} is the stiffness of the master and H_{cc} is the inverse stiffness of the slave.

using this idea, which have been experimentally verified in Eusebi and Van der Ham (1994). The algorithms all had problems with position drift, which was alleviated during that investigation. The drawback of these schemes is the complexity. The following scheme is rather simple and effective. It has been tested in practice using a transputer based digital control system (figure 5) with a sampling time of 1 ms.

A control scheme has been developed that gives true 1:1 force and position control and can cope with time delay (see figure 6). Stiffness controllers at each end take care of the local tracking behaviour and communication is done basically using the position (and force is fed back to adapt the master stiffness). During contact the stiffness on both sides is equal, yielding accurate control of the force, see figure 7. When the gripper is moving freely the stiffness of the glove is adapted such that a small force is felt to indicate the tracking error. More about the scheme is explained in the paper by Van der Ham, *et al.* (1994).

In case of time delay (1 second or more) the system reacts sluggish. Experiments over internet of a system

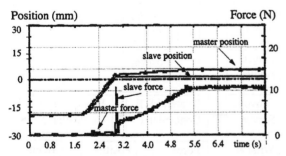

Fig 7. Bilateral position/force control of a 1 DOF master/slave system. The gripper moves toward a steel block, hits it at 3 seconds and applies a force during the rest of the experiment.

in Bologna and a system in Delft (with a time delay of 2 seconds) resulted in sluggish behaviour of the system. Though the system was still stable, the response time was confusing to the operator. Experiments with predictive control proved that the response time could be improved. Due to the complexity of the Unified Predictive Controller the maximum delay compensation that could be achieved was 40 ms due to memory problems (Nouwen 1994). The sampling time was 5 ms.

The glove carrier arm is controlled in a similar manner as the 1 DOF master. Using the Jacobian transpose J^T the cartesian forces are transformed to the joint forces of the robot. The Jacobian matrix is calculated relative to the base frame. This scheme has been tested on a kinematically redundant robot: 4 cartesian DOF and 6 joint space DOF (see figure 8). Due to the dynamics of the robot arm, certain resonance frequencies were filtered out using a low-pass filter at 10 Hz. The force reflection part, needed to feed back the tracking error, has not been tested yet.

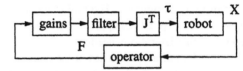

Fig 8. Force controlled master robot.

The remote robot will, in the beginning, be controlled by position only. Relying on the force being fed back by the gripper controller, force feedback will be incorporated later on in the same manner as with the 1 DOF slave system by using a stiffness controller (Craig, 1986). The position controller will use the cartesian position of the master arm and transform this to set points by calculating the inverse kinematics. The slave robot will usually be a kinematically redundant system (e.g. snake-alike manipulators). Therefore the analytic solution yields an infinite number of possible solutions. The set points for the controller will be close to each other, therefore a numerical algorithm has been developed, based on stiffness control, that will use the previous joint values as a starting point (see figure 9).

The numerical algorithm also allows the incorporation of joint and environmental constraints, thus solving the collision avoidance problem at the same time. The collision avoidance shuts off those joints

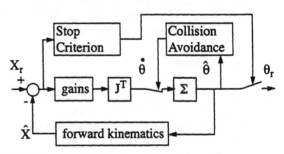

Fig 9. A numerical inverse kinematics solver.

which have reached their respective constraints. Only the joint constraints have been implemented until now, because the world model, which is contained within the autonomous control system, has not been coupled yet. Another problem is the conversion of cartesian constraints to adequate joint constraints.

For large changes in position the algorithm needs 19 ms on a INMOS T805 transputer, depending on how tight the constraints are set and the stop criterion. For small changes in position the algorithm needs 5 ms calculation time. The accuracy is set to 2 mm for each of the 4 cartesian degrees of freedom. In case of small movements, the numerical algorithm could be connected directly to the joint controllers, which will be investigated.

For stability reasons, the gains of the numerical algorithm have been chosen such that the maximum open-loop gain is lower than 1. This conservative approach ensures stability, but does not achieve the best performance possible. More research into stability issues is needed.

Due to the constraints and possible singularities, convergence can not always be guaranteed. Therefore a maximum number of iterations (10000) has been included within the stop criterion.

The problem of singularities also needs to be addressed. However, the stability does not suffer from this problem as it would using the Jacobian inverse. A number of heuristic rules solve this problem, but this approach is not flexible when using different robot arm configurations. A minimum value for the joint velocity may solve this problem. This is yet another aspect which needs to be investigated.

6. CONCLUSIONS

The kinematical configuration of the dexterous gripper is not equal to the human configuration. A mapping has been achieved which should give a natural feel to the operator, based on the assumption that the posture of the base finger joints is set before grasping.

A bilateral position/force controller has been tested that is stable in the event of time delay. Due to its adaptability, it behaves well during free motion and during constrained motion. In the event of time delay the system becomes sluggish. A simple form of predictive control may solve this problem.

The telemanipulator that is to be controlled is most probably kinematically redundant. A scheme to solve the inverse kinematics has been introduced that also copes with the joint constraints. A general solution to singularities and overall performance still needs to be found.

ACKNOWLEDGEMENTS

This research has been supported in part by the EC Teleman project, contract no. F12-0021.

REFERENCES

Anderson, R.J, M.W. Spong (1989). Bilateral control of teleoperators with time delay, *IEEE transactions on automatic control*, **Vol. 34**, pp. 494-501.

Craig, J.J. (1986). *Introduction to Robotics Mechanics and Control*, Chapter 8.9, Addison-Wesley Publishing Company, Reading Massachusetts.

Eusebi, A., A.C. van der Ham (1994). Experimental Verification of Bilateral Force/Position Control Schemes Using the Communication Network Analogy, *IEEE Conference on Decision and Control*.

Hannaford, B., L. Wood, D.A. McAffee and H. Zak (1991). Performance Evaluation of Six-Axis Generalized Force-Reflecting Teleoperator. *IEEE Transactions on Systems, Man and Cybernetics*, **Vol. 21, No. 3**, pp. 621-633.

Kazerooni, H., T.I. Tsay and K. Hollerbach (1993). A Controller Design Framework for Telerobotic Systems, *IEEE transactions on control systems technology*, **Vol. 1, No. 1**, pp. 50-62.

Kim, W.S. (1992). Developments of New Force Reflecting Control Schemes and an Application to a Teleoperation Training Simulator. *Proceedings of the 1992 IEEE International Conference on Robotics and Automation*, pp. 1412-1419.

Nouwen, H.J.O (1994). Adaptive Predictive Force/Position Control for Telemanipulation, *Proceedings of the Teleman Student Research Projects Congress*, pp. 133-138.

van der Ham, A.C., E.G.M. Holweg, W.Jongkind (1993). Future gripper needs in nuclear environments, *BNES proceedings of remote techniques for nuclear plant*, pp. 173-178.

van der Ham, A.C., G. Honderd, W. Jongkind (1994). A Bi-lateral Position-force Controller with Gain Scheduling for Telemanipulation. *IMACS International Symposium on Signal Processing, Robotics and Neural Networks*, pp. 282-285.

THE IMPORTANCE OF MOVEMENT FOR THE DESIGN OF TELE-PRESENCE SYSTEMS

C.J. OVERBEEKE & P.J. STAPPERS *

Laboratory for Form Theory, Delft University of Technology, Faculty of Industrial Design Engineering, Jaffalaan 9, 2628 BX Delft, the Netherlands. E-mail: c.j.overbeeke@io.tudelft.nl

Abstract. The experiments presented in this paper centre on closed-loop tasks and provide behavioural evidence about the importance of movement and interactivity for perception. Perception studies have too long been concerned with open-loop task, where the onlooker was forced to sit still, unable to interact with the scene he/she was looking at. The experiments concentrate on what humans need in tele-presence visualisations: movement parallax and depth perception, scaling the world by movement, and the trade-off between movement and image resolution.

Key Words. Movement, Teleoperation, Virtual reality, Resolution, Closed loops

1. INTRODUCTION

Recent years have shown a growing interest in the study of our senses as exploring perceptual systems (the approach initiated by Gibson, 1966, 1979), where the perceptual capabilities of human observers are studied while they are exploring or performing tasks involving perceptual and motor skills. These studies, recently dubbed 'active psychophysics' (Warren & McMillan, 1984; Warren, 1988; Flach, 1990), show much more potential for achieving measurements of human capabilities and the technical demands that must be satisfied to support them.

Two principles govern work at the Laboratory for Form Theory: the interplay between theory and implementation, each stimulating the other, and the attention for what people need to act in a (tele)presence environment, instead of for what machinery is able to do. Findings stress the importance of the distinction between open-loop and closed-loop tasks. In the former, the subject needs only to react, in the latter he receives feedback to his own actions, i.e., he can also explore. The theory of perception and action, as developed by J.J. Gibson (1979), is taken as a starting point, as our interest lies in what humans need to experience (tele)presence. Gibson's theory stresses the importance of active movement for perception. A technological consequence of this theoretical point led to the first application. In Experiment I a new system to get a 3D impression on a flat screen, was tested. This impression is realized by measuring the onlooker's head movements and synchronizing the movements of the camera to match them. In this way the onlooker gets the image corresponding with his viewing angle.

The second step was to investigate what sort of information onlookers need to get a 3D impression. First, how do people scale the world? How do they get information to be able to walk through a door or up a staircase? For this experiment a VR system was used, as all perceptual variables can be controlled in such a system. At the same time such a system allows the onlooker freedom of movement. Thus, in experiment II a navigation task, walking through a narrow aperture under VR conditions, was put to the test. Second, is movement really so important for perception, and even more important than the resolution of a screen? In experiment III the trade-off between movement and resolution was tested.

2. EXPERIMENT I: SPACE THROUGH MOVEMENT

The DVWS (Delft Virtual Window System) is based on movement parallax. Many 3D film and television systems present the viewer with a stereo pair of images, one for each eye. All of these systems are based on binocular disparity, i.e., on spatial differences. Monocular movement parallax provides the same information using differences over time. Movement parallax consists of a pattern of parallax shifts in an onlooker's field of view caused by the movements of the onlooker (Figure 1). This principle serves to create a 3D impression on a flat screen. An onlooker is sitting before a monitor. His/her head movements are recorded and fed through to a camera steering mechanism. The camera, recording a scene, thus makes the same movements. As the onlooker gets a view on the screen, corresponding with his viewing angle, he gets a 3D impression of the scene.

The system was put to the test. Only the major results of the experiment are reported here. Details can be found in Smets, Overbeeke, & Stratmann, 1987. Figure 2 shows the experimental conditions.

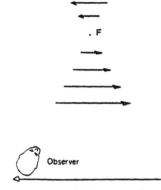

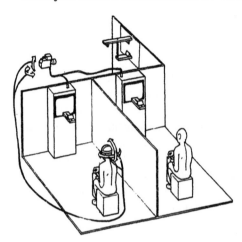

Fig. 1. When the onlooker moves from right to left looking at a point F (the fixation point), objects in front of F shift in the field of view against his/her movement, and objects behind F shift in the same direction.

Fig. 2. The experimental conditions used in experiment I.

2.1. Procedure

A camera records a wedge system, displayed on the two monitors. The frontal wedge is placed in front of the monitors. Subjects had to align this frontal wedge with the wedge apparently leaping out of the screen. Two conditions were compared: active (where the onlooker's head movements steered the camera) and passive (where the onlooker got the same image as the active one, without being able to steer the camera). In the 'real' condition the subject had to align real wedges (i.e., not depicted on a screen). All conditions were monocular. All subjects were students of Industrial Design Engineering. Each received a fee of US$ 5.

2.2. Results

Means and standard deviations (for active, passive, and real) are $\overline{X}_A = -1.70$ cm, $\overline{X}_P = 0.65$ cm, $\overline{X}_R = 0.71$ cm, $S_A = 1.8$ cm, $S_P = 5.30$ cm, $S_R = 0.83$ cm. The value of the mean of the active condition indicates that the wedge leaping out of the screen seems shorter. But that should be no problem, if the standard deviation (variability) is small. A systematic underestimation can be corrected. The standard deviation of the active

in front condition is small, as compared to the passive in front condition, where the subjects were unable to perform the task. None of the subjects in the active condition, however, found the task a strange one, which is an indication of the realism of the task.

2.3. Discussion

Results show that the coupling between head and camera movements is sufficient to enable an onlooker to make reliable depth estimates. The system has been patented world-wide (Smets, Stratmann, & Overbeeke, 1988, 1990). Several applications are being implemented in X-ray luggage inspection and endoscopy.

3. EXPERIMENT II: WALKING THROUGH A NARROW APERTURE

In the walking experiment, the subject's task was to walk in as comfortable a manner along a track of 12 meter length. Halfway the track was an aperture formed by two posts, which the subject was to avoid touching. For a narrow aperture, the subject would have to rotate his shoulders, the wider ones he could pass without any extra effort. This task has been used in perception studies (Warren & Whang, 1987), which found that the amount of shoulder rotation could be used as an index to the aperture's perceived width: shoulder rotation is maximal at very narrow apertures, and suddenly drops when aperture width is approximately 1.3 times shoulder width.This ratio seems to be generally applicable in navigation studies, a similar ratio is found to rule the comfortable passage of cars through gates (Shaw, Flascher, & Kadar, in press).

In this experiment changes in the perceived width-curve were studied for subjects performing the aperture-passing task in three conditions, as illustrated in Figure 3. One condition was 'real life', in which the subject performed the task as in the studies cited above; another was 'Virtual Reality', in which the subject performed the task in the same environment but with visual feedback from a matching simulated environment; the third condition was intermediate between the above two, in which the subject performed the task wearing a 'mock-up' of a VR simulator, imposing on him the same restrictions of field-of-view and of movement, but not those of image resolution, graphics quality, and time lag. With this set-up, the effect of two types of technical limitations of present-day VR simulators could be evaluated separately in two stages.

3.1. Apparatus

The simulator used was a W Industries (now Virtuality Group) Virtuality SU-1000 running the VGS graphics library. The head-mounted display (Visette) weighs 2.9 kg (!), and provides binocular 276*372 pixel images with a horizontal field of view of 90 degrees. Time

Fig. 3. Conditions in Experiment II: from left to right natural, mock-up, and VR.

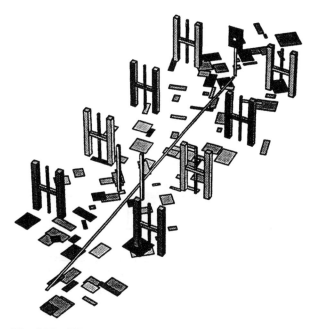

Fig. 4. The VR environment used in Experiment II.

lag was approximately one-fifth of a second, well below the 300 msec suggested for manipulative tasks by Funda, Lindsay, and Paul (1992), but well over the 100 msec seemed necessary by others. Image refresh was 50 Hz, image updates approximately 15 Hz. Range limitations due to the Polhemus sensor (1660 mm) were overcome by placing the entire 200 kg. simulator on a custom-built rails along the walking track, and moving it manually to follow the walking subject; displacements of the simulator were registered using an angular encoder. The mock-up consisted of a weighted belt to simulate the Visette's cable- belt, and a 2.9 kg helmet-mockup with 90 deg view-restrictor and ear mufflers (similar mock-ups have also been used in validating experiments with inverting goggles; see Dolezal, 1982). The aperture posts were two pieces of yellow plastic piping, 40 mm diameter, suspended from the ceiling. These posts were physically present in all conditions, so that even in the VR condition the subject would get tactile feedback if he failed to shoulder-rotate at a narrow aperture. The simulated environment, depicted in Figure 4, was optimized to produce as vivid a three-dimensional impression as possible with the available graphic means, based on principles of ecological optics (Gibson, 1979): the H-bars prominently exhibit (dis)occlusion, the rectangular random patches on the ground show texture and hue gradients. In the VR condition, the subjects' responses (shoulder rotations) were registered using a shoulder-attached second Polhemus sensor. In the 'real life' and 'mock-up' conditions these were registered optically using reflective markers. All subjects were students of Industrial Design Engineering. Each was paid ca. 10 US$ for each afternoon session. Six students participated in the pilot series, five in the full experimental series.

3.2. Procedure

Each session of the experiment consisted of a number of trials. In the 'free' and 'mock-up' conditions, they performed 9 repeat measurements * 8 aperture widths = 48 trials. In the 'VR' conditions 3 sessions * 6 repeat measurements * 8 aperture widths (students in the pilot series did only one VR session). Within each condition, trials were performed in randomized order. Subjects were instructed for each trial to 'wait until a sign was

given and then walk as comfortably as possible to the end of the marked track, without touching the aperture'. In the VR condition they had to wait at the end of the track, so the simulator could be halted and put in reverse. For each trial, the maximal angle of shoulder rotation during passage of the aperture was recorded.

3.3. Results

The repeated measurements allow a calculation of mean and standard deviation of the shoulder rotation at each aperture width, resulting in 'perceived aperture width' curves. The thickness of the curves indicated the standard deviation, the centre of the curve joins the means. These curves are shown for the three conditions in Figures 5 and 6 for three subjects, labeled M-1, M-2, M-3. Figure 5 depicts the comparison between the 'free' and 'mock-up' conditions. Responses in the 'free' condition are shown as a grey band in the background, responses in the 'mock-up' condition as separate mean/standard deviation markers in the foreground. Figure 6 depicts the comparison between 'mock-up' and 'VR' conditions in the same format. Even though the number of responses was quite large (all in all, subjects and simulator travelled over 30 km during the experiments!), variance within and between subjects complicate a quantitative analysis. Here only a qualitative description can be given at the hand of Figs. 7 and 8. (Full details and discussion of the responses can be found in Stappers, 1992.) The mean curves follow the same shape of the values from the literature, i.e., a base rotation level of 10 degrees at the widest apertures, rising to a large rotation at aperture widths around 1.3 times shoulder width, but with marked individual differences. In both conditions, standard deviations are largest around this value.

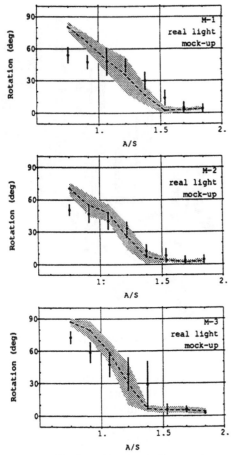

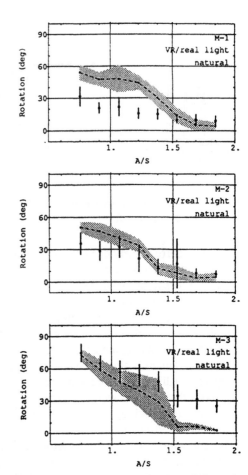

Fig. 5. Comparing results for 'free'(shaded) and 'mock-up'(symbols) conditions.

Fig. 6. Comparing results for 'mock-up'(shaded) and 'VR'(symbols) conditions.

3.4. From 'free' to 'mock-up'

See Figure 5. The mock-up worn in the 'mock-up' condition clearly has behavioural consequences. for all three observers, the mean curves obtained in the 'mock-up' condition lie to the right of those obtained in the real life condition, indicating subjects rotate more readily (take fewer risks) with the mock-up. Furthermore, at the narrowest apertures the 'mock-up' curve lies well below the real life curve, the maximal angle of shoulder rotation not exceeding 70 deg (not even 50 deg in the case of subject M-1). The total variance of the curves, however, did not consistently increase (for two subjects not depicted here it did even decrease). This last effect may have been due to training, as the real life trials preceded the 'mock-up' trials.

3.5. From 'mock-up' to VR

Figure 6 shows the behavioural effects of limited resolution and time delay added to the reduced field of view of the 'mock-up' condition. Again the global shape is retained fairly well. For two of the subjects, M-1 and M-2, maximal rotations get smaller. Only subject M-3 keeps making rotations up to 75 deg at the narrowest apertures. For all subjects the range of responses contracts and the standard deviations become more uniform over the curve, indicating that the action boundary is less pronounced.

3.6. Discussion

The three-step comparison shows clearly that some of the effects of VR on task behaviour are due to computer-graphics qualities like resolution, picture complexity, and time delay, others to interface qualities like field- of-view and encumbrance (weight of helmet and sensors), and that the effect of these two sources can be separated. Furthermore, observation showed that the subjects quickly trusted their ability to walk in the virtual environment, and made very little mistakes in navigating, e.g., there were very few collisions with the posts, subjects rotated at the appropriate moment.

4. EXPERIMENT III: MOVEMENT VERSUS RESOLUTION

The point this experiment wants to make is that resolution is much less important for the interactive tasks where immersive VR is typically brought to bear, (where the user can explore his environment by moving his head and body), than it is in classical computer graphics applications (where static detail becomes important because the user can only explore, by directing his gaze, over this single picture).

Fig. 7. Apparatus used in experiment III.

4.1. *Apparatus*

Subjects were fitted with a head mounted system containing a display and, depending on the experimental condition, a micro camera (Figure 7). The camera image is fed through a video-processor to manipulate aspects of the image stream, e.g. the number of grey values, pixels, and frames per second. The camera is a Panasonic WV-CD1 micro-camera. The camera head has a diameter of 17 mm and a length of 48 mm. It weighs approximately 20 g. The obtained PAL 625 lines video signal is manipulated with a Panasonic WJ-MW10 production mixer. The display is an electronic viewfinder of a Sony Video Hi 8 camera type CCD-V900E. The size of the screen is 11.0 mm x 8.2 mm and it weighs about 75 g. The helmet mounted system including camera, viewer and helmet, weighs 350 g. The visual angle obtained by the combination of the camera and the ocular lenses is 60 degrees. The enlargement factor is 1.

4.2. *Procedure*

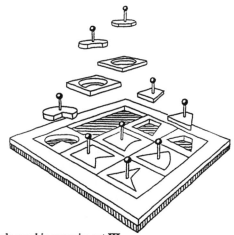

Fig. 8. puzzle used in experiment III

The subjects have to complete a specially designed puzzle, depicted in Figure 8. This puzzle excludes any learning by place, since the location of each piece was randomly varied throughout the trials. It was used with the explicit instruction and control that the pieces only be handled by the pegs as to prevent tactile exploration. Subjects were five volunteer students in Industrial Design Engineering, who had no previous experience in psychological experiments. None had an uncorrected visual problem.

Table 1 Design and data Experiment III. (t.o.: trial ended on time-out when no piece of the puzzle had been placed after 600 s)

movement	SS	PAL 625 lines	Mosaic 36×30	Mosaic 18×15
active	1	33	73	242
	2	35	141	259
	3	32	51	210
	4	24	190	t.o.
	5	27	88	t.o.
passive	1	56	143	t.o.
	2	51	596	t.o.
	3	41	153	t.o.
	4	48	531	t.o.
	5	101	600	t.o.
still	1	35	422	t.o.
	2	49	123	t.o.
	3	41	123	384
	4	97	472	t.o.
	5	56	436	t.o.

Table 2 Analysis of Experiment III. (SR = Spatial Resolution; MO = Movement; *: $p < 0.05$, **: $p < 0.01$).

Source	SS	df	MS	F
SR	388968.53	1	388968.53	29.26**
MO	140221.07	2	70110.53	5.27*
SR×MO	92785.87	2	46392.93	3.49*
Blocks	98595.53	4	24648.88	1.85
Residual	265844.87	20	13292.24	
Total	986415.87	29		

4.3. *Design and Results*

The design was simplified by leaving out the temporal resolution manipulation. The independent variables spatial resolution and movement were retained with identical levels. Design and results are shown in Table 1 and Table 2. The lowest spatial resolution condition was excluded from the ANOVA, since it contained a lot of null cells. It can be seen however that in the active condition three out of five subjects still can resolve the puzzle, a much better result than in both other movement conditions (passive and still).

The main spatial effect is significant (p<0.01). Movement is also significant (p<0.05). The interaction between spatial resolution and movement is significant

193

(p<0.05). With decreasing spatial resolution observers perform better when actively controlling the camera with their head movements as compared to conditions were the camera was passively moved or still.

4.4. *Discussion*

The results of this experiment shows that the added active movement of VR can compensate for losses in spatial resolution in a way that animated graphics cannot. The advantages of VR conditions match up well with similar results from medical prosthetics, tele-operation, and Gibsonian and Gestalt perception theory (a.o., Gibson, 1979). For instance, Bach-y- Rita (1972) found in his Tactile Visual Substitution System TVSS, in which digitized camera images are presented to a congenitally blind subject by means of an array of vibrating pins placed against the skin of his back, that no recognition takes place unless the subject himself controls the movements of the camera.

5. GENERAL DISCUSSION

Results of Experiments I to III fit in well with Sheridan's (a.o., 1992) three factor model of tele-presence, where three independent factors together add up the quality of presence realized by a tele-operator system. These three factors are (i) the extent of sensory information (a.o., resolution), (ii) the amount of control over sensors (called exploration and movement in this paper), and (iii) the ability to modify one's environment (interactivity). In the applied research the above experiments are also used for the development of non-immersive systems for tele-operation and non- immersive surgery using the Delft Virtual Window System (Overbeeke & Stratmann, 1988; Smets, Stratmann, & Overbeeke, 1988, 1990), in which movement parallax is produced by adapting the viewpoint of a real or virtual camera to match the displacements of the observer's head in front of the display. For several application areas, a.o., X-ray inspection, these easily outperform static binocular displays (Smets, 1992, in press).

6. REFERENCES

Bach-y-Rita, P. (1972) *Brain mechanisms in sensory substitution.* Academic Press. London, UK.

Dolezal, H. (1982) *Living in a World Transformed. Perceptual and performatory adaptation to visual distortion.* Academic Press. New York, NY.

Flach, J.M. (1990) Control with and eye for perception: Precursors to an active psychophysics. *Ecological Psychology*, 2(2): 83-110.

Funda, J. Lindsay, Th.S., and Paul, R.P. (1992) Teleprogramming: Toward delay- invariant remote manipulation. *Presence* 1(1).

Gibson, J.J. (1966). *The senses considered as perceptual systems.* Houghton Mifflin. Boston MA.

Gibson, J.J. (1979) *The Ecological Approach to Visual Perception.* Houghton Mifflin. Boston, MA.

Overbeeke, C.J., & Stratmann, M.H. (1988) *Space through Movement.* Doctoral dissertation, Delft University of Technology, Delft, the Netherlands.

Sheridan, T. (1992) Musings on telepresence and virtual presence. *Presence* 1(1), 120-126.

Smets, G.J.F. (1992) Designing for Telepresence: The interdependence of movement and visual perception implemented. In: *Fifth IFAC/IFIP/IFORS/IEA symposium on analysis, design, and evaluation of man-machine systems. (International Federation on Automatic Control)*, 9-11 June 1992. The Hague, Netherlands, p. 1-7.

Smets, G.J.F. (in press). Designing for telepresence. In J.M. Flach, P.A. Hancock, J. Caird and K.J. Vicente (Eds.), *The ecology of human-machine systems*, Erlbaum. Hillsdale NJ.

Smets, G.J.F., Overbeeke, C.J., & Stratmann, M.H. (1987) Depth on a flat screen. *Perceptual & Motor Skills*, 64: 1023-1034.

Smets, G.J.F., Stratmann, M.H., & Overbeeke, C.J. (1988) Method of causing an observer to get a three-dimensional impression from a two-dimensional representation. *US patent 4,7575,380.*

Smets, G.J.F., Stratmann, M.H. & Overbeeke, C.J. (1990) Method of causing an observer to get a three-dimensional impression from a two-dimensional representation. *European Patent 0189232.*

Shaw, R.E., Flascher, O.M., & Kadar, E.E. (in press) Dimensionless invariants for intentional systems: Measuring the fit of vehicular activities to environmental layout. In Flach, J.M., Hancock, P.A., Caird, J., & Vicente, K.J. (Eds.) *The Ecology of Human-Machine Systems.* Erlbaum. Hillsdale, NJ.

Stappers (1992) *Scaling the Visual Consequences of Active Head Movements: A study of active perceivers and spatial technology.* Doctoral dissertation, Delft University of Technology, Delft, the Netherlands,

Warren, R. (1988) Active Psychophysics: Theory and Practice. In: Ross, H.K. (Ed.) Fechner Day '88: *Proceedings of the 4th Annual Meeting of the International Society for Psychophysics*, p 47-52. Stirling, Scotland.

Warren, R. & McMillan, G.R. (1984) Altitude control using action- demanding interactive displays: Toward an active psychophysics. In *Proceedings of the 1984 Image III Conference, p 37-51, Air Force Human Resources Laboratory*, Phoenix, AZ.

Warren, W.H., & Whang, S. (1987) Perceiving affordances: Visual guidance of walking through apertures. *Journal of Experimental Psychology: Human Perception and Performance* 13(3): 371-383.

A LABORATORY EVALUATION OF FOUR CONTROL METHODS FOR SPACE MANIPULATOR POSITIONING TASKS

Eric F.T. Buiël and Paul Breedveld

*Delft University of Technology, Department of Mechanical Engineering and Marine
Technology, Laboratory for Measurement and Control (Man-Machine Systems Group),
Mekelweg 2, 2628 CD Delft, The Netherlands*

Abstract: A space manipulator is a lightweight robotic arm mounted on a space station or a spacecraft. If the manipulator is manually controlled from a remote site (teleoperation), the pictures from cameras in the neighbourhood of the manipulator are the only aids the human operator can use. Besides, manipulator dynamics and the possible presence of time delays in the control loop also complicate manual control.
In space manipulator positioning tasks, the manipulator hand (the end-effector) has to be positioned accurately in front of a known (physical) target. This paper evaluates four control methods to position a space manipulator by hand with a force activated input device with six Degrees-of-freedom. With two of these methods, forward and downward end-effector control, the operator controls the movements of the end-effector ('forward' resp. 'downward' refer to the direction in which the input device must be pushed to let the end-effector move forward in the camera picture shown on the interface console in the remote control room). With the two other methods, the target is the controlled object. The results of man-machine experiments clarify that subjects by far prefer end-effector control above target control. Furthermore, subjects prefer controlling the end-effector in downward mode instead of in forward mode.

Keywords: Teleoperation, Manual Control, Evaluation, Graphic Displays

1. INTRODUCTION

At Delft University of Technology (DUT), Department of Mechanical Engineering and Marine Technology, manual control of a space manipulator is an object of study (Bos, 1991). A space manipulator is a lightweight robotic arm mounted on a space station or a spacecraft. There, it performs specific maintenance and inspection tasks, e.g. the replacement of Orbital Replaceable Units (ORU's) containing scientific experiments (see Figure 1). There are two ways to fulfil these tasks under manual control:

1. <u>Extra Vehicular Activities (EVA)</u>
   An operator floating in the neighbourhood of the manipulator controls the manipulator movements manually.

2. <u>Teleoperation</u> (Sheridan, 1992)
   Here, the operator controls the manipulator movements from a remote location, e.g. a space station's manned module or a ground station on earth.

To the operator, teleoperation is much safer than EVA. By using this method, he does not need to go outside the spacecraft anymore; the manipulator movements can be controlled with the help of the pictures from cameras installed in the neighbourhood of the manipulator and the pictures from cameras mounted on the manipulator. All these pictures are shown on a monitor at the remote site.

To the operator, manual control of a space manipulator with the help of camera pictures shown at a remote location is a difficult task. At first, the lack of spatial information in these camera pictures is a major problem. Besides, task execution suffers from the manipulator dynamics: because of the lightly constructed links, the manipulator will be flexible. Finally, when the operator controls the manipulator on earth, time delays are introduced in the control loop. These delays are caused

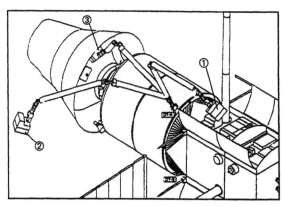

Figure 1 An activity of a space manipulator: the replacement of ORU's (Haman and Bentall, 1989)

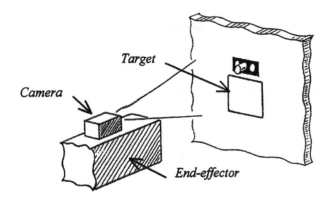

Figure 2 The space manipulator positioning task

by the transmission of control signals from earth to space and the transmission of camera signals from space to earth.

To help the operator overcome the mentioned problems, new operator aids are being developed at DUT. Current research is aimed at the development of a new man-machine interface for an elementary subtask of a space manipulator: the positioning task (see Figure 2). Before the manipulator is able to grasp and displace an object (e.g. a damaged satellite or an ORU), the manipulator hand (the *end-effector*) must be positioned accurately in front of the object (the *target*) without damaging it. To be able to fulfil this task under manual control, the operator has to take control over either the target or the end-effector (the *controlled object*). Here, all Degrees-of-freedom (Dof's) of this object (three translations and three rotations) need to be controlled. If this task is performed by teleoperation, and the distance between the end-effector and the target has decreased to less than a few centimetres, it is difficult to detect whether the end-effector and the target collide.

Breedveld (1995) has developed the concept for the new man-machine interface for space manipulator positioning tasks. At the display-side of the interface, a single camera picture is shown: the picture from a camera that is attached to the end-effector (see Figure 2). To assist the operator in deriving spatial information from this picture, a graphical camera overlay is added. At the control-side of the interface, a commercially available input device with six Dof's is used: the Spaceball®[1] (see Figure 3). By means of this device, the operator is able to control the translational and angular velocity of the controlled object. The *magnitude* of the force (resp. torque) applied to the Spaceball determines the magnitude of the object's translational (resp. angular) velocity. The *direction* of the applied force (torque) determines the direction of the object's translational (angular) velocity. So, if the user grasps the Spaceball as if he grasps a car's gear lever, he might feel it as if he grasps the controlled object (*virtual grasping*).

At the time the interface was being developed, the choice of the *control method* appeared to be a delicate one. The

---

[1] Spaceball is a registered trademark of Spatial Systems Inc.

control method represents the way the Spaceball movements in the remote control room are mapped to the manipulator movements in space. In this paper, the experimental outcomes of a comparison between four control methods are presented:

1. *Forward end-effector control*
2. *Downward end-effector control*
3. *Forward target control*
4. *Downward target control*

Detailed information about the work presented here can be found in (Buiël, 1994).

2. FOUR CONTROL METHODS FOR SPACE MANIPULATOR POSITIONING TASKS

2.1 Forward and downward end-effector control

With both forward and downward end-effector control, the end-effector is the controlled object (see Figure 3). The terms 'forward' and 'downward' refer to the direction in which the Spaceball must be pushed to let the end-effector move in the direction perpendicular to the camera lens (all other Spaceball Dof's are mapped to end-effector Dof's in accordance with the chosen direction). With forward end-effector control, the operator pushes the Spaceball forward to let the end-effector move in that direction (see Figure 3a). To do this with downward end-effector control, the Spaceball is pushed downward (see Figure 3b).

2.2 Forward and downward target control

In Figure 4 the principle of forward and downward target control is clarified. Now, with both methods, the target is the controlled object. Again, the terms 'forward' and 'downward' refer to the direction in which the Spaceball must be pushed to let the controlled object move in the direction perpendicular to the end-effector camera lens.

Obviously, it is not possible to control the movements of the target object in space directly. Instead, the operator controls the target's velocity relative to the end-effector. After the operator has set the desired magnitude and direction of that velocity, the manipulator controller computes the magnitude and direction of the end-effector velocity needed to realise the demanded target velocity. In this way, the operator feels as if he is controlling the movements of the target, but in fact still the end-effector is the controlled object. Because of this, in the end-effector camera picture, the objects in the neighbourhood of the target will always move in the same direction as the target (just like with end-effector control).

2.3 Discussion

To the operator, target control is totally different from end-effector control. With end-effector control, the perception of the end-effector movements may be compared to the perception of locomotion in daily life. According to Gibson (1979), the optical flow patterns in the retinal picture of the human eye form an essential cue

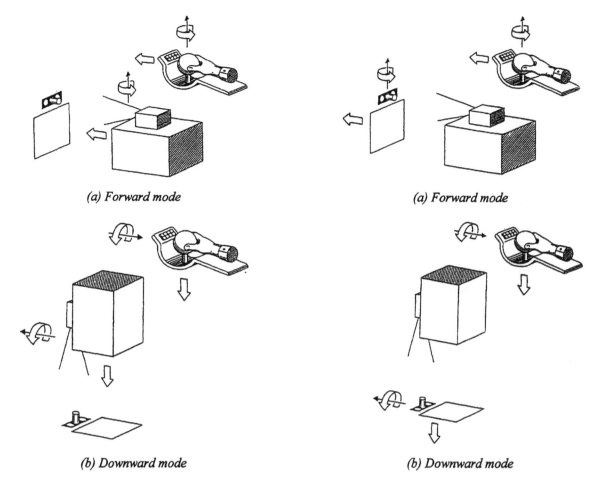

(a) Forward mode   *(a) Forward mode*

(b) Downward mode   *(b) Downward mode*

Figure 3 End-effector control   **Figure 4** Target control

for the perception of locomotion. If the operator utilises the flow patterns in the end-effector camera picture (i.e. the movements of the target and other objects visible in that picture) in a similar way, he might more or less become familiar with the remote environment and in some sense feel as if he is actually *moving* in space (telepresence, (Sheridan, 1992)). In that case, the choice between forward- and downward end-effector control (i.e. the choice of the *control mode*) might well depend on whether the operator feels as if he is moving forward or downward.

By comparison with end-effector control, target control might result in a lower degree of telepresence. Here, the operator '*stands still* on the end-effector and directs the target towards him'. In this way, target control corresponds more or less to remote control of a model aeroplane. At first sight, this method seems to be less usual in daily life. However, with target control the movements of the Spaceball in the remote control room correspond exactly to the target movements seen in the end-effector camera picture. In fact, the principle behind this method is the same as the principle behind the control of the movements of the arrow-pointer used in window-based software. In this software, the movements of the mouse input device correspond exactly to the movements of the arrow-pointer.

In order to find out which method is most easy to use for the human operator, two series of man-machine

experiments have been performed. The first series have been aimed at finding the favourable controlled object; the second series have been aimed at finding the favourable control mode.

3. CHOICE OF THE CONTROLLED OBJECT

To find the favourable controlled object, the members of a group of twelve subjects (all students in Mechanical Engineering, without any experience in tasks like the space manipulator positioning task) have all executed the same experiment in a simulator. This simulator has been designed specially for man-machine experiments with the newly developed man-machine interface. In the simulator, the movements of the six Dof European Robotic Arm[2] (van Woerkom *et al*, 1994) are simulated by means of a Silicon Graphics Indigo II Extreme graphical workstation. This computer animates the pictures from cameras attached to that manipulator in real time. In the simulator, subjects can perform a generalised positioning task in predefined circumstances.

3.1 The generalised positioning task

At the start of the generalised positioning task, the end-effector is at one metre distance from a particular target

---

[2] The European Robotic Arm (ERA) is developed at Fokker Space and Systems in Leiden, The Netherlands. At the end of the 20th century, it might become operational on the new global space station Alpha.

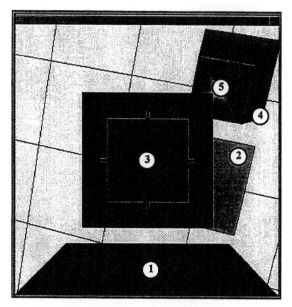

Figure 5 The animated end-effector camera picture
(① end-effector top-side, ② target, ③ frosted glass,
④ insert-box, ⑤ pyramid)

in space. To finish the task correctly, the subject has to position the front side of the end-effector straight in front of this target, without coming into collision with the target itself. While moving the Spaceball, subjects can watch the results of their control actions in the animated end-effector camera picture.

Figure 5 shows the animated end-effector camera picture as seen at the start of the task (① end-effector top-side, ② target). To assist the operator in deriving spatial information from this picture, a graphical camera overlay is added: the *Pyramid Display* (Breedveld, 1995). This overlay contains three graphical objects: the frosted glass (③), the insert-box (④) and the pyramid (⑤). To the operator, the frosted glass seems to be attached to the end-effector camera lens; the insert-box and the pyramid seem to be attached to the target. When the end-effector is very close to the target, the position and orientation of the end-effector relative to the target can immediately be perceived from the intersections of the frosted glass with the insert-box and the pyramid.

3.2 The experimental design

Each experiment lasted about three hours. With each control method, the subject could practise the generalised positioning task for approximately one hour. To compensate for learning effects, six subjects started with an one-hour-training for end-effector control; the others started with an one-hour-training for target control. At the end of each training, the subject's global task performance was measured by means of suitable performance measures (completion time, number of controlled Dof's, etc.). Finally, in a verbal interview, the subject was asked to express his or her opinion about the two practised control methods.

To be able to detect possible interaction between the two experimental variables (i.e. *controlled object* and *control mode*) the control mode was varied among the subjects.

Six subjects practiced both control methods by means of forward control; the others practiced both methods by means of downward control. In each of these groups three subjects started with end-effector control; the others started with target control.

No manipulator dynamics were implemented in the mathematical model of the space manipulator. The end-effector translational and angular velocities were limited to 0.1 m/s and 10°/s resp. (so, the minimal time possible to finish the generalised positioning task was 10 seconds). To finish the generalised positioning task, position-misfits needed to be reduced to less than 2 millimetres; orientation-misfits needed to be reduced to less than 1°.

3.3 Results

During the experiments with the two control methods, for most subjects target control turned out to be difficult. This is illustrated in Figure 6. For eleven subjects, the upper part of this figure contains the mean completion time and standard deviation of the last five task runs of the training for end-effector control (for subject 3 no data were available). The overall mean completion time is 68 seconds. The lowest part of the figure contains the same data for target control. Here, the overall mean completion time is 172 seconds. Subjects 1 and 5 found it very difficult to perform the positioning task with target control. During one hour of training, they never finished a task run correctly.

The detected performance differences indicate that the end-effector logically turned out to be the preferred controlled object. Eleven subjects said they preferred end-effector control to target control. Only subject 6 had no preference for a specific control method.

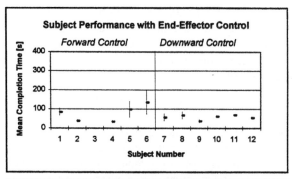

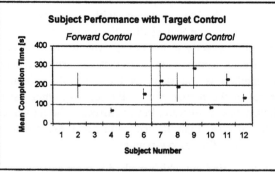

Figure 6 Subject performance during
the controlled object experiments

In their explanation for their preference for end-effector control, six subjects referred to daily life experiences. During the two training's it had appeared that end-effector control coincided with a well-known situation: the situation in which an object is moved in a stationary environment. From the explanations of four subjects, it can be concluded that they initially identified the end-effector as the moving object; they identified the target as (a part of) the stationary environment. Two other subjects even compared the end-effector with their own hand that is moving through their natural environment. These explanations support Gibson's theory of locomotion perception (Gibson, 1979).

With target control, in fact the target is the moving object and the end-effector top side is the stationary environment. Almost all subjects said they had found it difficult to imagine this. Since the target does not move relative to the background of the animated end-effector camera picture, it seemed that not only the target, but also the entire background was the controlled object! Especially at the start of the generalised positioning task (when the distance between the target and the end-effector is relatively large), this appeared very strange to many subjects. It seemed that a very large object (i.e. the target and the background) was controlled relative to a very small environment (i.e. the end-effector top side). This situation is hardly observed in daily life.

4. THE CHOICE OF THE CONTROL MODE

The results of the controlled object experiments clarified that end-effector control was preferred almost unanimously to target control. So, the control mode experiments all have been performed with end-effector control.

4.1 The experimental design

The experimental design of the control mode experiments was the same as the design of the controlled object experiments. A (new) group of six subjects was trained shortly for forward and downward end-effector control. Three subjects practiced forward end-effector control first; the others practiced downward end-effector control first. At the end of each training, the subject's global task performance has been measured. In a verbal interview, the subject was asked to express his or her opinion about the two practiced control modes.

4.2 Results

Figure 7 shows the global task performance at the end of each of the short training's for forward control and downward control. For each subject, the upper part of this figure contains the mean completion time and standard deviation of the last five task runs of the training for forward control. The overall mean completion time is 75 seconds. The lower part of the figure contains the same data for downward control. Here, the overall mean completion time is 38 seconds. So, just like in the previous experiments, a performance difference occurred between the two compared control methods. In accordance with this difference, five subjects said that they

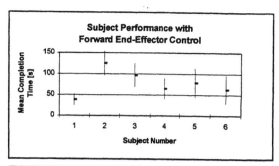

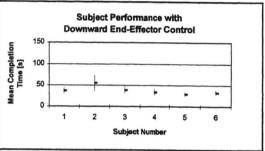

Figure 7 Subject performance during the control mode experiments

finally preferred downward control to forward control. Subject 1 had no preference for a specific control mode.

From the remarks made by the six subjects, three important advantages of downward end-effector control can be extracted:

1. Downward control 'fits' very well to the Spaceball's ergonomic design. Five subjects said that to them, the end-effector translation in the direction perpendicular to the end-effector camera lens is the main movement direction in the end-effector positioning task. While directing the end-effector in this direction, the orientation of the end-effector relative to the target can be corrected gradually (multivariable control). Also, the lateral and vertical distances between the end-effector and the target can be reduced gradually. For downward control, the main movement direction is mapped to the vertical central axis of the Spaceball; for forward control, it is mapped to the longitudinal central axis of the Spaceball. According to the subjects, the first mapping is somewhat easier to handle when performing multivariable control. This is because the Spaceball is grasped *from above* (see Figure 8). Then it is easier to combine some relatively small correction forces with a relatively large downward force than to combine these with a relatively large forward force. In the last case, mainly carrying out a small correction force in an upward direction is difficult.

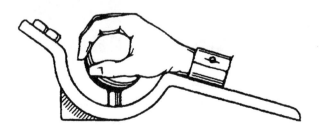

Figure 8 The Spaceball is grasped 'from above'

2. With downward control, the tabletop can serve as a control reference plane. In window-based software, the movements of the mouse parallel to the tabletop (the control reference plane) correspond to the movements of the arrow-pointer parallel to the front plane of the monitor (the movement reference plane). To the user, the tabletop is more or less equivalent to that plane. According to three subjects, the principle of downward control is the same. The top of the table on which the Spaceball is placed is the control reference plane; the plane parallel to the end-effector camera lens, the monitor plane, is the movement reference plane.

3. The animated end-effector camera picture looks like a downward view. At the start of each experiment, the participating subject was asked in which way he interpreted the animated end-effector camera picture. Five subjects said that they felt as if they were looking downward, when they saw the end-effector moving towards the target.

5. CONTROL BEHAVIOUR

According to the subjects, downward end-effector control is a very convenient and intuitive control method for the positioning task. To confirm this, six subjects have been further trained for this method (these subjects also participated in the control method experiments). At the moment they were able to perform eight successive task runs with sufficient and almost constant performance, the training was ended and their control behaviour was analysed. To attain the performance level mentioned, subjects had to perform eight successive task runs without any collisions between the end-effector and the target. Also, the standard deviation of the eight task run completion times had to be less than 3 seconds.

To most subjects, it took approximately 75 task runs to attain the desired performance level. At the end of the training, all subjects were able to finish a task run in approximately 22 seconds. From the analysis of control behaviour, it appears that the operator subdivides the run in two stages (see Figure 9). In the first stage, efficient multivariable control is used: while directing the end-

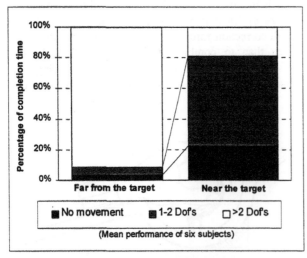

Figure 9 Mean number of controlled Dof's in two stages of the generalised positioning task

effector toward the target, the orientation of the end-effector is adjusted gradually (translations and rotations are controlled at the same moment). Most of the time three to six of the end-effector Dof's are controlled simultaneously. At the moment, the end-effector is less than 10 centimetres distant from the target, the operator switches to scalar control. Here, translations and rotations are controlled separately. On average, one or two end-effector Dof's are controlled at the same time.

6. CONCLUSIONS

The outcomes of the controlled object experiments discussed above clarify that end-effector control can make the operator feel as if his own hand is approaching a physical object in space. The degree of visual telepresence attained with this control method is by far sufficient for successfully performing space manipulator positioning tasks. However, the results of the control mode experiments show that the attained degree of telepresence is not the only criterion for selecting a suitable control method for teleoperation tasks. Other ergonomic aspects (like the Spaceball's ergonomic design) play also a major part in the choice for downward end-effector control instead of forward end-effector control.

ACKNOWLEDGEMENTS

This research is supported by the Dutch Technology Foundation STW (Utrecht, The Netherlands) and Fokker Space and Systems B.V. (Leiden, The Netherlands).

REFERENCES

Bos, J.F.T. (1991). *Man-Machine Aspects of Remotely Controlled Space Manipulators.* PhD Thesis. Delft UT, Dept. of Mechanical Engineering, Delft.

Breedveld, P. (1995). *The Development of a Man-Machine Interface for Telemanipulator Positioning Tasks.* Proc. 6th IFAC Symposium on Analysis, Design and Evaluation of Man-Machine Systems, Cambridge, Massachussetts (to be published).

Buiël, E.F.T. (1994). *A Laboratory Evaluation of four Controller Configurations for a teleoperated Space Manipulator.* Internal Report N472. Delft UT, Dept. of Mechanical Engineering, Delft.

Gibson, J.J. (1979). *The ecological approach to visual perception.* Houghton Mifflin, Boston.

Haman, R.J. and R.H. Bental (1989). *The Hermes Robot Arm - System Description.* Proc. 2nd Europe in Orbit Symposium, Toulouse.

Sheridan, T.B. (1992). *Telerobotics, Automation, and Human Supervisory Control.* The MIT Press, Cambridge, Massachusetts.

Woerkom, P.Th.L.M. van, A. de Boer, M.H.M. Ellenbroek and J.J. Wijker (1994). *Developing algorithms for efficient simulation of flexible space manipulator operations.* 45th Congress of the Int. Astronautical Federation, Jerusalem.

The Development of a Man-Machine Interface for Telemanipulator Positioning Tasks

Paul Breedveld

Delft University of Technology, Faculty of Mechanical Engineering & Marine Technology, Laboratory for Measurement & Control, Man-Machine Systems Group, Mekelweg 2, 2628 CD Delft, the Netherlands.

Abstract. Currently, the space manipulator ERA (European Robot Arm) is being developed in Europe. It will be used on the international space station Alpha. When the ERA is controlled in manual control, its movements are monitored with camera's, like the end-effector camera, which is used for positioning the 'hand' of the manipulator. At the Delft University of Technology in the Netherlands, research is done on the development of a man-machine interface for the ERA. Herewith, the movements of the manipulator are animated on a graphical workstation. The input device is a 6 degrees of freedom Spaceball controller, which is used to control the velocity of the end-effector. In this paper, the development of a novel graphical display is discussed, that improves the spatial observability of the end-effector camera picture. In this display, the spatial information is presented by means of intersections between 3-dimensional graphical objects. Also, a control method is described, that transforms the movements of the Spaceball into movements in the end-effector camera picture. Finally, some results of man-machine experiments with the display and the control method are given.

Keywords. Man/machine systems, Man/machine interfaces, Manual control, Telemanipulation, Telerobotics, Graphic displays.

1. INTRODUCTION

Currently, a *space manipulator* is being developed in Europe. The manipulator is called *ERA*, which is short for 'European Robot Arm'. It is a 10 m long anthropomorphic manipulator with 6 degrees of freedom (DOF's). Fig. 1 shows a schematic drawing of the ERA. The links of the manipulator are made of lightweight carbon fibre. Its joints are driven by electric brushless DC motors. The 'hand' of the ERA is called *end-effector*. On the end-effector front, a gripper is placed, which is used for grasping an object. In the early 21th century, the ERA will be placed on the Russian part MIR 2 of the international space station Alpha. It will be used for a variety of space activities. One of the most important activities will be the displacement of *ORU's* ('Orbital Replaceable Units'). These are containers with objects that must be moved from one part of the space station to another. Since the ERA is long, and since it has been made of lightweight materials, *flexibilities* are present when it moves. In the future, it is the intention that the ERA will be controlled from Earth. Then, *time delays* are present, due to the up- and down data transmission through space. More information about the ERA can be found in [Schoonejans, 1990].

Whenever possible, the ERA will be controlled in *supervisory control*. In that case, the human operator gives commands to the manipulator like 'put that there'. The manipulator performs these tasks automatically. The *manual control* mode is a 'backup control mode', which is used when the automatic controller is not capable to perform the task correctly. On the MIR 2, in general, direct vision is absent. Therefore, camera pictures will be used to get information about the position of the

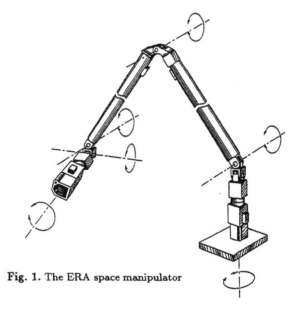

Fig. 1. The ERA space manipulator

ERA. When the end-effector must be positioned, e.g. for grasping an ORU, the picture of the camera, placed on the end-effector will be used. The operator watches the picture on a television screen. With a *control device*, the operator generates control signals that are sent to the manipulator.

In case of manual control, it is often difficult to estimate the 3-dimensional movements of the manipulator from a 2-dimensional camera picture. Furthermore, a space manipulator is a multivariable system with 6 DOF's. In general, it is difficult for a human operator to control such a system manually. The slow dynamics, the flexibility and the time delays make controlling even more difficult. At the Delft University of Technology, research is carried out on the development of a man-machine interface for manual control of a space manipulator like the ERA. The objectives of the project are to develop methods, that

1. improve the spatial observability of the camera pictures;
2. improve the manual controllability of a 6 DOF space manipulator; and
3. reduce the effects of the slow dynamics, the flexibility and the time delays.

Besides space manipulators, the results of the project might also be suitable for other applications, like endoscopic surgery, manually controlled manipulators undersea, and in nuclear industry [Sheridan, 1992].

The hardware facility consists of a fast Silicon Graphics Indigo II Extreme *graphical workstation* and a *Spaceball controller*. The graphical workstation is used to simulate the movements of the manipulator, and to animate the camera pictures; both in real-time. The Spaceball controller is a 6 DOF force operated control device that consists of a sphere on a base (see also Fig. 13). The sphere, or Spaceball, can be slightly translated and rotated in 3 perpendicular directions. The forces and torques the operator imposes on the Spaceball are transformed into translational and rotational velocities of the end-effector. The choice for end-effector velocity control and the selection of the Spaceball controller as control device for this project is motivated in [Breedveld, 1994]. Instead of velocity control, also end-effector *position control* could have been used. In [Kim, 1987], however, it is shown that position control is more suitable for *fast* manipulators with a *small* workspace, while velocity control is more suitable for *slow* space manipulators with a *large* workspace like the ERA.

The research of the project is phased as follows:

A. Activities in an operating point:
I. Manual control of an ideal manipulator;
II. Manual control of a flexible manipulator.
B. Activities along a track

Stage A is focused on positioning the end-effector, e.g. for grasping an ORU. Herewith, the demanded accuracy is high, but the risk of an unexpected collision between the manipulator and the environment is small, since the movements of the manipulator are small. Stage B is focused on moving the end-effector along a track, e.g. for displacing an ORU. Herewith, the demanded accuracy is small, but the risk of a collision is large. In phase I of stage A, objectives 1 and 2 mentioned above are investigated. Herewith, the manipulator dynamics are not implemented yet. In phase II, objective 3 is investigated. Herewith, the dynamics of the manipulator are modelled in an accurate real-time simulation model. The time delays will be compensated with *predictive displays* [Breedveld, 1992].

This paper focuses on the developments at the *display side* of the man-machine interface in phase I of the research. At the Delft University of Technology, a novel graphical display has been developed, that improves the spatial observability of the end-effector camera picture in a clear way. The development of this display is described in sections 2 and 3. In section 4, a control method is described, that transforms the Spaceball movements into movements on the screen. In section 5, some results of man-machine experiments with the display and the control method are given. The paper ends with the conclusions in section 6.

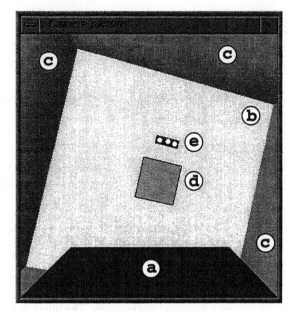

Fig. 2. The end-effector camera picture. *(a) End-effector (b) backplane (c) sideplanes (d) target (e) vision target.*

2. DEVELOPMENT OF THE PYRAMID DISPLAY

The end-effector camera picture
Fig. 2 shows a stylized impression of the end-effector camera picture in the case of the positioning task, animated on the graphical workstation. The end-effector camera is placed *on top* of the end-effector and gives a view of the environment. In the animated picture, the end-effector is represented as a block. Its front is visible at the bottom of the picture. The environment is simplified to five planes. The *backplane* represents the object to be grasped. The four *sideplanes* represent the environment around the object. On the backplane, the *target*, the place where the end-effector front must be positioned, is approximated with a square. Above the target, a *vision target* is visible. This is a black base plate with three white discs, from which the middle one is placed on an elevation (see also Fig. 13).

The vision target has the following function: When the end-effector is close to the object to be grasped, the vision target is clearly visible in the end-effector camera picture. From the size and the location of the three white discs, a computer program, called *proximity sensor*, accurately calculates the distance and orientation of the end-effector with respect to the target. In case of supervisory control, this information is used to position the end-effector on the target automatically. The slow dynamics and the flexibility of the manipulator, however, make automatic positioning difficult. In some cases, e.g. if an object is present between the end-effector and the target, or if the automatic controller is damaged, automatic positioning might even be impossible. Therefore, it is also possible to position the end-effector by hand.

Estimating spatial movements from the picture
In the last stage of the positioning task, the distance between the end-effector and the backplane becomes so small, that the four sideplanes get out of sight. Then, only the target and the vision target remain to give spatial information to the operator. Unfortunately, for a human, it is very difficult to estimate the position and orientation of the end-effector from the size and the locations of the three white discs. To assist the

Translations (mm)		Rotations (degrees)	
up	-153	yaw	12
right	-87	pitch	-12
forward	-237	roll	-21

Fig. 3. The position and orientation numbers.

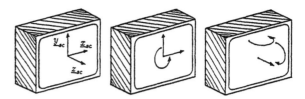

Fig. 4. The visualization of spatial movements on a flat television screen

operator, the proximity information can be used for drawing a *graphical overlay* on the picture that makes it more easy to position the end-effector. Such an overlay could for example consist of the position and orientation numbers that are calculated by the proximity sensor (Fig. 3). When all numbers are made equal to zero, the end-effector is positioned against the backplane, and the positioning task is completed. With this way of presenting the information, however, the operator constantly has to translate the six numbers to spatial movements on the display. This is a tiring activity that increases his mental load. In order to decrease his mental load, the risk exists, that the operator simplifies his task to just putting the numbers to zero one by one, without looking at the movements on the picture anymore. Then, the operators involvement with the real situation decreases, which can be dangerous in abnormal situations.

In fact, there are two basic ways in which the presentation of the proximity sensor information can be improved:

1. The way, in which the 6 numbers are displayed is improved, but they are still displayed *individually*.

2. The 6 numbers are not displayed individually, but *integrated* in a graphical overlay that makes it possible to estimate the distance and orientation of the end-effector at a single glance.

If the first, classical way is followed, the operator still has to translate six quantities to movements on the screen, and the risk above is still present. The second way, however, seems to lead to a more direct and natural way to present spatial information to the operator. This way will be followed here.

Note: Besides the end-effector camera overlay described in this paper, also two other end-effector camera overlays for the ERA have been developed. In the overlay in [Bos, 1992, p. 2], the first way above was followed to improve the presentation of the proximity sensor information: At the left and right side of the camera picture, the position and orientation misfits are displayed individually with six analogue dial indicators. In the overlay in [Ferro, 1992], both ways above were followed: the three translational misfits are displayed individually with three analogue distance scales; and the three rotational misfits are integrated in a figure that looks like the artificial horizon of a plane. This overlay is used at the moment to position the ERA [Schoonejans, 1990, p. 8].

As said before, it is very difficult for the operator to estimate distances and orientations from the movements of the vision target. Below, the cause of this will be explained. Consider Fig. 4: In the left part of the figure, a television screen with a coordinate system is drawn. The x_{sc} and y_{sc} axes lie in the screen. The z_{sc} axis is perpendicular to the screen. The spatial movements of a 3-dimensional object that is displayed on the screen, can be split up into two categories:

1. *Movements* parallel *to the screen.*

    These are (middle part of Fig. 4): translations

along the x_{sc} and y_{sc} axes, and rotations around the z_{sc} axis. Since these movements all lie *in* the screen plane, the operator can estimate them *directly* and accurately.

2. *Movements* perpendicular *to the screen.*

    These are (right part of Fig. 4): rotations around the x_{sc} and y_{sc} axes, and translations along the z_{sc} axis. Since these movements are *perpendicular* to the screen plane, the operator can only estimate them *indirectly* from movements *in* the screen plane.

In the end-effector camera picture, mainly the estimation of translational and rotational misfits *perpendicular* to the screen gives problems. In the last stage of the positioning task, rotational misfits perpendicular to the screen can only be estimated from displacements of the middle, elevated disc of the vision target with respect to the other two discs, and translational misfits perpendicular to the screen can only be estimated from size changes of the target and the vision target, caused by the perspective. Both estimations are not very accurate.

Below, the development of an overlay will be described, that uses 3-dimensional graphical objects, to visualize the translational and rotational misfits perpendicular to the screen in an integrated way. The overlay can also be used in the case of supervisory control, to *monitor* the movements of the end-effector.

Visualizing large rotational misfits perpendicular to the screen: the insert-box

The problems in estimating rotations perpendicular to the screen in the last stage of the positioning task, are caused by the fact, that the vision target is too *flat*. If its middle disc would stick out more, these rotations would be easier to detect. Following this line of thought, the visualization of rotations perpendicular to the screen plane can be improved by attaching a 'sticking-out' graphical object to the backplane. The longer the object, the more the spatial effect, the easier to detect its rotation. In general, many kinds of 'sticking-out' graphical objects are possible. In this case, a rectangular box with a square base and an open front is selected. This object is called *insert-box* and is shown in Fig. 5. It is a simple, recognizable object, which is easy to animate. Its open front is essential in combination with the 'frosted-glass' (see below). Furthermore, its angular base is very helpful for visualizing rotations parallel to the screen. To improve the sensation of depth, a rectangular grid is drawn in the box. The figure shows, that the rotational misfit of the end-effector can easily be detected from the orientation of the box.

Visualizing large translational misfits perpendicular to the screen: the frosted-glass

When the end-effector is moved towards the backplane,

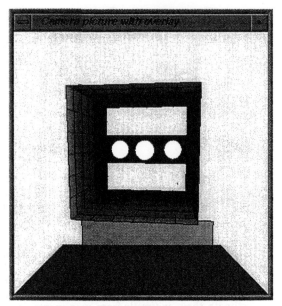

Fig. 5. End-effector camera picture with insert-box overlay in the last stage of the positioning task.

Fig. 7. End-effector camera picture with insert-box/frosted-glass overlay in the last stage of the positioning task.

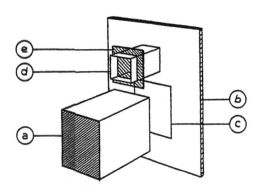

Fig. 6. Side view of the end-effector, the insert-box and the frosted-glass. (a) End-effector, (b) backplane, (c) target, (d) insert-box, (e) frosted-glass.

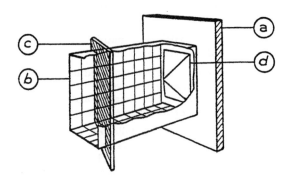

Fig. 8. Side view of the insert-box, the frosted-glass and the pyramid. (a) Backplane, (b) insert-box, partially cut away, (c) frosted-glass, (d) pyramid.

the distance between the end-effector front and the target decreases. When the distance becomes small, there will be a need for a more precise distance indicator than the size changes caused by the perspective. In combination with the insert-box, a *frosted-glass* visualizes this distance in a clear and natural way. The frosted-glass is a transparent square plane, 'fixed' to the front of the end-effector, like the sight of a gun. In the first phase of the positioning task, the frosted-glass is moved towards the insert-box. Then, it is used as a viewfinder. At a certain moment, it will *intersect* the box, as shown from aside in Fig. 6. Fig. 7 shows, that this intersection is clearly visible on the inside of the box. The rectangular grid in the box can then be used as a distance indicator, perpendicular to the screen, and the distance left to the backplane can easily be estimated.

Visualizing small rotational and translational misfits perpendicular to the screen: the pyramid

The combination of insert-box and frosted-glass is very useful for roughly positioning the end-effector in front of the backplane. However, at the end of the positioning task, when the distance from the end-effector front to the backplane is decreased to only a few millimeters, the small translational and rotational misfits left, are

hardly visible in the display. Then, it would be useful to have another aid, that more clearly amplifies the misfits. Again, it seems to be handy to visualize the misfits by means of an intersection of the frosted-glass with a graphical object attached to the backplane. It seems to be best to locate this graphical object at the base of the insert-box. Then, at the end of the positioning task, the object is located in the middle of the display, which is the place where the operator focuses his attention on. Small translational and rotational misfits perpendicular to the screen plane can be amplified strongly if a flat, tapering object, like a pyramid or a cone, with a top pointing towards the frosted-glass, is used. When the frosted-glass intersects such an object, a small translation perpendicular to the screen results in a large *size change* of the intersection and a small rotation perpendicular to the screen results in a large *shape change* of the intersection. In this case, the *pyramid* is selected; like the insert-box, a simple, recognizable object with an angular base, which is easy to animate.

The principle of the pyramid is shown from aside in Fig. 8 and from the front in Fig. 9 (since the vision target is covered by the pyramid it is not visible anymore). When the end-effector is very close to the backplane,

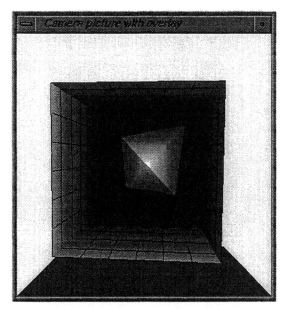

Fig. 9. End-effector camera picture with insert-box/frosted-glass/pyramid overlay at the end of the positioning task.

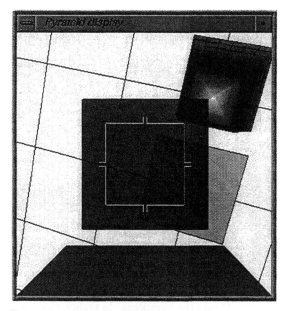

Fig. 10. The Pyramid Display.

the frosted-glass will intersect the pyramid. Then, the size of the intersection visualizes the distance left to the backplane. The larger the intersection, the smaller this distance. The shape of the intersection visualizes the rotational misfit perpendicular to the backplane. If a rotational misfit is present, the intersection is a trapezium or a kite, like in Fig. 9. If all rotational misfits are zero, the intersection is a square. When the intersection coincides with the pyramids base, the front of the end-effector is positioned against the backplane. The amplification of the pyramid can be changed by its height. The lower the pyramid, the larger the effects. A very large amplification, however, will result in a sudden appearance of the intersection and increase the risk of a collision.

Visualizing translational and rotational misfits parallel to the screen

With the pyramid, the end-effector can be positioned against the backplane accurately. However, it is still not clear, where exactly the end-effector should be positioned against the backplane, or consequently: where the intersection with the pyramid should be located on the frosted-glass. To solve this problem, on the frosted-glass, a square is drawn at the location where the pyramid should be at the end of the positioning task. The square is visible in Fig. 10. It has the same size as the pyramid base. When the intersection with the pyramid coincides with the square, the translational and rotational misfits are zero, and the positioning task is completed.

The Pyramid Display

Fig. 10 finally shows the resulting graphical overlay on the end-effector camera picture, called the *Pyramid Display*. To improve the sensation of depth, rectangular grids are projected on large unicoloured planes, and the front part of the graphical objects, which is closer to the viewer, is drawn lighter than the back part. The graphical objects have contrasting colors that make them clearly distinguishable. The grids are coloured black, the insert-box blue, the frosted-glass transparent black, and the square on the frosted-glass bright yellow. In combination with the grids, the 'holes' at the sides of the square make it easier to approach the insert-box

straight from the front. The pyramid is drawn bright red; an alarming color, that enforces the operator to decrease speed.

During try-outs with some subjects in an early stage of the research, the Pyramid Display turned out to be handy for positioning the end-effector. However, many subjects still had some problems with accurately positioning the end-effector by using the pyramid. Therefore, some improvements were made in the display. These improvements are discussed below.

3. IMPROVEMENTS OF THE PYRAMID DISPLAY

The shift

At the end of the positioning task, the operator estimates the distance left to the backplane from the size of the intersection of the pyramid and the frosted-glass. During the try-outs, however, this estimation turned out to be unreliable if a rotational misfit perpendicular to the backplane was present. If the rotational misfit was large, it could even happen, that the end-effector front collided against the backplane, while no intersection was visible. This phenomenon is caused by the fact, that, in reality, the *end-effector front* moves towards the target, while, with the Pyramid Display, the operator moves the *frosted-glass* towards the pyramid. The frosted-glass and the end-effector front differ in location. The left part of Fig. 11 shows the effect of this difference. The backplane and the end-effector are shown from aside, at the end of the positioning task. Due to the large rotational misfit, the bottom of the end-effector front collides against the backplane, while still no intersection is present.

To solve this annoying phenomenon, in the display, the difference in location is compensated by *shifting* the insert-box and the pyramid in a direction perpendicular to the backplane, as shown in the right part of Fig. 11. The shift is a function of the rotational misfit [Breedveld, 1994]. In the figure, the insert-box and the pyramid are shifted away from the backplane, and an intersection is present. If the end-effector has a rotational misfit in the opposite direction, the insert-box and the

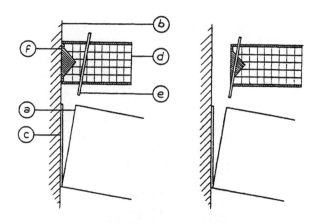

Fig. 11. Pyramid Display from aside, without (left) and with shift (right). *(a) End-effector, (b) backplane, (c) target, (d) insert-box, (e) frosted-glass, (f) pyramid.*

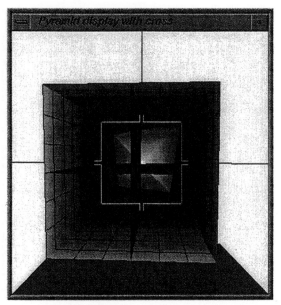

Fig. 12. Pyramid Display with cross at the end of the positioning task.

pyramid shift *into* the backplane. In reality, this would of course not be possible. However, the insert-box, the frosted-glass and the pyramid are just animated objects. In the graphical overlay, they are drawn with a *preference*: they are always visible, whether or not they are shifted into the backplane. Due to this, the human operator doesn't really notice the shift. With the shift, the distance can always be estimated correctly from the size of the intersection. Also, it is easy to see a collision coming: a collision occurs, when one of the four vertices of the intersection touches the base of the pyramid (Fig. 9).

Note: In the case of the ERA, the end-effector camera is placed *on top* of the end-effector, and the end-effector camera gives a view *alongside* the end-effector. Instead of an end-effector camera overlay, also a virtual view *through* the end-effector front could have been developed. With such a display, no shift is needed. In the positioning task, however, there is always a possibility, that the proximity measurement is wrong. Therefore, the operator cannot fully rely on the graphical display: the real end-effector camera picture must always be present for verification. In the Pyramid Display, the camera picture and the display are combined in one. This makes verification easy. With other kinds of graphical displays, like a virtual view through the end-effector, the operator will have to spread his attention between two or more pictures. This is less handy and less safe. Therefore, priority was given to the development of an end-effector camera overlay, instead of a virtual view through the end-effector.

The accuracy indicator
During some try-outs with shift, the positioning task was completed, if the six rotational and translational misfits left were within the desired accuracy of 1.5 mm and 1.5 °. To be sure about finishing the task correctly, first of all, most subjects manoeuvred the end-effector straight in front of the target. Herewith, by using the graphical objects in the display, they made all misfits as small as possible, except for the distance left to the backplane. Then, they approached the backplane carefully, till also this misfit reached the desired accuracy, and the positioning task was completed. Many of them complained, that they wanted to be more certain that the end-effector was positioned straight enough in front of the target. Some of them said, that it would be handy to have a kind of *accuracy indicator*, like a light switching on, that made them sure about having

reached all desired accuracies except for the distance left to the backplane. This advice was followed. A separate light, however, was not used, since this would make the displays more crowded. Since the operator focuses on the pyramid, this object was given the function of accuracy indicator. This was done like a traffic light: In general, the pyramid is coloured bright red. When all desired accuracies, except for the distance left to the backplane are reached, its colour changes to bright yellow. When the misfits are made even smaller, its colour changes to bright green, and the backplane can be approached safely. The yellow color functions as a kind of security margin. The subjects were very satisfied with the indicator. It turned out to be very helpful.

The cross
The pyramid turned out to be very handy for visualizing the distance left to the backplane. However, some subjects had problems with translating the shape of the intersection to the direction of the rotational misfit. Therefore, another graphical aid was added to the display, that visualizes small rotational misfits perpendicular to the screen more clearly. Like in the case of the insert-box, again, a 'sticking-out' graphical object seemed to be very suitable to visualize these misfits. In this case, a *cross*, consisting of two planes, perpendicular to each other was chosen. The cross is placed in the insert-box, as shown in Fig. 12. If a rotational misfit is present, the cross is clearly visible in the display. If the misfits are reduced to zero, the two planes of the cross are perpendicular to the screen plane, and the cross will not be visible anymore. So: the smaller the cross visible on the screen, the smaller the misfits. The cross turned out to be very helpful. However, it has the disadvantage, that it makes the display more crowded.

4. THE CONTROL SIDE OF THE INTERFACE

At the control side of the man-machine interface, six *control methods* have been developed, that transform the movements of the Spaceball into movements on the screen. This is done in such a way, that the movements of the hand of the human operator correspond with the

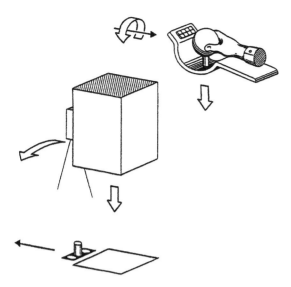

Fig. 13. Downward 'corrected' end-effector control

movements of the controlled object in the end-effector camera picture. This gives the operator the feeling 'as if he/she holds the object in the hand'. A Spaceball controller makes such an intuitive way of controlling possible, since it can be translated and rotated in three perpendicular directions. As an example, one of the six control methods is shown in Fig. 13. The figure shows the Spaceball controller, the target, the vision target and the end-effector with camera. With this control method, the operator experiences the task as positioning the end-effector downward against a floor. If the operator pushes the Spaceball downward, the end-effector will move downward, as shown in the figure.

More about the control methods can be found in [Breedveld, 1994]. An evaluation of the control methods by means of man-machine experiments can be found in [Buiël, 1994 & Buiël, 1995]. To give an impression of the suitability of the man-machine interface for positioning the end-effector, some results of the experiments are given below.

5. RESULTS

During the man-machine experiments with the control method of Fig. 13, 6 subjects had to position the end-effector several times with a desired accuracy of 1.5 mm and 1.5°. They all used the Pyramid Display with shift and accuracy indicator. The cross was not present. The dynamics of the manipulator were not implemented yet: the end-effector responded exactly to the velocity control commands of the subjects. The maximum velocity of the end-effector was 0.1 m/s, and the initial positions and orientations of the end-effector were varied randomly in such a way, that the distance to the target was always equal to 1 m. This made the minimum time required to complete the task equal to 10 s. After a relatively short training period of about 3.5 hours, the subjects were able to perform the task with surprisingly fast completion times between 19 and 23 s. The Pyramid Display and the control method proved to be very handy for positioning the end-effector.

6. CONCLUSIONS

In this paper, a graphical display is described, that improves the spatial observability of the end-effector cam-

era picture. Rotational misfits are visualized with an *insert-box*, a 'sticking-out' graphical object, fixed to the object to be grasped. The distance left to the object is visualized with a *frosted-glass*, fixed to the front of the end-effector. When the frosted-glass intersects the insert-box, the distance is clearly visible on the inside of the box. At the end of the positioning task, the small misfits left are visualized with a *pyramid*. When the frosted-glass intersects the pyramid, a small rotational misfit causes a large shape change, and a small translational misfit causes a large size change of the intersection. Man-machine experiments with the display showed its suitability for positioning the end-effector.

7. ACKNOWLEDGEMENTS

This research is supported by the Dutch Technology Foundation STW and takes place in cooperation with Fokker Space & Systems B.V., the European Space Agency ESA, and the National Aerospace Laboratory NLR.

8. REFERENCES

Bos J F T (1992), Aiding the Operator in the Manual Control of a Space Manipulator. *Proc. 5th IFAC Symp. on Analysis, Design and Evaluation of Man-Machine Systems, June 9-11, the Hague, the Netherlands, paper 1.3.3.*

Breedveld P (1992), Controlling a Space Manipulator from Earth with Predictive Display Techniques. *Proc. 11th European Annual Conference on Human Decision Making and Manual Control, Nov. 17-19, Valenciennes, France, paper 27.*

Breedveld P (1994), The Design of an Optimal Man-Machine Interface for a Manually Controlled Space Manipulator. *Report N-469, Delft University of Technology, Fac. of Mechanical Eng. & Marine Tech., Lab. for Measurement & Control.*

Buiël E F T (1994), A Laboratory Evaluation of four Controller Configurations for a Teleoperated Space Manipulator. *Report N-472, Delft University of Technology, Fac. of Mechanical Eng. & Marine Tech., Lab. for Measurement & Control.*

Buiël E F T, Breedveld P (1995), A Laboratory Evaluation of four Control Methods for Space Manipulator Positioning Tasks. *To be published in Preprints 6th IFAC Symp. on Analysis, Design and Evaluation of Man-Machine Systems, June 27-29, MIT, Cambridge, USA.*

Ferro D V (1992), Process and System for Remotely Controlling an Assembly of a First and a Second Object, *US Patent no. 5,119,305, 1992. Assignee: Aerospatiale Societe Nationale Indust., Paris, France.*

Kim W S, Tendick F, Ellis S R, Stark L W (1987), A Comparison of Position and Rate Control for Telemanipulations with Consideration of Manipulator System Dynamics, *IEEE Journal of Robotics and Automation, vol. RA-3, no. 5, pp 426-436.*

Schoonejans P H M, Andre G, Danan G (1990), The Hermes Robot Arm: Advances in Concepts and Technologies, *Proc. International Astronautical Federation IAF-90-025.*

Sheridan T B (1992), Telerobotics, Automation and Human Supervisory Control, *MIT Press, USA, ISBN 0-262-19316-7.*

HUMAN-MACHINE INTERACTION
WITH MULTIPLE AUTONOMOUS SENSORS

Steven A. Murray

*Navy Command, Control and Ocean Surveillance Center, RDT&E Division,
San Diego, California*

Security and inspection systems are becoming increasingly automated. Many such systems
include mobile platforms capable of autonomous sensing and analysis of the environment
from a multitude of perspectives. This increased automation shifts the responsibilities of
humans from active patrolling and inspection to passive monitoring of remote sensor
information. The operator brings perceptual and cognitive characteristics to this task which
need to be addressed in both system architecture and interface design if desired
performance reliability is to be achieved. A study is reported which examines the impact
of these characteristics on system performance.

Keywords: Human-centered design, human-machine interface, human reliability.

1. INTRODUCTION

The use of autonomous systems for remote sensing
and inspection has long been an important design
effort of system engineering. Autonomous sensing
can preclude the need for humans to physically patrol
large areas, such as factories or warehouses, and can
protect them from hazardous environments. In
addition, such systems do not fatigue, or vary
significantly in detection performance over time, as
humans often do.

Autonomous systems, however, are not foolproof.
Sensors can fail to detect important events, or can
report false alarms as a function of physical fault or
imperfect analysis algorithms. Human monitoring is
therefore included in most system designs to ensure
proper operation and improved signal classification
(e.g., Everett *et al.*, 1992). Furthermore, to realize
the best economic potential of such systems, a single
operator is often responsible for supervising several
remote platforms simultaneously.   This leveraging of
human presence presumes that significant events will
occur for only a fraction of the platforms at any one
time; that is, designers assume that "worst case"

events will still be within the response capabilities of
the human operator.

Controlling, or even monitoring multiple-platform
systems such as these can be complex if the areas
under surveillance are large or if many sensors are
employed. Because human performance tends to
degrade over time when monitoring systems with low
event rates (the "vigilance decrement;" Parasuraman,
1986), such automated systems usually provide for
operator cueing when a significant event occurs by
delivering an alert signal and/or information about the
event. This latter case can involve presentation of a
video image covering the area where the detection
was first triggered. The operator must confirm the
nature of the event, locate it in physical space (e.g., Is
it inside or outside the building? Which hallway or
room is it in? etc.), and initiate an appropriate
response.

Human-machine interfaces for this task almost
invariably provide an overall, two-dimensional
depiction of the surveilled area -- such as a building
diagram, displayed on a CRT -- so that the location
of the alerting platform can be determined and the

video image correlated with a point in physical space. Thus, the operator must rapidly map information between 2-D to 3-D representations of the environment. In addition, many sensors have panning capability, so the location of a sensor platform does not fully define the location of its image. Although some systems provide field-of-view markers to aid this task, the 2-D to 3-D information mapping must still be performed.

Operators can learn, over time, what to expect from sensor images of conventional security and monitoring systems (e.g., those with fixed, wall-mounted cameras), based on their known location and orientation in the surveyed area. The use of mobile sensor platforms, however, considerably complicates the operator's control task by removing the predictability of fixed sensor images. Each alert must be independently interpreted, based on the current location of the reporting platform(s).

The independent patrol patterns of autonomous platforms open the possibility of multiple signals arising from the same physical event due to overlapping fields of coverage (i.e., during those occasions when patrol patterns come into close proximity). Although such redundancy may complicate the operator's task, there may be sound reasons for desiring such overlaps in sensor coverage, such as increased detection reliability. Therefore, redundant reports may be expected to occur in some fraction of system events. In these situations, the operator must map each image in space, based on platform location, and must correlate information across images, to generate an integrated spatial model of the events being reported.

Development of multiple-platform, autonomous systems such as these, for both security and process monitoring, are underway at several government and commercial laboratories, including the Naval Command, Control, and Ocean Surveillance Center. These systems support both structured and unstructured patrolling by autonomous robots, which search for objects or events based on loosely-structured criteria (e.g., some combination of movement, contrast, or temperature thresholds). The wide variety of settings in which these systems need to function make stringent demands on automation capabilities; human presence is very much needed to ensure reliable performance and appropriate follow-up response. An information flow diagram is shown in Figure 1.

Depending on the degree of automation and the complexity of the task (e.g., building security with fixed cameras, or outdoor surveillance with mobile robots), the operator can spend most of the time in a passive role. All sensor control and raw data

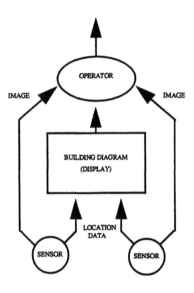

Figure 1. Processing and control diagram of multiple-sensor security system.

analyses are handled automatically, and the operator is required to interact with the system only when alerted. From this point, all subsequent performance is essentially manual, as the operator searches and interprets each display, takes the appropriate action (such as directing a response to the location of the event), and releases the sensors to resume surveillance. Control is, thus, more "traded" than "shared" (Sheridan, 1992, page 65).

Models of human supervisory control (e.g., Rasmussen, 1983; Sheridan, 1987) are not readily applied to tasks such as these, where the operator is essentially decoupled from the system until an alert effectively forces a change to exclusively manual control. A problem with this approach, as in many automated system designs (Woods, in press) is that the operator is required to perform at precisely those moments when the system can provide no further assistance, i.e., the sensors have done their job of detecting and relaying events. These moments are also those conditions (e.g., suspected intrusions, fires, etc.) for which prompt and correct action is essential.

Performance estimates for such complex systems often rely on extrapolations of operator models generated from much simpler, single-sensor applications. A consequence of this approach may be an overly-optimistic assessment of what the human-machine system can accomplish, assignment of too many systems to the control of a single person, or inadequate human-machine interface design. Different interface modeling and design approaches may be required to support operator performance in such one-to-many control situations. An initial examination of human performance capabilities in such settings was therefore conducted to demonstrate the limitations of single-sensor models for predicting operator performance, and to identify certain task variables in need of special interface support.

2. PERFORMANCE STUDY

The selection of an industrial security system as the experimental setting was dictated by practical need; a research effort was required to support the development of prototype systems which used multiple autonomous robots as sensing and reporting platforms. The task nevertheless illustrates general human capabilities to rapidly assess (pictorial) sensor information from a number of distributed sources and to integrate it into a single model of the environment, which should have wider application.

Three task characteristics were selected for initial study: the number of displays which had to be monitored (corresponding to the number of remote sensor systems), the amount of information to be interpreted (i.e., the number of images which were presented at any one time), and the complexity of the information (i.e., whether the images depicted separate events or a common event from overlapping sensors).

2.1 Hypotheses

The number of displays used for this experiment was limited to three. Each display represented a potential source of task-relevant information from a remote sensor, i.e., the presence of an image indicated that a sensor had detected something significant, and the operator was therefore compelled to examine the image, if only to confirm a false alarm. Although not all images contained a target (for this application, a target was a simulated human "intruder"), all targets were clearly visible in those images where they were presented. Target figures were all replications of the same model, i.e., the figures all looked the same but were positioned at different locations within the environment. Because the major influence on search time is the size of the search set (Hyman, 1953; Scanlan, 1977), operator response time was predicted to increase with larger numbers of displays and with greater numbers of target figures.

Task complexity was manipulated by controlling image redundancy. High redundancy was defined as multiple images of the same object, i.e., each display showing a different perspective of the same target figure. Low redundancy was defined as a separate object in each image. It was hypothesized that the effort involved in correlating similar scenes, to resolve the number of actual targets present, would be a more complex task than resolving images containing distinct targets and, therefore, high-redundancy conditions would require more

processing time. This is also in accord with a prediction by Vickers (1970) of increased reaction time as a function of reduced inter-stimulus discriminability.

It was further hypothesized that additional cognitive processing demands would be required for this task if the operator had to map images from unpredictable platform locations (and viewpoints) for each trial, i.e., if the operator had to interpret images from mobile platforms. This additional workload should be evidenced by poorer performance for trials imaged from mobile sensor platforms than trials imaged from fixed-position platforms.

To make the task as realistic as possible, and to control for individual speed-accuracy tradeoff strategies (e.g., Wickens, 1984), subject instructions for this experiment emphasized accuracy; operators were told that response accuracy was more important than speed. Response times were therefore expected to be more informative than error rates as a dependent measure.

2.2 Method

Subjects. Six volunteers from the laboratory staff (four males and two females, aged 26 to 32) were used as subjects. All participants were familiar with the task environment used in the simulation.

Stimulus preparation. An indoor setting -- a single, open-bay building -- was selected as the task environment. This setting was already modeled on a Silicon Graphics computer system, and had been previously used for virtual environment applications. A set of static scenes was generated from this model by moving through it with a simulated sensor platform (about four feet above ground level) and capturing images at various locations and along various directional bearings. Images were monochromatic, and approximately 9.0 x 7.5 cm in size. A human figure was modeled in the simulation to serve as an "intruder." Copies of this figure were placed at varying locations in a subset of these images to serve as visual targets. No more than one target was contained in any given image. Images were displayed on a Silicon Graphics Indigo computer with a 43 cm (diagonal), high-resolution (1280 x 1024 pixel) monitor.

A paper diagram of the building was provided for each trial, which depicted major features of the interior (e.g., furniture, doors) and which contained the location of the sensor(s) which had generated the associated video image(s). An example of such an image - diagram pair is presented in Figure 2. The

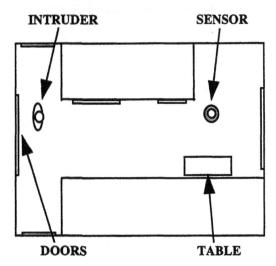

Figure 2. Simulated video image of intruder with diagram of corresponding object locations

illustration shows an image (from a single sensor) of a figure standing at the end of a hallway, in front of a door. Subjects used these diagrams to record their task responses, i.e., by marking locations where they believed intruders to be, based on the images they were given.

Design. A repeated measures design was used, which consisted of two conditions for Sensor Mobility (i.e., fixed-position platforms versus platforms which were free to change position between trials), three levels of Number of Displays (i.e., one, two, or three displays, presented simultaneously from a corresponding number of simulated sensor platforms), three levels of Number of Figures (i.e., how many images actually contained a target figure), and three levels of Redundancy (i.e., HIGH, where all displays showed a common target, MEDIUM, where some displays showed a common target, and LOW, where each display showed a unique target).

A decision was made that each image would contain, at most, a single target figure. This necessarily resulted in an incomplete design (i.e., some cells of the 2x3x3x3 factorial were not used). Target detection across displays was the performance of interest. If a full factorial design had been employed, some displays would have contained multiple targets, allowing the task to be completed by straightforward

counting; the desire to avoid this confounding behavior led to the design approach described.

Procedure. The nature of the security monitor job was explained to the subjects using standardized verbal instructions. Response accuracy was emphasized by explaining that erroneous interpretations would result in time being wasted by response forces being dispatched to the wrong location. A set of training trials was then administered, which contained at least one example of every condition to be encountered in the actual experiment. A trial began when the images appeared on the monitor, and ended when the subject had marked the position(s) of the intruder(s) on the paper diagram; the experimenter then pressed a key which recorded elapsed time for the trial. Subjects were always told whether they were dealing with fixed or mobile platforms, and how many sensors were active in the environment. Other conditions were randomized. Every trial had at least one image with a target figure in it; i.e., there were no completely "false alarm" trials.

2.3 Results

Main effects from an analysis of variance were all significant and were in expected directions. Because the incomplete nature of the design cells prevented calculation of a complete ANOVA , a multiple regression analysis was also performed to gain an understanding of the simultaneous action of these variables.

ANOVA effects. The results of the sensor mobility manipulation showed that responses to images from mobile sensor platforms took approximately forty percent more time than responses to images from fixed-position sensors; $F(1,5) = 8.97$, $p < .001$. Response time also increased as a function of monitoring increasing numbers of displays ($F(2,10) = 126.31$, $p < .001$) and as a function of the number of target figures shown on those displays ($F(2,10) = 95.69$, $p < .001$). The redundancy manipulation showed that system operators took longer to process multiple independent images than they did to process redundant ones ($F (2,10) = 77.11$, $p < .001$). Mean response times for subjects are presented in Table 1.

Multiple regression. A forward stepwise regression analysis was performed separately for the response time results of the fixed-platform and the mobile-platform conditions; it is unlikely that these two sensor positioning schemes would ever be designed into the same system. Results showed that levels of image redundancy and the number of figures shown on the displays contributed the most to response time performance for both types of systems. The number of displays which the operator had to monitor did not

Table 1. Mean response times results, by condition

		mean response time(sec)
Sensor Mobility	fixed	8.91
	mobile	12.15
Number of Displays	1	5.96
	2	9.89
	3	11.61
Number of Figures	1	6.97
	2	10.59
	3	14.01
Redundancy	high	8.53
	medium	12.45
	low	16.59

account for enough additional variance to be included in either regression equation. For the fixed-platform condition:

Response Time (sec) = 2.440 + 0.478 (Redundancy) + 0.328 (Number of Figures)   (1)

R^2 for this equation was 0.5157, $F_{(2,117)} = 62.762$, $p < .001$. For the mobile-platform condition:

Response Time (sec) = 2.035 + 0.477 (Number of Figures) + 0.343 (Redundancy)   (2)

R^2 for this equation is 0.5342, $F_{(2,117)} = 67.083$, $p < .001$.

3. DISCUSSION

This experiment succeeded in highlighting the relative importance of selected system and display variables to operator performance, and provided some indication of the sensitivity of performance to manipulations of those variables.

3.1 Analysis

In keeping with the priorities established for the security monitor's job, response time proved to be an effective performance measure. Subjects appeared to trade response time for some threshold level of accuracy. More complex systems, however, could demonstrate even longer latencies or significant error rates (i.e., in determining numbers and locations of intruders). A modern industrial security system, for example, may have as many as 125 displays under the control of a single operator. The most important consequence of such data is that they demonstrate a

potential bottleneck on total system performance; the system can only be as fast and accurate as the operator, who is the final filter on input data and the only initiator of system action.

The most significant factor influencing task performance was the use of autonomous, mobile sensors, i.e., platforms whose physical location (and viewpoint) could change from one trial to the next. Clearly, information from such sensors took longer to interpret than information from fixed-position sensors. The flexibility and expanded sensor coverage afforded by using such an autonomous system therefore, may come at a price in operator workload.

The experiment showed shorter response times for trials with redundant images. For a given number of images, subjects were apparently able to determine common objects more rapidly than unique ones. This was not a trivial task, however, as different viewpoints used for redundant images were made at different (apparent) distances from the target figures, as well as different viewing angles. These changes resulted in different effective fields of view, and thus to changes in both aspect and relative size of the target figures, and to shifts in the contents of scene backgrounds.

Subjects used in the study were familiar with the visual environment of the simulation, and knew both the building layout and the locations of its contents (e.g., doors, windows, shelves, etc.). This result, therefore, could have been a function of subject experience with the particular environment used in the experiment, or could have been obtained by some process that worked equally well with any set of redundant images. The results of the sensor mobility manipulation would seem to support an explanation based on the inherent redundancy of the images. If familiarity with the environment were the essential factor, this manipulation would probably not have shown such a large difference in performance, as all images -- fixed or mobile -- contained portions of the "familiar" background. This conclusion is indirectly supported by other research (Thorisson, 1993), that used both reaction time and eye movements to determine that subjects could extract three-dimensional information from multiple two-dimensional images using a feature search process; mental reconstruction of a scene in three-dimensions was not necessary.

3.2 Application

Results of this study provide some guidelines for predicting human-machine performance for systems involving multiple, autonomous sensors. The rapid increase in response time for even the modest levels of manipulations used here is cause for concern, especially when newer systems are planned with larger numbers of sensors and are designed for operations in cluttered environments.

Operator activities in systems like these do not match the common functions typical of supervisory control (Sheridan, 1992, chapter 1). Certainly, the closed feedback loops found in many human-machine control systems are not present. Nevertheless, the operator has a central control function, as it is the operator who filters and transforms sensor data to produce the final system output.

Woods (in press) has long advocated the need for human-machine design which supports the extraction of task-relevant meaning from input data, rather than the mere delivery of data to the operator, i.e., to design for information extraction at the whole task level. Like the computer applications which Woods addresses, each image from a sensor platform acts as one "keyhole" into a much larger data space. It is the coordination of multiple views, or "keyholes," into a single picture of this space that is not supported by the design approaches examined here.

Additional measures could be exploited toward this end, to improve the human-machine interface and, thereby, to enhance system performance. Providing additional visual cues (such as directional lines from each sensor) on the diagram display, for example, could resolve sensor views by identifying overlaps where those lines crossed. This approach would still require operator processing, however, and consume additional response time, especially for systems using higher numbers of sensors. Another approach would be to provide an inhibitory feature to the display of such redundant events (Sheridan, 1992, page 289), whereby only one alert is provided regardless of the number of overlapping contacts. This would require additional computer processing, however, to automatically detect such redundancy. Both of this design concepts are readily testable, and additional investigations are being initiated to measure their effects.

System applications continue to emerge which require extensions to models of human-machine interaction, and which motivate empirical measurement to establish and scale critical variables. Multiple, autonomous sensor systems are one such application. Solutions to these problems can, in turn, enhance human-machine design for a variety of other engineering needs, as well.

ACKNOWLEDGMENTS

This research was supported by an Office of Naval Research contract N0001493WX24310AA, under the administration of Dr. Harold Hawkins.

REFERENCES

Everett, H.R., Gilbreath, G.A., and Laird, R.R. (1992). Multiple robot host architecture. *NRaD Technical Note 1710.*

Hyman, R. (1953). Stimulus information as a determinant of reaction time. *J. Exp. Psychol.*, **45**, 423-432.

Parasuraman (1986). Vigilance, monitoring, and search. In: *Handbook of Perception and Human Performance* (K.R. Boff, L. Kaufman, and J.P. Thomas, Eds.), Vol. 2, Chap. 43, pp. 1-39. John Wiley, New York.

Rasmussen, J. (1983). Skills, rules and knowledge: signals, signs, and symbols, and other distinctions in human performance models. *IEEE Trans. Systems, Man and Cybernetics*, **SMC-133**, 257-267.

Scanlan, L.A. (1977). Target acquisition in realistic terrain. In: *Proceedings, 21st Annual Meeting of the Human Factors Society* (A.S. Neal and R. Palasek Eds.). Human Factors Society, Santa Monica, CA.

Sheridan, T.B. (1987). Supervisory control. In: *Handbook of Human Factors* (G. Salvendy, Ed.), Chap. 9, pp. 1243-1268. John Wiley, New York.

Sheridan, T.B. (1992). *Telerobotics, Automation, and Human Supervisory Control.* MIT Press, Cambridge, MA.

Thorisson, K.R. (1993). Estimating three-dimensional space from multiple two-dimensional views. *Presence*, **2(1)**, 44-53.

Vickers, D. (1970). Evidence for an accumulator model of psychophysical discrimination. *Ergonomics*, **13**, 37-58.

Wickens, C. (1984). *Engineering Psychology and Human Performance.* Charles E. Merrill, Columbus, OH.

Woods, D.D. (1988). Coping with complexity: The psychology of human behavior in complex systems. In: *Mental Models, Tasks and Errors* (L.P. Goodstein, H.B. Andersen, and S.E. Olsen Eds.). Taylor & Francis, London.

Woods, D.D. (in press). Towards a theoretical base for representation design in the computer medium: ecological perception and aiding human cognition. In: *The Ecology of Human-Machine Systems: A Global Perspective* (J. Flach, P. Hancock, J. Caird and K. Vicente, Eds.). Erlbaum Associates, Hillsdale, NJ.

DYNAMIC, SIMULATION-INTEGRATED, INTELLIGENT VISUALIZATION: METHODOLOGY AND APPLICATIONS TO ECOSYSTEM SIMULATION

Jeff Yost, Michael M. Marefat, and Jinwoo Kim

Dept. of Electrical and Computer Engineering, University of Arizona

Abstract: A simulation is only useful if its data is understood by the user and it has increased usefulness if the data could be used dynamically to adjust simulation focus and steer it. This paper proposes a method which enables intelligent interrogation of a large database for spatial and temporal understanding, and which also provides a dynamic and flexible environment for simulation-integrated visualization. A language is developed for defining and detecting semantically significant events based on interesting data patterns or trends. A programmable dynamic finite state machine based approach is used to develop the dynamic and flexible control. The dynamic finite state machines are automatically generated by enabling the users to define semantically interesting events (changes in data) and sequences of reactions to these events. This approach enables the generated behavior to be modifiable without the need to reprogram the system. The methodology developed uses simulation and visualization systems to create an intelligent problem solving environment.

Keywords: Visualization, Dynamic Behavior, Flexible Automation, Man/machine Interaction, Detection systems, Decision making, Simulation, Events

1. INTRODUCTION

With the development of parallel and distributed systems along with the ability to efficiently program these systems, simulations now exist that can process large scale databases. Although advances in the above areas have made the use of simulations a more valuable part of scientific experimentation, one area is holding back the efficiency of a simulation environment. This area is data visualization and human-directed feedback based on observation.

Data visualization is necessary to interpret and understand the data resulting from the simulation. Data understanding is the ultimate reason for simulations, without it the simulation results are useless. Allowing human feedback provides a way of tailoring a simulation to a user's specific needs to produce desired results. Efficiency can be increased by providing a method for the human feedback to be intelligently captured and integrated.

There have been many advances in visualization systems which create more realistic and faster displays, however there is still a lack for visualization systems which are intelligent enough to understand data, selectively reduce data, appropriately visualize data of importance and focus the simulation scenario. Databases such as those in Geographical Information Systems (GIS) and ecological management are so large that interpreting and understanding the data spatially and temporally in a brute force manner is not feasible. Regions of interest occur in parts of the data, but with a large database some of these regions may be overlooked. The rest of the data is highly redundant. In addition, the magnitude of the data running simulations over and over to tweak simulation parameters or to change whole simulation scenarios. Re-focusing the simulation as the process progresses allows the simulation/visualization environment to become a problem solver.

In many cases it is likely that only small amounts of data are truly interesting or useful. Most scientists interacting with data know that information of importance is episodic - it tends to be clumped in space and time. This assumption is key in the development of the methodology. An agent that could interpret the needs of the user and could understand what data are interesting, and then limit the size of data and appropriately visualize it, would greatly increase the efficiency. The agent would allow the simulation to use a database as large as the computing facilities allow and the user would still view all of the spatially and temporally important data.

Two main components are needed to implement such an agent. First is a language that will allow a user to define any pattern or trend in data as a significant event. Secondly, a system that can transform the event definitions and reactions into a control structure that will make the system behave accordingly is needed. Flexibility is needed by the language and control system so any data deemed as interesting can be defined and the system can be modified to change its operation without having to reprogram it.

2. PREVIOUS WORK

Very little work in the area of integrating simulation and visualization into a single environment has been done. Two systems were found which show links between simulation and visualization. RSYST (Lang, *et al.*, 1991) was developed to integrate scientific calculation and visualization into one system. The goal of RSYST was to reduce the turn around time between calculations and visualizations and then the appropriate feedback to the simulation. Another system developed to combine computation and visualization into one framework was GRASPARC (Brodlie, *et al.*, 1993). This system was designed to allow simulation steering and to allow the simulation to be backed up in time. Conceptually GRASPARC provides a scientist with an environment to search for solutions to problems using a simulator and a visualizer combined.

Neither of the above two research works mentioned how to solve the problems associated with visualization and examination of a very large dynamic database, that is too large to view, but in both systems it was necessary to investigate the data before decisions on how to change the simulation parameters or models could be made. In these scenarios, the visualization lacks the necessary intelligence, is not fully integrated, and is not dynamic.

3. SYSTEM ARCHITECTURE

The goal of the agent is to make a simulation run more effective and efficient for the user. The agent will accomplish its goals by monitoring the simulation database for interesting events which upon detection will be communicated by a user-defined reaction. Figure 1 shows the system architecture that employs the agent. The agent is broken down into three main parts.

First is the data monitor which evaluates event definitions, which are stored in the definition module of the agent, using the database. Once an event is detected, the data monitor triggers the reaction

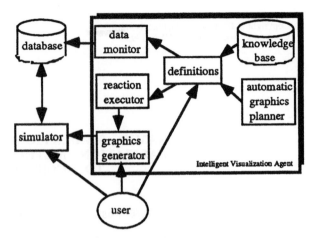

Fig. 1. Abstract System Architecture.

executor. The reaction executor, depending on the reaction definitions which come from the definition module, sends commands to the simulator to change parameters and/or generate specified graphics. The user is allowed to interact with the system in three ways. First the user can define events and reactions. Secondly, the user can interact with the visualizer. The third link allows the user interaction with the simulator. The agent being described can be implemented with different types of simulators and visualizers so the exact types of interactions will vary. Event definitions can be defined by the user as mentioned and by means of artificial intelligence. A knowledge base of relevant events and an automatic graphics generator can be employed. The development of this second method has been left for future work.

4. SEMANTIC EVENTS

One of the most important parts of the system is the significant event definition language. Large scale simulations will be operating on a variety of data types and with data having varied semantic meanings. The combination of data defining an event in one simulation environment may be completely different in another environment. Even within a particular simulation each user might be interested in different events (combinations of patterns or trends). Event definition therefore must allow any pattern and trend in the data to be labeled as an event. These event definitions will then be represented internally so that the system can use the representation along with the data to infer the occurrence of an event.

The event description language consists of *predicates, functions, connectives, variables* and *constants*. From a syntactic view point, an *event* is specified by a formula which describes it as qualifications over predicate terms and connectives combining term (Charniak and McDermott,1985). With the combination of the predicates and functions, a user can define the semantic meaning of significant events.

Predicates. Relational predicates compare a data variable or a function to other data. Six predicates are provided; *LT, LE, EQ, GT, GE.*

Functions . The functions needed are divided into two categories: statistical and calculus. Five statistical operators all of which take one parameter in the form of data variable have been defined. They are *MIN, MAX, AVG, TOTAL* and *STD-DEV*. Two calculus operators are provided: *SLOPE* which calculates a spatial derivative and *RATE-OF-CHANGE* which computes a temporal derivative.

Connectives. Three connectives are provided: *AND, OR* and *NOT*. For significant event descriptions, the connectives include a set of predicates or other connectives as arguments.

To describe events, there also needs to be a class of variables which are arguments of predicates and functions and include information such as: name of variable, file name to extract relevant data, geometric data (x,y,z-coordination) and sampling time, etc. Using the above variables and event description scheme, a user can define an significant ecological event in following way:

Ecological variable specification
rainfall: file=rainfall00.dat,
   coord[100,345,560,890,0,0]
infiltration: file=infiltration03.dat,
   coord[450,705,260,490,0,0]

Significant events specification
EVENT1: [AND (GT(MAX(rainfall),13.3),
     LT(AVG(infiltration),21.0))];

Interpretation : Event1 is defined as the maximum rainfall value of given geometric subspace is greater than 13.3 and the average amount of infiltration of given geometric subspace is less than 21.0.

A parser reads the descriptions of significant event and generates a parsed-tree for later evaluation. For instance, figure 2 shows the parsed-tree for a given event specification:

AND[GE(MAX(rainfall1),23.1),
  OR[EQ(infiltration1,15.2),GE(rainfall2,7,32)],
  LE(AVG(rainfall3,2.12)]

When the parsed-tree is evaluated the predicates which are objects read data file and return the evaluation result to the connectives. The connectives aggregate the truth value of the elements and report to the higher connective if there is. Since event detection often requires processing of various information, the evaluation scheme of the parsed-tree can be well-adapted to "information fusion". For processing data and combining facts, the entities of the tree can employ more complicated algorithms such as Neural Networks, signal processing algorithms and Possibility models.

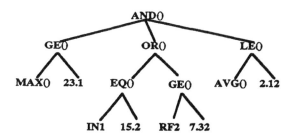

RF:rainfall
IN:infiltration

Fig. 2. Parsed-tree for event specification. The tree provides a systematic method for evaluating events.

5. CONTROL STRUCTURE

The term behavior generally describes the processes that connect perception to action. Thus, behavior senses the changes in the environment and takes an appropriate action based on what was perceived. A combination of behaviors is also a behavior, and in this way complex behavior can be achieved by combining generic simple behaviors (Draper, *et. al*, 1994). In general, the goal is to dynamically and intelligently control the integrated behavior of the system for different scenarios.

An intelligent system must recognize the significant events and change its behavior as appropriate. Thus the sequence of actions (behaviors) actually executed depends upon the sequence of semantic events detected. To interact with the system, the user identifies the processes which constitute a behavior and for each possible event of semantic interest describe the system's reaction. To achieve the necessary flexibility and integration, the control structure is developed as a programmable finite state machine (Ehrman, *et. al*, 1994).

A finite state machine (FSM) is a system with discrete inputs and outputs, and at anytime it is in a particular state out of a given number of states. Each state will produce a certain behavior (Draper, *et al.*, 1994) or basic action (i.e. monitoring the data or displaying a 2D graph), which determines how the system reacts to inputs or changes. The current state of an FSM is determined by the previous state and the inputs. Ultimately each current state is a summarization of the initial state and all past inputs or changes. States are entered and exited through state transitions which correspond to arcs in the FSM. These state transitions are implemented as a transition table which gives the next state of the FSM depending on current inputs and the current state. These transitions encode occurrences of events or inputs by the user. Hence, both external (user) and internal (data driven) events are accommodated in a unique and consistent manner.

An FSM is formally described by a 5-tuple (Q, Σ, d, q_0, F) (Hopcroft and Ullman, 1979). Q is the set of all possible states, Σ is the set of inputs, d is the transition function, q_0 is the initial state, and F is the set of final states. For our programmable FSM, Q will be replaced by P where P is a group of one or more states that form primitive FSMs. The system has an expandable set of generic primitive FSMs which will be instantiated and combined to generate the desired control structure. These primitive FSMs (which include primitives to create graphics, monitor data, and allow user interaction) can be assembled in almost any combination. Since the composition of the states and the transitions is not fixed, the combined FSM generated from the primitive generic FSMs can be tailored to the particular tasks the system is trying to achieve. See section 5.1 for more on primitives. F will be replaced by f which will be only one state. User input, event definitions, and rules for assembling primitive FSMs will determine the transition function. Three types of transitions will cause the FSM to move from one state to another. The first type is an event in the data. The second is an automatic done signal by the current state when its has finished its task, and finally a user caused transition. q_0 will always be the monitor data state, and the final state, f, will also be the monitor data state.

The programmable FSM allows for the needed flexibility and dynamics by the inherent properties of finite automata and by its programmability. These properties have led to a system with behavior that is determined by simulation data and that is modifiable with having to reprogram.

5.1 *Primitive FSMs*

For modularity, code reusability, implementation and testing, and control reasons, states that perform only the most basic behavior are used, but then grouped these states into primitive FSMs so that they could be referred to by higher level behaviors. These primitives will then be used to build the FSM, and three categories of primitives are needed for a complete control system. The categories are system, graphical, and user primitives. System primitives will take care of communicating with the database and simulation. Graphical primitives will displays the graphics, and user primitives will allow the user to interact with the system once it has started. Figure 3 shows an example primitive FSM along with a script that describes how the primitive operates. The primitive shown displays a three-D graph.

Monitoring and User Interaction. Two important primitive FSMs that need to be included in every implementation of the system are the monitor data and wait primitives. The monitor data performs as expected by searching the database while the simulation is running for specified events. Upon event detection the control is transferred to another state which will begin to handle the event. The wait primitive is almost as important. Within the wait state the user is allowed to interact with the system. Manipulation of the graphics and event definitions is allowed.

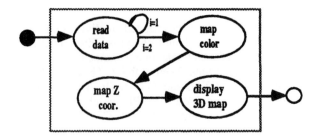

3D FSM

inputs: data_variable for color mapping,
           data_variable for 3D,x1,x2,y1,y2
output: 3D map

3D FSM script

states: read data, map color, map Z coor.,
       and display 3D map;
processes: read, map to color, map to Z,
         and display;
events: read done, map done, display done;
variables:i=0;

```
while read data(variable_i, x1, x2, y1, y2) {
        i++;
        RUN read data;
        EVENT read done && i=1 GOTO
                read data;
        EVENT read done && i=2 GOTO
                map color;
}
while map color(data variable_0) {
        RUN map to color;
        EVENT map done GOTO map Z coor;
}
while map Z coor(data variable_1) {
        RUN map to Z;
        EVENT map done GOTO
                display 3D map;
}
while display 3D map() {
        RUN create;
        EVENT create done GOTO end;
}
```

Fig. 3. 3D primitive FSM. Primitives combine basic functionality to create a simple behavior.

6. IMPLEMENTATION OF PROTOTYPE

The prototype of the system was developed with an object-oriented programming language (C++) on a distributed computing environment. The visualization system used with the agent was developed with IRIXGL and Motif libraries. Also all of the user interfaces used Motif libraries. Two workstations were employed: a SPARC 10 workstation for watershed model simulation and SGI (Silicon Graphics INDIGO2) for data monitoring and visualization.

The control system was implemented to interact with four types of displays. The displays are a 2D colormap, 3D colormap, a histogram, and a hydrograph which is an x-y plot of water runoff versus time. The number of display types can easily be increased if needed, but for the particular simulation being integrated the above four are sufficient. Reactions input by the user and certain rules are used to construct FSMs which behave as desired. The method to inputting reactions is given below.

```
(Event: (event name)
    [Show: (list of graphics and parameters)
    (Simulate: (new simulation parameters)
    NoWait
    Subevents: (list of sub-events)]])
```

The key word **Event** connects the appropriate event definition with reaction. The specifications given after the event definitions are commands that should be executed upon the detection of the event. **Show** indicates that the following graphics should be displayed. **Simulate** causes the given parameters to be sent to the simulation. The simulation parameters should be given as pairs of parameter and value. **NoWait** prevents the simulation to be halted while the user investigates the event. By default the simulation halts. More on this is included in section 5. Sub-events are defined the same way events are defined, and then placed after the **Subevents** command. A sub-event allows different events to be monitored for given that the main event has been detected. Their use provides more control over the visualization with less run time interaction by the user.

An example reaction is given below:

```
(Event: e1
    Show: 2Dgraph,histogram)
(Event: e2
    Show: 3Dgraph
    Simulation: (new parameters)
    (Subevents: Event: e3
                Show: 2Dgraph
                NoWait)
    (Subevents: Event: e4
                Show: 3Dgraph))
```

where e1 - e4 are event definitions. Also note that for simplicity the exact parameters needed for the graphics have been left out. Figure 4 shows the resulting FSM. At first the rest of the FSM from the simulate primitive could be completed, but the user will be allowed the option to open other graphics in the wait state and those graphics will also need to be updated if a simulation is started. In the wait state, the user has the option to restart the simulation of the displayed data or to return to monitoring the data. Figure 5 shows the results of the detection of e1. The control system causes two displays to be generated. The top display shows a 2D colormap of a water map, and the bottom shows a hydrograph of the point on the 2D map marked with an 'X'.

The communication environment is based in a UNIX-based message passing system: PVM (parallel virtual

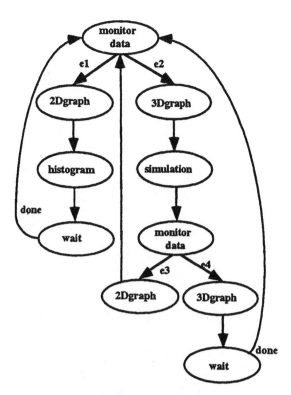

Fig. 4. FSM created from example in section 6. Each node is a primitive fsm combined to create create a complex behavior.

machine). PVM is a powerful and flexible software system which allows a network to be used as a single computational resource, a virtual machine. Before the system starts, user specifies multiple events and variables that he/she might be interested in, but at the monitoring stage of FSM, only subset of events are evaluated against incoming data.

During the operation, the intermediate simulation data is continuously transferred to the SGI computer, where the event agent monitors the data. If a pre-defined event specification satisfies, the visualization agent displays the data on the SGI screen. New simulation commands and parameters can also be sent to the SPARC 10 to modify the simulation.

7. CONCLUSIONS AND FUTURE WORK

Use of an agent which detects significant events and controls reactions to the events greatly enhances the capabilities of a simulation/visualization environment. Allowing current data to effect future results as dictated by the user gives rise to better scientific experimentation. Decision-making based on simulation results can be done with more confidence because interesting data that occurs anywhere spatially or temporally will be detected. A third enhancement supports more realistic modeling with real-world data. This results again from the ability to more efficiently view the data.

Section three indicates an area where more intelligence can be added to the agent. The use of a knowledge base of event/reaction definitions decreases the

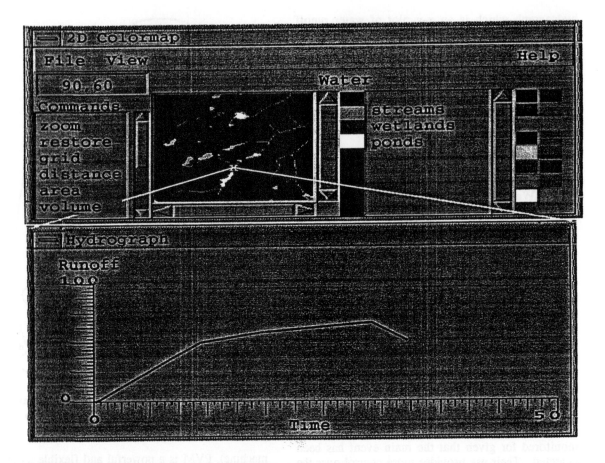

Fig. 5. Graphics generated by agent. A 2D colormap and hydrograph are shown. The 'X' marks the point on the colormap where the runoff data for the hydrograph was taken. The white lines extending from the 'X' to the hydrograph help distinguish where the data comes from.

qualifications need by the user to effectively apply the agent. Complex definitions can be created by a research group and then used by others. The automatic graphics generator frees the user from always having to define the graphics to be displayed, and can use intelligence to create more informative displays. These informative displays will not only show what event has taken place, but will show why the event has occurred and how it will change in the near future.

ACKNOWLEDGMENTS

This work has been supported in part by a National Science Foundation high performance computing grant, ASC-9318169. The provided support is gratefully appreciated.

REFERENCES

Brodlie, K. *et al.*, (1993). GRASPARC- A Problem Solving Environment Integrating Computation and Visualization. *IEEE Visualization '93 : Proceedings* , pp. 102-109.

Charniak, E. and D. Mc Dermott (1985). *Introduction to Artificial Intelligence*, pp. 321-343. Addison-Wesley Pub. Co., Menlo Park, CA.

Draper, B. *et al.*, (1994). Integration for Navigation on the UMass Mobile Perception Lab. *AIAA/NASA Conference on Intelligent Robots in Factory, Field, and Service.* pp. 473-482.

Ehrman, J., M. Marefat, J. Yost (1994). Intelligent Visualization Agents and Management of Large Scale Ecosystem Simulations. In Proc. of *Annual International Conference on Geographicsal Information Systems - Dcision Support 2001.* Toronto, Canada, Sept. 1994.

Hopcroft, J.E. and J.U. Ullman (1979). *Introduction to Automata Theory, Languages, and Computation.* Addison-Wesley, 1979.

Lang, U., R. Lang and R. Ruhle, (1991). Integration of Visualization and Scientific Calculation in a Software System. *IEEE Visualization '91 : Proceedings* , pp. 268-273.

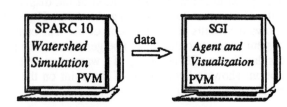

Fig. 6. Prototype System Configuration.

220

INTERACTION RELATIONSHIPS: A PARADIGM FOR THE DESCRIPTION OF VISUAL MANIPULATION

Manfred Tscheligi

Vienna User Interface Group
University of Vienna
Lenaugasse 2/8
A-1080 Wien
mt@ani.univie.ac.at

Abstract: Visual manipulation means the transmission of the intentions of the user to the computer by means of visual elements. A structural description is introduced to offer a model for understanding and describing visual user interfaces. The model supports the development of user interfaces based on a general interpretation of a visual element not constrained by standard look and feel standards. The user interface is modelled by changes in the visual presentation. The visual presentation is composed of visual elements. The dynamics of the visual presentation can be described by interaction relationsships initiated by manipulators.

Keywords: User Interfaces, Interaction, Model, Direct Manipulation

1. INTRODUCTION

Visual elements are the main constituent of modern user interfaces, which embody the visualization of the application state and the basis for the input actions of the user (Draper, 1986). Perceivable affordance has to be expressed by the appearance of visual elements. Every internal state change should be reflected at the presentation surface to offer the user an actual manipulation basis.

In this paper a structural description of visual manipulation is introduced to offer some means for the understanding and description of visual user interface behavior. Visual manipulation extends the term direct manipulation, which is often used in the context of manipulating predefined interaction techniques like scrollbars or menus. The structural description is based on general interpretation of visual user interface objects not constrained by predefined interaction techniques. This forms the basis to support the design of Non Standard User Interfaces.

The identification of structural properties is beneficial for the understanding of the phenomena, relationsships and elements of the information exchange between user and computer. This identification of structural properties supports the foundation of conceptual model of visual human computer interaction and act as a meta level which influences all acitivities during the development of the user interface (design, specification, implementation, runtime architecture and support).

Whereas traditional approaches to user interface representation are influenced by high level sturctural models (e.g. the language model according to Foley, 1980) visual manipulation dialogues lack such a conceptual framework. This missing frame of understanding leads to a high complexity and usage problems for the designer of modern dialogues.

In this paper such a structural framework is presented which is based on a basic interpretation of visual user interfaces, behavioral abstractions to describe user interface dynamics and the explixit integration of semantic feedback as an inherent feature of the interaction relationship based description approach.

2. RELATED WORK

The structural characterization of visual manipulation is influenced by several types of modelling approaches: interaction paradigm characteristics,

221

structural or interaction models for modern human computer commnication, design specification techniques and implementation oriented representation techniques within user interface tools.

Modern visual oriented user interfaces are based on the model world approach often circmscribed by the term direct manipulation (e.g. Draper, 1986; Shneiderman, 1983). For this paper the general abstraction of direct manipulation resulting in visual manipulation is the user controlled alteration of visual elements forming the constituents of the user interface.

Structural models describe the general process of human computer interaction and define the phenomena, relationships and elements of the information exchange between user and computer (Hix and Hartson, 1987). Interaction models are another term used in this context. Often conversational style oriented structural models are used for model world based dialogues. A more recent approach is the surface model (Took, 1990). Took defines the surface as abstract, persistent graphical structure, on which user interface actions take place. The surface is similar to the visual world of visual elements used in this work. The main emphasis of the interaction relationship approach used here lies more on the dynamic behavior of manipulations than on the graphic structure. More formal oriented interaction models are the family of PIE models (based on the use of mathematics to describe principles and properties of human computer interaction) and the interactors model (Duke, et al., 1994), which is based on reference models used in computer graphics. Duke gives an ecellent overview of these two types of models and their usage for user interface design.

Design specification techniques are used to the communication and documentation in earlier stages of interactive system development (Hix and Hartson, 1993). Specification techniques have to transfer the general concepts of an interaction model within a notation used throughout the design. A prominent example is the User Action Notation (UAN) (Hartson, et al., 1990). The specification notation used in this paper is used for the demonstration of the interaction relationship approach but can also be used for conceptual design activities.

A lot of different approaches exist in the domain of User Interface Management Systems. Several types of user interface languages (Myers, et al., 1992) are in existence. Some systems try to generalize the notion of direct manipulation to handle non standard user interface objects (Myers, et al., 1990) or to provide alternative application views. One of the main motivations for the model presented in this paper is the difficulties user interface designers face in understanding the general model of the interface implicitly incorporated in these user interface languages. By providing a more conceptual characterization the designer could be supported in a much better way.

3. VISUAL MANIPULATION PRINCIPLES

The main feature of visual manipulation is the alteration of a visual representation within the user world. The whole visualization of the application functionality (user world) is composed by visual elements. All changes regarding the state of the application are triggered by user controlled changes of visual elements which are the fundamental manipulation units. These alterations become the initiating determinant of state changes. A visual element is defined by its shape (e.g. circular, rectangular, triangular, pattern, color), its position (in n-dimensional space) and its appearance (the look of the visual elements in a particular situation). The definition of the shape also includes the determination of the size. Together these attributes form the characteristics of a visual element.

The characteristics of a visual element has to be changed to get an new visual configuration. The alteration takes place by means of interaction relationships defined between visual elements. Interaction relationships define the interaction behavior of the participating elements. The interaction behavior results in an altered visualization configuration.

The manipulation of visual elements is primarily carried out by using the concept of manipulators. Manipulators embody available physical or logical input devices. On the other end of the interaction relationship there are the visual elements acting as the primary alteration target. This relationship between a manipulator and an element is designated as primary interaction relationship and the manipulated element is the primary element. Only one manipulator is allowed to alter the characteristics of one element at a given time but there are no restrictions regarding the concurrent usage of different manipulators for different elements.

Visual elements cannot be modelled without taking into account other visual elements. All elements are constituents at a visual world where changes in the characteristics of one visual element cause effects in the characteristics of other visual elements. This sort of dependencies is modelled by secondary interaction relationships. One primary change can cause more than one secondary changes. So the primary element gets the role of a manipulator for other elements. Secondary changes are also able to trigger other secondary interaction relationships and manipulators can also be influenced by a secondary change. The interaction relationship model is illustrated in Figure 1.

A typical example for a secondary relationship is can be illustrated by the usage of a trash can for a delete operation. Altering the position of some object which has to be deleted is a primary change carried out by a manipulator (usually the cursor) until the element is on top of the trash can. The movement of the object is the primary change and the object is the primary element. The movement is initiated by the interaction relationsship between manipulator (visualized by the cursor) and the object. The delete operation (the disappearance of the object) results from the

secondary interaction relationship defined between the primary element (the object) and the secondary element (the trash can). All manipulations presented within the visual user world can be modelled by these two types of interaction relationships.

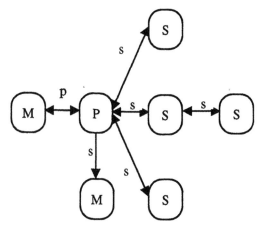

M: manipulator   p: primary relationship
P: primary element   s: secondary relationship
S: secondary element

Fig. 1: Interaction Relationships

4. INTERACTION RELATIONSHIPS

The interaction relationship determines the visual changes which are made during the interaction of two visual elements. The characteristics of an element determine the visual presentation and therefore the role of the element within the visual configuration. By varying the characteristics an element gets different presentations. Different presentations can have different interaction relationships to other visual elements. So the actual presentation takes part in determining the behavior of the visual element. With this feature of the model semantic state changes (visualized by an altered characteristics) influence subsequent element behavior. Caused by these different sets of interaction relationships, the different characters Conceptually the different combinations of shape, position and appearance establish the sides of a polyhedron

Different visual elements may have a different number of interaction sides. In its simplest form an element has exactly one interaction side. This means that interaction relationships are independent of element characteristics.

The actual presentation of an element determines the actual interaction surface of an element at a given time. The actual interaction side guides the interaction at interaction time. The interaction time determines the time two elements interact. The interaction relationship drives the visual changes to be made. Visual changes lead to changed characteristics which may have some consequences to the actual interaction side and therefore result in an altered interaction behavior. The two partners of an interaction relationship are called initiating element and manipulated element to get a direction of an interaction.

According to the result of an interaction relationship the following classes of interaction relationships can be identified:

-   no changes of the participating elements
-   visual change of the manipulated element (-->)
-   visual change of the initiating element (<---)
-   visual change of both elements (<-->)

The first case is equal to the absence of an interaction relationship. The movement of a visual element (alteration of the position) by the mouse is an example for the second case. The mouse is modelled as a manipulator and is represented by means of a mouse cursor. The change of the position results from the interaction relationship between manipulator and visual element. This interaction relationship does not influence the presentation of the manipulator. In the opposite direction (case 3) only the manipulator gets an altered characteristic (e.g. the color of the cursor) but the manipulated element is not changed. However, the altered presentation of the cursor can have some influence on the following manipulations by an altered set of interaction relationships for the changed presentation. Finally (case 4), the manipulator is represented by a new color and the manipulated element is moved.

5. PRIMARY INTERACTION RELATIONSSHIPS

Exactly one visual element (primary element) is influenced by a primary interaction relationship. Primary manipulations are the starting point of any visual manipulation. The initiating element is the manipulator.

So far the notion of the interaction time has not been discussed. The interaction time determines the elements whose actual interaction sides (depending on the current presentation) interact. For primary interaction only the visual correlation of two visual elements is necessary. Here, defined parts of the interacting elements have to correspond in position. The process of primary interaction can be summarized as follows:

-   external events (mouse movements, mouse button activations) cause changes in the manipulator characteristic
-   the alteration of the characteristic determines the actual interaction side of the manipulator
-   the manipulator visually touches other visual elements while changes of the manipulator position take place
-   the actual presentation of the touched element (visual correlation) determines the actual interaction side of this element
-   the manipulator interacts with the visual element involved in the visual correlation
-   if an interaction relationship exists between both interaction sides the visual change specified in the relationship takes place
-   the new characteristic potentially influences the actual interaction side and activates other interaction relationships (the interaction is continued as long as the visual correlation takes place)

To illustrate the mechanism of primary interactions the following example will be used. A visual element (rectangle) should be moved with the mouse. The change of the position takes place while the mouse button is pressed (only a one button mouse is assumed, but the concept is valid also for more buttons). The appearance attribute of the primary element has two possible values: normal and inverse. The movement is only possible if the element is in its inverse representation (the element is prepared for moving by changing to an inverse representation). With a single click on the element the representation switches from normal to inverse. A double click causes the return to the normal representation. The assignment of the characteristic to the interaction sides for the manipulater element is shown in Figure 2. Shape and appearance are described by example. The characteristic of the manipulator is influenced by the physical actions of the user. The actual presentation of the manipulator gives the user appropriate feedback of the current behavior. In this example the position of the manipulator is insignificant for the determination of the interaction side.

Figure 3 shows the situation of the visual element to be moved (primary element). In the example this is a rectangle defined by x and y. Figure 4 shows the specification of the necessary interaction relationships determining the interaction behavior of these two visual elements. The actual interaction results is initiated by visual correlation. Caused by the physical activation of the mouse button the actual interaction side of the manipulator changes. As long as the visual correlation holds, the corresponding interaction relationships are active. If there are no interaction relationships defined (e.g. a pressed mouse button does not interact with a not inverse primary element) nothing happens. The presentation of the manipulator could also reflect the ongoing alteration of the element´s position. For this only slight changes have to be made.

6. SECONDARY INTERACTION RELATIONSHIPS

The primary change of a visual element is the source for additional changes of other elements. These additional elements are the secondary elements and the relationships are designated as secondary relationships. Secondary visualization elements altered may again be the reason for additional alterations. Manipulators can be secondary elements, too.

With the concept of secondary relationships, for example, the usage of so called tools to alter other visualization parts can be modelled The primary relationship manipulates the tool and the tool is connected via a secondary relationship to the target element. Here, the primary element gets the role of the manipulator.

The interaction relationship between the new manipulator and the target element is handled as before.

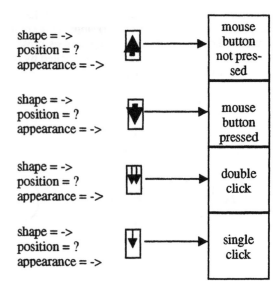

Fig. 2: Manipulator Mousecursor

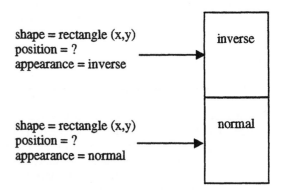

Fig. 3:. Primary Element Rectangle

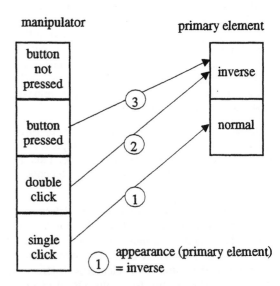

Fig. 4: Interaction Relationships

224

Secondary interactions are activated by additional forms beside explicit visual correlation. The valid activation conditions are summarized as follows:

- visual correlation between two visual elements
- entry of a special interaction side (the condition is applied to the interaction relationships of the new interaction side)
- leave of a special interaction side (the condition is applied to the interaction relationships of the old interaction side)
- changes to the characteristic of the element (changes without an immediate change of the interaction side)

This results in the following typical interaction process (for illustration purposes the initiation is done by visual correlation as before):

- primary manipulations result in changes of the characteristic of the primary element
- the visual changes potentially results in the change of the itneraction side
- due to successful evaluation of conditions interaction relationships to actual interaction sides of other elements become active (the restriction to actual interaction sides guarantees that the user realizes these changes)
- the changes defined in the activated interaction relationships take place
- a new element characteristic potentially activates other interaction relationships

If secondary interaction relationships are used, one has to define the primary element of an interaction relationship to specify the initiating element.

To illustrate the usage of secondary interaction relationships the previous example is extended. In addition to the moveable visual element a tool element is used. With the help of the tool element it is possible to alter the shape of the moveable element. The change takes place at the beginning of the visual correlation. The tool element shows the possible transformation by means of the appearance attribute. In the example case the object can be a rectangle, a triangle or a circle. The tool element is the primary element to transform the shape and the secondary interaction relationship takes place with the moveable object (secondary element). The opposite direction is excluded with an appropriate entry in the interaction relationship definition. As before, the position of the moveable object can be changed too. The changes of the tool element are carried out by a double click. This causes different effects in the secondary interaction.

7. SEMANTIC FEEDBACK

The integration of necessary application knowledge within direct manipulative interaction is often stated as being a problem within modern visual manipulation specification. Semantic feedback is necessary to inform the user during active manipulation. The next example explicitly illustrates the modelling of semantic feedback within the paradigm of interaction relationships. Application semantics are incorporated within the presentation by

the visual appearance to have its effect for further interaction relationships and to show the user what state the element is in.

A file, represented by a file symbol, has to be deleted. It is possible to delete the file only if the file is not protected from deletion (semantic feedback). This results in two possible presentations, which can also be changed by the application semantics. The delete operation is carried out by dragging the visual element to a trash can element (visual correlation). If the file is not protected the execution of the operation will be presented to the user by an inverse trash can (semantic feedback) and the file symbol subsequently disappears. Afterwards the trash can returns to the normal presentation. If the file is protected no visual change is done during the dragging activity and the file cannot be deleted. The activation/deactivation of the delete protection is done out by a single click from the user. Figure 5, Figure 6, Figure 7, Figure 8 and Figure 9 contain the necessary specifications.

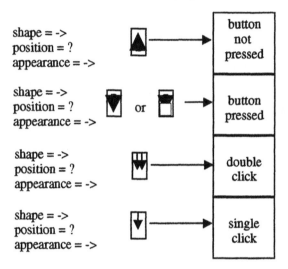

Fig. 5: Manipulator Mousecursor

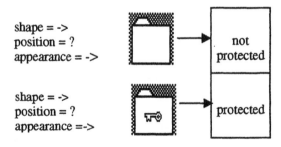

Fig. 6: File Symbol

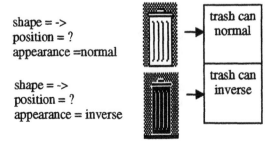

Fig. 7: Trash Can

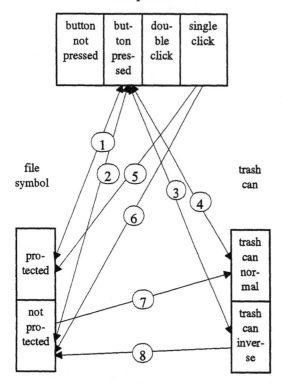

manipulator

button not pressed	button pressed	double click	single click

file symbol

trash can

① ② ⑤ ③ ④ ⑥ ⑦ ⑧

| pro-tected |
| not pro-tected |

| trash can nor-mal |
| trash can inver-se |

Fig. 8: Interaction Diagram

① position (file symbol) =
position (manipulator)
appearance+shape (manipulator) =
type: primary

② position (file symbol) =
position (manipulator)
appearance+shape (manipulator) =
type: primary

③ position (file symbol) =
position (manipulator)
appearance+shape (manipulator) =
type: primary

④ position (file symbol) =
position (manipulator)
appearance+shape (manipulator) =
type: primary

⑤ appearance+shape (file symbol) =
not protected
type: primary

⑥ appearance+shape (file symbol) =
not protected
type: primary

⑦ appearance (trash can) = inverse
type: secondary
trigger: file symbol, visual correlation

⑧ position (file symbol) = (0,0)
appearance(trash can) = not inverse
type: secondary
trigger: file symbol, visual correlation

Fig. 9: Interaction Relationships

8. SUMMARY AND CONCLUSION

The paradigm concentrates on the individual dependencies between the visual constituents of a user interface. Visual elements are the basis of visual manipulation and can actively as well as passively take part in an interaction relationship. The collection of all interaction relationships results in an interaction language for a concrete user interface instance. The description supports prototyping because the connection between two elements can be modelled independently of the rest of the system. The description formalism used to demonstrate the features of the model is declarative in its nature to free the designer from thinking about low level implementation details. Further effort is necessary to finalize this specification approach and integrate it into a design oriented prototyping environment.

REFERENCES

Duke D., Faconti G., Harrison M., Paterno F. (1994). Unifying Views of Interactors. In: *Proceedings of the Workshop of Advanced Visual User Interfaces (AVI´94)*, pp. 143-152, ACM Press, New York.

Hartson H. R., Siochi, A. C., & Hix, D. (1990). The UAN: A User-Oriented Representation for Direct Manipulation Interface Designs. *ACM TOIS, Vol 8, Nr. 3*, pp. 181-203.

Hix D., Hartson H. R. (1987). A Structural Model for Hierarchically Describing Human-Computer-Dialogue. In: *Proceedings of the INTERACT´87*, pp. 695-700, North Holland.

Hix D., Hartson H. R. (1993). *Developing User Interfaces: Ensuring Usability Through Product & Process.* Wiley, New York.

Draper S. W. (1986). Display Managers as the Basis for User-Machine Communication. In: *User Centered System Design* (Norman D. A., Draper S. W., Eds.), pp. 339-352, Lawrence Erlbaum Associates, Hillsdale.

Foley J. D. (1980). The Structure of Interactive Command Languages. In: *Methodology of Inter-action* (Guedj et al (Eds.), pp. 227-234, North Holland.

Myers B. A., et. al. (1990). Garnet: Comprehensive Support for Graphical, Highly Interactive User Interfaces. *IEEE Computer, Vol. 23, Nr. 11,* pp.71-85.

Myers B. A. (1992). *Languages for Developing User Interfaces.* Jones and Bartlett, Boston.

Shneiderman B. (1986). Direct Manipulation: A step beyond programming languages. *IEEE Computer, Vol. 16, Nr. 8,* pp. 57-69.

Took R. (1990). Surface Interaction: A Paradigm and Model for Separating Application and Interface. In: *Proceedings CHI´90 Human factors in Computing Systems*, pp. 35-42, ACM Press, Ney York.

INFORMATION MANIPULATION ENVIRONMENTS:
AN ALTERNATIVE TYPE OF USER INTERFACES

Sabine Musil

Georg Pigel

Manfred Tscheligi

Vienna User Interface Group
University of Vienna

Abstract: Information manipulation environments (IME) provide the designer with a design framework for non-standard user interfaces that meet the demands of today's users for intuitive, visual and useable applications. They are a new type of user interfaces, which go beyond the classical graphical user interfaces (GUI) and are thus applicable to more application areas, especially for areas where traditional WIMP, Look&Feel Guide oriented user interfaces do not suit well like process control, consumer products etc. After discussing shortfalls of current user interface solutions, we will introduce the concepts of IMEs and then will illustrate those with examples from three very different application areas.

Keywords: User Interfaces, Non-WIMP Paradigm, Metaphors

1. INTRODUCTION

Designing user interfaces has become a complex and critical task. Application areas are becoming wider, users are more diverse than ever and so the "one-for-all" user interface is not applicable at any time. Graphical user interfaces have been a milestone in the development of user oriented interfaces between humans and the computer. The WIMP (Windows, Icons, Mouse and Pointing) paradigm in connection with the desktop metaphor proved useful in making it easier for users to fulfill their tasks. But not everything will be carried out on the desktop and not every user is the "typical office worker" for whom the desktop metaphor was originally intended and designed. So it is time to move on and to start thinking on how to improve and develop further the idea that stood behind graphical user interfaces in the first place: facilitate problem solving. A new set of guidelines and metaphors (Green and Jacob, 1992) has to be found.

New kinds of user interfaces are under development and are published under various names, like for example Non Standard User Interfaces, Next Generation User Interfaces and Non Command User Interfaces. What they all have in common is that they share the Non WIMP paradigm, which means that these interfaces are not constrained to those four elements. We distinguish three classes of Non WIMP interfaces: virtual reality, augmented reality and interaction environments.

Virtual reality interfaces try to bring the human and the computer together IN the computer. A computer generated world is provided and the user can explore it by being a part of it. The illusion can be assisted by immersive interaction devices like stereoscopic glasses and data gloves, but very often good three dimensional graphics and a three dimensional input device are enough.

Augmented reality interfaces walk down the path in the opposite direction. They try to integrate the computer in the everyday world of the human. Ubiquitous computing - the other well known term - demands that users shouldn't be aware of the fact that they use a computer and are supported by electronical devices. The "Active Badge" system (Hopper, *et al.*, 1993) or the DigitalDesk (Newman and Wellner, 1992) are good examples for this interaction paradigm.

Interaction environments stand somewhere in the middle of these two. They tend to use conventional technology, but still try to extend and/or replace the traditional desktop. They are not constrained by any look and feel guide, which is supported by the fact that they are user interfaces for non standard problems usually not supported by style guides. Dealing with information of any kind is their main concern. This can be program information, monitoring information or office information, for every type of information a specialized solution can be found. The Information Visualization Architecture (Robertson, *et al.*, 1992) is one prominent example for such interaction environments.

Information Manipulation Environments (IMEs) fall within the category of interaction environments. In the following we will explain this term in detail and describe the various features of this type of user interface. The rest of the paper is dedicated to examples for IMEs, which will illustrate this type of user interface. We will provide different kinds of IMEs to show the huge applicability of IMEs.

2. INFORMATION MANIPULATION ENVIRONMENTS

Encouraged by the fact that there is a strong demand for alternative user interface solutions that go beyond the traditional WIMP desktop, we have developed a concept for a specific kind of interaction environment, which we employ for most of our user interface projects. We have called this kind of user interfaces Information Manipulation Environments (IMEs), as we consider the most important aspect of a user interface to be the accomplishment of a task, which almost always means manipulating some kind of information in order to achieve a goal. IMEs are metaphor oriented, user specific, direct manipulative, highly visual and application oriented.

Metaphor orientation of user interfaces has until now mostly meant using the desktop metaphor. But this metaphor is not sufficient for all application areas and so the designer has to think of alternative metaphors that are more suitable for a specific problem, which is a task that will be supported in IMEs. Not every problem is typically solved on the desktop and sometimes applications seem to be pressed into the corset of windows, trashcans and clipboards. For applications that are not used by typical office workers who are most familiar with the concepts of a desktop, it makes sense to think about other metaphors, which possibly are more suited to their experiences and to the task they have to solve. Popular examples of alternative metaphors include the book metaphor used in hypermedia tools (e.g Apple's HyperCard), the fisheye metaphor (Sarkar and Brown, 1992) and the filter metaphor (Stone, *et al.*, 1994).

The use of metaphors is important because they provide a good means to transport new concepts to users via some other concepts from a familiar domain (Indurkhya, 1992). Metaphors are a cognitive map for the orientation in the functionality jungle of today's software systems. Users can reuse well known concepts to accomplish something new or to remember a specific task. This fact makes metaphor oriented user interfaces easier to learn and easier to use. Nevertheless metaphors have to be chosen very

carefully and to be tested in early stages (Musil and Tscheligi, 1994; Carroll, 1988; Erickson, 1990).

Today's user interfaces are mostly oriented towards the "average" user with some experience in using computers, some experience in the application domain and an North-American cultural background. Again, this is not a good thing to do in many situations. IMEs are user specific in the way that they take care of the background of the target user group, take into account the fact that users not always are familiar with the problems they have to solve with the computer or that users often do not want to use the computer at all. On the other hand there are users who are too familiar with the problems to accept computer assistance or who are too fond of the bits and bytes of a computer to want more than a cryptic command language. And for this broad range of users different user interfaces are necessary. We think it is highly necessary for a good interaction environment to be tailor made for a specific user group or at least to be adaptable for different kinds of users. Apple (Halfhill, 1994) seems to have realized this fact, too, as their new operating system AtEase is adaptable for specific users.

Direct manipulation has become an important part of graphical user interfaces, but often it is not exploited to all its possibilities and mostly used only for advertising novel products. In the IME oriented interpretation of the term choosing from a menu is not everything that can be done via direct manipulation. We use more direct forms of interaction techniques to support the feeling of engagement. Real direct manipulation is an integral constituent of any information manipulation environment. Permanent visibility of the objects to manipulate and a common visualization for input and output are necessary to achieve this kind of interaction. Furthermore the operations on the objects have to be immediately reversible, they must result in an altered configuration or representation of the objects (in order to detect that something has happened) and the operation must be executable directly by the user, indirectly by the program or via the alteration of visual elements that influence other visual elements.

IMEs are also highly visual. Visual information is processed best by most humans. Many reasons have been given for that fact. But still a lot of today's graphical user interfaces are at most equipped with nice designs. Adding on backgrounds and icons here and there doesn't make an interface visual. IMEs make use of visual elements, which is a general term for any concept expressed by visual means (more than standard user interface widgets), in order to take away mental workload from the user and give a more intuitive interaction environment. Visual elements convey the structure and order of things and aid the user's perception and cognition of information. They direct the user's focus and attention and thus make it easier to navigate within a system. Furthermore they are the prerequisite for direct manipulation and immediate feedback on user actions.

Last, but not least, IMEs are application specific. For every problem situation, specific solutions have to be found. This does not imply that the wheel has to be reinvented all the time, but the most suitable solution has to be worked out. Well fitting and often alternative interaction techniques have to be discovered, interaction devices determined and the like. IMEs do not enforce technology where it is not appropriate and do not call for innovations where existing solutions can be applied, but they demand and encourage careful choice. For example, we wouldn't propagate the use of stereoscopic goggles for a three dimensional text editor where reading is the most important task. HMDs today do not provide the necessary resolution to accomplish this.

3. EXAMPLES

We will illustrate the concepts of Information Manipulation Environments with three examples. One is N/JOY, an integrated office application for casual users that makes use of the office building metaphor and works completely without menus. The functionality of N/JOY is totally encapsulated in different types of objects. The second is from the realm of parallel programming. In InHouse a parallel program is visualized, its behaviour monitored and the results are visualized. The metaphor of a hotel represents the parallel system. Our last example is a user interface for the tourism sector. Users can select the features of their desired holiday in a virtual supermarket.

3. 1. N/JOY - A World of Objects

N/JOY (Tscheligi, et al., 1991) provides the user with an overall intuitive interaction philosophy to realize the coexistence of wordprocessing, business graphics, database, spreadsheet and necessary operating functions like copying or printing. The general aim was to give the user a single, consistent environment to solve the necessary office tasks. Our intention was not to overwhelm users with a simple combination of the functions coming from the different parts of an integrated software system but to cut down the functions according to the users' needs. As in real life there are no artificial boundaries between related sets of functionality.

The system is designed according to a real life metaphor. The real life office behavior was used as an analogy during the whole design activity, an approach named "design by symmetry" by Erickson (Erickson, 1990). So we arrived at an extension of the desktop metaphor by using objects of an office environment.

The user enters a building containing several offices. Behind the entrance the user enters a room containing all the objects that are needed to carry out intended tasks. Among other things it can contain more doors, one of them being the exit door, leading out of the building. A room forms the background for all other objects and can be equipped with objects according to the special requirements and tasks of the user. The user is able to create as many rooms as necessary and is able to change the working environment by simply opening the door to the other room. Objects can be transferred from one to another by dragging the objects to the door. Thus specifically configured objects can be moved to other places throughout the building and can be used there.

N/JOY is highly visual, because it is composed of various visual objects, realizing the different functional features of the system. This results in an explicitly visually represented model world, where the user directly manipulates well defined objects with manipulation primitives forming a visual interaction language. There is no distinction between a manipulated world and a manipulator world (usually menus). We totally omit menus. Furthermore we explicitly visualize normally hidden system concepts is the cut and paste function. Selected text can be copied and becomes a movable clipping object that can be used for any part of the system where text is needed.

N/JOY is application specific in the way that it is extendable via the catalogue, which contains all available objects. The user can "order" things that are delivered immediately. If additional tools are introduced by later enhancements they can be simply integrated into the old catalogue. Flexible structure is very important in office tasks. The demand for flexibility is also met with the structuring concept of interleaves for folders. Users can put interleaves wherever they want in a folder.

N/JOY is also user specific as it can be parameterized. A central object in this respect is the document. The document consists of pages and can be organized in folders. There are two different types of folders: folders and archive folders. Both have the same functionality but with different presentations for two diffferent cultures. Folders correspond to the type of folders typically used in the US. Archive folders correspond to the European type of folders.

Another example for the user orientation via parameterization is the pencil. A pencil acts as a representative for a set of text attributes. Many pencils can lie around the desk. To change the text style the user clicks on a pencil. The active pencil is shown inclined. The conceptual model of activating a pencil is taking it into the hand. As long as the pencil is active the same text attributes are in use. This improves the mode of interaction used by common text processors, where the system continues to write with the attributes of the character before the text insertion position.

N/JOY is also highly direct manipulative, because all functional features of the system are embodied in interaction relationships between various objects. The interaction relationship is initiated by a small set of manipulation actions. Each interaction has a defined meaning depending on the interacting objects. Take for example the interaction between a copier and a wastebin. Either one may take the role of the operand or the role of the operator. Marking the copier and activating the wastebin initiates the function of the wastebin. The copier which is the operand is deleted. Marking the wastebin and activating the copier copies the wastebin. This results in symmetric interaction behaviour.

A further contribution to the user orientation is N/JOY's capability to allow a high degree of flexibility to configure the working environment according to their needs and preferences. An unlimited number of rooms can be created and objects can be arranged in the room without any constraints. An office can be configured like the user's office in real life, having it disorganized with things lying around or neatly organized in named cupboards. Multiple instances of any object can exist simultaneously either with the same behaviour or parameterized with different settings. So the above mentioned set of pencils can be installed or there are some printers representing different physical print devices. With this feature, which enables the user to prepare tools in advance, the user won't be distracted during task execution by permanent cycles of tool adjustment. The possibility of parameterization is one prerequisite to work without menus.

3.2. InHouse

InHouse is a user interface for a parallel monitoring system for students. Parallel programs are recorded as the user wishes to in order to analyze their runtime behaviour in a later phase. The monitoring tool consists of three phases. First of all there is the object specification phase, where the users see a visualization of the parallel program they want to analyze and can select the items of interest to be monitored. This is done on the one hand to keep the trace file small and on the other hand to give the users a feeling of what their programs are like. Next, the instrumented parallel program runs on the parallel system and a tracefile is produced containing the runtime behaviour. This tracefile can be analyzed in the third phase, the visualization phase. Various displays show different aspects of the runtime behaviour and a lot of them can be viewed dynamically like a videotape.

The target user group of this monitoring tool were students who are not familiar with the concepts of parallel programming. So we were not satisfied with the current solutions (e.g. Heath and Etheridge, 1992) of how to present the results of such a monitoring process, which were mainly traditional visualization techniques like Gantt charts or Kiviat diagrams presented in a window based system. The biggest problem was that many of these diagrams were showing different kinds of information but using the same technique. For a novice to parallel concepts they are not much more than a nice coloured pattern of rectancles or free forms. This led us to the conclusion that we had to think about a different solution and arrived at deciding to implement an IME.

InHouse is first of all metaphor oriented. We have taken up the house/hotel metaphor for representing the parallel system. The number of rooms in the hotel corresponds to the number of processors on the

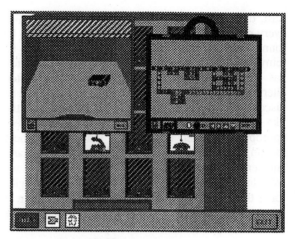

Fig. 1: A typical InHouse screen

system and the number of suitcases in a room is equal to the number of processes that will be executed on this processor during runtime. Suitcases contain puzzle pieces which are a visual representation of the code. All the other necessary functionality is centered around this metaphor, e.g. a remote control is used to regulate the dynamic displays and this can also be found in a hotel. We have tried several metaphors like a tree, a school or a prison, but after some evaluation we finally settled on the hotel metaphor (Musil, et al., 1994).

InHouse is also highly visual. We have tried to avoid text as far as possible and rely completely on the intuitiveness of the graphics and colour coding. A user test has shown that this is possible, at least for parallel system not having more than 16 processors and not much more than six processes per processor. A small set of visual elements was enough to convey the whole functionality. These visual elements were consistently used throughout the whole program, both in the selection and in the visualization phase. Figure 1 shows a typical screen from the selection phase. The visual elements were of extreme importance in the visualization phase as they helped determine what was shown in a single display. For example, the state of a processor is represented by a picture of a room window that has eyes and a mouth. Depending on the state of the eyes (closed, half-open and open) and the mouth (open or closed) the four states idle, waiting, active and communicating could be visualized. The users immediately knew that they were watching a processor as there was a room window graphics shown and it was even possible to determine from the colour which processor it was.

InHouse is also user oriented and application specific. We chose a real world metaphor, which everyone is familiar with, because we are dealing with novice users with various backgrounds. Novice users deal better with such metaphors (Vaananen, 1994) than with abstract ones. We assisted the comprehension of the concepts by providing a very realistic "model" of the hotel, which would have been best implemented with a really three dimensional environment. Nevertheless we have chosen a two and a half dimensional, highly visual, mouse regulated interface running on a large screen to be most suitable for our purposes, because our users are students and very often only have access to so called standard equipment with a pointing device and a large screen. In order to see more parts of the parallel system (the hotel) at once, which was declared a must not only by parallel programming experts, but also by many future users, we use multiple windows.

InHouse is last, but not least direct manipulative in the way that it allows users to select the objects directly and that some objects can be freely arranged. Selections have immediate feedback in that all objects affected by that selection change their graphic representation. Rearrangements and other visual adaptations like zooming are also immediately visible. Navigation within the system is also accomplished by direct manipulation. Every object has a region for selection and one for opening. Upon clicking on the open region a new window displaying the contents pops up and can be closed with a button. These region are intuitive for the users. For example on the suitcase there is a lock to unlock the suitcase and the rest of the suitcase is the selection region.

3.3. The Holiday Supermarket

In the last years automation spread wider than could ever be imagined. Even sectors selling services shift to selling self-services. Banks, which are trying to replace their clerks by ATMs, are a good example for this trend. Also travel agencies consequently will replace their employees by some kind of automated service. But it was due to the special conditions in the banking sector that the automatation of banking services was such a success. Cash machines for example have a horrible user interface, but they are the only way to get cash outside the limited opening times. Travel agencies cannot rely on similar conditions. Users cannot be forced to use automated services in travel agencies, they must prefer using the system to talking to a human clerk. It must be fun to use the automated service. So the basic requirements on the design of automated services of travel agencies were general understandability of what the system is good for, extreme ease of use, and attraction of potential users.

To solve this design problem we came up with the holiday supermarkt (Musil, *et al.*, 1995; Musil, *et al.*,

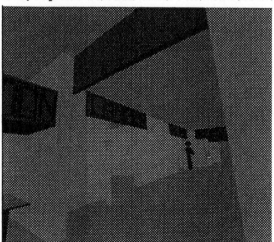

Fig.2: A typical aisle of the holiday supermarket

Fig.3: In the sports department

1995b), which allows users to put together their dream holiday by selecting features in the form of various objects that are a symbol for a certain feature. Users go through the supermarket (see Figure 2) and take these objects off the shelves and collect them in their trolleys: For example, the users head for the sports department and take out a tennisracket to indicate that they want to play tennis (see Figure 3). It is also possible to put items back onto the shelves or to exchange objects. At the cash desk the users finally get a list with suggestions for holiday trips fulfilling their desires. This holiday supermarkt encorporates all the points typical for an IME:

The holiday supermarket is metaphor oriented. The customers of travel agencies (in the following simply called users) use this system only occasionly and will not be willing to invest any effort in learning how to use this system. They have to understand it without further help. The best way to reach this goal is to reuse knowledge from a different domain. The vehicle that allows this transfer of information is a metaphor, the supermarket in our case. This supermarket metaphor performed well in user tests. First of all, no explanation how to understand the metaphor was needed. It also fits fine into the context of a travel agency. As users buy nearly everything in a supermarket they will very likely accept the fact that holidays can be obtained there, too. In addition shopping is very contrary to work, which is a basic requirement for an automated travel agency services system. Furthermore there is no evident mismatch between the source domain and the target domain, which makes this metaphor very easy to understand.

The design of this user interface is user specific. The holiday supermarket has to attract a very broad user group. The only feature all potential users have in common is that they can be assumed to like to spend their money, else they would not visit a travel agent. From this point of view a user interface based on a supermarket is an optimal choice. Besides that people visiting a travel agency are out for some fun and entertainment and generally will not appreciate anything which reminds them of their work. So some VR techniques like moving around in a 3D building were included to give the whole interface more the feeling of an arcade game than of a serious "worklike" application.

The holiday supermarket is highly visual with direct manipulation. There are no icons to click on which indirectly initiate any commands or selections out of menus. Everything the users do is done directly with the artifacts displayed (and every artifact existing in ths system is displayed visually) on the screen. This makes this user interface much more intuitive to use. Every user can be assumed to know that one has to put artifacts into one's trolley in a supermarket, but it is not so sure, that every user would know that one has to doubleclick on icons to put them into action.

This solution is application specific. The implementation of a supermarket as an interface to a computer application cannot claim any generality for being a generic solution for a wide range of applications. The holiday supermarket interface is an appropriate solution for the given problem. Furthermore there is a very broad user group in Western European countries which has to be attracted and fascinated by an arcade game like interface. We are convinced that the holiday supermarket would not prove a good interface for people who organize holidays professionally, i.e. travel agents. The holiday supermarket is a non standard interface design for a specific application, not an off the shelf solution, which is very likely not to be accepted for this application.

The holiday supermarket as a metaphor oriented, user specific, direct manipulative and application specific user interface design is a very fine example for the successfull application of the IME concept.

4. CONCLUSIONS

In this paper we have introduced an alternative type of user interfaces called Information Manipulation Environments. We have pointed out the most important features and have illustrated the concepts with three examples. Future work will include further prototyping and implementation of various IMEs. We furthermore work on a design process for IMEs that especially focuses on the metaphor orientation of the interaction environments.

REFERENCES

Carroll, J.M. (1988). Interface Metaphors and User Interface Design. In: *Handbook of Human Computer Interaction* (M. Helander (Ed.)), Elsevier Science Publishers, Amsterdam.

Erickson, T.D. (1990). Working With Interface Metaphors. In: *The Art of Computer Interface Design* (B. Laurel (Ed.)), pp. 65-73, Addison-Wesley, New York.

Erickson, T.D. (1990b). Interface and the Evolution of Pidgins. In: *The Art of Computer Interface Design* (B. Laurel (Ed.)), pp. 11-16, Addison-Wesley, New York.

Green, M. and R. Jacob (1992). SIGGRAPH Workshop Report: Software Architectures and Metaphors for Non-WIMP User Interfaces. *Computer Graphics*, **25**(3), pp. 229-235.

Halfhill, T.R. (1994). Apple's High Tech Gamble. *Byte*, December 1994, pp. 50-70.

Heath, M.T. and J.A. Etheridge (1992). Visualizing the Performace of Parallel Programs. *IEEE Software*, **8**(5), pp. 29-39.

Hopper, A., A. Harter and T. Blackie (1993). The Active Badge System. In: *INTERCHI '93 Conference Proceedings* (S. Ashlund, K.Mullet, A. Henderson, E. Hollnagel and T. White (Eds.)), pp. 533-534, ACM Press, New York.

Indurkhya, B. (1992) Metaphor and Cognition. Kluwer Academic Press, Amsterdam.

Musil, S. and M. Tscheligi (1994). InHouse: An Information Manipulation Environment for Monitoring Parallel Programs. In: *CHI '94 Adj. Proceedings* (B. Adelson, S. Dumais and J. Olson (Eds.)), pp. 135-136, ACM Press, New York.

Musil, S., V. Giller and M. Tscheligi (1994). InHouse: Designing a Metaphor Based Interaction Environment. *VUIG Technical Report Series*, **TR 94/15**.

Musil, S., G. Pigel and M. Tscheligi (1995). The Holiday Supermarket. To appear in: *Proceedings of ENTER'95*, Springer Verlag, Wien..

Musil, S., G. Pigel and M. Tscheligi (1995b). Der Urlaubssupermarkt. To appear in: *Proceedings of Softwareergonomie'95*, Springer Verlag.

Newman, W. and P.A. Wellner (1992). A Desk Supporting Computer-based Interaction with Paper Documents. In: *CHI '92 Conference Proceedings* (B. Bennersfield and J. Bennet (Eds.)), pp. 587-592, ACM Press, New York.

Robertson, G.G., S.K. Card and J.D. Mackinlay (1993). Information Visualization using 3D Interactive Animation. *Communications of the ACM*, **36**(4), pp. 57-71.

Sarkar, M. and M.H. Brown (1992). Graphical Fisheye Views of Graphs. In: *CHI '92 Conference Proceedings* (B. Bennersfield and J. Bennet (Eds.)), pp. 83-91, ACM Press, New York.

Stone, M.C., K. Fishkin, and E.A. Bier (1994). The Movable Filter as a User Interface Tool. In: *CHI '94 Proceedings* (B. Adelson, S. Dumais and J. Olson (Eds.)), pp. 306-312, ACM Press, New York.

Tscheligi, M., F. Penz, and M. Manhartsberger (1991): N/JOY - The Word of Objects. In: *Proceedings IEEE Workshop on Visual Languages*, pp. 126-131, IEEE Computer Society Press.

Vaananen, K. and J. Schmidt (1994). User Interfaces for Hypermedia: How to Find Good Metaphors? In: *CHI '94 Adjunct Proceedings* (B. Adelson, S. Dumais and J. Olson (Eds.)), pp. 263-264, ACM Press, New York.

PATTERN BASED HUMAN-MACHINE SYSTEM

T. Vámos, K. Galambos

Computer and Automation Institute, Hungarian Academy of Sciences
H-1111 Budapest, Lágymányosi u. 11., Hungary

Abstract: The unclear knowledge on issues where logical structure is not known, and the closed word assumption fails, can be treated by pattern metaphor that reflects the brain representation of knowledge. Recognition of patterns is a man-machine interaction process best supported by those means that are close to the user's practice. Visual representation of input and output have different requirements. The applications of different methods of graphic art are analyzed from the point of view of easy parallel data input and cognitive psychology of knowledge acquisition.

Keywords: human brain, human-centred design, meta-level knowledge, multidimensional systems, pattern recognition, physiological models, visibility

1. INTRODUCTION

Designing a system, having any relation to both agents, man and machine, we start with considering representations of knowledge in the two agents. The computer representation is mostly a data base, having some, possibly flexible, structure. The mind representation is the great problem of brain research and cognitive psychology. Not getting into not yet consolidated disciplinary details we can use those views that are more or less accepted. The distributed mechanisms of the brain combined with localizations, the lessons of evolutionary development, suggest the pattern model. The model started with the memory and response organization of the most primitive living creatures. This is a model only, not the real picture of complexity of the mind; in man-machine interactions, however, we should be satisfied with such simplified models, especially if they work in a limited extent and validity.

The pattern model is one of the oldest, just because of its apparent justification. It returns in different disciplines and naming, like idea in ancient philosophy, Gestalt in psychology, or connectivist structures in computer science, objects in programming. A further consequence of considerations based on the pattern model was the suggestion to see the continuity of representation from the pattern-like memory ensembles to the abstracted schemes of reasoning, like logic (Vamos, Koch and Katona, 1994). The term metapattern is used to express pattern relations and dynamics. The extension of the pattern concept from the original, unstructured visual metaphor to disciplinary methods of inference has a practical outcome, too: looking for flexible connections and mixtures of structured and unstructured representation schemes. The metaphor of visual patterns covers loosely coupled traces of information on certain impressions, situations, objects, faces. Foreground presence, intensity of the pattern is changing, due to relevance. An object in a room can have a rather vague trace but can be the most remarkable one, if it received a special role in the scenario of the events.

The visual pattern analogy is used in two ways: finding a suitable computer representation of pattern-like information, and looking for such graphic representations that can bridge the knowledge representation differences of man and machine. The methods are proved in three different projects. All three relate to subjects where human judgment has a primary role, the knowledge has no well-consolidated, rule-based structure.

2. COMPUTER REPRESENTATION

The computer representation fits best to the general concepts of databases and object-oriented programming. A pattern, which in our projects reflects a diagnosis of a disease, a legal case or economical situation of a certain country, has a descriptive part, the *body*, containing all available data, i.e. the *label* of individual data, their specific *value*, the *range* of the values, *conversion* calculation of the values to scaling of the system, the *relevance* of the datum from the point of view of the pattern and the *uncertainty* of the measurement, estimation of the value. E.g. in economy, the GNP/capita is a label of a datum, $ 7500/year is a given value, the range of medium income countries can be between $ 5000 to 10.000/year, the relevance of the datum from the point of view of economical status is high (1.0 in a 0-1 scale), the uncertainty of the estimate can be 10 %, i.e. the reliability of the datum 0.9 in a 0-1 scale.

The file of the pattern has a *head* with the *name* of the pattern, a calculation *algorithm* for defining distance to other members of the pattern class, and an estimate for the completeness, (*exhaustiveness*) of the pattern description in the body. In the cited example, the pattern name is the economical situation of Ireland, the algorithm can be a linear combination of different local prices compared to prices in other countries, the exhaustiveness of the status description can be estimated to 0.8, on a 0-1.0 scale.

The *tail* of the file provides the connections to other related patterns. It lists the names of these *patterns*, the types of *relationships* (see further) and the *intensity* of these relations. In this example the list contains those countries that have a relevant economical influence on the situation of Ireland, those patterns of world politics that change the priorities of the economy, etc.

The *relationships* are some week generalizations of the usual inference schemes, reflecting our view on the nature of metapatterns. *Evocation* is a weak logic, means that the patterns usually appear together or follow each other, like interest rates and investment patterns. *Assessment* changes the viewpoint, a pattern strengthens or weakens the relevance and scaling values of the other, like unemployment and crime. *Reordering* changes the internal structure (if some exists) and the relevance values, like a catastrophic weather period does in the pattern of foreign trade.

Transformation is another logical relationship: a replacement of a pattern with another one due to the arising of a third pattern. The pattern of the economy of a country can change completely due to dramatic historical patterns, like war, occupation or liberation.

The relationships can be time dependent, in this case methods of temporal logic can be included into the representation. Most time dependencies have a fuzzy and varied character, like long- range and short range effects, interventions having different time delay, nonlinear behavior in time with fuzzy phase transitions, and this means that time dependency has a pattern characteristic, too. Changes in different levels of an education policy can demonstrate these long range effects, their different behaviors in time. Other modalities can be represented similarly by patterns, especially by using the ternary pattern relations, where the third pattern represents the modality type, connecting the first and the second pattern.

Two types of patterns are defined, *standard* and the *individual*. Standard pattern is the more general one in a conceptual hierarchy, like a prototype of an average European economy. Individual is related more to the specifics within a class, like Ireland, in our case. Due to the multiple imbedded nature of concepts a standard can have an individual role in an other relation and vice versa. The distinction is practical because of the inheritance relations within one conceptual class.

This representation includes several data and types of relationships that are more or less human estimates, can not be given by exact measurements or reliable statistics. In all these cases man-machine interaction is supposed. The most difficult task is the determination of the distances among patterns, those which express similarities and differences. One of the advantages of the pattern representation is the expression of the relativity of these relations. The distances are calculated in two ways. The first borrows its methods from the pattern recognition discipline, finding suitable algorithms for clustering. The simplest is the linear combination of values, weighed by relevance and uncertainty estimates. This is the way of finding separating hyperplanes among the pattern clusters. The human interaction has an apparent possibility: visualization of the clusters from different points of views (i.e. selecting different two or three dimensional views), and tuning the relevance values for a refined separation.

The other method is borrowed from signal processing. The pattern bodies are treated as signatures. General similarities and differences are given by the usual code distance definitions; the procedure selects relevant looking, similar code groups in otherwise different patterns and peculiar code groups, indicators in generally similar patterns. No surprise: this is the obvious way of human identification and discrimination, i.e. another offer for appropriate visual representation in the man-machine interaction. In this representation the choice of standard and individual patterns has an important role: the code differences are apparent in these relative comparisons.

3. THREE PROJECTS USING THE METHOD

We have an experience of more than a decade with the application of the shortly described method. At least three generations of software and refinement of other tools can be marked, but the basic idea had not to be changed during this long period of experimentation and dramatic progress of all related technologies. Not only computer hardware, software, computer science made a revolutionary progress, providing us with more and more powerful tools, but in the most advanced project, medical instrumentation and diagnostic methods survived the same changes, in instrumentation from translumina-tion of newborn's head by an incandescent lamp to sophisticated brain mappers, positron emission tomography (PET), and in communication from traditional blackboard discussions to ongoing, picture-based regular transatlantic teleconsultations.
The projects are reported in details (Vamos, Koch and Katona 1994). The first is a diagnostic and habilitation system for discovery and compensation of brain injuries of newborn babies, using the flexibility of the first few months of brain development after the birth. Patterns are diagnostic statements and habilitation procedures. This project helped to recover thousands of babies who were sentenced for a lifelong mentally and physically handicapped state. The second project is a sociological-legal problem of decisions on custody lawsuits where patterns are situations seen by different experts, people involved. The third project investigates patterns in economy, especially patterns of long range cultural, social, historical attitudes influencing the responses of different countries, to similar phenomena of change. This has a particular importance in the changing patterns of the late Soviet Empire countries in Central and Eastern Europe.

4. GRAPHIC TOOLS

Graphics is the most important communication medium of man-machine interaction in a system, based on patterns, applying visual analogy. Though communication between the user and the computer system is an interactive procedure, input and output have different features and requirements. The time schedule of application is different, too: a session starts mostly with massive data input and later the user ponders more on output representations of the processed data, reasoning. Psychologically, input is more related to the original forms of representation of phenomena, i.e. the natural visual forms of the investigated patterns, output is more to the ab-straction of the evaluation, i.e. the very few regular configurations, like circular or rectangular diagrams, which represent normal, standard states, and suitable deformations for easy recognition of the similarities and differences, characteristic features. Abstraction characterizes those more sophisticated representations as the Chernoff faces (Chernoff, 1973) or the harmonic representations of Andrews (1972). These latter look to be challenging but did not find much application because of the deviation from the simplicity principle and therefore due to more subjectivity in evaluation-recognition. The out-put representation task is similar to that used in multivariate statistical analysis, and that is the reason why until now efforts were concentrated more on specifics of input graphics.

Graphics has two important features that are utilized. The first is its inherent parallel nature. A figure can represent a complete pattern with the conceptual part, details, attributes (Lohse, et al., 1994). This means that appropriate figures can replace the whole file of pattern description. If figures of patterns are suitably designed, the selection of the characteristic pattern figure is equivalent with the complete data input operation, which is the most boring part of human tasks in computer usage. In our system this happens in two steps. The first is the usual icon-type menu; previously fixed icons represent mostly the standard patterns. The second step is the real art of graphic pattern design, and we cooperate with artists in this task. The individual pattern (it was emphasized that the distinction between standard and individual pattern is an arbitrary pragmatic decision) is a figure that can be varied. Variation is an adjustment of the standard data to the individual values, i.e. the standard data are handled as defaults, and the deviations have to be corrected. This correction happens by several ways depending on the convenience of the usage. Keyboard input, adjustment on scroll bars, changing width, coloring, changing dimensions of figure components are all possible interaction means. These interactions should happen automatically simultaneously on the figure and in the database. Warning of adjustment of some relevant features, contradictions have special signals (blinking). All these are devices for fast quasi-parallel input, bypassing the boring, dull query files. The method is more qualitative than quantitative, coming closer to the fuzzy nature of patterns and parameter estimates. The direct numerical representation, like positioning in the Paintbrush applications, combining scroll bars with numerical data, and other means of parallel portrayal for more fuzzy visual attraction and more precise data control are available. Enhancing help functions is another way for solving this ambivalence.

Designing the two steps, i.e. the separation of the hierarchically organized data and those which are treated unbiased in respect to their hypothetical causal, logical structure is a trade off of the knowledge engineer. This is a problem which has to be revised at each application; data input in each questionnaire has a certain linear structure which hides the experience and hypothesis about the graph representable structure of the data. A device is given to see the situation like the real expert does, parallel, receiving impressions on a first and consecutive glance. How many information should appear parallel (Tufte, 1983; Tufte 1990), this is a problem of the two agents' characteristics: how much can be well observed by the user, and how much can be represented suitably on the computer screen?

The other advantage of this sketch representation is a cognitive feature. Associations occur to us by concatenations of mind representations, i.e. primarily visual ones. Fortunate sketches can stimulate association activity of the user.

The task is difficult. For each subject different sketches are needed, best recognizable by the user, emphasizing the relevant features, like the talented cartoonist does. Similarly to the cartoonist requirement, the sketch should be made of a minimal number of rather simple lines but sufficient to express all details of the pattern. Density of the figures, composition for best representation of relations is another hard choice. In future man-machine communications this can evolve a special artistic profession like many other previous ones of applied graphic art. Cognitive psychology is deeply involved, experimentation with the best user stimulating representation and interaction methods.

Animation is used mostly in time dependent processes but it can be helpful in representation of multidimensional spaces, sweeping through different views, zooming special details. The use of original pictures received by instrumentation (aerial, radiographic, microscope pictures, etc.) is an other way, we apply it in the medical project.

5. SOME GRAPHIC APPLICATIONS IN THE ABOVE PROJECTS

The three projects demonstrate three different representation variants. The medical project offers figures of motion problems used regularly in teaching the diagnostic methods. The inputs of individual data can be given on selected points of the general baby figure. Motor functions attracted simple forms of animation. Other diagnostic features have conventional representation possibilities on anatomical and radiographic pictures. More abstraction is needed in the representation of mental status but attention, reaction have also direct figurative metaphors.

Interesting task was the representation of social characteristics in the custody project. Nevertheless the impression on the physical and psychological environment could be visualized in easily alterable sketches. Resulting patterns, closer to general judgements, i.e. on higher conceptual level returned to the icon method, the process here was in some respect opposite to the diagnostic one, diagnostics starts with a few general hypotheses, i.e. on a higher conceptual level, the judiciary process should avoid as much prejudice as possible, i.e. follow a bottom up procedure, from alterable details to conceptual level of judgement.

The project in economy lies furthest from the visual representation. The graphic fantasy has a very important role in developing ensembles of regularly used smaller icons. It is very important to use the most conventional habitual representations as much as possible because the purpose of the representation is the facilitation of association, fast response.

Different means should be offered for different users. Further a possibility can be given for changing the means if the users find it boring, the attraction of the first glance is over. The illustrations of this paper show a small selection of a comprehensive set, experimented for different users. Fig. 1 and 2 represent extreme values of data in the sociological project and in the project in economy. This representation suits the conceptual frame of the Repertory Grid Analysis knowledge acquisition method, used in our projects. The attribute poles of concepts are portrayed, and in between a scroll bar, changing width or coloring sets the fuzzy values. Fig. 3 is a small part of the same but with different figures for the poles. Fig. 4 is a motor function representation in the medical project. The arrows can be used for input of different data on detailed specification. The figure can be applied in animation, too; this represents the motion types more impressive. Fig. 5 is a general pattern from the sociological project, representing different levels of bad human relations. As it can be seen the art of representation is a problem of the artist's wit, the psychological and conceptual skills of the knowledge engineer and of the habits, taste of the users. The system design should provide a variety for challenging the user's satisfaction.

REFERENCES

Andrews, D. F. (1972). Plots of high-dimensional data. *Biometrics,* **28**, 125-36.

Chernoff, H. (1973). The use of faces to represent points in K-dimensional space graphically. *Journal of American Statistical Association,* **68**, 361-368.

Lohse L. G., Biolsi K., Walker N., Rueter H. H. (1994). A Classification of Visual Representations. *Communications of the ACM,* **37**, 36-49

Tufte, E. R. (1983). *The Visual Display of Qualitative Information.* Graphic Press, Cheshire, Conn.

Tufte, E. R. (1990) *Envisioning Information.* Graphic Press, Chesire, Conn.

Vámos, T., P. Koch, F. Katona (1994). A strategy of knowledge representation for uncertain problems: modeling domain expert knowledge with patterns—to appear in *IEEE Trans. on SMC*

Social workers' pattern

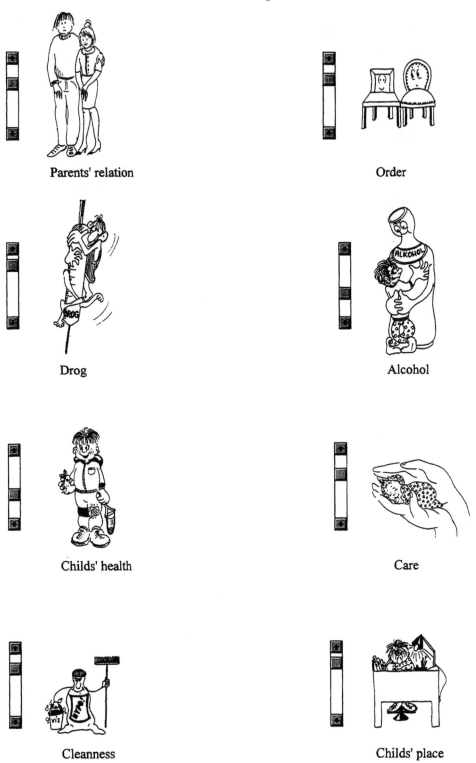

Parents' relation

Order

Drog

Alcohol

Childs' health

Care

Cleanness

Childs' place

Fig. 1. Social workers' pattern

Socio-economical pattern

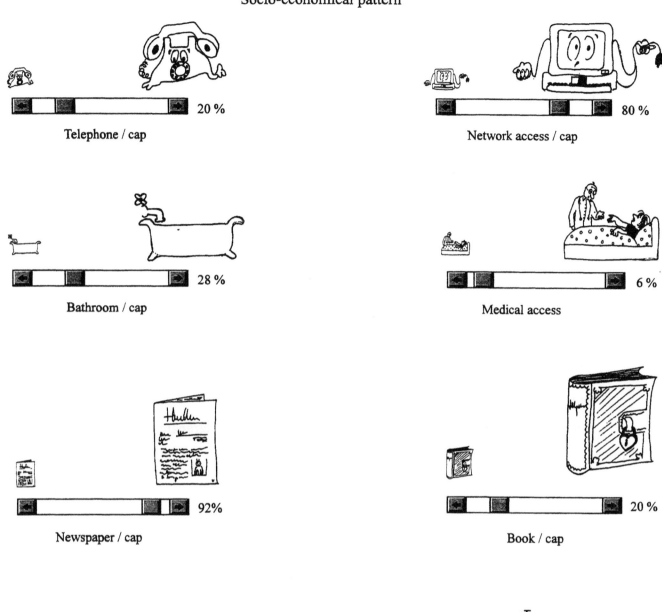

Telephone / cap — 20 %

Network access / cap — 80 %

Bathroom / cap — 28 %

Medical access — 6 %

Newspaper / cap — 92%

Book / cap — 20 %

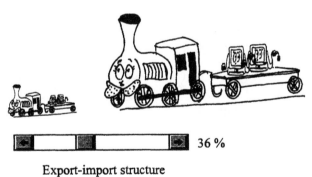

Export-import structure — 36 %

Socio-economical indicators
medium-low developed country

Fig. 2. Socio- economical pattern

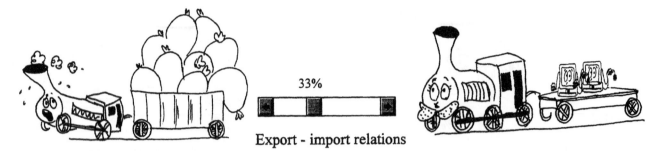

Fig. 3. Export- import relations

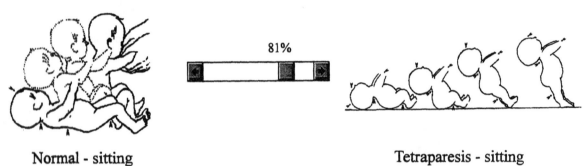

Normal - sitting

Fig 4. Tetraparesis - sitting

Tetraparesis - sitting

Human relations

phase 1   phase 2   phase 3   phase 4   phase 5

Fig 5. Human relations

IN-FLIGHT APPLICATION OF 3-D GUIDANCE DISPLAYS:
PROBLEMS AND SOLUTIONS

E. Theunissen[*]

*Delft University of Technology, Faculty of Electrical Engineering,
P.O. Box 5031, 2600 GA Delft, NL*

Abstract: Perspective flightpath displays can be used to present spatially integrated guidance information. Position and angular errors can be obtained from a single image, and the successive presentation of images produces additional cues. These latter cues, are influenced by the resolution, accuracy, and update-rate of the position information. This paper discusses the mechanisms behind these cues, the resulting information requirements, problems which can occur when using GPS or DGPS for the determination of the 3D position, and possible solutions.

Key Words: Aerospace engineering; flight control; guidance systems; man-machine systems; navigation; displays

1. INTRODUCTION

Perspective flightpath displays present spatially integrated information about the future desired trajectory. They can be used to provide the pilot with spatial and navigational awareness (Dorighi, *et al.*, 1993), and to present guidance information to fly a three- or four-dimensional trajectory (Grunwald, 1984; Wickens, *et al.*, 1989; Theunissen, 1993). The perspective presentation of the flightpath allows the pilot to synthesize information about position and angular errors (Theunissen, 1994), the trajectory preview allows an error neglecting control strategy and a certain amount of anticipatory control to be applied, and the range of error gains resulting from the perspective projection allows adaptable closed-loop control (Theunissen and Mulder, 1994). The dynamic nature of the information, allows the determination of error-rates and under certain conditions the extraction of temporal range information. Since this information is inferred from the successive presentation of images, the data needed to generate these display formats must meet certain requirements with respect to resolution, noise, and update-rate. In a flight simulator it is easy to satisfy these requirements, but in a real aircraft problems may occur as a result of the limited accuracy and update-rate of the required information. For a proper design of the display format, and the application of techniques to achieve the required information update-rate, a thorough understanding of the cues resulting from the dynamic nature of the displayed information is required.

This paper addresses the mechanisms responsible for the generation of visual motion cues, the cues required for the extraction of temporal range information, the resulting requirements which must be met when using the concept in a real aircraft, problems which can occur due to system limitations such as limited information update rates of attitude and position, and solutions to compensate for these deficiencies. In December 1994 Delft University of Technology has successfully demonstrated the in-flight application of a perspective flightpath display using Differential GPS information. The laboratory aircraft, a twin engine business jet, performed several curved approaches with the pilot obtaining the required information from the perspective flightpath display.

2. GUIDANCE & NAVIGATION

The Navigation Error (NE) of an aircraft consists of a Positioning Error (PE) and a Flight Technical Error (FTE). The PE is the difference between the true position of the aircraft and the position reported by the positioning system. The FTE represents the difference between the desired position of the aircraft and the position reported by the positioning system. The navigation performance requirements are based on statistical knowledge about the PE, and an allowed margin for the FTE. The pilot is only aware of the FTE, and a change in PE will be perceived as a change in FTE. To satisfy the navigation performance requirements, the guidance tasks requires keeping the current and future FTE below a certain threshold.

2.1 *Perspective guidance displays*

To generate a perspective presentation of the flightpath, 3-D world-space coordinates must be transformed to 2-D device coordinates. The transformation from the World Reference Frame (WRF) to the Ego-centered Reference Frame (ERF) requires the specification of a viewpoint, a viewplane-normal, and a view-up vector. A detailed description of these transformations can be found in computer graphics literature (Hearn and Baker, 1986). After the coordinate transformations, a perspective projection on the viewplane is performed, which requires the specification of a viewing volume. Equation 1 presents the relation between the coordinates of an object in 3-D space and it's position on a 2-D viewplane.

$$X_{screen} = \frac{screensize * X_{object}}{2d * \tan(\frac{fov}{2})} \qquad (1)$$

The parameter d represents the distance from the viewpoint to the plane perpendicular to the viewing vector in which the object lies. X_{object} represent the x coordinate of the object in the 3D ERF, X_{screen} represent the x coordinate of the object on the viewplane, *fov* indicates the field-of-view used for the projection, and screensize the size of the screen.

2.2 *Information processing*

A major difference between command displays such as the Flight Director (FD), and perspective flightpath displays such as the Tunnel-in-the-Sky, is that the former is based on the presentation of a weighted sum of position and angular errors

and error rates, whereas the latter presents an abstraction of the real world, and thus is based on position and attitude. Figure 1 presents a typical control scheme for a FD for lateral control, and Figure 2 for a perspective flightpath display. In these Figures, H_{disp}, H_p, and H_c represent the transfer functions of respectively the display, the pilot and the aircraft.

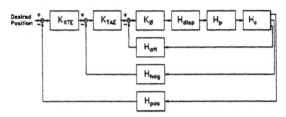

Figure 1 FD information processing

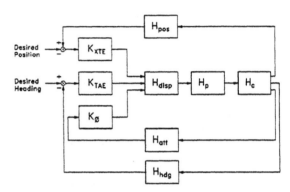

Figure 2 3D display information processing

The transfer function of the display can be used to include the effects of display-latency, update-rate, and resolution. The transfer functions H_{att}, H_{hdg}, and H_{pos} in the feedback loops represent the systems used to determine attitude, heading, and position and can be used to include the effects of update-rate, latency, resolution, accuracy, and noise of these systems. The attitude, heading, and position error gains are represented by respectively K_{att}, K_{hdg}, and K_{pos}. With a perspective flightpath display, the error gains are a function of the perspective design parameters. The relation between position- and angular errors, and the resulting distortions of symmetry of the flightpath can be described as a function of the distance to the specific elements of the flightpath and the geometric field of view used for the projection (Theunissen, 1994).

To avoid information conflicts, visual stimuli obtained through the perspective flightpath display must be compatible with visual stimuli from the outside world and the motion cues obtained through the vestibular system. In order for the pilot to believe the FD, the commands must have <u>a certain degree</u> of consistency with the other information available. The fact that a FD command is not required to have a one-to-one relation with any other perceivable cue, allows the algorithms driving the FD to be adapted to achieve an optimal control-display compatibility. McRuer, *et al.* (1971) describe an analytical approach for the design of a FD which satisfies both pilot-centred, and guidance and control requirements. With a perspective flightpath display, the information is not combined into a single parameter. The pilot is able to synthesize the information required for aircraft guidance from the display, and perform the integration of the information required to derive a guidance command. With a well designed perspective flightpath display, following a 3D trajectory is fairly easy (Grunwald, 1984; Wickens, *et al.*, 1989, Theunissen, 1993). During the trajectory following task, position and angular errors result in a distortion of the natural symmetry of the perspective representation of the flightpath, and provoke almost spontaneous control reactions.

If, for some reason, the cues presented by such a display are not fully consistent with outside-world cues, the confidence level of the pilot will decrease.

3. MOTION

The relation between the Ego-centered Reference Frame (ERF) and the World Reference Frame (WRF) is determined by the position and orientation of the aircraft. A change in position results in a translation of the ERF in the WRF, and a change in orientation causes a rotation of the ERF. Whereas a rotation of the viewpoint only results in a translation of the visual scene, a translation of the viewpoint causes motion perspective. This provides a strong cue for three-dimensionality, allows the perception of relative distances, velocities, and locations as well as the direction of movement of these objects (Wickens, *et al.*, 1990), and thus gives the pilot a sense of aircraft velocity. Furthermore, under certain conditions, the dynamic nature of the visual scene allows the pilot to extract temporal range information.

3.1   *Perception of egospeed*

With a perspective flightpath display, most visual motion cues are presented through the movement of the cross section frames of the tunnel toward the observer. The mechanisms responsible for the generation of these cues can be divided into optical edge rate and global optical flow rate. Optical edge rate is defined as the speed at which texture elements pass a given point in the subject's field of view (Warren,

1982). With a perspective flightpath display, optical edge rate is determined by the distance between the successive frames.

Optical flow rate of texture elements within the visual field is determined by the following equation (Gibson, *et al.*, 1955):

$$Flowrate = \frac{V}{h} \sin^2(elevation)\cos(azimuth) \qquad (2)$$

where *elevation* and *azimuth* refer to the optical coordinates of a texture element and V/h is a global scaling factor determined by the speed V and the altitude h of the observer which indicates the rate of expansion of the visual field. Warren (1982) named this scaling factor global optical flow rate. He hypothesized that the perception of egospeed scales with global optical flow rate. In case of a perspective flightpath display, global optical flow rate is determined by the geometric field of view and the tunnel size.

During experiments in which the tunnel dimension was varied to present the pilot with different error gains (Theunissen, 1993), subjects mentioned the apparent lower velocity when flying the larger tunnel, and the very high velocity when flying the smallest tunnel, confirming the contribution of optical flow rate to the perception of egospeed. Due to the inverse relation of perceived egospeed with tunnel size, optical flow rate will only contribute below a certain threshold in the distance to the tunnel walls.

3.2   *Three-dimensionality*

Optical flow rate is an important cue for three-dimensionality. To illustrate this, it is necessary to examine the velocity gain and introduce the concept of virtual acceleration. Equation 3 presents the formula for the velocity gain, which is defined as the velocity of an element on the display, divided by the relative velocity between the observer and the element.

$$G_v = \frac{x * screensize}{2d^2 \tan(\frac{fov}{2})} \qquad (3)$$

The parameter d represents the distance from the viewpoint to the plane perpendicular to the viewing vector in which the element lies, x indicates the distance between the central display axis and the element, *fov* the geometric field of view which is used for the perspective projection, and *screensize* the size of the screen. Because velocity cueing is obtained through the movement of the tunnel frames, x can be substituted by *(tunnelsize/2)-XTE*, which yields: From Equation 4 it follows that a linear decrease

$$G_v = \frac{screensize * (\frac{tunnelsize}{2} - XTE)}{2d^2 * \tan(\frac{fov}{2})} \qquad (4)$$

in distance towards the observer results in a quadratic increase in velocity gain.

Since the velocity gain is a function of the distance to the observer, the perspective projection of the elements in the 3-D world on the 2-D display results in a virtual acceleration of these elements, which generates a feeling of three-dimensionality. By using the relation between the distance to the object and the position of the object on the screen of Equation 1, and substituting tunnelsize/2 for X_{object}, Equation 4 can be re-written as:

$$G_v = \frac{2 * X_{screen}^2 * \tan(\frac{fov}{2})}{(\frac{tunnelsize}{2} - XTE) * screensize} \qquad (5)$$

From Equation 5 it follows that the velocity gain is proportional to the square of the distance from the central projection axis. Figure 3 presents an example of a perspective flightpath display, and Figure 4 graphically represents the velocity gain for the tunnel in Figure 3, calculated for a discrete number of points on the display. Figure 5 presents the representation of the flightpath in the presence of a cross-track error, and Figure 6 presents the velocity gain for the tunnel in Figure 5. As can be seen from Figure 6, a position error results in an asymmetry of the optical flow field.

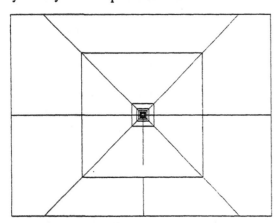

Figure 3 Flightpath, XTE=0

The virtual acceleration is a function of the distance to the central projection axis, and can be obtained by multiplying aircraft velocity with the derivative of Equation 5 which yields:

$$A_v = \frac{4 * X_{screen} * \tan(\frac{fov}{2})}{(\frac{tunnelsize}{2} - XTE) * screensize} * V \qquad (6)$$

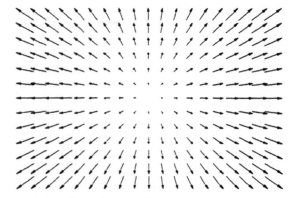

Figure 4 Velocity gain for Figure 4

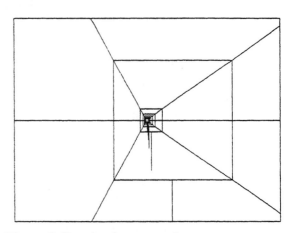

Figure 5 Result of cross-track error

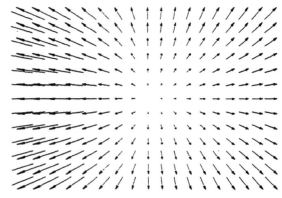

Figure 6 Velocity gain for Figure 5

Thus, the acceleration of an object on the screen is proportional to the distance from the central projection axis. The compellingness of the motion perspective is related to this acceleration. If the acceleration is below a certain threshold, it will not contribute to the feeling of three-dimensionality.

3.3 Temporal range information

Temporal range judgements are based on global optical flow rate, which must exceed a certain threshold to allow accurate estimates to be made. Temporal range information is often used to determine the moment to initiate certain anticipatory control actions. Various studies addressed the perception of temporal range information for flight control. Advani, et al. (1993) investigated the initiation of the landing flare where temporal range information can be perceived from the size of the runway image and its rate of expansion. Kaiser and Mowafy (1993) studied the extraction of temporal range information from the angular position of an object and the angle's temporal derivative. They report that temporal range judgements exhibit non-veridical scaling. Theunissen and Mulder (1994) studied the relation between the moment an error-correcting control action is initiated and several parameters. One of these parameters was a prediction of the remaining time before crossing a tunnel wall, which is referred to as the Time-to-Wall Crossing (TWC). Their results illustrate that the initiation of the control actions is based on a certain combination of position- and angular errors and error rates, and suggest that the TWC represents temporal range information used by the pilot to maintain a temporal distance from the tunnel wall.

4. INFORMATION REQUIREMENTS

For the presentation of guidance information, the actual aircraft position, attitude and heading must be determined. The typical processing of this information for a FD is presented in Figure 1, and for a perspective flightpath display in Figure 2. It is likely that in the future augmented GPS, and in the proximity of a reference stations Differential GPS (DGPS) will be used for the determination of the 3D position. In the previous section it was demonstrated that the effectiveness of visual cues presented by the perspective flightpath display is related to the tunnel size. Due to the need to reduce the error gain as a result of the resolution and accuracy of position information, a trade-off between the tunnel size dictated by the error gain, and the tunnel size required for adequate cueing may be required. This section discusses some of the problems, which must be solved when using GPS as a source for 3D position information for perspective guidance displays.

4.1 *Update-rate*

To achieve a smoothly animated display, the information update-rate must exceed a certain threshold. Information update-rates in the order

of 20 to 30 Hz prove to be adequate. As a result of the limited bandwidth of the carrier tracking loop in GPS receivers (typically about 16 Hz), these receivers output position information at an update-rate below that required for smooth animation. In case it is impossible to oversample the position information, inter- or extrapolation techniques are needed to increase the position information update-rate. However, interpolation introduces latency, which reduces the stability of the control loop due to a decrease in phase-margin. Thus, interpolation is only acceptable in case the position update-rate is sufficiently high. With extrapolation, the prediction, which is based on position data and models which use other elements of the state vector such as velocity, attitude, and heading is inevitably accompanied by a prediction error which is corrected at each new position update. However, these corrections can be perceived as a sudden change in FTE, and introduce a noise component in the optical flow field with the same frequency as the position information update-rate. Therefore, the prediction algorithm must apply some form of error smoothing to avoid a distortion of the dynamic cues. An in-depth discussion of such prediction techniques is beyond the scope of the paper, however, more information can be found in Mulder (1992). For the in-flight testing of the Tunnel-in-the-Sky display, a Kalman predictor was used.

4.2 *Resolution and accuracy*

As indicated previously, a change in PE is perceived as a change in FTE. Due to the successive presentation of images, a change in along-track PE is perceived as a fluctuation in egospeed. The magnitude and the frequency distribution of the PE determine the influence of positioning noise on aircraft control. To prevent a contribution of positioning noise on the guidance commands presented by a FD, a low gain for the FTE is selected. With the perspective flightpath display, a low gain for the FTE can be obtained by selecting a large tunnel. However, a larger tunnel increases the minimum trajectory preview distance and only reduces the average position error gain. The actual error gain is a function of the distance to the tunnel wall. A further disadvantage of a large tunnel is that temporal range information and an acceptable magnitude of virtual acceleration are only available in a very asymmetric configuration. The lack of these cues in the symmetrical configuration, and the increased minimum preview distance are likely to affect pilot performance. To compensate for the lack of this information, display augmentation concepts, such as the integration of a flightpath predictor, can be used to provide the pilot with additional guidance information. Another method to reduce the effect of positioning noise on the perception of the FTE, is to apply some form of filtering on the position information. Research is needed to determine the extent to which filtering is possible without influencing control behaviour. For the in-flight evaluation of the DELPHINS Tunnel-in-the-Sky display, DGPS was used, which allowed a high position error gain.

5. CONCLUSION

This paper discussed the cues for ego-motion, three-dimensionality, and temporal range information, which can be obtained from a perspective flightpath display. It was shown that for a successful presentation of these cues, more stringent requirements on the resolution, accuracy, and update-rate of position information as compared to FD displays apply. When the positioning system does not meet all these requirements, prediction techniques can be used to increase the information update-rate, while a reduction in error gain and the application of filtering techniques can be used to reduce the effect of noise. To achieve the full potential of perspective flightpath displays for aircraft guidance, accurate position information is required with a high update-rate.

ACKNOWLEDGEMENTS

The author would like to thank D. van Willigen for his support of the DELPHINS Tunnel-in-the-Sky project, D.J. Moelker, B.C.M. Blesgraaf, and S.P. van Goor for developing and implementing the Kalman predictor, C. Dam and K. van Woerkom for aircraft modifications and J.A. Mulder for performing the in-flight experiments.

REFERENCES

Advani, S., van der Vaart, J., Rysdyk, R. and Grosz, J. (1993). What Optical Cues Do Pilots Use to Initiate the Landing Flare. In: *Proc. of the AIAA FST Conference*, pp. 81-89, Monterey, CA.

Dorighi, N.S., Ellis, S.R. and Grunwald, A.J. (1993). Perspective Format for a Primary Flight Display (ADI) and its Effect on Pilot Spatial Awareness. In: *Proc. of the Human Factors and Ergonomics Society 37th Annual Meeting*, pp. 88-92, Seattle, WA.

Gibson, J.J., Olum, P. and Rosenblatt, F. (1955). Parallax and perspective during aircraft landings. *American Journal of Psychology*, **68**, pp. 372-385.

Grunwald, A.J. (1984). Tunnel Display for Four-Dimensional Fixed-Wing Aircraft Approaches. *Journal of Guidance*, **7**, pp. 369-377.

Hearn, D. and Baker, M.P. (1986). *Computer Graphics* Prentice-Hall, ISBN 0-13-165598-1.

Kaiser, M.K., and Mowafy, L. (1993). Visual Information for Judging Temporal Range. In: *Proc. of Piloting Vertical Flight Aircraft*, pp. 4.23-4.27, San Francisco, CA.

McRuer, D.T., Weir, D.H., and Klein, R.H. (1971). A Pilot-Vehicle Systems Approach to Longitudinal Flight Director Design. *Journal of Aircraft*, **8**, pp. 890-897.

Mulder, M. (1992). *Aviation Displays and Flightpath Predictors*. Master's Thesis, Delft University of Technology, Faculty of Aerospace Engineering.

Theunissen, E. (1993). A Primary Flight Display for Four-Dimensional Guidance and Navigation: Influence of Tunnel Size and Level of Additional Information on Pilot Performance and Control Behaviour. In: *Proceedings of the AIAA Flight Simulation Technologies Conference*, pp. 140-146, Monterey, CA.

Theunissen, E. and Mulder, M. (1994). Open and Closed Loop Control with a Tunnel-in-the-Sky Display. In: *Proc. of the AIAA FST Conf.*, pp. 32-42, Scottsdale, AZ.

Theunissen, E. (1994). Factors influencing the design of perspective flight path displays for guidance and navigation. *Displays*, **14**, pp. 241-254.

Warren, R. (1982). Optical transformation during movement: Review of the optical concomitants of egomotion *Technical Report AFOSR-TR-82-1028*. Bolling AFB, DC: Air Force Office of Scientific Research (NTIS no. AD-A122 275).

Wickens, C.D., Haskell, I. and Harte, K. (1989). Ergonomic Design for Perspective Flight Path Displays *IEEE Control Systems Magazine*, **9**, pp. 3-8.

Wickens, C.D., Todd, S. and Seidler, K. (1990). *Three-Dimensional Displays: Perception, Implementation, Applications*. University of Illinois, Savoy, IL.

PERCEPTION OF FLIGHT INFORMATION FROM EFIS DISPLAYS

Ruud J.A.W. Hosman and Max Mulder

Delft University of Technology, Faculty of Aerospace Engineering
P.O. Box 5058, 2600GB Delft, The Netherlands
and
Erik Theunissen

Delft University of Technology, Faculty of Electrical Engineering
Mekelweg 4, P.O. Box 5031, 2600 GA Delft, The Netherlands

Abstract: Pilot's perception of variables presented on the Electronic Flight Instrument System, EFIS, has been investigated. A stimulus response technique has been used to determine the accuracy and speed of the perception process. By varying the exposure time of the stimuli, it has been shown that the perception of the variable magnitude is faster and more accurate than the perception of the first derivative or rate of that variable. Results of experiments on roll and pitch attitude perception, the influence of scale division, and the perception of the indicated airspeed are shown.

Keywords: Displays; Man/machine interface; Human Factors; Cognitive systems.

1.INTRODUCTION

The presentation of flight information using instruments started about 90 years ago with the first flight of the Wright Flyer. Since then, flight instruments developed concurrently with the increasing complexity of the pilot's task. In this ongoing development, a real breakthrough was the introduction of the Electronic Flight Information System, EFIS, around 1980. Early displays used 2D presentations of flight variables. With the introduction of fully electronic displays, however, the much larger potential of 3D presentations could be realized. This potential has been demonstrated by evaluations of 3D display formats, as for instance the tunnel in the sky, in flight simulators and laboratory aircraft.

The high-precision navigation systems and terrain data bases are the basic elements necessary for the 3D synthetic vision displays currently under development. These 3D display formats are expected to be introduced on transport aircraft in the next decade. It is anticipated that such displays will help the pilot to improve his situation awareness. This improved situation awareness is called for to further improve the safety record of the air transport industry. Some questions have to be solved, however, before a successful introduction of 3D displays is possible.

In a 3D presentation, as for instance the tunnel in the sky, not all required flight information can be displayed or perceived with the required accuracy. Speed relative to the synthetic world can be derived from the changing 3D view but is not equal to and should not be perceived to be as accurate as the presently displayed indicated airspeed. Altitude may also be estimated from the 3D display but the altitude tape or altimeter still proved superior accuracy. So at least 2D indicated airspeed and altitude indicators have to be present resulting in a mixture of 3D and 2D presentations. Many questions are raised in developing these displays. How does the pilot perceive the indicated variables, or which variables are necessary to perform a certain flight task, etc. To make the future application of 3D synthetic flight displays successful, additional knowledge on how the visual system derives the information from 2D and 3D displays is required.

At the Delft University of Technology the pilot's perception of 2D presentations of individual variables and the perception of 3D presentations of integrated flight information is under investigation. The main objective is to determine the specific characteristics that influence the perceptual behavior such as perception accuracy and perception speed. The research is based on stimulus response tasks, as well as on pilot-in-the-loop control tasks. This paper will discuss the research on pilot's perception of 2D presentation of flight information.

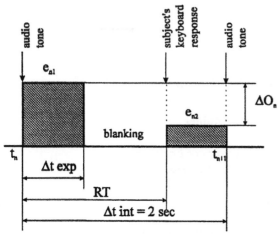

Fig. 1. One sequence of a measurement run

Three stimulus response experiments will be presented. These experiments were intended to establish roll and pitch attitude perception from the artificial horizon, the influence of the scale division on the perception accuracy and the effect of conventional and modern presentations on the perception of indicated airspeed.

In all the experiments, not only the perception accuracy and perception speed of the presented variable but also the first derivative or rate of that variable is studied. Quite large differences between the perception of the presented variable and its rate are found.

2. EXPERIMENTAL TEST PROCEDURE

As mentioned earlier, a stimulus-response technique was chosen to obtain information about perception accuracy and response time. During the experiments, the stimulus (the presented variable or its rate) is displayed to the subject for a limited duration, Δt_{exp}.

During the experiments, each run consisted of approximately 100 presentations at fixed time intervals. The sequence during one interval was at follows (see Fig. 1).

At the beginning of an interval, a new value e_{n_1} of the stimulus was presented, the event being marked by an audio tone in the subject's headset. After observation, the subject was required to respond by - pressing the appropriate key of the keyboard (see Fig. 2). The response magnitude is designated by ΔO_n. Immediately after the response the error value:

$$e_{n_2} = e_{n_1} - \Delta O_n \qquad (1)$$

was shown on the displays, thus giving the subject an immediate knowledge of the result.

The exposure time Δt_{exp} could be varied and was set either at a constant value by the experimenter prior to each run or was extended till the response of the subject. In the first condition the stimulus e_{n_1} was made to disappear at the end of the preset exposure time. In that case, subjects were required to give responses only after the termination of the exposure, while responses during the exposure time were neglected by the computer program in all test runs with limited exposure time. At the start of the next interval, a new value of the stimulus was generated. This new value was obtained in several ways.

In the first experiment, a new value of the stimulus was generated by adding a quantified sample i_n of a noise signal to the latest (n-1) value of the error value e_{n_2}:

$$e_{n_1} = e_{(n-1)_2} + i_n \qquad (2)$$

The standard deviation of the noise signal from which the sample i_n was taken and the scaling of the display were chosen such that the number of discrete values and the practical range of i_n, e_{n_1}, e_{n_2}, ΔO_n during the experiments corresponded to the range and number of keys on the keyboard device.

In later experiments the discrete stimulus magnitude, e_{n_1}, was sampled from a string of numbers in random sequence. The 25 different values of the numbers in this string correspond with the keys (zero $\pm$ 12) of the keyboard. The distribution of the values in the string was either uniform or normal.

During each run, the variables i_n, e_{n_1}, e_{n_2}, ΔO_n and the subject's reaction time, RT, were recorded. After the run the following values were calculated: the number of times a stimulus value j was shown, ns_j, the sum of observation errors per stimulus value j, $n e_j$, the mean value and variance of i_n, e_{n_1}, e_{n_2}, ΔO_n, the reaction time RT and a score parameter Sc defined by:

$$Sc = \sigma_{e_{n_2}}^2 / \sigma_{e_{n_1}}^2 \qquad (3)$$

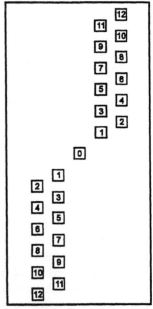

Fig. 2. The keyboard for the pitch attitude and indicated airspeed perception task

The measurements for experiment 1 and 2 were performed in a low noise room. A CRT-display (Textronix monitor 604) was mounted in an instrument panel in front of the subject displaying the artificial horizon. A refresh rate of 250 Hz and an update rate of 50 Hz was used.

The measurements for experiment 3 were performed in the research flight simulator at Delft University. A 14" SVGA monitor was used to present the indicated airspeed indicators. The symbol generation was based on the Delphins Display Design System, see (Theunissen, 1993). The refresh and update rate of the display was 55 Hz.

Two keyboards were used; one with the keys positioned laterally for the roll attitude perception tasks and one with the keys positioned vertically for the pitch attitude and indicated airspeed perception tasks (see Fig. 2).

3. EXPERIMENTS

Experiment 1: Roll attitude and rate perception

Subjects were required to make accurate and quick estimates of roll attitude and rate presented on an artificial horizon display (see Fig 3). Exposure times were systematically varied in both attitude and rate measurements. The values of Δt_{exp} used for attitude perception were 40, 60, 80 and 100 msec and for rate perception 100, 150, 200, 300 and 400 msec. An approximate normal distribution of the discrete stimuli was used. For each measurement run, a new randomized sequence of the stimuli was used. The standard deviation was 10 degrees for the attitude stimuli and 8 degrees/sec for the rate stimuli.

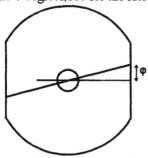

Fig. 3. The artificial horizon with indication of the roll angle φ

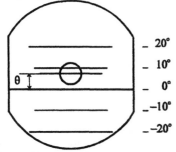

Fig. 4. The artificial horizon with the "pitch ladder" indicating the pitch attitude

Threepilot-subjects participated in the experiment and performed their measurements after extensive training. All of the measurement runs (4 attitude and 5 rate conditions) were performed 5 times by all subjects. For more details about the experiment see (Hosman and van derVaart, 1988).

Experiment 2: Influence of a scale division on perception accuracy

In the previous experiment no scale division was used. This experiment was designed to investigate the influence of a scale division on the perception process. The presentation of a so called "pitch-ladder" on the artificial horizon (see Fig. 4) was used to analyse the changes in perception accuracy due to the incorporation of a scale on the presentation of a variable. Both pitch attitude and rate perception were incorporated in the experiment. The same exposure times as in experiment 1 were used. An approximate uniform distribution of the stimuli values was applied to obtain a constant probability of all stimuli values. The range of pitch attitude extended from -20 to +20 degrees while pitch rate magnitudes varied between -10 and +10 degree/sec. Three pilot subjects participated in the experiment. Training was extended until stable performance of the accuracy and response time was reached. All measurement conditions (2 x 4 attitude and 2 x 5 rate) were performed 5 times by the subjects.

Experiment 3: Perception of indicated airspeed

In this experiment, attention was focussed on the conventional and new presentation formats of indicated airspeed. Since the introduction of the Electronic Flight Instrument System, EFIS, two new presentation formats for the indicated airspeed have been introduced: the analog moving tape and the digital presentation. In this experiment the conventional dial airspeed indicator is compared with the moving tape and the pure digital presentation of indicated airspeed. The perception of both airspeed magnitude and rate (or aircraft acceleration) were investigated. To complete the comparison, a fixed tape presentation was added to the experimental configurations for airspeed magnitude perception. In case of airspeed rate perception, a moving tape with the airspeed trend vector was incorporated in the display design. Thus in both parts of the experiment four display formats were investigated (see Fig. 5).

For the perception of indicated airspeed two exposure times were used; Δt_{exp} = 50 msec and Δt_{exp} = RT. In the latter case no blanking was applied.

For the perception of airspeed rate, the exposure time Δt_{exp} had to be adjusted to the particular

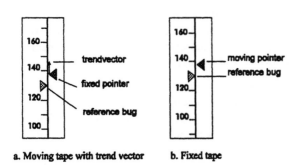

a. Moving tape with trend vector   b. Fixed tape

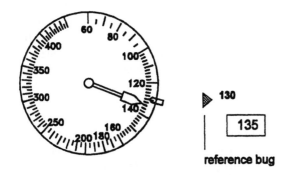

c. Dial indicator   d. Digital indicator

Fig. 5. Four indicated airspeed indicators

characteristics of the presentation format. For the dial and moving tape instrument, the values of the exposure time were: Δt_{exp} = 0.5 sec, 1.0 sec. and equal to the response time RT. For the digital presentation, 1.0 sec and RT were used, while for the moving tape with trend vector Δt_{exp} was either 50 msec or RT. The interval time was adjusted to the exposure times.

An approximate normal distribution of the discrete stimuli values for airspeed and its rate was used. Stimuli values corresponded with ± 12 keys. In each measurement run, 116 stimuli were presented in a random sequence. The range of indicated airspeed was 130 ± 16 knots, while the range of airspeed rate was 0 ± 3.89 knots/sec (± 2 m/sec^2). The 16 knots and the 3.89 knots/sec were equally divided over the 12 key intervals.

Three subjects, one pilot and two students, participated in the experiment. Subjects passed an extensive training before they performed the measurements runs.

4. RESULTS

In the following discussion the dimensionless score parameter, Sc, is used as a performance index to demonstrate the effect of the experimental conditions on perception accuracy. In addition, the influence of the experimental conditions on the response time, RT, is shown. Standard deviations presented in the figures are those from the mean values of the individual measurement runs. Although the perception accuracy as a function of stimulus magnitude can be obtained from the measurements,

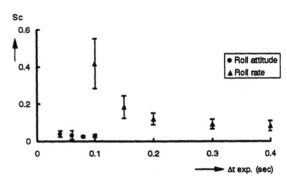

Fig. 6. Sc for roll attitude and rate perception as a function of exposure time

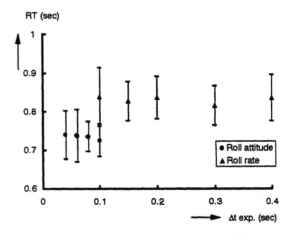

Fig. 7. RT. for roll attitude and rate perception as a function of exposure time

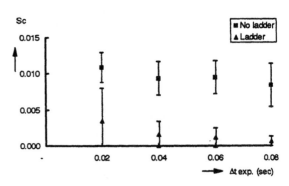

Fig. 8. Sc for pitch attitude perception as a function of exposure time and scaling

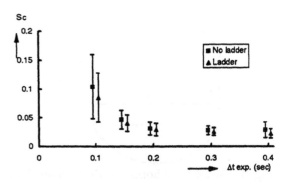

Fig. 9. Sc for pitch rate perception as a function of exposure time and scaling

it is byond the scope of this paper and will not be discussed.

Experiment 1

Fig. 6 presents the score, Sc, for roll attitude and roll rate. Roll attitude perception can be performed at a much shorter exposure time than roll rate perception. Typical values are 50 msec for attitude and 300 msec for rate perception. Roll attitude perception is more accurate than rate perception. For long exposure times the standard deviation of the perception error is approximately 1.5 degree for roll attitude and 2.0 degree/sec for roll rate. Response time for attitude perception is approximately 100 msec shorter than for rate perception (see Fig. 7).

Experiment 2

Fig. 8 and 9 present the score for pitch attitude and rate as a function of exposure time. The magnitude of the score parameter is smaller than for experiment 1 due to the different range and distribution of the stimuli. Clearly, including the "pitch ladder" in the scale improves the attitude perception accuracy. The standard deviation of the perception error at long exposure times decreases from 1.0 degree to 0.3 degree. In contrast, the perception accuracy of the pitch rate does not change with the presentation of the pitch ladder. In both cases, at long exposure times, the standard deviation of the perceived pitch rate is 0.9 degree/sec. The difference in required exposure time between attitude and rate perception corresponds to the results of experiment 1. The response time for pitch attitude and rate perception was not influenced by introducing the pitch ladder. The differences in the response time for attitude and rate perception (see Fig. 10) correspond to those of experiment 1.

Experiment 3

The results of experiment 3 are presented in Figs. 11 - 14. As shown in Figs. 11 and 12 the perception of the indicated airspeed from the digital indicator has the best accuracy and the shortest response time (standard deviation of perception error 0.22 knots

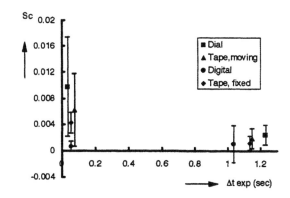

Fig. 11. Sc for airspeed perception as a function of exposure time

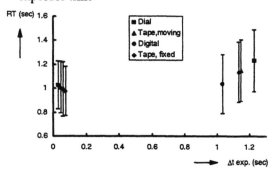

Fig. 12. Response time for four airspeed indicators as a function of exposure time

and mean response time 1.03 sec). The fixed tape gives slightly better results than the moving tape indicator. Subjects obtained the lowest performance with the dial indicator (standard deviation of perception error 0.35 knots and mean response time 1.23 sec). The differences in the score parameter and the response time as a result from the indicators are not significant.

The influence of the indicators on airspeed rate perception is presented in Figs. 13 and 14. The accuracy and speed of perception with the trend vector is superior to the other indicators (standard deviation of perception error 0.04 knots/sec and mean response time 1.04 sec). Depending on the exposure time, the response time for airspeed rate perception is 1 second or more shorter than for the dial and tape indicator. The perception accuracy for

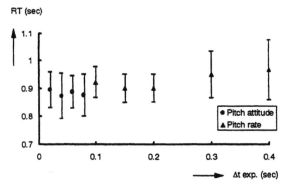

Fig. 10. Response time for pitch attitude and rate perception

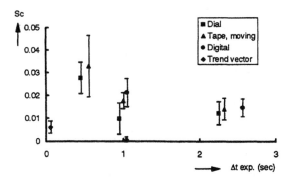

Fig. 13. Sc for airspeed rate perception as a function of exposure time

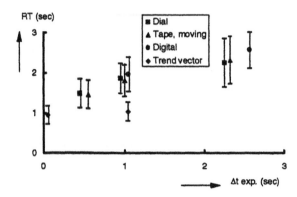

Fig. 14. The response time of airspeed rate
    perception as a function of indicator and
    exposure time

the dial indicator is slightly better than for the
moving tape indicator. Perception of airspeed rate
with the digital indicator proved to be possible for
exposure times of 1 second and larger.

5. DISCUSSION AND CONCLUSIONS

From the results of the experiments, it appears that
both the magnitude of a presented variable (roll and
pitch attitude and indicated airspeed) and its first
derivative or rate, can be perceived. For all variable
and display combinations it was shown that the
perception of the magnitude of the presented
variable is, as expressed with the score parameter,
more accurate than the perception of its rate. In
addition the required exposure time and the response
time for the perception of the variable magnitude are
shorter than for its rate.

For all variable and display combinations it turned
out that the variable magnitude can be perceived
after an exposure time of approximately 50 msec.

The required longer exposure time for the perception
of the first derivative of the presented variable
corresponds well with the characteristics of the
visual system. In Van der Grind et al. (1986) it is
shown that some time delay occurs in visual
perception of motion due to the fact that the stimulus
has to pass a motion receptor field on the retina
before the associated neural circuit can generate an
output. Time delays of 50 msec to 2 sec depending
on stimulus velocity are reported. This is confirmed
by the results of the experiments in this paper. To
obtain an accurate perception, stimuli had to be
exposed to the subjects from 300 msec for attitude
rate perception up to 1500 msec for indicated
airspeed rate perception. In addition it was found
that the response time for rate perception is
approximately 100 msec up to 2 sec longer than the
perception of variable magnitude. The longer time
required to perceive the rate of a presented variable
has direct consequences for human operator control

behavior in a closed-loop control task. (Hosman and
van der Vaart, 1988).

As could be expected, the addition of a scale division
as the "pitch ladder", improves the perception
accuracy as expressed by the score parameter,
without increasing the required exposure time or
response time. The increase of the stimulated area
of the retina by the pitch ladder did not have any
influence on the perception of pitch rate in
experiment 2. Neither the score parameter nor the
required exposure time and response time were
significantly changed.

The comparison of analog and digital presentation of
indicated airspeed demonstrated that the most
accurate and fastest perception of airspeed
magnitude is obtained with the digital presentation.
Perception of airspeed rate is improved dramatically
by adding the trend vector to the moving tape
indicator. This corresponds well with the results for
variable magnitude perception. With the addition of
the trend vector on the indicated airspeed
presentation the variable rate perception is changed
to variable magnitude perception, the length of the
trend vector. This result confirms that, although the
visual system has motion detection circuits that are
able to differentiate visual position, perception of
variable rate is a difficult process.

Surprisingly, the results of experiment 3 show that
perception of indicated airspeed rate from the digital
presentation is possible. The required exposure time
and the response time are slightly lager than for the
analog displays.

The presentation of indicated airspeed on modern
Primary Flight Displays combining the moving tape
indicator with the trend vector and a digital indicator
corresponds well with the experimental results
discussed.

6. REFERENCES

Grind, W.A. van der , Koenderink, J.J. and Doorn,
    A.J. van (1986). The distribution of human
    motion detector properties in monocular visual
    fields. *Vision Research*, **26**, 797-810.

Hosman, R.J.A.W. and Vaart, J.C. van der (1988).
    Visual-Vestibular Interaction in Pilot's
    Perception of Aircraft or Simulator Motion.
    AIAA Flight Simulation Technology
    Conference. Atlanta, Ge. AIAA paper 88-
    4622-CP.

Theunissen, E (1993). D^3S: The Delphins Display
    Design System. AIAA Flight Simulation
    Technology Conference. Monterey, CA. AIAA
    paper 93-3556-CP.

THE RELATIONSHIP OF BINOCULAR CONVERGENCE AND ERRORS IN JUDGED DISTANCE TO VIRTUAL OBJECTS

Stephen R. Ellis, Urs J. Bucher
and Brian M. Menges

*NASA Ames Research Center,
Moffett Field CA, 94035 USA*

Abstract: Errors in judged depth of nearby virtual objects presented via see-through helmet mounted displays are shown to be linked to changes in binocular vergence. This effect varies measurably across the subjects examined and correlates with the magnitude of the subjects' individual depth judgment errors. The relationship is demonstrated by visually superimposing computer-generated virtual objects on physical backgrounds in a manner similar to that suggested for some practical applications. Suggestions for improved control of virtual objects under these viewing conditions are briefly discussed.

Keywords: see-through display, virtual objects, stereo vision, computer-aided work, binocular convergence

1. JUDGED DISTANCE TO VIRTUAL OBJECTS IN THE NEAR VISUAL FIELD

The perceptual cues to space have been classically divided into either static or dynamic or classified either as monocular or binocular information sources. More recent analyses of depth perception focusing on the behavioral affordances of vision have usefully reclassified the classical depth cues into three categories, those important with respect to personal space (2m ~ 1-2 eye heights), those relevant for action space (3-30 m ~ 2-20 eye heights), and those relevant for vista space (>30m ~ >20 eye heights). (Cutting and Vishton, in press).

This reclassification of sources of information concerning the spatial layout surrounding a viewer is particularly useful since it focuses attention on what vision is to be used for in each of these distinct regions. Depending on the category of relevance, different cues to depth have varying importance. In particular binocular convergence and accommodation play roles mainly relevant for personal space associated with coordinated, manipulative activity. Understanding the interaction of these physiological responses and depth perception will have growing importance as head mounted displays of virtual objects are introduced into the workplace.

Head or panel mounted see-through displays of conformal, computer generated imagery have been used in aircraft cockpits for many years as heads-up displays (Weintraub & Ensing, 1992) or as helmet mounted sights, i.e. the Honeywell sights. But these applications have almost universally presented users with virtual images at the far end of action space or into vista space. More recent applications of such displays are designed to present to their users spatially conformal computer generated virtual objects for medical and manufacturing applications (Rolland, 1994; Janin, Mizell & Caudell, 1993; Azuma & Bishop, 1994). This work has focused attention on precise calibration of the displays and also on perceptual phenomena that might degrade performance even in well calibrated systems.

Previous reports have indicated that indeed such phenomena are observable. The present two experiments are designed to investigate their causes and practical implications. In particular, previous observations (Ellis & Bucher, 1994) have shown that optical superposition of a virtual object on a physical backdrop changes its judged position. In particular, if a physical surface is introduced at the judged depth of the stereoscopic virtual image constituting the virtual object, the virtual object is judged to be closer to the observer. This effect is enhanced by slowly moving the physical surface. Since a change in rendering to make the virtual image completely occlude the backdrop did not affect its judged depth and since the motion of the backdrop is likely to have attracted visual attention and binocular convergence (Ellis & Bucher, 1994), it was concluded that the change in judged depth was not due to the occlusion. Rather, it was suspected that the effect was due to an increase in binocular convergence associated with the physical object. The following experiment is a direct test of this hypothesis using an unobtrusive, nonius technique to detect convergence.

2. MEASUREMENT OF CHANGES IN STATIC CONVERGENCE DURING VIEWING OF VIRTUAL OBJECTS

2.1 Methods

Stimuli

A stereoscopic virtual image of an upside down, axially rotating (2 rpm) tetrahedron was presented at a distance of 108 cm away from the subjects' eyes by a head-mounted see-through display. This display was operated under normal room illumination and was an improved, previously described device (Ellis & Bucher, 1994). One diopter accommodative relief was provided for the virtual image only. Display resolution was better than 5 arcmin.

The depicted size of the presented virtual object was randomly scaled from 70 to 130% of its nominal size for each trial preventing subjects' use of angular size as a depth cue. The wire-frame tetrahedron had a nominal 10 cm base and 5 cm height. The width of the wire frame lines was about 9 arcmin. The lines of the wire frame an all other computer generated lines had a luminance of about 65 cd/m^2. and were seen against a 2.9 cd/m^2 gray cloth background placed 2.2 m from the subject. Occasional variations in the depicted depth of the tetrahedron which were not analyzed further were also introduced to insure that the subjects did not notice that the same depicted depth was repeated. But these additions proved to be unnecessary due to large and variable, experimentally induced perceptual effects that influenced the judged depth of the target.

A physical surface that was introduced along the line of sight to the tetrahedron was provided by a rotating checkerboard made for foamcore.The checkerboard was a disk 29 cm in diameter with 5 cm black and white checks having either 1.3 cd/m^2 or 17.8 cd/m^2 luminance.

Subjects

Five men and one woman with stereo resolution of better than 1 arcmin as measured with the stereo vision test on an Orthorater participated in the experiment. Some subjects had vision corrected by contact lenses or glasses and were able to wear their corrections during the experiment. Subjects' ages ranged from late teens to late '40's and included laboratory personnel as well as paid subjects recruited by a contractor at Ames.

Task

The first part of the subject's task was to mechanically place a yellow-green LED (about 20 cd/m^2) pointer under the nadir of the slowly rotating, wireframe tetrahedron, which had a size randomly selected for each trail. After aligning the pointer, the subjects were presented with two sets of nonius lines just flanking the tetrahedron. These lines were then adjusted to equal visual directions on each side of the tetrahedron by moving the lower left and right segments (See Figure 1). The second part of the task involved a second adjustment of the pointer to the tetrahedron's depth after a slowly, irregularly rotating, (~2 rpm) opaque checkerboard was introduced along the line of sight to the tetrahedron. . The tetrahedron was presented a second time at the same depicted depth in this new configuration but the experimental variations generally concealed this fact from the subjects so that they believed each trial, with or without the checkerboard, involved a potentially different depicted depth.

After the second judgment of the tetrahedron's depth, the nonius lines were flashed briefly (ca 250 msec) next to the tetrahedron while the subjects fixated it. Then the subjects made a forced choice indicating whether the upper or lower pair of the flashed nonius lines were closer. The eye assignments of each segment of the nonius lines were randomly selected so the that meaning of the alternative possibilities in terms of convergence or divergence varied randomly across the trials. The assignment of the lower part of the left nonius line and the upper part of the right line to one eye and the other upper-lower pair to the other eye, produced a differential effect doubling the relative misalignment for any given vergence change. The subject reported whether the upper or lower segments of the paired nonius lines were closer.

Three different experimental conditions were used. In the "on" condition the checkerboard was mechanically introduced at the judged depth of the virtual tetrahedron object. For the "in front" condition the checkerboard was introduced 30 cm in front of the judged depth. In the control condition the second judgment was a replication of the first judgment in that the subject made a second judgment of the depth of the virtual image. But this time instead of aligning the nonius lines, the subject made the forced-choice judgment of the nonius lines configuration without the addition of the checkerboard. The control thus was identical to the experimental conditions except the checkerboard was not introduced into the line of sight. This control, thus, provides a check on the subjects' judgment bias or changes of their convergence during the experiment. Each condition was repeated 15 times for each subject in a randomized block design in which blocks of 5 replications of each condition were repeated. The 6 possible orders of the 3 conditions were distributed randomly across the 6 subjects in the experiment.

The change in judged distance of the virtual object, the tetrahedron, was analyzed in a single factor re-

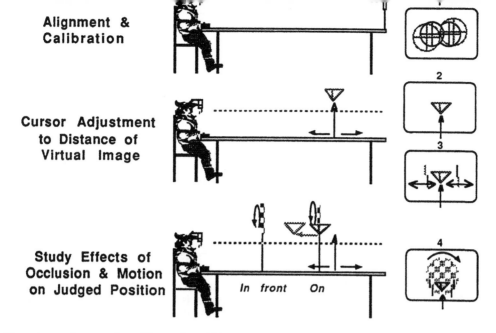

Figure 1. Experimental procedure illustration. Top: alignment, magnification, and interpupilary adjustment, Middle: Pointer and nonius lines adjustment, Bottom: A variety of testing conditions are shown, the "on" or " in front" placement of the rotating checkerboard and the flashed nonius lines.

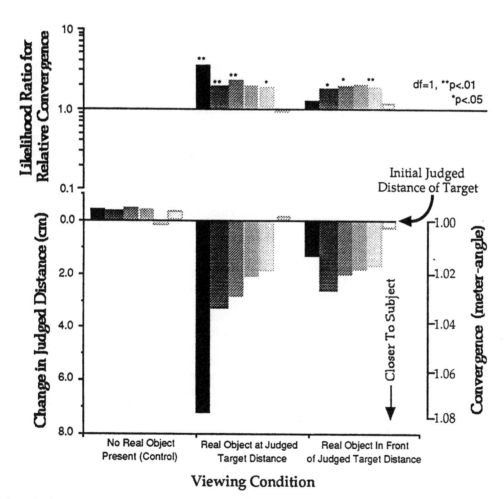

Figure 2. Results for 6 subjects sorted for magnitude of change in judged distance in the "on" condition.

peated measures ANOVA. Chi-square analyzes were conducted on each individual subject's distribution of judgments of convergence/divergence for each of the 3 experimental conditions. Taking the control condition as a baseline, the relative strength of convergence could be measured by a likelihood ratio of the probability of convergence in each experimental condition divided by the probability of convergence in the control.

2.2 Results

Single factor repeated measures analysis of the effect of superposition of the checkerboard and virtual images replicated the previous observations that the virtual object was moved closer to the viewer ($F_{(2,10)}$= 7.549 $p < 0.01$). Individual data are shown in Figure 2. This effect was somewhat stronger for the "on" condition than for the "in front" case and varied in strength across the 6 subjects. One subject interestingly showed essentially no effect.

Table 1 Frequency of convergence and divergence during depth judgments of virtual objects

	Converg.	Diverg.	Likelihood ratio of convergence to divergence
On	84	21	1.58
In front	70	35	1.32
Control	53	52	--

The cause of this individual subject's variation is illuminated by considering all subjects' tendency to relatively converge during judgment of the depth of the virtual object in the presence of the checkerboard. This tendency is summarized for the experiment in Table 1 which displays the frequency of convergence or divergence indicated by the nonius judgments for all subjects in the three experimental conditions (Chi-square = 20.37, df=2, p< .001). The control case shows an almost perfect 50:50 break while the other two conditions show decided convergence, the "on" condition being somewhat stronger.

Each subject's individual likelihood ratio of convergence was computed as the ratio of the probability of convergence in an experimental condition to the probability of convergence in the control. These ratios are plotted in Figure 2 for each subject. A 2X2 Chi square contingency was computed to compare the distribution of convergence and divergence for each experimental condition to that of the control condition. This was done separately for each subject. Statistically significant distributions are indicated by asterisks in Figure 2.

2.3 Discussion

The individual subject's results in Figure 2 are sorted by the size of the change in the judged position of the virtual object for the "on" condition. These results can then be compared with the likelihood ratio of relative convergence. As is clear from the figure, the two measurements are almost perfectly correlated across the subjects. The only subject not to show a displacement of the virtual object caused by the checkerboard, also is the only one to show essentially no relative convergence. The results for the "in front" condition show a weaker apparent displacement of the virtual image but also show a correlation convergence. The correlation of relative convergence with magnitude of displacement for the "on" and "in front" conditions is across subjects and conditions, r= 0.894, df=10; (t=6.31, p < .002). The results generally support the supposition that the change in judged depth could be due to a change in convergence, but the mechanism underlying this change remains to be clarified.

One possibility is that the effect observed here is the reverse of the inward shift of accommodation with infinity-collimated virtual image displays discussed at length by Roscoe (eg. 1991). In his case users of these displays do not fully accommodate to infinity, but remain focused somewhat closer. The reflex coupling of accommodation and vergence would be expected to change the vergence as well. In the cases we investigate in the present experiment, subjects might be accommodating beyond 1 diopter when the small, relatively poor quality, ~20/100 acuity, virtual object is presented. Their accommodation and vergence could be then brought closer by the insertion of the real surface which provides a much bigger and sharper stimulus to accommodation and to vergence through the accommodation vergence reflex. Another view of this change may be that a fixation disparity present during viewing of the virtual object is removed by the physical target.

Further experimentation will test these alternatives and examine possible relationships to perspective vergence (Enright, 1991) or proximal vergence (Cuiffreda, 1992) which appear to indicate that higher-level spatial interpretations of the visual image can simulate the vergence system. The results from the present experiment, while indicating that there is a clear oculomotor response associated with the error in judged depth, do not resolve the question of causation since the vergence change could produce the change in judged depth through disparity reinterpretation just as easily as a perceptual interpretation of occlusion could produce a vergence response through proximal vergence. This question can only be experimentally resolved if a viewing condition can be found that would differentiate the lower level from higher level convergence cues.

The present results were observed with a static eye point and can be expected to change when significant head-movement coupled motion parallax is introduced. But it is important to realize that many of the possible new applications of head-mounted see-through displays involve situations in which viewing will be relatively static and for which weight considerations might suggest the use of monocular displays. The results of this experiment indicate that such displays should have a variable focus control and probably should be used with a bore-sighting procedure in which focus is adjusted to a reference target so as to correct for any errors in depth due to inappropriate vergence. Alternatively, the computer generated targets used with such displays could be binocularly generated with distorted disparity to correct their spatial perception, pushing them away from the viewer.

REFERENCES

Azuma, R. & Bishop, G. (1994) Improving static and dynamic registration in an optical see-through HMD. Proceedings of SIGGRAPH '94, July 24-29, Orlando, Fl.

Cuiffreda, Kenneth J. (1992) Components of clinicalnear vergence testing. *Journal of Behavioral Optometry,3*,1,3-13.

Cutting, J. E. & Vishton, P. M (1995) Perceiving layout and knowing distances: the integration, relative potency an d contextual use of different information about depth. In *Handbook of Perception and Cognition*, Vol. 5., W. Epstein a& S. Rogers, eds., Academic Press. (in the press).

Ellis, S. R. (1991) Nature and origin of virtual environments: a bibliographical essay. *Computer Systems in Engineering, 2*, 4, 321-327.

Ellis, S. R. (1994) What are virtual environments? *Computer Graphics and Applications, 14*, 1, 17-22

Ellis, Stephen R. and Bucher, Urs J. (1994) Distance perception of stereoscopically presented virtual objects superimposed by a head mounted see through display. *Proceedings, 38th Annual Meeting of the Human Factors and Ergonomics Society*, Santa Monica CA,

Enright, J.T. (1991) Paradoxical monocular stereopsis and perspective vergence. In Ellis, S. R., Kaiser, M. K., & Grunwald, A. J. (eds). *Pictorial Communication in Virtual and Real Environments*. London: Taylor and Francis.

Foley, J. M. (1993) Stereoscopic distance perception. In Ellis, S. R., Kaiser, M. K., & Grunwald, A. J. (eds). *Pictorial Communication in Virtual and Real Environments*. London: Taylor and Francis.

Janin, A.L.; Mizell, D.W.; & Caudell, T.P.(1993) Calibration of head-mounted displays for augmented reality applications Proceedings of IEEE VRAIS '93, Seattle, WA

Ritter, M. (1977) Effect of disparity and viewing distance on perceived depth. *Perception and Psychophysics, 22*, 4, 400-407.

Rolland, J. P., Ariely, D. & Gibson, W. (1994) Towards quantifying depth and size perception in 3D virtual environments. to appear in *Presence*

Roscoe, S. (1991) The eyes prefer real images. In Ellis, S. R., Kaiser, M. K., & Grunwald, A. J. (eds). *Pictorial Communication in Virtual and Real Environments*. Taylor and Francis London: Taylor and Francis.

Weintraub, D. J., & Ensing, M. (1992) *Human factors issues in head-up display design: the book of HUD*. Wright Patterson AFB, Ohio: CSERIAC.

PILOT PERFORMANCE AND WORKLOAD USING
SIMULATED GPS TRACK ANGLE ERROR DISPLAYS

Charles M. Oman[+] , M. Stephen Huntley, Jr.[*], and Scott A. Rasmussen[+]

[+]Man Vehicle Laboratory, Department of Aeronautics and Astronautics
Massachusetts Institute of Technology, Cambridge, MA USA

[*] Cockpit Human Factors Program, DTS-45
Volpe National Transportation Systems Center, Cambridge, MA

Abstract: The effect on simulated GPS instrument approach performance and workload resulting from the addition of Track Angle Error (TAE) information to cockpit RNAV receiver displays in explicit analog form was studied experimentally (5 display formats, 6 pilots, 20 approaches each) in a Frasca 242 light twin aircraft simulator. Inter subject differences in ability to use the displays were found, but sliding and tilting pointer TAE formats significantly improved intercept and tracking performance measures. Determination of wind correction angle was simplified. Workload scores were not significantly influenced by display format, but depended on approach geometry and phase.

Keywords: man/machine interfaces, displays, manual control, simulators, aircraft control, flight control, air traffic control, multiloop control, mental workload, global positioning systems, navigation systems.

1. INTRODUCTION

Satellite based navigation systems and a new generation of microprocessor based cockpit avionics are revolutionizing air traffic control world wide. In the USA, many transport and military aircraft are now equipped with Global Positioning System (GPS) based area navigation (RNAV) computers or flight management systems, which are used for supplementary enroute and oceanic navigation. Research is underway to develop differential GPS systems with the horizontal and vertical accuracy and integrity needed for precision instrument approaches so that the aging VOR and ILS navaids can be phased out. Meanwhile, the horizontal accuracy of ordinary non-differential GPS receivers (100 m) is sufficently good that the Federal Aviation Administration encourages their use for less demanding non-precision approaches, employing conventional altimetry for descent. Pilots are now permitted to fly most existing non-precision approaches using GPS as the primary reference. FAA has also begun to certify new approaches specifically designed for GPS equipped aircraft. This initiative is particularly important for the general aviation (GA) community, since non-precision approaches to thousands of new airports will eventually be possible. Because GPS approach waypoints can be arbitrarily positioned, non-traditional approach geometries can be employed to improve obstacle clearance, or reduce noise and air traffic congestion. GPS RNAVs have flexible electronic displays, updatable databases, and many more operating modes than traditional VOR, DME, ILS, and ADF equipment. The new RNAVs can potentially make instrument flying both easier and safer, provided that the human factors aspects have been properly considered at the design stage.

GPS RNAVs for civil aircraft must meet minimum performance and display standards established by the FAA (TSO C-129, and RTCA/DO-208). TSOed units are now available from some manufacturers. In most GA aircraft, instruments are of traditional type, and panel space is limited. Hence, GPS receivers are typically stand alone devices which occupy a radio or instrument slot. Only a small LCD or CRT display and a limited set of control buttons and knobs are practical. Since the GPS cross track error (XTE) information functionally replaces that from VOR, XTE is typically converted to an analog signal, and sent to an existing course deviation indicator (CDI), usually on a VOR head or Horizontal Situation Indicator (HSI), as shown in Fig 1. Alternatively, a simulated CDI needle can be displayed on the RNAV itself (Fig. 2). As with VOR driven CDIs, the pilot always flies "toward" the needle to center it, but needle sensitivity is in linear, rather than angular units, and is scheduled: +/- 5 nm full scale while enroute, increasing to 1 mile during initial approach, to 0.3 miles 2 miles before final approach, and returning to 1 mile if a missed approach is flown.

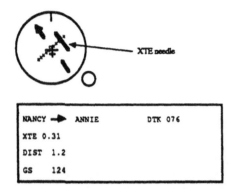

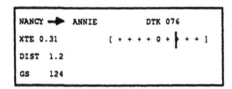

Fig. 1. (Top): Analog XTE on Horizontal Situation Indicator CDI. (Bottom): Alphameric data as they appear on nearby GPS receiver display.

Fig. 2 Analog XTE CDI on GPS Receiver Display

Maintaining an aircraft on course centerline using XTE information is a demanding multiloop manual control task The pilot's stick position controls the third derivative of XTE, so the combined pilot-aircraft system is unstable unless the pilot properly monitors and controls bank angle and heading, in addition to XTE. In a crosswind, the pilot must use a cut-and-try technique to determine the proper wind correction angle. The pilot must base his control movements in part on the rate of change of XTE to avoid large oscillations across the course centerline, particularly during the critical stages of the final approach. Judging XTE needle movement is not so easy, since the pilot also must sequentially scan attitude, heading, airspeed and altitude instruments. XTE needle sensitivity is increased on final approach to help the pilot see its movement and null small errors. However, XTE must then be even more frequently scanned, reducing the time available to perform other tasks, and increasing workload. Any side task which delays instrument scan impairs tracking performance. Inflight studies have demonstrated a direct relationship between CDI sensitivity and pilot workload, and an inverse relationship with XTE during non-precision approaches (Huntley, et al, 1991) . In the parlance of

manual control, the pilot's ability to create outer loop lead is determined by effective instrument scanning delay (Clement, et al, 1968). Since it is even more difficult to sense the derivative of XTE using a numeric (digital) display, TSO C129 requires GPS RNAVs to present XTE in analog form.

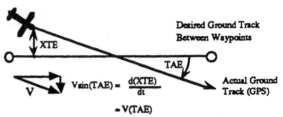

Fig. 3 The derivative of Cross Track Error (XTE) is proportional to Track Angle Error (TAE)

The capability of GPS RNAV systems to determine the direction of the aircraft's ground track with only a brief delay is potentially of importance for aircraft manual control. Since the desired heading is known, it is possible to compute "track angle error" (TAE), the difference between the desired track and the actual track. As shown in Fig. 3, TAE is mathematically proportional to the derivative of XTE, the important manual control variable. If TAE were directly displayed to the pilot, XTE could be less frequently scanned, and the pilot's performance might improve. There are several ways this could be done. One is to incorporate TAE information into inner loop attitude commands. However, many GA aircraft lack the necessary flight director equipped attitude indicator. A second possibility is to derive a TAE based "course to steer" - a heading flight director command. However, this ideally requires an additional indicator on the primary heading display not available on existing instruments. Another option is to use TAE to estimate future XTE - a "quickened XTE" display. However, information on present XTE is then lost. The fourth and perhaps simplest alternative is to present TAE information on the GPS RNAV itself in analog form. It makes sense that if TAE were explicitly displayed, pilots might learn to take advantage of it. TSOed GPS RNAVs are required to have *at least* a numeric display of TAE. TSO C-129 suggests that "the use of non-numeric XTE data integrated with non-numeric TAE data into one display may provide the optimum of situation and control information for the best overall tracking performance". However, the TSO did not suggest the format of such displays, and no experimental data has yet been available as a guide to FAA or manufacturers.

The purpose of this flight simulator research project was to investigate the fourth alternative, and see how pilots used TAE information when it was presented in several different formats. Which format do pilots favor, and why ? To what extent does the addition of TAE information allow pilots to quantitatively improve their approach performance, or reduce workload ? If pilots must look away from their primary instruments to monitor TAE on the GPS receiver, is some of the potential advantage lost ?

2. METHODS

2.1 Displays

The TAE GPS receiver display formats evaluated were:

1) <u>Separate TAE and XTE sliding pointer displays</u> (2 versions). This format[1] , shown in Fig. 4(top) added a TAE sliding pointer display beneath a conventional, "fly to" XTE CDI. The TAE pointer was a triangle, located just beneath the XTE needle, and using the same "ten dot" (123 pixel wide) scale. When the triangle was centered, TAE was zero. Full scale TAE triangle deflection was set at ± 90 degrees, since this is the maximum useful course intercept angle. Which way should the TAE triangle move in response to a roll command ? One alternative is to have the triangle move in the <u>same</u> direction as the stick roll command. This is easy to remember, and has the advantage that both the needle and the triangle appear on the <u>same</u> side of the display when converging with course. This version was therefore referred to as "Triangle/Same". However, a concern was that this makes the TAE triangle a "fly from" display. Since the XTE display above it is "fly to", this version apparently violates the well known human factors "command-response consistency" guideline. So we also evaluated a second version of this format, where the sign of the triangle movement was reversed. This version was referred to as "Triangle/Opposite", and is shown in Fig. 4(middle).

2) An <u>TAE/XTE sliding/rotating pointer integrated display</u>. In this format, shown in Fig. 4(bottom), the horizontal displacement of the arrow was proportional to XTE and the tilt angle was equal to TAE. This way, dimensional correspondence was preserved for the linear and angular variables. The sign of the pointer rotation was chosen so the arrow always moved in direction of tilt. In practice, the display appeared much like a "mail slot view" of a track up moving map display, where the arrow corresponded to the desired track, and the CDI scale a a downward looking view of the aircraft's wings. If the pilot adopted this "inside out", aircraft centered frame of reference, interpretation of this display was very intuitive This format was referred to as the "Track Vector" display.

To assess the value of explicit TAE information, these three displays were experimentally compared to an "XTE only" receiver display, shown in Fig.2. All four formats required the pilot to frequently look over to the GPS receiver for XTE/TAE information, so a fifth format was also included, in which XTE was presented along with heading information on an HSI, and only alphameric information was presented on the receiver, as shown in Fig. 1. The HSI was 70 cm from the pilot's eye, and 9.5 cm beneath the attitude indicator. The GPS receiver display was created on

---

[1] originally suggested by G. Lyddane, an FAA pilot, National Resource Specialist for Flight Management Systems, and an author of TSO C-129.

an high resolution LCD display, located 35 cm (27 deg) to the right of the HSI, and which subtended approximately 10 degrees of horizontal visual angle. A consistent set of generic alphameric data was presented on all 5 displays: last and next waypoint, desired track (DTK), numeric XTE, groundspeed (GS), and distance (DIST) to waypoint. Numeric TAE was shown only on TAE displays. In all approaches, the pilot had to monitor DIST, and if a turn at the next waypoint was required, initiate a dead reckoning, standard (3 deg/sec) turn at the appropriate point to intercept the next leg. Waypoints automatically sequenced when the aircraft crossed a line bisecting the angle between the inbound and outbound legs.

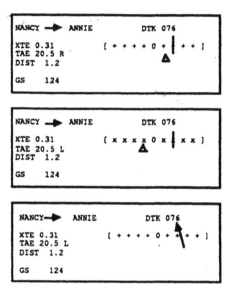

Fig. 4: Three GPS receiver TAE display formats studied: Top: Triangle/Same; Middle: Triangle/Opposite; Bottom: Track Vector.

2.2 Subjects , Sessions, and Experiment Design

Six multiengine, instrument rated pilots, were recruited locally. Total flight time averaged 1967 hr. (range 750-3387 hr.), and included an average of 73 hr. (10-213 hr.) actual instrument, 125 hr. (30-370 hr.) of simulated instrument., and 78 hr. (4-240 hr.) of time in various simulators. Multiengine experience averaged 498 hr. (22- 1500 hr). They had flown an average of 4 (0-14) approaches and 26 hr. (0-82 hr) in the past month, and an average of 6.3 hr. (0-28 hr) in the week preceeding the experiment.

Each pilot flew a total of 20 approaches, four with each of the five display formats, and two different types of approach geometries. Each approach required about 15 minutes, so the experiment was conducted in two ten-approach sessions on separate days. Pilots were given a written and oral briefing on the displays and the experiment procedures. Each day, they flew the simulator and practiced with the displays until they felt familiar with them. They then flew 3-4 complete practice approaches with the different displays to asymptote practice effects, and then flew 10 test approaches. To minimize confusion between the two triangle formats, pilots flew with only one of the triangle formats each day. Half the subjects flew with the triangle/same first. The presentation order of triangle displays was thus blocked, but for the three other formats was randomized and balanced within sessions.

2.3 Aircraft Simulation and Approach Geometries

Pilots flew a fixed base, light twin engine flight simulator (Frasca International, Inc. Urbana, IL, Model 242). Aircraft dynamics (furnished by Frasca) resembled a Piper Aztec. Nongaussian, patchy disturbances (Jansen, 1981) were added about the three aircraft attitude axes, independently. The ratio of roll/pitch/yaw mean disturbance was 15/5/1 times, respectively. The disturbances qualitatively resembled moderate-severe turbulence, and required the pilot to closely monitor the attitude indicator to maintain control. Additional networked computers created the GPS displays, altitude dependent wind, and collected data. Wind was always a 45 degree left or right head wind with respect to the final approach heading, but strength varied from 35 kts at 3100 ft above ground to 15 kts at the surface, using a power law atmospheric model. On many legs, up to 14 degrees of heading correction was required. Pilots knew the wind direction varied, but were not told that only two relative wind directions were used.

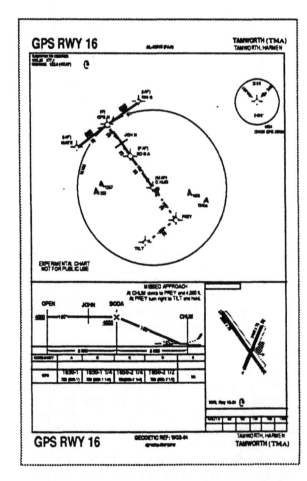

Fig. 5 Example of the NOS style GPS "T" approach plates used. Top: plan view. Bottom: elevation view, minima, and fictitious airport diagram.

Eight approach charts were employed, each with a different final approach heading and required altitudes, so that pilots could not memorize the numbers. Half the charts used a GPS "T" geometry, (Fig. 5). The aircraft was initialized 0.5 nm upwind and abeam of the initial approach fix (IAF), located at one end of the top of the T. The pilot was required to intercept the initial approach leg, and fly five miles to an intermediate approach waypoint (IF) at the center of the "T", maintaining 3100 ft. above ground level (AGL). At the IF, the pilot was instructed to turn 90 degrees, and fly five miles to the final approach fix (FAF). A waypoint 2 miles before the FAF showed where XTE CDI sensitivity changed from ±1 nm to ±0.3 nm. The pilot was to lower the flaps and landing gear just before the FAF, and after passing it, fly five miles to the missed approach point (MAP) while descending to the 750 ft. minimum descent altitude (MDA). At the MAP, the pilot was to retract gear and flaps, and climb back to 3100 ft. AGL, flying to a first missed approach fix directly ahead five miles away, make a second 90 degree, level turn and then fly five more miles to a missed approach holding fix (MAHF). (Each chart had 2 or more IAFs, but the one used was always on the same side of the runway as the MAHF, so all turns were made in the same direction on any given run.) The turns permitted us to study performance while intercepting the subsequent leg. The initial, intermediate, and second miss legs were flown at constant altitude, and provided opportunities to measure tracking performance.

The remaining approaches used a "Crooked T" geometry[2] which required the pilot to also make a 45 deg. turn while initiating descent at the FAF, and then fly a two mile descending dogleg before turning back to the runway heading. There was a minimum crossing altitude at the dogleg waypoint. Since these were descending turns made with 0.3 nm CDI sensitivity, the Crooked T approaches were expected to be much more difficult to fly.

During the approach and missed approach pilots were required to perform the usual checklist items, such as turning on and off fuel pumps, and tuning the radio to a frequency found on the chart and announcing their position. All pilots were instructed to fly the approach and the missed approach at 120±10 kt., to fly as close to course centerline as possible, to maintain altitudes within 100 ft., but never to descend below the MDA.

2.4 Workload, Display Preference and Approach Performance Metrics

Immediately after each approach, pilots were asked to rate the overall workload on a 10 point modified Bedford workload scale, which emphasized spare attention (Roscoe and Ellis, 1990; Huntley, et al 1993). They were asked to describe any errors made,

---

[2] The "T" approaches resembled an approved GPS approach at Oshkosh, WI. The "Crooked T" geometry was hypothetical.

and then to rank order the 6 legs of the approach from easiest to hardest. After the second test session, a questionnaire was administered which required the pilots to subjectively rank the 5 displays using several different display preference scales. These included ease of interpretation (EOI), effect on flight path control accuracy (FPA), and overall preference (OP). In addition, each pilot was asked to indicate relative preference between individual pairs of displays on a ± 7 point scale. The scores from these 10 pairs were summed using a tournament method, and ranked by display, to yield a second measure of overall preference, based on direct "head to head" comparisons (HTH).

XTE, TAE, altitude, airspeed, pitch and roll attitude performance parameters were sampled continuously at approx. 1 Hz by computer. Aircraft position data from each of the 120 approaches were rotated to a common southerly final approach heading, east/west reversed where appropriate, and ground tracks were compared by display. The combined track records were also used to retrospectively separate the approach into a series of 13 segments of varying lengths, chosen to isolate the various intercept, tracking, turning, descending and climbing phases of the approach. Mean, standard deviation, and RMS values of all six performance parameters were computed longitudinally along 13 different approach segments, and analyzed using Systat v.5.2 (Systat, Inc., Evanston, IL).

3. RESULTS

3.1 Pilot Display Evaluations

Based on pilots debriefing evaluations, TAE displays appeared to have the following <u>advantages</u> over the XTE only formats:

d) When intercepting a new leg, pilots could choose an appropriate intercept TAE, and then reduce it in several steps as they approached course centerline, so as to avoid overshooting.

a) While tracking along a leg, an offset of the triangle or a tilt of the vector allowed pilots to detect and anticipate the magnitude and direction of slow changes in XTE.

b) Pilots found that they could distinguish the "diverging" and "converging" XTE/TAE pointer configurations at a glance, and react appropriately. For example, when the triangle/same pointers were on opposite sides of center, or when the track vector was tilted away from center, corrective action was immediately needed.

c) When tracking, it was possible to immediately determine the cross wind correction angle without using a cut-and-try approach. Pilots noted the heading when TAE equaled zero and many chose to set the heading indicator "bug" to this value, and simply make small left-right course corrections by flying one side of the bug. While training, pilots

found it was possible to adopt a loop separation control strategy, using bank angle to control TAE and then TAE to control XTE, while effectively ignoring heading. However, closed loop control of the TAE pointer required frequent scan of the TAE display. Most pilots found this difficult or inappropriate to do in turbulence, because the attitude indicators required so much attention, and instead relied on familiar attitude and heading control strategies, using TAE to command an appropriate heading.

d) If the XTE indicator was off scale, an appropriate indication on the TAE pointer reassured the pilot that XTE would soon be on scale again.

The following TAE display deficiencies were noted:

1) Pilots said that small TAE offsets seemed more easily detected using either of the triangle displays than the vector display. The graphical display resolution was identical, but the track vector pointer showed staircasing (aliasing) when it was nearly vertical, and there was no vertical reference mark.

2) Three pilots reported confusion while interpreting the track vector display. There appeared to be two reasons for this. The first was that the format resembled a conventional CDI, not a moving map display, and there was no explicit aircraft symbol in the center. Several pilots said they had "difficulty remembering which was the airplane and which was the track". The second was that the vector would suddenly "flip" by 90 degrees during turns. Although the triangle display also flipped position during turns as the aircraft crossed the turn bisector, with the track vector display the pilot needed to mentally rotate his frame of reference when the vector flipped in order to maintain the "inside out" map interpretation. Several pilots found they could not always do this.

3.2 Pilot Display Preferences

Post-session questionnaire data clearly indicated a preference for the triangle/same display over either of the other two TAE displays. At the same time, the rankings underscored the relative importance to the pilots of having XTE information in the primary instrument scan area, rather than alone on the GPS receiver. Results are shown in Table 1. A statistically significant effect of display was found for the HTH, FPA, and EOI scale rank scores from the individual subjects (Friedman rank ANOVA, $p < 0.04$). An OP scale display effect was found at the $p < 0.06$ level.

On the Overall Preference (OP) and Head-To-Head (HTH) comparison scales, pilots consistently preferred the HSI display over the "triangle/same" TAE display. However, in terms of summed rank scores, the margin was slight. When asked to make head to head comparisons, half the pilots preferred the triangle/same display, two preferred the HSI display, and one judged it a tie.) Directly comparing the "triangle/same" and "triangle/opposite" versions, 4 pilots preferred the former, and only one the latter.

All pilots always ranked one or more of the TAE display formats above the XTE only on both scales, so the consensus was clear that TAE information was subjectively useful.

Table 1. Pilot Display Preference Ranks by display, using 4 different scales (see text). Rank = 1 is best.

| Display | Display Preference Scales | | | |
Format	OP	HTH	FPA	EOI
Δ/Same	2	2	1	2
Δ/Opposite	3	4	2	5
Vector	4	3	4	3
HSI	1	1	3	1
XTE only	5	5	5	4

For accurate flight path control, pilots preferred the triangle/same display, though the three subjects (1, 4 and 5) who actually tracked most consistently (see below) ranked the HSI first in this respect. Four of the five pilots said the HSI was the easiest of the displays to interpret, though three of the four cited long-standing training and experience with the HSI format as one reason for this preference.

Four of the six pilots said they never referred to the numeric TAE information at all, since it was not obvious how to interpret the L/R TAE indication, and because the analog TAE pointer was available.

3.3 Influences on Subjective Workload

Average Bedford subjective workload was 3.5 out of 10 on T approaches, and increased to 4.5 on Crooked T. Subject 2 had the highest average workload (6.2) and Subject 1 the lowest (3.0). ANOVA of modified Bedford Workload scores revealed significant effects between subjects ($F(5,108)=29$; $p<.0001$), and T and Crooked T approach types ($F(2,108)=28.4$; $p<.0001$). The average workload scores by display type were: triangle/same=3.8; triangle/opposite=3.9; track vector=4.0, HSI=3.9; XTE only=4.5. However, adding display to the ANOVA did not produce a significant effect. No trends were found by sequential approach or session number, suggesting training had asymptoted practice effects. Ranking workload scores within subjects did not reveal a display dependent effect. It was concluded that display effects, if they exist, must be small compared to approach geometry and inter subject effects.

The pilots ranked the approach legs in order of decreasing difficulty: 1) long/dogleg final, 2) short final, 3) intermediate leg, 4) first miss , 5) initial leg, and 6) second miss leg. Ranking was identical for both approach geometries, and the concordance of workload rankings within geometries was statistically significant (Friedman ANOVA, df=5; p <0.0001)

3.4 Display Effects on Track and Performance

The reoriented aircraft ground tracks are shown superimposed in Fig. 6, staggered by the five display

types. The Crooked T tracks can be distinguished by the dogleg after the FAF. Several approaches where pilots made gross errors are apparent, particularly for the XTE only display (upper right). The frequency of such errors was noted to be less with the HSI and triangle/same displays (left and third from left).

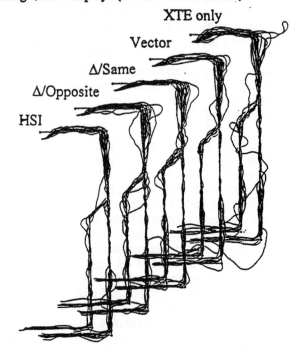

Fig. 6. Reoriented ground tracks by display type.

When the simulation began, the pilot's first task was to intercept the initial approach leg. Crosswind (always from the right in Fig. 6) tended to steepen the intercept angles of the non-TAE display approaches. Pilots flying with the triangle/same and vector displays had smaller downwind biases, as shown in Fig. 7. The difference in average tracking bias was significant by display ($F(4,90)=7.8,p<.001$) and subject. ($F(5,90)=7.8$; $p<.001$). However, the display effect was not significant for the second miss leg, possibly because pilots knew the wind direction by then, and the workload on this leg was lower, so pilots may have paid more attention to nulling XTE.

For the tracking portions of the initial (miles 2-4), intermediate (miles 1-3), and second missed approach (miles 1-4.5) legs, ANOVA showed significant effects of display ($F(4,329)=2.7,P<0.03$) and subject ($F(5,329)=14.8,p<0.0001$) in the combined data for standard deviation of XTE and also standard deviation of TAE (display: $F(4,168)=4.9$, $p<.001$; subject:$F(5,168)=11,p<.0001$). This was consistent with the notion that pilots were controlling TAE. For this and other approach segments tested, a significant subject by display interaction term was found, suggesting that certain subjects learned to make better use of the TAE information than others. For most segments, the HSI or Triangle/same displays generally ranked best, followed by the track vector display. Average display effects were smaller than subject effects. The standard deviation of XTE was consistently lower on the final approach leg, as the aircraft approached the MAP. The standard deviation of XTE on final approach (0.09 nm) with the "XTE only" display was quantitatively similar to that obtained in a comparable flight study (0.06 nm), in a Beech 55 Baron, where XTE also was displayed on a separate CDI (Huntley, 1993).

Inter subject differences in XTE and TAE tracking performance appeared to relate in part to inner loop attitude control. Two subjects (2 and 6) consistently showed larger values of standard deviation of pitch and roll attitude, airspeed, and altitude, suggesting that their effective attitude instrument scanning delay was longer. There was no clear effect of display format in the longitudinal axis on pitch, airspeed, and altitude, but for the lateral/directional axis, ANOVA revealed a significant effect of both · subject ($F(5,449)=59, p<.0001$) and display ($F(4,449)=3.9$, $p<.004$) on the variation in roll attitude. No correlation of performance with low recency or experience was found, except for pilot 2, whose recent instrument time was only in helicopters.

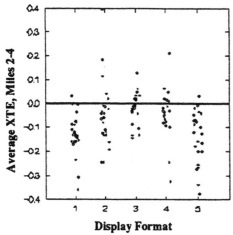

Fig. 7. Average XTE for miles 2 - 4 of Initial Approach Leg, by Display Type. Key: 1=HSI, 2=Δ/opposite, 3=Δ/same;4=vector; 5=XTE only.

4. CONCLUSIONS

Results showed that even under turbulence conditions requiring diligent attitude instrument scan, the addition of analog TAE information to a GPS receiver XTE display significantly improved approach leg intercept and tracking performance measures, probably by allowing the pilot to predict XTE changes. Determination of wind correction angle was greatly simplified. Pilots chose not to use the numeric TAE information provided. Thus, the study quantitatively supports the FAA TSO-C129 recommendation for analog TAE displays.

Certain pilots were better able to improve their performance with TAE displays than others. Of the three TAE display formats evaluated, a sliding TAE pointer located beneath the XTE CDI, moving in the same direction as aircraft bank, produced the largest performance improvement, and was preferred by the pilots for flight path control. However, control tests with HSI displaying heading and XTE (but not TAE) on the primary instrument panel showed that the performance advantages of adding TAE to the

receiver display were partly lost because of the need to widen the pilot's scan. The simultaneous display of XTE on the HSI and XTE/TAE information on the GPS receiver has not yet been tested.

The integrated XTE/TAE "track vector" display evaluated in this study had the advantage that pilots could visualize it as a track up moving map. However, it was found to be less useful for small TAE corrections because it lacked a vertical reference line. The display resembled a conventional CDI, rather than a moving map, and pilots occasionally had difficulty maintaining the map interpretation, particularly during turns when a cognitive mental rotation was required to reinterpret the display after waypoint changeover. The addition of a fixed, central aircraft reference symbol and the use of a surrounding outline box may improve this display.

Although Bedford scores were sensitive to approach geometry, no consistent effect of display format on workload was found. It is possible that our pilots adjusted their performance criteria so that workload remained approximately constant, and performance varied instead.

5. ACKNOWLEDGMENTS

We thank G. Lyddane, S. Robinson, D. Hannon, T. Carpenter-Smith, J. Bastow, F. Sheelen, J. Giurleo, and J. Turner, who all made important conceptual or technical contributions, and also our subjects. Protocol was approved by IRBs at MIT and DOT/Volpe. Supported by MIT CTS/DOT-Volpe Contract DTRS-57-92-C-0054 TTD#27A.

6. REFERENCES

Clement, W.F., Jex, H.R., Graham, D. (1968). A Manual Control-Display Theory Applied to Instrument Landings of a Jet Transport. IEEE Trans. Man Machine Sys. 9:4 93-110

FAA (1994) Technical Standard Order C129:Airborne Supplemental Navigation Equipment Using the GPS. FAA, Washington, DC.

Huntley, M.S. Jr. (1993) Flight Technical Error for Category B Non-Precision Approaches and Missed Approaches Using Non-Differential GPS for Course Guidance. Report DOT-VNTSC-FAA-93-17

Jansen, C.J. "Non-Gaussian Atmospheric Turbulence Model for Flight Simulator Research." AIAA J. Aircraft 19 (1981): 374-379.

Roscoe, A.H. and Ellis, G.A., (1990) A subjective rating scale for assessing pilot workload in flight. Tech. Report TR90019, RAE Farnborough, UK.

RTCA Special Committee 15. (1991) Minimum Operational Performance Standards For Airborne Supplemental Navigation Equipment Using Global Positioning System, DO-208 July, 1991, 1140 Connecticut Avenue, N.W., Washington, DC.

CAB-SIM: A COMPUTER CODE TO ANALYZE ERRORS OF COMMISSION FOR PROBABILISTIC RISK ASSESSMENT

A.P. Macwan*, F.J. Groen*, and P.A. Wieringa*

*Laboratory for Measurement and Control, Dept. of Mechanical Engineering, Delft University of Technology, Mekelweg 2, 2628 CD Delft, The Netherlands

Abstract. Intentional errors, especially, errors of commission are becoming increasingly important in safety analyses. The Human Interaction Timeline (HITLINE) methodology aims at analyzing and incorporating these errors into probabilistic risk analysis (PRA). A computer coded version of the HITLINE, called HIT-SIM, was developed using an object-oriented approach. An important component of HIT-SIM is CAB-SIM which simulates operator errors. Another component HIT-MERGE was developed to perform merging of resultant sequences. CAB-SIM can also be used in dynamic simulation techniques as a model of intentional operator errors.

Key Words. Probabilistic Risk Assessment, Human Reliability, Human Factors, Error Analysis, Safety Analysis, Computer Programs, Simulation

1. INTRODUCTION

Intentional errors, also referred to as cognitive errors, especially errors of commission, are receiving attention within risk and reliability assessment(Gertman, 1992). The methods currently used to perform human reliability analysis (HRA) are aimed at estimating a human error probability (HEP) for non-performance of a required task within a specified time-window. Thus, these methods are found inadequate to handle different types of errors, only perform limited analyses of dependencies and coupling factors among actions taken at different times, and lack treatment of dynamics of operator-plant interaction.

Recent research has led to development of simulation-based and analytical methods to analyze different errors, as well as models to generate these errors. Simulation based techniques include the continuous event tree technique (Devooght & Smidts, 1992) and dynamic event tree techniques, such as DYLAM(Cojazzi et al., 1993) , DETAM(Siu & Acosta, 1994), ADS(Hsueh & Mosleh, 1992). Analytical methods include the HITLINE technique(Macwan & Mosleh, 1994), the application of which is described in this paper.

Developments in modeling errors include enhancing existing models, such as THERP(Swain & Guttmann, 1983), SLIM(Embrey et al., 1984). Other researchers have focussed on developing cognitive models of operator behavior. These include CES(Woods et al., 1987), COSIMO(Cacciabue et al., 1992), and CO-COM(Hollnagel, 1993). The latter are too detailed for implementation in probabilistic framework, and are not intended for exhaustive error analysis.

2. BRIEF DESCRIPTION OF HITLINE

The objective of the HITLINE methodology is to systematically analyze causes and consequences of intentional operator errors. Errors are defined as deviations from operating procedures. For implementation to nuclear power plants, emergency operating procedures have been analyzed.

The HITLINE methodology consists of three steps. The first step is the screening of information about system availability, instrument failure, and operating crew related information, for analysis of intentional errors. Step 2 is the analysis, that is carried out by constructing HITLINEs. A HITLINE is a sequence of operator actions, including procedural instructions and deviations from the procedures. These deviations are generated by the CAB Model (Cause- based Behavioral Model). A HITLINE is, in principle, an event tree with multiple branches at each node. In the last step, results from HITLINEs are evaluated with respect to impact on the plant, and subsequently incorporated into the risk assessment.

One of the problems with forward branching tree techniques is the rapidly developing size of the tree(Siu, 1994). In developing the HITLINE, size is managed by

1. truncation, using probability thresholds of sequences;
2. termination of a sequence with known hardware configuration that the operator cannot change;
3. merging branches with similar impact in terms of operator errors.

3. HITLINE CODES

This paper describes the development of the computer code for the HITLINE analysis. Implementation of HITLINE into a computer code was done using an object-oriented-like approach. Coding was done in ANSI C(Kernighan & Ritchie, 1988). Three modules were used to accomplish three different objectives. The first module, HIT-SIM, is used to construct the HITLINEs. The CAB Model for generation of errors was implemented in the module CAB-SIM. During development of sequences, merged branches were stopped from further simulation. The merging operation was subsequently performed by the module HIT-MERGE.

3.1. *HIT-SIM (HITline SIMulation)*

HIT-SIM is based on codes used for dynamic reliability such as DYLAM and ADS, which generate dynamic event trees. These codes couple the stochastic behavior of hardware systems with deterministic behavior of process variables, and generate sequences of events in time. Besides handling stochastic behavior of hardware systems, these codes are capable of handling operator models. However, these codes have not incorporated the types of operator errors analyzed in the HITLINE methodology. Furthermore the operator error models they use are too simple.

The dynamic event trees can get very large because of the exponential growth of the number of sequences. The codes usually use truncation, i.e., cutting off sequences with very low probabilities. They also cut off sequences when the process conditions reach unacceptable levels. However, they are not capable of performing the merging operation that is an important feature of the HITLINE methodology. The user of HITLINE may merge a given branch. The code then terminates the branch, and labels it as "merged by user". Additionally, the user is also given an option to terminate branches that were judged to have guaranteed consequences. The code also ends these sequences and labels them as "terminated by user".

3.2. *CAB-SIM (CAB model SIMulation)*

At every branch point, HIT-SIM calls the CAB-SIM which generates information about operator actions necessary for construction of the HITLINE. CAB-SIM essentially simulates the CAB Model. However, some of the the mapping tables of the CAB Model were modified in order to simulate the model as a computer code. Examples of such modifications are discussed in section 4.

The inputs to the CAB-SIM are of three types: plant-related, which include process conditions, hardware status and alarms; procedure-related, which include characteristics of the procedural step and the system or parameter addressed by the step; and operator-related, including perceptions, expectations, etc. These inputs are processed through CAB-SIM as described in section 4. The outputs of CAB-SIM include execution directives about which procedure and which step to follow, and probabilities of branches needed by HIT-SIM. The actual execution in terms of which system to check or change status of, is done by another module HIT-EXEC (HITline EXECution).

3.3. *HIT-MERGE (HITline MERGEr)*

As explained earlier, the codes for dynamic event tree generation are incapable of merging sequences. Merging involves removing the sequence that is being merged and redistributing the probabilities over the sequences starting from the point where it was merged. An initial attempt showed that it would require lots of effort for the HIT-SIM to do such an operation. Thus, merging is accomplished after the HITLINE analysis is completed. HIT-SIM produces all the sequences including the ones which were terminated or merged by the user. A program, HIT-MERGE, was developed for the purpose, which reads the sequences generated by HIT-SIM and produces a condensed number of sequences after merging.

4. CAB-SIM IMPLEMENTATION

The most important part of HIT-SIM is CAB-SIM which generates operator errors at each branching point using the CAB Model.

4.1. *Object Representation*

The implementation of CAB-SIM, uses an object-oriented representation of the plant and emergency operating procedures (EOPs). The representation of the plant is done using three types of objects: systems, parameters and alarms. The information that is stored in these objects, is generated by CAB-SIM itself, or by other models within HIT-SIM.

The system objects represent the hardware components like pumps, valves et cetera. An important attribute of the system objects is the hardware status perception, in which CAB-SIM's perception about availability is stored. The parameter objects represent the process parameters such as pressure and temperature. In these objects, the values and trends calculated by the thermal-hydraulic models and instrument failures are stored. Attributes of the alarm objects include the status of the alarms, and related instrument failures.

The representation of the EOPs is done in EOP objects, which contain a number of EOP step objects. The EOP step objects contain the results of a classification of the steps that is performed prior to the simulations. HITLINE includes guidelines on such classification, and includes aspects such as type of EOP step, number of logical conditions, etc.

4.2. Simulation of Cognitive Processes

The CAB Model consists of simplified cognitive processes. Three processes are simulated:

1. Selection of systems and parameters
2. Comparison of expected and observed process behavior
3. Generation of internal diagnosis of initiating event

These steps are discussed below.

4.2.1. Selection of systems and parameters.

The selection of systems and parameters is included to simulate the fact that the operators cannot and do not continuously monitor the many parameters and systems associated with the plant. This is in accordance with the limited working memory concept(Woods et al., 1987).

In the CAB-SIM model, selection is done based on the operator's internal diagnosis state and the EOP-step being performed. For each accident class analyzed, a mapping table provides associated parameters and systems.

Selection based on EOP-step also consists of a mapping of EOP-step to parameters and systems. For each step, a list of parameters and systems that are addressed by it, is specified through pre-simulation evaluation of the EOPs.

4.2.2. Comparison of expected and observed process behavior.

This step is included to account for the expected courses of process parameters, based on the operator's understanding of the plant. The comparison of expected and observed behavior is done for every selected parameter.

In CAB-SIM, expectations about process behavior are defined in terms of trends. The range of trends is divided in three, partly overlapping, subranges: trend up, trend down, and trend zero, see Fig. 1. The input from the plant model is also divided in three ranges

The expectations are generated based on:

1. Internal diagnosis from the previous operator-plant interaction.
2. Execution of the previous EOP-step.

When a mismatch between expectation and observation is found, additional operator actions are executed. These are limited to checking instrument failure and verifying status of hardware systems, and result in an updating of CAB-SIM's perception of instrument and

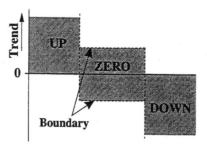

Fig. 1. Definition of trend areas.

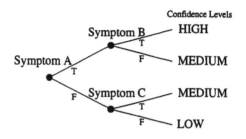

Fig. 2. Example of a diagnostic tree.

hardware state. Actions which change hardware states such as turning a pump off are not included here. These are included in HIT-EXEC.

4.2.3. Generation of internal diagnosis of Initiating Event.

During an incident situation, the operators combine their knowledge about process behavior and the information received from the plant, to arrive at a diagnosis about the root cause of the situation. This internal diagnosis has significant influence, as it affects selection of parameters as well as expectation about plant behavior.

The CAB Model consists of mapping tables in which combinations of symptoms result in associated confidences for the different accident classes. Thus, the model limits itself to a limited number of predefined accident classes. The approach was modified for implementation in CAB-SIM, and consists of the following steps:

1. Assignment of confidence levels
2. Generation of global diagnosis state
3. Update of global diagnosis state

For assignment of confidence levels, three levels are defined: low, medium and high. The assignment of confidence levels to the accident classes is done using predefined symptoms. For each accident class, a binary tree structure, like the one shown in Fig. 2, is predefined. The nodes of the tree are formed by pre-selected (diagnostic) symptoms. Each node has two branches, one for presence and one for absence of the symptom. At the end of the search, a confidence level is determined. CAB-SIM searches through these tree structures for each accident class.

Table 1 Global diagnosis state as a function of confidence levels

Occurences			Global Diagnosis
Low	Med	High	
≥ 0	0	0	None
≥ 0	1	0	Exclusive
≥ 0	> 1	0	Proportional
≥ 0	≥ 1	1	Proportional
≥ 0	0	1	Exclusive
≥ 0	≥ 0	> 1	Confusion

Table 2 Updated global diagnosis state as a function of previous and current states.

Global Diagnosis State		
Previous	Current	Updated
Unknown	any	Current
Exclusive	Unknown	Previous
"	Proportional	Prev./Curr.
"	Exclusive	Current
Proportional	Unknown	Current
"	Proportional	Current
"	Exclusive	Current
Confusion	any	Prev./Curr.
any	Confusion	Current

After confidence levels have been found for each accident class, they are combined to generate a global diagnosis state. The global diagnosis state is an additional component that was not defined in CAB Model. It can be one of the following:

- *None:* Indication for no accident class
- *Proportional:* Weak indications for multiple accident classes
- *Exclusive:* Strong indications for one accident class
- *Confusion:* Strong indications for more than one accident classes

Table 1 shows how the global diagnosis state is found as a function of the confidence levels.

In CAB-SIM, memory is accounted for by combining the current diagnosis state with the previous one, resulting in an updated diagnosis state. This helps to model phenomena such as fixation on one diagnosis or recovery from a misdiagnosis(Macwan & Mosleh, 1994). In most cases, updating is entirely based on the previous and current global diagnosis state. The updating rules are presented in Table 2. In two cases, further evaluation of diagnosis states is necessary:

- When the previous state was "exclusive", and the current is "proportional", the updated state is "exclusive" if the associated accident class again has a medium or high confidence level. Otherwise, it is taken as proportional.
- When the previous diagnosis was "confusion", but the symptoms have been indicating exactly one accident class for two consecutive CAB-SIM runs, the updated global diagnosis state changes to "exclusive". In other cases, the diagnosis remains "confusion".

The updated diagnosis state is used as CAB-SIM's internal diagnosis state.

4.3. *Quantification*

For generation and quantification of operator actions, CAB-SIM uses a scheme in which the type of procedural instruction is mapped to a predefined set of deviations from the instruction. Translation of the results of the cognitive processes into numbers is done through the use of Performance Influencing Factors (PIFs). First, a set of PIFs is assigned as a function of different inputs from the plant, procedure and operator, as explained below. Subsequently, this set is used to retrieve a set of error factors, which are used to calculate probabilities for the selected operator actions.

4.3.1. *Performance Influencing Factor Assignment.*
CAB-SIM uses the same PIFs as the original CAB Model. Basically, the PIFs are divided into scenario-independent (ScI) and scenario-dependent (ScD) PIFs.

The ScI PIFs do not change considerably during the evolution of an accident. Thus, their values can be assigned as a boundary condition. A more detailed discussion can be found in (Macwan & Mosleh, 1994). The ScI PIFs include:

- Crew training and experience
- Crew confidence
- Relative experience of reactor operator and senior reactor operator
- Recent experience with faulty signal(s)

As the name indicates, the scenario-dependent PIFs do change during the course of an accident situation. Hence, values must be assigned to them each time the model is called. This is partly done through the use of lookup tables, and partly through the use of rules as described below.

The EOP related ScD PIFs are looked up in tables and include the following:

Phase of EOP: The EOPs used in nuclear power plants are divided into five phases: Verification, Diagnosis, Reverification, Rediagnosis and Recovery(Macwan & Mosleh, 1994). A sixth phase called Success was added to simulate successful performance of a step, in order to limit the number of sequences.

Type of instruction: This PIF is kept the same as for CAB Model.

Number of logical conditions: This PIF is kept the same as for the CAB Model.

The other ScD PIFs, all binary, are plant related and

Table 3 Error factor lookup table for reverification phase.

OD_1	PD_1/PD_2	α_s	α_{proc}	α_{scl}
0	0	0.995	0.98	0.1
0	1	0.985	0.85	0.9
1	0	0.990	0.95	0.1
1	1	0.980	0.90	0.9

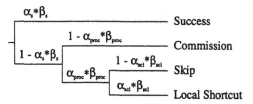

Fig. 3. Quantification scheme for reverification phase.

operator related, and are assigned by rules as follows:

Value of critical parameters: Is set to 1 for when a selected parameter exceeds constant, predefined boundary values.

Trend of critical parameters: Is defined and set similar to the previous PIF.

Instrument Failure: This PIF is set to 1 if for any system or parameter, an instrument failure has occured and is not detected.

Operator-related PIFs simulate the mental state of the operator and are implemented as follows:

Operator Diagnosis: This PIF is set to 1 when the global diagnosis state is "exclusive" or "proportional".

Memory of recent actions: This PIF is set to 1 when perception state is "unavailable" for no system, but "verified available" for at least one system.

Perceived Importance: When any system is selected based on both the diagnosis and EOP step, this PIF is made 1.

Perceived consequences: This PIF is not implemented in CAB-SIM.

Expectation mismatch: This PIF was not explicitly mentioned in CAB Model. It is set to 1 when a mismatch between expected and observed behavior occurs for any parameter.

4.3.2. Error Factor retrieval. The PIF values are used to determine a set of error factors through the use of lookup-tables, see Table 3. In this table, OD_1, PD_1 and PD_2 represent three PIFs, the α-s are the error factors(Macwan & Mosleh, 1994). These tables are defined for each EOP phase separately, with the phase of EOP serving as the table selection criterium. CAB-SIM transforms the binary PIF values into an index number that is unique for each PIF combination. With this number, a set of error factors is retrieved. The current version of CAB-SIM does not use the last two PIFs related to EOPs, because they are assumed not to influence the intentional error modes.

4.3.3. Weight calculation. The weights for the different operator actions are found by filling the error factors into quantification schemes like the one shown in Fig. 3. These schemes are developed such that the weights of all actions always add up to 1. A quantification scheme is available for each EOP phase since each phase has an associated set of errors(Groen, 1995).

Additional calculation is necessary when weights are assigned to global short cut branches, which represent a premature entry into an EOP associated with an accident class. Shortcuts are generated for each accident class that is part of CAB-SIM's internal diagnosis. The probability is distributed over the accident classes using a weight factor 0.8 for ones with high confidence and 0.2 for ones with medium confidence.

4.4. Output generation

The results of the CAB-SIM model need to be communicated to HIT- SIM, to keep track of how many HITLINE branches to generate and with what probabilities. The main piece of information for HIT-SIM is the weight of the action, to calculate the probabilities for the generated sequences of events. Further, the contents of some of CAB-SIM's internal registers, like diagnosis state and perceptions, are passed to HIT-SIM. These are used as inputs for the next time CAB-SIM is called by HIT-SIM.

Additionally, information is also given to HIT-EXEC execute the appropriate step. These instructions include the EOP step plus the error mode directives. The EOP step is determined by CAB-SIM based on the current EOP step and the type of action, and tells HIT-EXEC which place in the EOPs to go to. For example, skipping a step is simulated by increasing the EOP step by 1. For a shortcut, HIT-EXEC enters the EOP for the associated accident class.

The error mode directives consist of four flags that guide HIT-EXEC in performing the steps. These are:

Execute: If this flag is not set, HIT-EXEC does not perform the step but returns to the simulation controller. This is necessary in case the error consists of a jump to a new position in the EOPs, so that CAB-SIM can be executed before entering the new step.

Delay: If this flag is set, HIT-EXEC uses longer than nominal execution time for the step.

Misdiagnosis: This flag can be set for EOP steps which ask the operator to perform a diagnosis. If this flag is set, HIT-EXEC performs the step incorrectly. For a step related to diagnosis of the simulated accident class, HIT-EXEC does nothing, while for a step related to other accident classes, HIT- EXEC chooses the accident class of the step as a diagnosis.

Change diagnosis: If this flag is set, HIT-EXEC leaves the current EOP, and enter another one, without performing any steps. The EOP to jump to depends on the (rediagnosis) step for which the change of diagnosis was made, and is defined in HIT-EXEC.

These directives are sufficient to let HIT-EXEC implement all the errors that are generated by CAB-SIM.

5. RESULTS

The programs described above were implemented on a hypothetical system consisting of two tanks with one coolant loop through each, of which one tank had internal heat generation. First-order differential equations were used to model heat transfer through the coolant loops. Two possible leaks were defined. Operating procedures similar to emergency operating procedures (EOPs) used in nuclear power plants were developed for the system. A simple risk assessment model was developed. The detailed description of the system, as well as the results of screening process, HITLINE analysis and their incorporation into event and fault tree are explained elsewhere (Macwan et al., 1995).

6. CONCLUSIONS

The HITLINE methodology for analysis of operator intentional errors has been implemented as a computer code HIT-SIM. In this form, it can be used to analyze and incorporate these errors into probabilistic risk assessments. HIT-SIM has the potential for applications in several industries. CAB-SIM has successfully been applied as an operator error model in dynamic simulation techniques and seems a good alternative for the ones currently in use.

ACKNOWLEDGEMENT

Part of the work was performed under the project "Human Performance in Fault Management Tasks" under the Dutch Ministry of Social Affairs (contract 220917), and Dutch Ministry of Health and Environment (proj.nr. 711221). The opinions, findings, and conclusions expressed are those of the authors and do not necessarily reflect those of the ministries.

7. REFERENCES

Cacciabue, P.C., et al. (1992). A Cognitive Model in a Blackboard Architecture: Synergism of AI and Psychology. in *Reliability Engineering and System Safety*, **36**, pp. 187- 197.

Cojazzi, G., P.C. Cacciabue, and P. Parisi. (1993). *DYLAM-3: A Dynamic Methodology for Reliability Analysis and Consequence Evaluation in Industrial Plants; Theory and How to Use*, ISEI/SER 2192/92 Joint Research Center, Ispra, Italy.

Embrey, D.E., P.C. Humphreys, E.A. Rosa, B. Kirwan, and K. Rea. (1984) *SLIM-MAUD: An Approach to Assessing Human Error Probabilities Using Structured Expert Judgment*, NUREG/CR-3518, U.S. NRC, Washington, D.C.

Devooght, J., and C. Smidts. (1992). Probabilistic Reactor Dynamics-I. The Theory of Continuous Event Trees. in *Nuclear Science and Engineering*, **107**, p. 229.

Gertman, D.I., et al. (1992). INTENT: A method for estimating human error probabilities for decision based errors. in *Reliability Engineering and System Safety*, **35**, 127-136.

Groen, F.J. (1995). CAB-SIM: a cause based operator model for use in dynamic PSA for nuclear power plants, M&R A-713, Delft University of Technology.

Hollnagel, E. (1993). *Human Reliability Analysis; Context and Control*. Academic Press, London.

Hsueh, K.S., and A. Mosleh. (1992). An Integrated Simulation Model for Plant/Operator Behavior in Accident Conditions, UMNE 92-004, University of Maryland.

Kernighan, B.W., and D.M. Ritchie. (1988). *The C Programming Language, Second Edition*. Prentice Hall.

Macwan, A., and A. Mosleh. (1994). A Methodology for Modeling Errors of Commission in Probabilistic Risk Assessment. in *Reliability Engineering and System Safety*, **45**, 139-154.

Macwan, A., P.A. Wieringa, and A. Mosleh. (1994). Quantification of multiple error expressions in following emergency operating procedures in nuclear power plant control rooms. in *Proceedings of PSAM-II*, San Diego, pp. 066-15 - 066-20.

Macwan, A., F.J. Groen and A. Mosleh. (1995). Analysis of errors commission and other intentional errors using HITLINE methodology: Application to a two tank system. To appear in *Proceedings of Workshop on Advanced Human Reliability Models: Theoretical and Practical Challenges*, Stockholm.

Siu, N., and C. Acosta. (1994). Dynamic Event Trees in Accident Sequence Analysis: Application to Steam Generator Tube Rupture, *Reliability Engineering and System Safety*, **41**, pp. 135-154.

Swain, A.D., and H.E. Guttmann. (1983). *Handbook of Human Reliability Analysis with Emphasis on Nuclear Power Applications*, NUREG/CR-1278, U.S. NRC, Washington, D.C.

Woods, D.D., E.M. Roth, and H.E. Pople, Jr. (1987). *An Artificial Intelligence Based Cognitive Environment Simulation for Human Performance Assessment*, NUREG/CR-4862, U.S. NRC, Washington, D.C.

METHOD TO IDENTIFY COGNITIVE ERRORS
DURING ACCIDENT MANAGEMENT TASKS

V. Gerdes

Delft University of Technology, Department of Mechanical Engineering,
Laboratory for Measurement & Control, Man-Machine Systems Group, The Netherlands
Detailed information for correspondence:
N.V. KEMA, BU-PCL, Utrechtseweg 310, 6812 AR Arnhem, The Netherlands
Phone +31 85 563034, Fax +31 85 515456

Abstract: A cognitive task decomposition is described that comprises components such as
diagnosis and planning, as well as alternative thinking modes. A dynamic model of the
cognitive process describes how the components interact. Then "error evaluation diagrams"
are developed that represent potential erroneous outcomes of different cognitive levels. The
combined results of these diagrams construct a cognitive error classification. This consists of
a set of basic error types that can be applied to the different components. The error
classification is turned into a flowchart that helps to systematically identify cognitive errors.

keywords: Cognitive systems, classification, decision making, error analysis, human
reliability, human supervisory control, identification, modelling errors, operators

1. INTRODUCTION

Errors in the "thinking process" of operating crews
have been shown to contribute significantly to the
overall risk of industrial plants (Kauffman et al.,
1992). These thinking errors are also referred to as
"cognitive errors". At this moment the methods
available to identify and analyze such errors are
scarce and premature (Gertman, 1991; Dougherty,
1992). In cognitive psychology, a human is modeled
as an information-processing system. From this,
cognitive errors are defined here as: *Errors in the*
intention-formation process. The intention-formation
process comprises everything between noticing that
there is a need for a task (detection) and the actual
performance of the physical interactions with the plant
to achieve this task (execution).

This paper presents a method to identify cognitive
errors (MICE) during accident management tasks.

MICE is developed in four steps: cognitive process
configuration - error evaluation diagrams - cognitive
error classification - identification flowchart. Each
step is described in the sections below.

2. DESCRIPTION OF MICE

2.1   Cognitive process configuration

First, the configuration of an accident management
task is described. The task is to adequately manage an
accident scenario that requires conscious problem-
solving thinking. Various cognitive models have been
developed (Rasmussen, 1976; Woods, *et al.*, 1990;
Cacciabue, *et al.*, 1992). Gerdes (1994) presents an
overall task decomposition that focuses on the
cognitive phase. This is shown in Figure 1, followed
by a short explanation of the different ingredients
(modified from Gerdes, 1994).

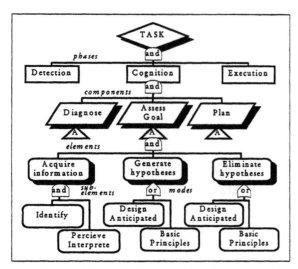

Fig. 1. Cognitive task decomposition (Gerdes, 1994). The structure below the middle "A"-triangle applies for "Diagnose" and "Plan" as well.

For the overall task, three **phases** can be distinguished, not necessarily separated or sequential:
- **Detection** Notice the need for a task.
- **Cognition** Generate the intention and strategy to carry out a task.
- **Execution** Physically perform the actions.

The detection and execution phases are not analyzed in more detail in this study. Within the cognitive phase three **components** are distinguished:
- **Diagnose**   Assess the current system state.
- **Assess Goal**   Define the target (desired) state.
- **Plan**   Specify a strategy to accomplish the transition from current state to target state.

For each component, three **elements** are required:
- **Acquire information**

Typically, various information-"items" are required, originating from several sources.
- **Generate hypotheses**

Find a set of candidate hypotheses. Often, there are several "close-enough" solutions.
- **Eliminate hypotheses**

Eliminate some of the hypotheses according to some set of criteria.

On the lowest level of Figure 1, two **sub-elements** for acquiring information are presented:
- **Identify**

Assess what information is relevant to acquire.
- **Perceive/Interpret**

Sense and comprehend this information. These two functions are taken together in one sub-element for practical reasons; failure of these functions can hardly be distinguished in practical applications.

For the generation and elimination elements two **modes of thinking** for solving a query exist:
- **Design Anticipated (DA-mode)**

The DA-mode should cover all events that occurred in the past as well as all events that have been identified before they actually occurred. It makes use of external sources such as written procedures and oral in-

structions and of internal sources such as long-term memory and learned heuristics.
- **Basic Principles (BP-mode)**

The BP-mode should cover all events not covered or not resolved by the DA-mode ("beyond design accidents"). It makes use of the mental representation of the fundamental physical relations of the process. These two thinking modes also relate to concepts described by Wickens (1992) and Reason (1990). A more detailed description of these ingredients can be found in Gerdes (1994).

The interaction of the phases, components and elements described above is depicted in Figure 2.

The cognitive process model

After the need for a task is detected, the current system state is assessed by the diagnosis component. This component starts with a search for situation-relevant information. From this point a three-loop process starts:

1. The acquired information is used to generate hypotheses about the current system state. How this generation of hypotheses takes place is not depicted in figure 2 (the DA and BP thinking modes). If no hypothesis can be generated, an additional search for information may be warranted, thereby reiterating the generation process.

2. In the elimination element the hypotheses generated can be checked against hypothesis-specific information (also available from the acquired information). If no hypothesis is considered acceptable ("all hypotheses eliminated"), another run of hypotheses generation can be carried out. If this generation run doesn't provide new hypotheses, the first loop may be entered again.

3. Another possibility is that the elimination element cannot make a choice between alternative hypotheses about the current system state ("multiple-hypotheses-retained"). This may also warrant additional, more decisive, information that reiterates the elimination process.

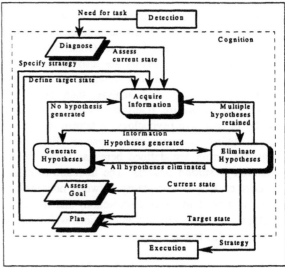

Fig. 2. Process model of accident management task

274

These three loops are active until one hypothesis is retained by the elimination element. This hypothesis becomes the solution for the query about the current system state. Based upon this current state a target state has to be assessed by the "assess goal" component. The three-loop process to assess what is desirable and what is not is analogous to the previous described process for the diagnosis component. If successful, the three-loop process results in a target state that is used, together with the current system state, to define a strategy that accomplishes the transition between these states. For this planning component again the three-loop process is carried out, that, if successful, results in a strategy that can be executed.

What can go wrong?

Obviously, one can get stuck in one of these loops. This means that no hypothesis is generated at all, all generated hypotheses are eliminated, or multiple hypotheses remain indecisive. Another possibility is that one of the elements provides a wrong result, so, not enough or wrong information, wrong hypothesis generated, or wrong hypothesis retained. All these erroneous possibilities can be represented by the sub-elements and thinking modes of the cognitive elements, as described in Figure 1. This results in the error evaluation diagrams of section 2.2.

2.2 Error evaluation diagrams (EED)

The error evaluation diagrams (EEDs) are derived by evaluating the "not-wrong-right" results of every factor in the cognitive task decomposition. In Figure 3 the need and the possible results of the cognitive phase are evaluated. The final result "Omission: no cognition" is inserted in the cognitive error classification right away. The result "Commission: to components" indicates a wrong cognition phase that is evaluated in more detail at a component level.

For the three components of the task decomposition analogous EEDs apply. These are not presented here because of lack of space. Of the component EEDs both the omission and commission results are evaluated in more detail at the element level. For the

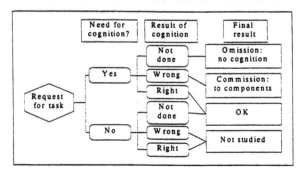

Fig. 3. Cognitive phase error evaluation diagram

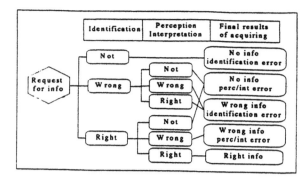

Fig. 4. Information error evaluation diagram

"Acquire information" element the EED is pretty straightforward. The final results and their derivation can be easily traced back in Figure 4.

This is not so simple for the generation and elimination elements. These two elements have a high level of dependency and cannot be treated separately. Figure 5 presents the EED of the generation and elimination elements, build from the possible results of the DA and BP modes of thinking.

The generation element of this EED is pretty straightforward; both modes of thinking can have no, a wrong, or a right result. In Figure 5 the DA mode of thinking precedes the BP mode. Because they are true alternatives (OR-gate in Fig. 1) and because there is a natural preference for the less strenuous DA mode of thinking (Wickens, 1992; Reason, 1990), this mode is represented first. This has no further consequences for the progression of the EED. The "wrong" result of a generation thinking mode signifies that *one or more* wrong hypotheses are generated. Analogous for the "right" result.

The results of the thinking modes of the elimination element are to be interpreted as a result of "hypotheses retained". This means that "None" indicates that no hypothesis is retained, hence, that all hypotheses available from the generation element are eliminated. "Multiple" indicates that more than one hypothesis was generated but that this set is not reduced to one or none by an elimination thinking mode. The "Wrong" and "Right" results of the elimination element indicate that *only one* wrong or right hypothesis is retained.

The EEDs of Figures 4 and 5 apply to the bottom level of Figure 1. The EEDs derived so far are the intermediate step to arrive at the cognitive error classification.

2.3 Cognitive error classification

The final results of the EEDs are arranged into a cognitive error classification. First steps in this

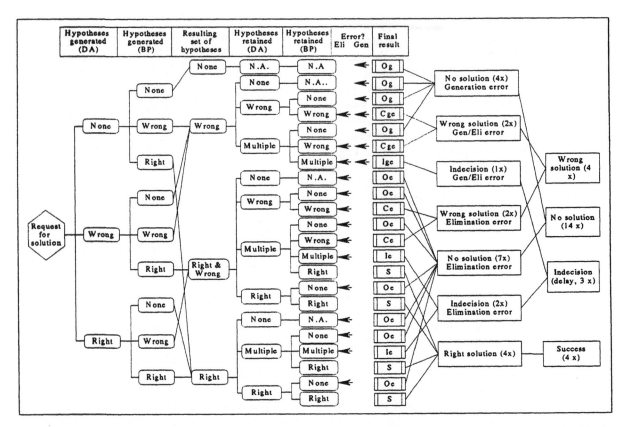

Fig. 5. Error evaluation diagram of the generation and elimination elements. The "hypotheses retained" blocks indicate the elimination element. eli=elimination, gen=generation. N.A.=Not Applicable. The codes are as follows: O=Omission, C=Commission, I=Indecision, S=Success, g=generation, e=elimination.

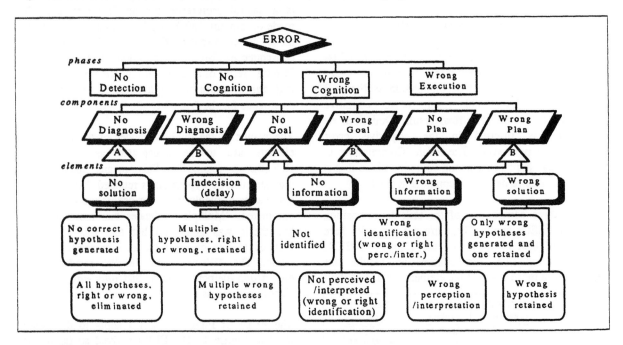

Fig. 6. The cognitive error classification. The structures below the "A" and "B"-triangles apply for the corresponding other triangles as well.

direction were presented in Gerdes (1994). The final classification is presented in Figure 6.

The "No Cognition" at the phase level stems directly from **Figure 3**. For the other phases "No Detection

(omission)" and "Wrong Execution (slips)" are shown for the sake of completeness, although these are not cognitive errors. The "Wrong Cognition" refers to the "Commission: to components" result in Figure 3.

At the component level in Figure 6 "No ..." and "Wrong ..." errors exist, resembling the omission and commission results of the component EEDs. Evaluating these errors in more detail reveals the element level in Figure 6. The "No Solution", "Indecision (delay)", and "Wrong Solution" errors are directly available from Figure 5. The six basic generation/elimination errors they comprise can also be derived from the six results displayed in Figure 5 (excluding "Right solution"). Something similar goes for the "No Information" and "Wrong Information" errors in Figure 6, that originate from Figure 4.

It is concluded that the cognitive error classification comprises 11 basic errors; 6 basic generation/elimination errors and 4 basic information errors at the lower level, and 1 omission of cognition at the higher level. The 4 basic information errors can both result in a "no..." and a "wrong..." component error. If the 10 lower basic error types are extended to all the 3 components, a set of 42 different errors results (3 x 14; the information errors count twice). So, including the omission of cognition, the total cognitive error set contains 43 different errors, built from 11 basic error types.

2.4   Identification flowchart

Based on the previous sub-sections a flowchart is devised. The flowchart can be used for incident analyses to identify and classify cognitive errors that may have been influential. The flowchart is shown in Figure 7, where it is assumed that there is a need for cognition. The chart guides to errors at the phase, component, element and sub-element level, thereby covering the full cognitive error classification of Figure 6. The two modes of thinking are not embodied in the flowchart. To assess whether it was a DA or BP mode error one has to investigate the specific circumstances in detail. However, it is very difficult to attribute an error either to the DA or to the BP mode of thinking; one can only suspect, given the prevailing influencing factors (with proper procedures, training, frequent experience, direct instruction, etc., a DA solution becomes more viable).

The flowchart is constructed in such a way that multiple errors within one incident can be dealt with. These multiple errors must be independent because an erroneous result of an element that mostly depends on previous erroneous outcomes in the cognitive process should not be called an error! A wrong generation of hypotheses due to wrong or insufficient information cannot be attributed to the generation element, but this should be classified as an information error. Similar arguments count for other errors as well: the primary error is decisive. This pitfall is settled by the "Related to previous error(s)?" question in the flowchart. The "ELEMENTS" box may be skipped for various reasons such as time constraints of the investigation (dotted line). In that case the flowchart identifies and classifies errors at a component level.

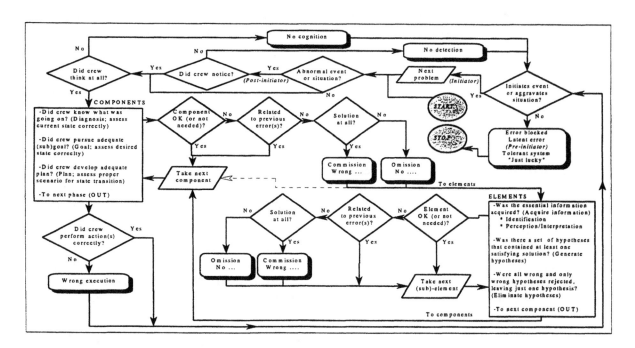

Fig. 7. Cognitive error identification flowchart

277

3. CONCLUSIONS AND PERSPECTIVE

3.1 Conclusions

This paper describes the transition from a cognitive task decomposition into a cognitive error classification. Furthermore, this cognivtive error classification includes the Basic Principles mode of thinking explicitly. Previous cognitive models and error classifications focused mainly on the Design Anticipated mode of thinking (Reason, 1990; Gertman, 1991).

The method to Identify Cognitive Errors (MICE) essentially consists of two main products;
- a four-level cognitive error classification that contains 43 different errors, build from 11 basic error types, and
- a flowchart to identify cognitive errors that results in a decomposed representation of, possibly multiple, errors.

For retrospective incident analyses application of the flowchart is effective. For prospective analyses (risk analyses, design evaluations, etc.) a more global survey of all possible cognitive errors is required. For this, all the 43 different cognitive errors of the classification should be checked for relevancy.

The other products that underlie MICE;
- a four-level cognitive task decomposition combined with a cognitive process model, and
- six error evaluation diagrams that evaluate erroneous results at all levels of the cognitive task.

The results of a MICE analysis can be interpreted with these two underlying products. This more fundamental insight gives direction to an effective root-cause analysis.

3.2 Perspective

Because human errors are dependent on the context in which they occur, classifying the errors is not enough to analyze them. To obtain insight into the reasons *why* a particular error occurs, the circumstances that provoke this error should be evaluated. In the field of human reliability the relevant circumstances are generally represented by a set of influencing factors such as training, recent experience, time pressure, crew issues, etc. Until now, the selection and definition of influencing factors for a certain situation has been controversial. A structured classification of pertinent factors that influence human cognitive performance will be presented elsewhere.

For every cognitive error a specific set of influencing factors wiil be designed, presented in a mapping diagram that describes which factors are most relevant for which errors. The overall method to identify and analyze cognitive errors then consists of three units (Gerdes, 1994);
- a classification of cognitive errors (presented in this paper),
- a classification of influencing factors, and
- a mapping diagram that combines these two classification.

Experiments that justify some issues and case studies that illustrate the application of the method are planned. MICE has been applied to several real cases. The results of this pilot study are very promising.

REFERENCES

Cacciabue P.C., Decortis F., Drozdowicz B., Masson M., Nordvik J. (1992) COSIMO: A Cognitive Simulation Model of Human Decision Making and Behavior in Accident Management of Complex Plants. *IEEE Transactions on Systems, Man, and Cybernetics*, **22, no. 5**: pp 1058-1074

Dougherty E.M. (1992) Context and Human Reliability Analysis. *Reliability Engineering and System Safety* **41**, edt. Apostolakis, G.E. Elsevier Science Publishers, ISSN 0951-8320: pp 25-47

Gerdes V.G.J. (1994) Towards a Method to Analyze Cognitive Errors during Accident Management: A taxonomy of cognitive errors. *Proceedings of the XIII. European Annual Conference on Human Decision Making and Manual Control* (Annual Manual), Espoo 13-15 June 1994, Finland.

Gertman D.I. (1991) INTENT: A Method for Calculating HEP Estimates for Decision based Errors. *Visions. Proceedings of the Human Factors Society 35th Annual Meeting*, San Fransisco, California, Sept. 2-6: pp 1090-1094

Kauffman J.V., Lanik G.F., Spence R.A., Trager E.A. (1992) *Operating Experience Feedback Report - Human Performance in Operating Events*. Office for Analysis and Evaluation of Operational Data, NUREG-1275, Vol. 8, US-NRC, Washinton D.C.

NUREG/CR-3114, (1982) *Workshop on Cognitive Modelling of Nuclear Plant Control Room Operators*. Proceedings of workshop, Dedham, Massachusetts, August 15-18, 1982

Rasmussen J. (1976) Outlines of a Hybrid Model of the Process Operator. In: *Monitoring Behaviour and Supervisory Control*, Sheridan and Johannsen (eds.), New York: Plenum Press.

Reason J. (1990) *Human Error*.-Cambridge University Press, ISBN 0-521-30669-8: 302 pp

Wickens C.D. (1992) *Engineering Psychology and Human Performance*. HarperCollins Publishers Inc. ISBN 0-673-46161-0: 560 pp

Woods D.D., Pople H.E., Roth E.M. (1990) *The Cognitive Environment Simulation as a Tool for Modeling Human Performance and Reliability*. NUREG/CR-5213, US-NRC, Washington D.C.

MODELLING ORGANISATIONAL FACTORS OF HUMAN RELIABILITY IN COMPLEX MAN-MACHINE SYSTEMS

W. van Vuuren
T.W. van der Schaaf

Safety Management Group
Graduate School of Industrial Engineering and Management Science
Eindhoven University of Technology
P.O. Box 513, Pav. U-42
5600 MB Eindhoven, The Netherlands

Abstract: This paper will give an overview of two research projects that were carried out in the Dutch chemical process and steel industry. The main goal was to show the importance of modelling organisational factors of human reliability in complex man-machine systems. Too often only human and technical factors are seen as the main contributors to industrial safety. The two research projects have shown that organisational factors cannot be neglected and that an insufficient analysis of incidents (accidents and near misses) is a prime reason for not acknowledging these organisational factors.

Keywords: chemical industry, human reliability, modelling errors, organisational factors, steel industry.

1. INTRODUCTION

Research into the causes of accidents in complex man-machine systems has always played an important role in the improvement of these systems. Such research used to look only at safety, but nowadays aspects as system reliability, quality and environment play an important part too. If the causes of accidents are subdivided into three groups (technical, human and organisational), it appears that at first investigators only looked at *technology* to take corrective actions. Later they became interested in *human behaviour* as a cause of failure, and only this last decade the importance of *organisational and management factors* as a cause of incidents has been accepted (Reason, 1991). However, this growing societal, industrial and scientific interest in the organisational causes of system failure still has not lead to any widely accepted explanatory theoretical model. Neither are there any practical tools that can be used in finding, describing, classifying and correcting

these organisational factors. The few real attempts that have been made till now came from researchers with a psychological background (Wagenaar, *et al.*, 1995), or from projects concerning technical risk analysis.

Because of this lack of research, many questions are still unanswered. In this paper the focus will be on the following two research questions:

1. How significant is the contribution of organisational factors, relative to human and technical factors, as a cause of safety related incidents?

The two projects that will be discussed in this paper were started with the expectation that this contribution is significant and should not be neglected. However in an overview given by Wagenaar (1983) the predominance of the human factor (80 to 100%) in accident research is clearly demonstrated. This leads to the second research question:

2. Why does almost all in-company accident research show us the opposite, namely that only human and technical factors are important?

It was predicted that it takes more than just a global investigation to come up with organisational causes. Technical and human types of failures are rather easy to detect, because the time between these failures and the real incident is usually very limited. Using the distinction made by Reason (1990) these failures would be called: 'active errors', because the effects of active errors are felt almost immediately. Organisational failures on the other hand are normally made much earlier in time and consequences lie dormant within the system, waiting for the right circumstances to become evident. Reason (1990) would call this type of failures: 'latent errors'. Because of this time delay organisational failures are much harder to detect. Therefore, in one project also the way of investigating incidents was examined. It is also very important that management acknowledges these organisational factors. Management can easily influence the way incidents are investigated, and thus the outcome of these investigations. Therefore, in these projects the investigations were done by an independent outsider.

To answer the research questions two projects were carried out by the Safety Management Group of the Eindhoven University of Technology. The first project was carried out in the Dutch steel industry and concentrated on the first research question. The second project was carried out in the Dutch chemical process industry with a main focus on the second research question. Besides answering the two research questions, the data collected in the two projects were also used to develop tools to describe and classify the organisational factors.

Both projects had practical as well as scientific aims. The practical aim was to implement a near miss reporting system to improve the safety level by gaining more insight in the causes of incidents. The following definition of a near miss was used: "A near miss is any situation in which an ongoing sequence of events was prevented from developing further and hence preventing the occurrence of potentially serious (safety related) consequences (Van der Schaaf, et al., 1991)". This practical part will not be discussed in this paper. This paper will only focus on the scientific aims of the two research projects.

2. STEEL INDUSTRY PROJECT

2.1 Plant profile

The first project was carried out in the Dutch steel

industry (Van Vuuren, 1993). For practical reasons, such as the considerable size of the total steel plant, only the coking plant was asked to participate in the project. The coking plant is the producer of cokes for the steel making process. The plant also purifies the gas that is originated by the coking process to keep the air emissions well below the levels permitted by the law, and to regain valuable substances from the gas. There are approximately 300 employees working in this coking plant, divided in two groups: a production and a maintenance group. Although the coking plant is only a part of the total steel plant it can be seen as a single plant, working independently from the other parts.

2.2 Aims of the research project

The scientific aims were to use the collected data:
1. to answer the first research question: how significant is the contribution of organisational factors, relative to human and technical factors, as a cause of safety related incidents?
2. to develop tools to describe and classify the organisational factors.

2.3 Method

Because implementing a near miss reporting system was the practical aim, near misses were used as input for reaching the scientific aims. Near misses are usually estimated to occur one or two orders of magnitude more frequently than actual accidents. So by using near misses, it is possible to collect a lot of information within a short period. The assumption is made that near misses have the same causes as real accidents.

The near misses were reported voluntarily in the setting of confidential 'critical incident interviews'. This technique is based on the 'critical incident technique' (Flanagan, 1954), a technique originally developed to trace the 'critical job requirements', for purposes like training and selection of personnel. In a similar way it is possible to look at the critical activities or decisions in the development of a near miss. The interviews were carried out on a confidential basis, to make the threshold to report and to talk about the near misses as low as possible. A selection was made to get an equal number of 'maintenance' and 'production' near misses.

After interviewing the people involved in the near miss, the information was used to build a 'causal tree' (see fig. 1). Causal trees, derived from fault trees, are very useful to present the critical activities and decisions in a chronological order, and also show how all the different activities and decisions are logically related to each other. For near misses the causal trees are divided in two parts: the 'failure side' (left side in fig. 1), which gives an overview of all the activities that have lead to the failure, and the 'recovery side' (right side in fig. 1), which gives an overview of the activities that prevented it from developing into a real accident. The importance of recovery will be discussed by Van der Schaaf (1995).

By using causal trees it becomes clear that there is never one single cause of a near miss, but that it is always a combination of many technical, organisational and human causes. The 'root causes' of the near miss, which are found at the bottom on the failure side of the causal tree, are the main products of the analysis. These root causes should be used to decide which preventive measures to take.

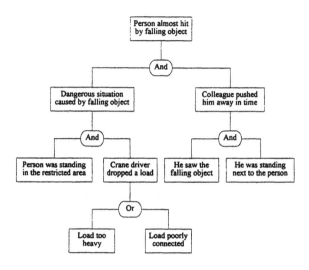

Fig. 1. Example of a causal tree describing a near miss.

To be able to take these measures, it is first necessary to classify the root causes. Without classifying the root causes, the information will be too descriptive and therefore too unclear to use. By classifying the root causes only a limited number of technical, organisational and human categories remain. As a starting point the Eindhoven classification model (Van der Schaaf, 1992) was used (see fig. 2). This model, originally developed in the chemical process industry with the main focus on the human part, had to be adjusted to the steel industry, and also an extension of the organisational categories was needed.

To adjust and extend the classification model the technique of analytic generalisation was used (Yin, 1991). Analytic generalisation uses the following iterative nature of 'explanation building':

* make an initial statement or an initial proposition;
* compare the findings of an initial case against such a statement or proposition;
* revise the statement or proposition;
* compare the revision to the facts of the second, third or more cases; and

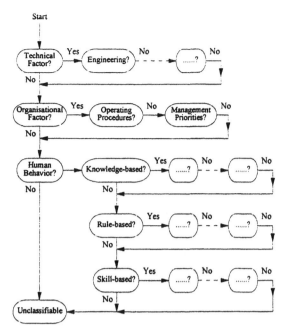

Fig. 2. The Eindhoven classification model (simplified form).

* repeat this process as often as needed.

In this research project the Eindhoven classification model was used as the initial 'statement'. Instead of using an initial case, 15 near misses were investigated and used as input for the first revision process. After the first revision of the model, five new near misses were investigated, and the model was revised again. This process was continued until the model did not change any more.

2.4 Results and discussion

After 25 near misses the classification results stabilised and enough data were collected to answer the first research question. By describing 25 near misses, 164 root causes were identified and classified according to the adjusted classification model. The distribution of the root causes is presented in fig. 3. Based on these results it can be concluded that the organisational causes are important and should not be ne-

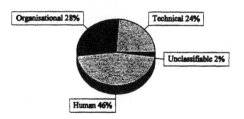

Fig. 3. Distribution of 164 root causes in the steel industry project.

glected. In this situation they are even more frequent than the technical causes.

The second aim was to develop tools to describe and classify the organisational factors. Critical incident interviews and causal trees were used for the description of the near misses. The company policy so far had been to use standard forms to analyse their incidents. But in this way only predefined questions are asked. By using critical incident interviews and causal trees a reconstruction of the incident takes place, and the person investigating the incident is free to ask any question that can help him to make this reconstruction. In this way the description will be much better and more comprehensive information is gained about the causes of the incident.

The second tool that was developed was the adjusted and extended classification model. To adjust the existing model to the steel industry only a limited redefinition of the categories was needed. It was necessary to extend the Eindhoven classification model on the organisational part. With the two original categories 'operating procedures' and 'management priorities' it was not possible to classify all organisational root causes. After the process of analytic generalisation, four descriptive organisational categories were added to the two already existing ones. With the following six descriptive categories it was possible to classify all the organisational root causes:

*   *Management priorities*: refers to any de facto pressure by management to let production prevail over safety. Especially maintenance personnel often have to do their work during the short planned production stops.
*   *Responsibilities*: refers to the inadequate allocation of responsibilities to persons. Especially in an emergency situation this can lead to the omission of vital actions.
*   *Operating procedures*: refers to the (inadequate) quality of procedures, not whether they are followed or not! Important are the completeness, accuracy and ergonomically correct presentation of operating procedures.
*   *Transfer of knowledge*: refers to the lack of care for education about situational knowledge, which is not included in the operating procedures. Especially new personnel has to be informed about things like system characteristics, known dangers and the existing operating procedures.
*   *Bottom-up communication*: refers to all the bottom-up signals that do not lead to any sort of reaction. Especially for safety and motivational reasons it is very important to listen to ideas or complaints from the workfloor.
*   *Culture*: refers to the inaccurate way of acting by a group. Groups within a company are used to do certain tasks 'their way', and not the prescribed way.

3. CHEMICAL PROCESS INDUSTRY PROJECT

3.1 Plant profile

The second project was carried out in the chemical process industry (Van Vuuren, 1994). The total plant, with a total number of employees comparable to the coking plant, was asked to participate in the project, including all the contractors who do most of the maintenance work.

The main products of this plant are Propylene Oxide (PO) and Tertiary Butyl Alcohol (TBA). Hundreds of everyday commodities, ranging from cosmetics to antifreeze and furniture cushions to car bumpers, contain PO or PO derivatives. TBA is used primarily to produce the gasoline component Methyl Tertiary Butyl Ether, which has good octane improving properties and facilitates the production of unleaded petrol. Because both raw products are extremely flammable, a high safety level on the workfloor is one of the top priorities of management.

3.2 Aims of the research project

For this project too, the practical aim was to implement a near miss reporting system. The scientific aims were to use the collected data:

1. to answer the first research question in a different domain.
2. to answer the second research question: why does almost all in-company accident research show us that only human and technical factors are important?

Also the refinement of the tools described in the previous chapter was considered as an important aim of this project.

3.3 Method

For this project the same approach as in the steel plant was used. Critical incident interviews, causal trees and classification of the root causes were used to collect the data and to answer the research questions. New to this project was the explicit attention for the way the company was used to analyse their incidents, and the results of these analyses. This was necessary to answer the second research question. As stated before, an insufficient level of analysis of incidents was expected to be a prime reason for the low percentage of organisational failure found in most of the companies. To check this expectation, two very recent accidents which had already been investigated by the company, were used for the project. For these accidents the investigation was done all over again, the way described in the previous chapter. After the new investigation a comparison of the results was made.

3.4 Results and discussion

During this project 24 near misses were investigated, which resulted in 138 root causes. The distribution of the classified root causes is presented in fig. 4. This leads to the same conclusion as in the first project, that the contribution of organisational factors is significant and deserves attention during incident analysis.

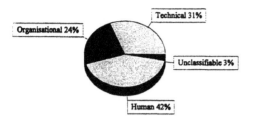

Fig. 4. Distribution of 138 root causes in the chemical process industry project.

To answer the second research question two serious accidents, already investigated by the company, were investigated all over again. The total results for the two accidents are summarised in table 1.

Table 1 Results of the comparison.

Types of root causes	Original results (number of root causes)			New results (number of root causes)		
	Techn.	Org.	Human	Techn.	Org.	Human
Accident 1	1	0	2	4	4	4
Accident 2	0	1	4	1	2	6
Total	1	1	6	5	6	10

In the original investigation eight root causes were identified, six of which were classified as human causes. Only one root cause was classified as organisational. During the new investigation 21 root causes were identified. 24% were classified as technical causes, 48% were classified as human causes and 28% were classified as organisational causes. This distribution is comparable to the distribution in fig. 4.

It was not possible to use more than two accidents for answering the second research question. The accidents should have happened very recently to obtain valid information. Only two accidents could satisfy this criterion. In spite of this limited number, the results of the comparison between their original investigation and the new approach convinced management to use another technique for incident investigation. Their previous way of investigating was shown to be insufficient. The results were incomplete and only covered the top of the causal tree, which mainly consisted of human and technical failures. Therefore causal trees based on critical incident interviews lead to better results than standard accident forms, because now also organisational failures become visible.

4. CONCLUSIONS

At the beginning two research questions were stated. After carrying out the two research projects described in this paper, it can be concluded that the results met the expectations. Based on the first research question, both projects were started with the expectation that the contribution of the organisational factors would be significant. The percentages found (28% in the steel industry and 24% in the chemical process industry) confirm this expectation. To make the best use of this information, the classification model was successfully extended to six descriptive organisational categories. With these categories it was possible to classify all organisational categories.

For answering the second research question, the way of analysing incidents was investigated. The expectation was that companies do not investigate deeply enough to identify organisational factors. The results of the second project have confirmed this expectation. Only the top of the causal tree was considered during the original investigation. This top mainly consists of non-organisational factors. During the new investigation not only more root causes, but also organisational root causes were found. Causal trees were demonstrated to be very useful to describe incidents in an orderly, comprehensive way.

Incident analysis as described in this paper is very useful for risk management in complex man-machine systems. By only using predictive models like Fault Tree Analysis, a unique opportunity to learn from accidents or near misses is missed. Especially for the low probability, high consequence risks it is crucial to learn as much as possible from near misses in order to prevent real accidents.

Investigating near misses also demonstrated the importance of 'human recovery'. Each year an innumerable number of dangerous situations are prevented from developing into a real accident by human recovery. Therefore, during the design of a new system attention should be paid not only to the prevention of possible failures, but also to the promotion of recovery possibilities. If it is not possible to prevent all types of initial failures, one should try to build in possibilities for timely recovery during the design phase. Timely recovery depends on the observability and reversibility of the emerging unacceptable effects (Rasmussen, 1986). This observability largely depends on the properties of the human-machine interface. Reversibility largely depends on the dynamics and linearity of system properties.

5. FOLLOW UP RESEARCH

At the Eindhoven Safety Management Group a four-year research project, which is part of the EC-funded Human Capital and Mobility program "Human Error Prevention", has been started with the following aims:

1. to develop and test an *explanatory model* for the organisational causes of safety-related incidents in complex technical systems, in order to be able to propose more effective and more efficient measures to increase safety performance.

2. to develop and test a number of *tools*, based on this model, to recognise, describe, classify and interpret these organisational causes and suggest preventive counter measures.

The classification model described in this paper only uses *descriptive* categories for organisational root causes. With these categories it is possible to classify the root causes, but that is only the start. Ideally, the categories should serve as a link between the root causes and preventive counter measures. For this an explanatory model has to be developed. As stated in the introduction, our research group aims at a contribution of industrial engineering and management science models and theories to build such a model.

Parallel projects are also carried out in other domains (energy production, medical care), and for other performance aspects of man-machine systems (quality, environmental control, and system availability).

ACKNOWLEDGEMENT

The presentation of this paper was supported by a CEC Human Capital and Mobility Network on Human Error Prevention.

REFERENCES

Flanagan, J.C. (1954). The Critical Incident Technique. *Psychological Bulletin*, **51**, 327-358.

Rasmussen, J. (1986). *Information processing and human-machine interaction*. Elsevier Science Publishers, Amsterdam.

Reason, J.T. (1990). *Human Error*. Cambridge University Press, Cambridge.

Reason, J.T. (1991). Too little and too late: a commentary on accident and incident reporting systems. In: *Near miss reporting as a safety tool*. (Schaaf, T.W. van der, D.A. Lucas and A.R. Hale, (Ed.)), 9-26. Butterworth Heinemann, Oxford.

Schaaf, T.W. van der, D.A. Lucas and A.R. Hale (eds.) (1991). *Near miss reporting as a safety tool*. Butterworth Heinemann, Oxford.

Schaaf, T.W. van der (1992). *Near miss reporting in the chemical process industry*. Ph.D thesis, Eindhoven University of Technology.

Schaaf, T.W. van der (1995). Human recovery of errors in man-machine systems. *This volume*.

Vuuren, W. van (1993). *Near miss reporting in the Dutch steel industry (in Dutch)*. Master thesis, Eindhoven University of Technology.

Vuuren, W. van (1994). *Near miss management in the chemical process industry (in Dutch)*. Internal report, Eindhoven University of Technology.

Wagenaar, W.A. (1983). *Human error (in Dutch)*. Inaugural lecture. University of Leiden.

Wagenaar, W.A., J. Groeneweg, P.T.W. Hudson and J.T. Reason (1995). Promoting safety in the oil industry. *To appear in Ergonomics*.

Yin, R.K. (1991). *Case study research: Design and methods*. Applied Social Research Method Series, Sage Publications.

EXPERIMENTAL INVESTIGATION ON MENTAL LOAD AND TASK EFFECT ON SYSTEM PERFORMANCE

Z.G. Wei, A.P. Macwan and P.A. Wieringa

*Laboratory for Measurement and Control, Faculty of Mechanical Engineering
Delft University of Technology, Mekelweg 2, 2628 CD Delft, The Netherlands*

Abstract: Tasks in supervising a system, whether performed by operator or automation, affect system performance. When operators perform tasks, they experience a mental load. This paper presents an investigation into the two phenomena, performance and mental load. Task effect was computed by conducting simulations without operators. Experiments were conducted with operators supervising partially or fully automated subsystems. Mental load was subjectively assessed by operators. The results show that performance and mental load were affected by the number and locations of automated subsystems, hence the complexity of the system. In addition, performance didn't always improve and mental load didn't always decrease with increased automation.

Keywords: Automation, tasks, performance, human factors, human perception, mental load.

1. INTRODUCTION

Automation has been motivated by the desire to reduce human workload, make system operation easier and increase system safety and efficiency. It shifts the human operator's role from manual to supervisory control. Supervisory control refers to a system where the closed-loop control tasks are performed by automation, while the human primarily monitors the automated system (Sheridan, 1992). However, if automation is not applied properly, there may be negative effects (Bainbridge, 1987). In the design of large complex processes, task allocation plays an important role in deciding which control tasks are allocated to humans and which to automation (Levis, *et al.*, 1994). Tasks which can be performed by both human and automation are often allocated using guidelines that consider strengths and limitations of humans and automation (Price, 1985).

Clearly by its design or definition, a task influences system operation and output, irrespective of whether it is performed by human or automation. According to Stassen *et al.* (1990), when a task is performed by a human operator, the operator experiences a mental load. The primary purpose of the investigation reported here is to explore whether and how the mental load is related to the task's influences on system output. For example, some tasks may evoke a high mental load but do not have significant effect on the system output. On the other hand, other tasks may induce a low mental load, but may have a strong effect on the system operation.

The approach that has been carried out in our investigation is described below. Some additional definitions are presented here to clarify the description that follows. Task Mental Load (TML) refers to the mental load that an operator experiences when he performs a particular task. Overall Mental Load (OML) is defined as the mental load experienced by the operator when he performs all tasks to supervise a system. System performance is measured by the mean error between actual outputs and desired objectives, integrated over time. Task configuration implies task allocation with respect to human and automation. Task Effect (TE) is defined as effect on system performance related to the control task of responding to a set-point

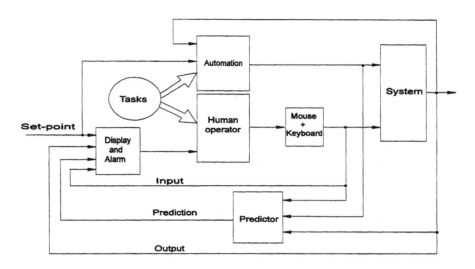

Fig. 1. Schematic of experimental system.

request for a specific subsystem.

An experimental system consisting of first-order subsystems was used to carry out the investigation. The effect of the tasks to control the system on performance was computed by analyzing a series of simulated experiments without human intervention. These simulated experiments provided a reference performance. Experiments were also conducted with human operators who were asked to control a system with certain combinations of tasks. System outputs and human control inputs were recorded. For each experiment, TML and OML were assessed subjectively by the operators themselves. From the data collected, the effect of different combinations of manual tasks on system performance and human perception of mental load was addressed. Then, the relation between performance and OML was discussed. Last, the correlation between TML and TE was analyzed.

2. EXPERIMENTAL SET-UP

The experimental system is shown in Figure 1 and was adapted from one developed by Stassen et al. (1993). It

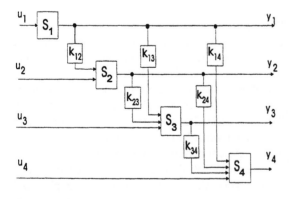

Fig. 2. Block diagram of 4 subsystems and their connections.

was made by using a graphic language, LabVIEW (National Instruments, 1993) installed on a Pentium PC. The simulated system consisted of 12 first-order subsystems. This number was chosen based on the following considerations: (1) If the number is smaller than 8, the system complexity is too low for observable differences in operator performance (Stassen et al, 1993); (2) With 12 subsystems, an adequate man-machine interface could be built on the 17 inch display that was used. With 16 subsystems, used by Stassen et al., the size of the icons were too small, making it difficult to operate the system. All subsystems were connected in a complete forward manner with constant coupling coefficients. They were sequentially numbered from 1 to 12. Thus, the output of *subsystem i* was connected to *subsystems i+1, i+2,...,12*, and meanwhile multiplied by the value of the corresponding coupling coefficient. Figure 2 shows a system with four subsystems. Each subsystem could be controlled automatically or manually as the experiment required. A proportional-integral, PI, controller was employed to execute automatic control.

To each subsystem, a set-point request could be presented at random instances in time. The operator's task was to generate an input for a subsystem, by means of the mouse or keyboard, and to bring the subsystem's output to its new requested set-point or to maintain it at its old set-point. In general, the operator had as many control tasks to perform as the number of manually controlled subsystems. The parameters that were changed among the experiments were the number of manually controlled subsystems and their locations.

Data collected from every subsystem included the value of the input given by the operator, the time at which it was given, and the subsystem output. For each subsystem, the mean error of the absolute difference between the actual output and the desired value, ME_i, was taken to be the measure of performance (van der Veldt, 1984), and was computed as follows:

$$ME_i = \frac{\sum_{k=0}^{M} \left| Y_i^k - R_i^k \right|}{M} \qquad (1)$$

where Y_i^k is the output of *subsystem i* for the computing period k, and R_i^k is the set-point request for *subsystem i* for the computing period k. M is the number of computing periods. $i = 1, 2, ..., N$ represents the ith subsystem, N being the total number of subsystems.

The mean of the performance, ME_{ov}, of all subsystems was taken as the performance of the overall system:

$$ME_{ov} = \frac{\sum_{i=1}^{N} ME_i}{N} \qquad (2)$$

As mentioned earlier, number and locations of manually controlled subsystems were changed as required by the design of the experiments. From a set of all possible combinations, 20 were selected. The number of manual subsystems was chosen to be 1, 2, 4, 6, 8 and 12.

19 experimental sessions were carried out by each subject, in addition to one session with full automation which was conducted without operators. Each session lasted 15 minutes. The subjects completed all experimental sessions in four half-day sessions.

Six subjects, all males with an average age of 22.8 years, who were students at Delft University of Technology following a curriculum for mechanical engineer served as operators. Each subject performed eight training trials of 5 minutes each. During training, the subjects got familiarized with the system, and got acquainted with estimating mental load. The students were paid a fixed amount of money after completing all requirements.

3. ANALYSIS OF TASK EFFECT AND ASSESSMENT OF MENTAL LOAD

The experiment was conducted in two parts. One was a computational system analysis of task effect on the overall system performance. Another was an experimental investigation on mental load when human operators control the simulated system.

3.1 Evaluation of task effect on system performance

One approach to evaluate the influence of a control task on system output is as follows. All subsystems are automatically controlled. Set-point requests of identical amplitude are given to individual subsystems at the same time instance, and the overall system performance is computed. For example, to evaluate the influence of *control task i* on system performance, a unit set-point request is given only to *subsystem i*. The process is repeated for each subsystem.

In an alternative approach, *subsystem i* is not automatically controlled, while all others are automated. Subsequently, during the trial, there is no response to the set-point request. Then, the system performance is computed. This process is repeated for each subsystem.

In using both approaches, dependencies among tasks can be analyzed, by replacing *subsystem i* with a combination of subsystems. The extended analysis is being carried out for a system with 4 subsystems and will be presented elsewhere.

3.2 Investigation on mental load

For this part of the analysis, human operators were asked to operate the system for different task configurations. Following each trial, they were asked to rate TML and OML. A subjective rating technique was used to assess the mental load. A unidimensional rating scale, called Rating Scale Mental Effort, or RSME, (Zijlstra, 1993), was used as a measure of mental load.

In other studies, a unidimensional rating scale was found to achieve results similar to a multidimensional scale, such as the NASA-TLX (Hart and Staveland, 1988), and was found simpler to use (Veltman and Gaillard, 1993; Vidulich and Tsang, 1987).

The RSME was presented to subjects on a vertical scale. The right side of the scale has statements related to mental load such as "not effortful", "a little effortful' and "very effortful" etc., while the left hand side includes numerical values from 0 to 150. The subjects were presented with a piece of paper which included the definition of mental load, the RSME, and a record table. Immediately after a session, subjects were asked to put a mark on the scale where they assessed the mental load to be. They assessed TML for each manually controlled subsystem and OML for the whole system.

4. RESULTS

During the experiment, three types of data were collected from each subject for each session. These were TML for the manual tasks of the session, OML for the whole session and ME for each subsystem. For each session, ME_{ov} was computed using Equation 2.

Figures 3 and 4 show ME_{ov} and OML, respectively, versus the number of manually controlled subsystems, averaged for all subjects. Multiple points for, 4, 6 and

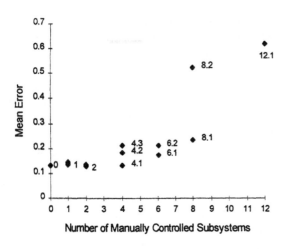

Fig. 3. System performance vs. number of manually controlled subsystems.

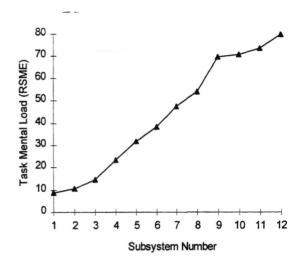

Fig. 6. Task mental load for each control task when all subsystems were manually controlled.

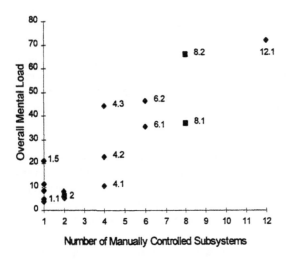

Fig. 4. Overall mental load vs. manually controlled subsystems.

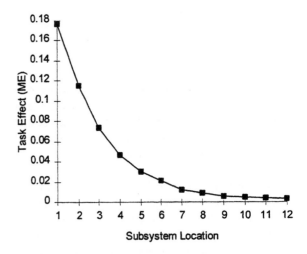

Fig. 7. Task effect for each subsystem on system performance with all subsystems automated.

8, indicate different locations of subsystems that were manually controlled. When 1 or 2 subsystems were manually controlled, the difference of ME_{ov} among different locations is not visible because of the scale of the plot. For each session, system performance was computed and OML perceived by operators when they controlled the whole system was estimated. Thus, to the system performance of every session there corresponds a value of OML. According to the session number, the relationship between ME_{ov} and OML, both averaged for all subjects, could be plotted as shown in Figure 5.

The task mental load for each subsystem is shown in Figure 6. TML for each subsystem was obtained from the session when all subsystems were controlled manually. TML was averaged for all subjects.

The TE for each subsystem on system performance when all subsystems were automatically controlled is

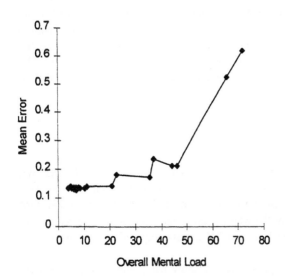

Fig. 5. Performance vs. overall mental load.

plotted in Figure 7. TE was evaluated using the approach described in Section 3.1 and was measured by system performance.

5. DISCUSSION AND CONCLUSIONS

5.1 Experimental system

In investigations on mental load, continuous manual tracking task of controlling a first- or second-order system is employed (see, for example, Backs *et al.*, 1994; Rouse *et al.*, 1993; Gopher and Braune , 1984). The experimental system presented in this paper consists of 12 first-order subsystems. These subsystems could be connected in different ways as long as the overall system is stable. To simplify, a complete forward connection was adopted. Thus, a 12th-order system was used in the experiment. Moreover, the experimental system could be modelled by a linear state-space equation, with the output of each subsystem as system state, from which a practical process might be simulated.

5.2 Observations from experimental results

From the results shown in Section 4, the following points could be observed:

(1) *System performance.* From Figure 3, first, it is noted that for the same number of manually controlled subsystems, different system performance has been achieved. This is indicated, for example, by points 8.1 and 8.2. Second, it can be seen that even for different numbers of manually controlled subsystems, the same performance may be achieved (see, for example points 4.2 and 6.1). Third, comparing points 4.3 and 6.1, it is found that despite more subsystems being manually controlled, system performance may not decline.

(2) *Overall mental load.* In Figure 4, similar observations as those for system performance can be made.

The observations in Figures 3 and 4 are due to the task configuration of the experimental system. Each subsystem has different numbers of inputs. Despite the number of manually controlled subsystems being the same, different subsystems have different number of coupling inputs which have to be compensated by the operators. This indicates that the location of manually controlled tasks in the system, not only the number of them, has effects on system performance as well as on mental load. The location reflects the task characteristics, and the complexity.

(3) *Relationship between OML and system performance.* In Figure 5, it is seen that system

performance varies only slightly for values of the RSME below 20. This indicates a threshold for low mental load situations and that mental load may vary without detectable change in performance (Backs *et al.*, 1994; Yeh and Wickens, 1988). When the RSME is larger than 50, ME varies largely and performance becomes poor. This indicates that despite the subjects having invested large mental load, system performance did not improve. In another continuous manual control experiment, Backs *et al.* (1994) observed a similar relation between mental load which was assessed based upon a rating scale and system performance which was measured by the root mean square error of tracking. They found that when subjective mental load increased, system performance declined. This is similar to the well-known inverted "U hypothesis" that if a human operator invested too much mental load, his performance will decline (see, for example, Sheridan, 1992).

(4) *Task mental load and task effect on system performance.* In Figure 7, the effect of coupling of subsystem 1 on other subsystems (2 to 12) is shown by the point corresponding to *subsystem 1*. However, in Figure 6, the mental load related to corresponding effect of *subsystem 1* on *subsystem 2* is shown by the point corresponding to *subsystem 2*. Similar observation applies for each subsystem. As seen in Figures 6 and 7, TML is inversely related to TE. A set-point request on the first subsystem produced the largest system error, but the subjects perceived the least mental load to control this subsystem. This could be explained as follows. If each subsystem is controlled by a local controller, despite being a first-order system, the forward coupling of all subsystems causes *subsystem i* to have a $(N+1-i)^{th}$ order behaviour. As the order of a subsystem increased, ME for the overall system also increased. However, for *subsystem i*, the operator had i inputs to compensate. When operators controlled *subsystem i*, the control task had more demand than that of *subsystem i-1*. Thus, TML was in a reverse order of TE. These observations correspond to those made by Gopher and Braune (1984) in a tracking experiment. They observed that estimates of task demands correlated highly with an index based upon the processing characteristics of tasks.

5.3 Conclusions

According to the above observations, the following, concerning the experimental investigations, can be concluded:

- It is not always true that as more tasks are automated, performance will improve. It is also not true that as more tasks are allocated to human operators, a higher overall mental load will be perceived by them.

- The task complexity for different tasks is not identical even when they are of the same type.
- An optimum between mental load and system performance depends not only on the automated task number, quantitative, but also on the task characteristics, qualitative.
- When too much task effect is automated, the operator may still have a high mental load, thus affecting operator performance and system performance. If too much mental load is reduced by automation, the operator has low mental load, but he may perform tasks with high effect. Thus, system performance may be adversely affected.

6. PRACTICAL IMPLICATIONS

The approach proposed and conclusions presented in this paper may be useful in the evaluation of task allocation strategies for optimal system operation. For task allocation, not only the question of what type of tasks is allocated to human or automation, but also which task in the same type is allocated to human or automation is important. Hence, it will affect system performance as well as operator's mental load. While improvement of system performance may demand automation of certain tasks, reduction of operator's mental load may require other tasks to be automated. The approach proposed here can help to achieve a balance of the two.

ACKNOWLEDGMENT

The authors would like to thank Mr. Hans Andriessen for his help in building the simulated system and discussing the analysis results.

REFERENCES

Backs, R. W., A. M. Ryan and G. F. Wilson (1994). Psychophysiological measures of workload during continuous manual performance. *Human Factors*, **36**, pp. 514-531.

Bainbridge, L. (1987). The ironies of automation. In: *New Technology and Human Error*, (Rasmussen, Duncan and Leplat (Eds.)), John Wiley, pp 271-283.

Gopher, D. and R. Braune (1984). On the psychophysics of workload: why bother with subjective measures?. *Human Factors*, **26**, pp. 519-532.

Hart, S.G. and L.E. Staveland (1988). Development of NASA-TLX (Task Load indeX): Results of empirical and theoretical research. In: *Human mental workload*. (P. A. Hancock and N. Meshkati (Eds.)), North- Holland, Amsterdam, pp.139-184.

Levis, A. H., N. Moray and B. S. Hu (1994). Task decomposition and allocation problems and discrete event systems. *Automatica*, **30**, pp.203-216.

National Instruments Corp. (1993), *LabVIEW for Windows -User Manual*.

Price, H.E. (1985). The Allocation of Functions in Systems. *Human Factors*, **27**, pp. 33-45.

Rouse, W.B., S.L. Edwards and J.M. Hammer (1993). Modelling the dynamics of mental workload and human performance in complex system. *IEEE Trans. Syst., Man, and Cybern.*, **23**, pp. 1662-1671.

Sheridan, T.B. (1992), *Telerobotics, automation and human supervisory Control*. The MIT Press, Cambridge, Massachusetts.

Stassen, H.G., J.H.M. Andriessen and P.A. Wieringa (1993). On the human perception of complex industrial processes. Presented in the 12th IFAC World Control Congress, Sydney, Australia.

Stassen, H.G., G. Johannsen and N. Moray (1990) Internal representation, internal model, human performance model and mental workload. *Automatica*, **26**, pp. 811-820.

Veldt, R.J., van der (1984). Looking ahead supervisory control. In *Proceedings of 4th European Annual Conference on Human Decision Making and Manual Control*. pp. 249-262.

Veltman, J. A. and A. W. K. Gaillard (1993). Measurement of pilot workload with subjective and physiological techniques. In *Proceedings of Workload Assessment and Aviation Safety, British Royal Aeronautical Society*, pp.3.1-3.13.

Vidulich, M. A. and P. S. Tsang (1987). Absolute magnitude estimation and relative judgment approaches to subjective workload assessment. In *Proceedings of the Human Factors Society-31st Annual Meeting*, pp.1057-1061.

Yeh, Y.Y, and C.D. Wickens (1988). Dissociation of performance and subjective measures of workload. *Human Factors*, **30**, pp. 111-120.

Zijlstra, F.R.H. (1993). *Efficiency in Work Behavior*. Ph.D. thesis, Delft University of Technology, Delft University Press, Delft, The Netherlands.

DEVELOPING SAFETY CULTURE - A NEVER ENDING PROCESS

Per Holmgren

RELCON AB
Box 1288
172 25 SUNDBYBERG, SWEDEN

Abstract: Companies with a need for a high level of safety must continuously develop their safety culture. The reason is that all changes and enhancements are made inside the companies culture. This paper discusses development of the safety culture by deliberate changes in attitudes and communication among the employées.

Keywords: Cultural aspects of automation, Communication, Human error, Human factors, Human-machine interface, Human perception, Man/machine interaction, Management systems, Safety

1. INTRODUCTION

Most safety and risk assessments have suggestions on how to increase the safety by different plant changes.. Those changes varies from minor changes in procedures to a complete reorganisation of the plant staff. Many improvements are made in systems and procedures, i e hardware" related improvements. These enhancements are good, and of course absolutely necessary. However, these changes are made inside the company's current culture. An example is the introduction of a new procedure. This new procedure is introduced into the current company culture. It is known that the culture or climate in a company is of vital importance for safety. Therefore, you can say that the changes are as effective as the culture of the company allows it to be.

Fortunately, the culture is not a static state, and therefore it is possible to deliberately introduce changes in the culture and way of thinking.

This paper does not try to distinguish or define culture. The main issue is to discuss and investigate "phenomena" like awareness, attitudes and communication, which have a large impact on the culture.

The paper discusses how one can work and develop the issues mentioned above in companies which are working in areas where a high level of safety is necessary. The nuclear power industry is the basis for the discussion in the paper, but the ideas and models are valid for all kinds of operations.

If one really would like to make a difference in peoples attitudes, awareness and communication it is probably not so effective using "ordinary" memory education. The point is that you cannot tell someone to be motivated, neither have a checklist for communication. For example, the knowledge about efficient communication is something that you needs to distinguish for yourself.

Lets look at an example from sport: A football player need to run to the left when the ball comes from a certain direction. If he wants to succeed, then there is no time available for taking out a note from his pocket and read what to do. To be a good football player, he must act more or less automatically, using his football instinct. If he try to remember what he should do he will also be too slow. Instead he needs to have an insight (even do he might not call it that) of where to run in a certain situation. He has distinguished that for himself through training. All his training has given him certain distinctions of where to run at the field when the ball comes from a certain direction.

Those who are world class sportsmen are always in training to become better. They are continuously trying out different ways of doing different moments or techniques in their sport - they are curios and they listen to different people from the perspective that these people can bring them forward in reaching their goals. They are listening for coaching and trying out different things that their coaches are saying.

So what does sport has to do with safety?

There are some similarities - the fact that they are all human beings and also that in times of an accident scenario the operator (or whoever that is involved) needs to be on top of their ability to manage the situation - like an athlete in a contest. You can compare the industry which need s an extraordinary safety in order to survive, with a large tournament in sport. If you are not good enough - if you make too many mistakes - you are out. Therefore there is of course a need for continuous efforts in developing important issues related to safety management.

Three major contributions to safety are declared in one of the IAEA guides (INSAG-4). These are *a questioning attitude, a rigorous and prudent approach* and *communication*. How can these factors be influenced in order to develop the safety?

It has already been stated that issues like the three above need to be like an insight for the person himself, in order to be really effective. This is true, especially in stressful situations. This paper has a discussion on how to achieve this insight and presents also a model that has been found very valuable for peoples awareness and questioning attitudes.

So the question is: What is a questioning attitude - and how can we achieve a new dimension in that area?

Let's start with the reactions in some different cases of incidents or accidents. How do people react in accident scenarios - and how can people get educated and trained to be more effective in their daily work and in accident sequences?

2. PEOPLES REACTIONS IN UNEXPECTED EVENTS

Usually one says that when an accident occur it is not a result of just one component that has failed or one manual action that was not properly performed. More likely it is a combination of incidents. Another way of saying this is that during the development of an accident scenario there are a lot of opportunities for people to mitigate the accident scenario. In the Chernobyl accident for example, there were many occasions when one single person could have stopped the incident sequence to become an accident by just standing up and communicate the uncertainties and unsafe handling that was present before the accident. Maybe someone did - but he was not strong enough to get through and convince the person in charge. Or the person in charge was not present to the actual situation. What the person in charge was present to was his own idea of the plant state, which in fact was a fantasy. Though his actions, listening and communication where in line with his experienced reality, it became devastating for the plant. He was living in a fantasy and his actions were correlated to that. You can compare it with the example of trying to drive in New York with a map over Stockholm.

You can see the same picture in a lot of different accidents and incidents. People are acting in a certain way, that is in line with "their world", while they are in fact living in a fantasy. It seems like people stop to investigate and question. The interesting question here is not why it is that way - maybe human beings are trained from birth to keep one answer as the truth . The interesting question is: How can the ability to investigate or question our own "realities", in situations when we really need it, be enhanced?

It would hardly be effective for people to walk around and question everything they do. However, by questioning in certain moments they can stop incidents from developing into accidents. By the way, usually people do stop accidents from occurring. By questioning, people have a better chance to be aware of what is really happening. People are present to their own experience, which is not always the same as the reality. And their actions are correlated to what they are present to. So, if people are "living in a fantasy", then their actions will be in line with that.

Figure 1 presents a model for peoples behaviour during unexpected disturbances or events. This model can be used for a single person as well as a team or an organization. By clarifying the status (in which phase they are) in the organization (or as one single person), the appropriate action to take, can be identified.

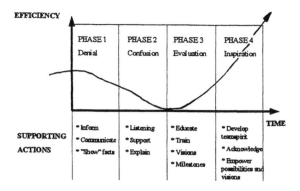

Fig. 1. The disturbance model

The different steps in the model are something everyone have been involved in???. If you really try the model, you can see that during your working day several disturbances will follow this route. It is nothing good or bad about this unexpected event - it is just the way life occurs. One advantage with the model is that, by asking yourself and other people in which phase you (they) are, one can actually see the most effective action to be taken, to shorten the time between phase one and phase four. Also, by just asking, you might see for yourself phase 1 and at the same moment phase 1 no longer have a "grip on you".

2.1 Phase 1 - Denial

In the first stage the unexpected event is introduced and people deny it. "This is not happening - it can't be true". This reaction is, as described before, not questioning. The unexpected events (disturbances) can be small or large. One example is when you unexpectedly are running out of gasoline. I know from my own and others experience that the immediate response is something like "it is impossible, I filled up yesterday!". One can distinguish the same pattern in most incidents that have been investigated. A look at the Chernobyl disaster again. The chief engineer decided a few moments after the blow-out that the core was still on place - even though there was a huge crater in the floor, graphite was spread in the reactor hall and the radiation meters went crazy. He wouldn't dare to question his own reality, because the alternative of a core spread around the plant and in the air was too horrible. Still, his decision made a few moments after the disaster was communicated as the truth. In many hours that decision, taken just a few seconds into the accident, was the base for all the following actions. The words from the chief engineer were widely spread - in fact all the way to Gorbatchov. The reality seemed to be as the chief engineer had said and everyone acted in line with that.

In this first phase of the model people often get upset and it seems like people get stopped from acting. It is like they are not participating in reality - they have no power to act effectively. Earlier, it was the example of running out of gasoline - instead of doing something about the problem you are sitting there thinking about "it cannot be - I filled up yesterday". You can call it to be a spectator in life. You are watching it - but you can't really do something about it, because you are not in the game. You are not playing.

An effective way of dealing with this situation is to communicate facts to the person that is denying the reality. You might have to tell this person over and over again until he (or she) really listen and sees it for himself (herself). In an organization you inform people about facts and make them aware of reality.

2.2 Phase 2 - Confusion

When people have gone through the first phase in the model, then the next phase is confusion. In this phase, people have no real power to act and they get less self confident. If this goes on for some time they are developing bitterness and they can blame someone/something as the reason to why they are in this position. The effective actions to take in the confusion phase are to explain, listen and give support. It is really important that the person who is in the confusion phase truly experiences that someone is listening, allowing the confused person to express everything he (she) want to. It is very important that confused persons get free to say everything so that they are ready to go on. The third phase starts immediately, when they confusion phase is completed.

2.3 Phase 3 - Evaluation

In the third phase, the person (or persons) starts to act again. They reassess and analyse. Appropriate actions to take during this phase is to educate and train, and to create visions and milestones.

2.4 Phase 4 - Inspiration

"We have a lift off". This is how you can characterise the fourth phase. Phase 4 has inspiration and a hunger for new ways and new ideas. The person who has been a spectator is now in the game and acting. Corresponding actions are to continue developing team spirit, acknowledge and reward, and continuously support and develop possibilities, visions and goals.

2.5 Conclusions of the disturbance model

The disturbance model can be used for both single persons, teams and organizations. It is valid both for short and long time perspectives. The accidenttime perspective is usually very short, but one can still use the model. If a company reorganizes, one can still use the model, but the time perspective will be much longer. Within the company itself, it can take different times for different groups. In the worst case, there are people that get stuck in phase 1 and phase 2. The disturbance model makes it possible to see where the person or team are and what you can do about it. Often, people try to soon to create visions and possibilities, but it does not work until the person has gone through the denial and confusion phases (phase 1- and 2).

The model can be used to deliberately get people more aware of the real situation.

The disturbance model has been used in training programs and seminars. Almost all participants could find out something about who they have been in certain situations and, why projects didn't succeed. The model usually reveals the phase in which their group or company resides for the time being. This information in turn give a new way of looking and acting upon things.

3. CONCLUSIONS

Investigations of different kinds of incidents in the nuclear power industry shows that the chain of incidents are following the "disturbance model".

Usage of a model, like the disturbance model, requires much more than just putting on a slide. The model really needs to be distinguished by the participants. Sometimes, a model can take hours to penetrate in detail. However, discussing, thinking and practising the model, makes it possible to deliberately cause a large difference in the three issues suggested by the IAEA guide, to be the major contributors to safety. There are also many methods and models from other disciplines (management, philosophy etc.) that haven't even started to be used in the nuclear power industry. There should be a lot to gain from these other disciplines.

A learning process involving a method or model which deliberately causes insights in what it is to be human, will promote a continuous safety development.

REFERENCES

Dougherty Ed (1992)
   Context and Human Reliability Analysis.
   Reliability Engineering and System Safety 41 (1993) 25-47
Granqvist Bo (1993)
   Inspiration ach resultat
Groth Michael (1994)
   "Skiftnyckeln".
   Stockholms Universitet,
   Psykologiska Instutitionen
IAEA (1991)
   INSAG-4 Safety Culture
   IAEA. Vienna 1991
Medvedev Grigory (June 1989)
   Chernobyl Notebook.
18220199 Moscow NOVY MIR in Russian No. 6 1989 pp 3-108

FUZZY-CONTROLLED-DRIVER-SIMULATION FOR EXHAUST EMISSION TESTS

K. Pfeiffer and R. Isermann

*Technical University of Darmstadt, Institute for Automatic Control,
Landgraf-Georg-Str. 4, D-64283 Darmstadt, FRG*

Abstract: A *dynamical engine test stand with a real engine and a drive-line simulation* is a capable tool for exhaust emission analysis. The driver behavior needs to be analysed to design a *fuzzy-controlled-driver-simulation*. The aim of the multivariable system *driver-simulation* is to imitate and to improve the human behavior. A comparison between a skilled human driver and the results of the driver-simulation is presented and discussed.

Keywords: Fuzzy control, velocity control, engine dynamometer, multivariable control

1. INTRODUCTION

Standardized exhaust emission test cycles are specified by law to get the type approval of a vehicle. Today exhaust emission analysis must be performed on roller dynamometers with real vehicles and skilled human drivers. Human drivers are able to follow pretended velocity-trajectories in a small range of tolerance. Nevertheless individual driving habits have a significant influence on fuel consumption and exhaust emissions. The test results are usually not reproducible because of the different and time variant behavior of human drivers, but they perform the imposition of the stringent test pattern. The statistical and systematical errors can be avoided by using a proper *driver-simulation*.

Fuzzy logic offers a notation of heuristical driver knowledge in form of linguistic rules, that can be transformed to a working control algorithm. Therefore a fuzzy-controlled-driver-simulation was investigated to imitate the human driver.

A *dynamical engine test stand* with a real engine and a drive-line simulation is a powerful tool to optimize the engine's transient behavior and also for exhaust

emission analysis. Therefore the driver-simulation is tested on a dynamical engine test stand.

2. DYNAMICAL ENGINE TEST STAND

Conceptionally the structure of the test stand may be divided into the plant with different hardware modules, some software modules and a user interface. Fig. 1 gives an overview of the main components of a dynamical engine test stand.

By real-time simulation of a vehicle's longitudinal dynamics with manual gearbox the dynamical load torque at engine's flywheel is calculated. This load is expected to be very similar to the load an engine has when built into a real car.

A fast dynamometer, which allows full electric inertia simulation, is applied to charge the engine with the simulated torque. It consists of a DC-motor with a 6-pulse current converter, a flat belt drive to reduce the moment of inertia and for speed adaption, a torque measuring shaft and a flange. The flange is directly connected to the combustion engine.

A torque controller is necessary to compensate the dynamometer's eigendynamics and for reduction of the disturbances by the combustion engine. Finally a combination of inertia compensation together with a PI-controller is used for torque control. Therefore statical deviations are avoided and a fast reaction on changing engine torque is achieved.

Fig. 2 shows the simplified mechanical model of the drive-line with interfaces to road, driver and tested engine. Main parts of the model are clutch, gear box, transmission shaft, wheel and vehicle body. Besides the linear components the model includes some nonlinear elements (e.g. aerodynamical drag, wheel slip and disk clutch).

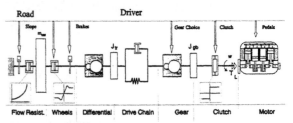

Fig. 2 Simplified mechanical model of a vehicle drive-line

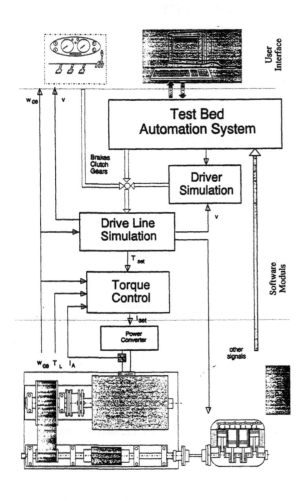

Fig. 1 Components of the dynamical engine test stand

2.1 Real Time Simulation of the Drive-Line

The simulation of the drive-line's rotational dynamics forms a central part of a dynamical engine test stand. The dynamic load torque at engine's flywheel is calculated as a function of driver commands, road condition and actual speed of the engine. To ensure real-time computation a simplified vehicle model is used for simulation, concerning inertia of the car, aerodynamic drag, rolling resistance, road gradient, wheel slip, torsion of the transmission shaft, gear shift with gearing losses and friction effects due to the clutch. Vibrations caused by other elements in a real drive-line (e.g. toothed gearing) show higher oscillation frequencies which are not of interest.

The *disk clutch* is considered by a Coulomb friction characteristic. Slipping clutch is assumed if the torque M_{cl} exceeds the maximal transferable torque $M_{cl,max}$, which depends on the actual clutch pedal position. In this case the load torque M_L at engine's flywheel is equal to $M_{cl,max}\text{sign}(\Delta\omega)$. The clutch continues to slip until engine speed ω_{mot} is equal to clutch disk speed ω_{cl}. Then the clutch sticks and load torque M_L is equated with M_{cl} and the clutch disk speed ω_{cl} is equated with ω_{mot} until M_{cl} exceed $M_{cl,max}$ again. This implies a change of model structure in case of sticking clutch.

The simulation model results in a set of ordinary differential equations with nonlinear coupling. There are many solution techniques for problems such as these but in case of real-time simulation the aspect of computation time is very important. Therefore the Heun-integration method was chosen for this specific problem.

2.2 User Interface

The user interface allows interaction with an operator. To consider the actions of a human driver a car cockpit has been included which interacts with the drive-line simulation. The car cockpit features acceleration, brake, clutch pedals, and gear stick as well. These components are directly connected to one VME-bus computer which controls the dynamometer by steadily calculating the states of the vehicle model. The acceleration pedal is directly connected to the tested engine. Thus operators may drive the engine in a way similar to the conditions in a real car. This tool allows the comparison between the actions of human drivers and a driver-simulation.

3. PROBLEMS AT DRIVING EMISSION TESTS

Driving exhaust emission tests are a multivariable control problem. The main problem is to follow pretended vehicle speed-trajectories in a range of tolerance by minimization the exhaust emissions. Exhaust emission test cycles can be separated into different modes of operation with different control strategies. The different modes of operation result in various manipulated variables and in single-input / single-output or multiple-input / multiple-output control loops.

Table 1 Control Loops dependent on the Modes of Operation

Mode of Operation	Controlled Variable	Manipulated Variable
Vehicle Startup	v_{car} ω_{mot}	acc. pedal clutch pedal
Gear Shifting	v_{car} ω_{mot} gear choice	acc. pedal clutch pedal gear stick
Cruise	v_{car}	acc. pedal
Increasing Velocity	v_{car}	acc. pedal
Decreasing Velocity	v_{car}	acc. pedal break pedal clutch pedal

v_{car} vehicle velocity
ω_{mot} engine speed

The most complex control task is the vehicle startup- and the gear-shifting control as multivariable control loop. Jerking while switching from these control loops to the single-input / single-output loops must be avoided. A pumping acceleration pedal has to be avoided, too.

4. FUZZY CONTROLLED DRIVER SIMULATION

A fuzzy controller is a set of IF-THEN rules. The premises contain linguistic statements on the input variables and the conclusions contain statements on the output variables. The linguistic statements of the IF-part are obtained by fuzzyfication of the numerical input values, the statements of the THEN-part are defuzzyficated to numerical output values. The basic idea is to obtain the rule base by interviewing experts (skilled human drivers).

4.1 Knowledge Acquisition

Remind the behavior of human drivers. They show a different behavior depending on the modes of operation. They use information about the desired velocity and acceleration during driving these tests. They also use information about the control error which is the *velocity difference*. The velocity difference is given for a sampling time k by

$$e(k) = v_{ref}(k) - v_{car}(k) \qquad (1)$$

Human drivers also use information about the change of control error which can be described as an *acceleration difference*.

$$\dot{e}(t) = \dot{v}_{ref}(t) - \dot{v}_{car}(t) \qquad (2)$$

The discrete-time version of this equation is given by

$$\Delta e(k) = \Delta v_{ref}(k) - \Delta v_{car}(k) \qquad (3)$$

First human drivers make a starting guess of the manipulated variable depending on the modes of operation and the desired velocity and/or the desired acceleration. This can be interpreted as feedforward control of the desired vehicle velocity (pilot control). Then they change the control output based on the velocity- and acceleration difference. It seems that they act like a PI-controller. Remind a PI-controller is given by

$$\Delta u(k) = K \Delta e(k) + \frac{K}{T_I} e(k) \qquad (4)$$

4.2 Fuzzy Knowledge Based Controller

The control error e(k) and the change of control error Δe(k) are chosen as input variables for the fuzzy knowledge based controller (FKBC). They are interpreted as linguistic variables *velocity difference* and *acceleration difference* with attributes defined as fuzzy subsets. Piecewise linear membership functions are chosen for fuzzyfication. A rule base for each mode of operation is created to imitate the human driver. An 'expert system' embedded in the FKBC decides which rule base needs to be activated. Because of computational time aspects the attributes of the linguistic output variables are defuzzyfied offline, by replacing the fuzzy set of each attribute by a singleton at the centre of gravity of the individual fuzzy set.

The FKBC has at most four output variables. These variables are the acceleration-, clutch and break pedal, which are feedback controlled and the gear-stick, which is feedforward controlled.

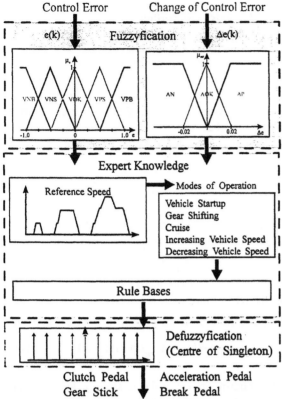

Control Error   Change of Control Error

Fig. 3 Structure of the fuzzy-controlled-driver-simulation

4.3 Fuzzy Sets

The *velocity difference* is described by the attributes *"velocity ok (VOK)"*, *"velocity negative small (VNS)"*, *"velocity negative big (VNB)"*, *"velocity positive small (VPS)"* and *"velocity positive big (VPB)"*. The membership functions are shown in Fig. 4.

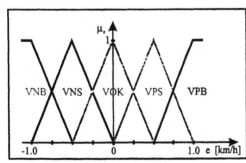

Fig. 4 Fuzzy sets of velocity difference

The linguistic variable *acceleration difference* is described by the attributes *"acceleration positive (AP)"*, *"acceleration negative (AN)"* and *"acceleration ok (AOK)"*, see Fig. 5.

The FKBC creates an output value Δu(k) which can be described by the linguistic variable *"change of position"*. This linguistic variable is described by the attributes *"nothing (NO)"*, *"negative small (NS)"*,

"negative medium (NM)", *"negative big (NB)"*, *"positive small (PS)"*, *"positive medium (PM)"* and *"positive big (PB)"*. Remind that there are different output values (acceleration-, clutch- and break pedal). The *change of position* needs to be combined with the correct output value dependent on the modes of operation

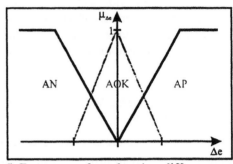

Fig. 5 Fuzzy sets of acceleration difference

4.4 Rule Base

Each rule base subsists of fifteen rules because of the five attributes of the linguistic variable *"velocity difference"* and the three attributes of *"acceleration difference"*. The whole driver simulation includes 75 rules divided into five rule bases.

The FKBC consists of rules of the form

IF *e* is LE AND Δe is LΔE THEN Δu is LΔU

where LE, LΔE and LΔU are linguistic values on the input and output values with the fuzzy sets presented in chapter 4.3 defined on the domains E, ΔE and ΔU.

For example consider now the two different rule bases for cruise control and increasing vehicle velocity. In these cases a single-input / single-output control loop with the acceleration pedal as manipulated variable is obtained.

Table 2   Rule base for cruise control

change of control output		acceleration difference		
		AN	AOK	AP
velo-city diffe-rence	VNB	NM	NS	NO
	VNS	NS	NO	NO
	VOK	NO	NO	NO
	VPS	NO	NO	PS
	VPB	NO	PS	PM

Table 3 Rule base for increasing velocity

change of control output		acceleration difference		
		AN	AOK	AP
velo- city diffe- rence	VNB	NM	NS	NO
	VNS	NO	NO	PS
	VOK	NO	NO	PS
	VPS	NO	PS	PM
	VPB	PS	PM	PB

The degree of fulfilment of the premise of each rule is calculated by combining the membership functions $\mu_{LE}(e)$ and $\mu_{L\Delta E}(\Delta e)$ into one membership function $\mu_{ant}(e,\Delta e)$

$$\forall e,\Delta e: \mu_{ant}(e,\Delta e) = \min(\mu_{LE}(e),\mu_{L\Delta E}(\Delta e))$$

(5)

The crisp output value Δu is computed by the simplified algorithm for singletons as a weighted mean value for all fired rules.

$$\Delta u(k) = \frac{\sum_{i=1}^{n} \mu_{ant,i}(e,\Delta e)\, u_{R,i}}{\sum_{i=1}^{n} \mu_{ant,i}(e,\Delta e)}$$

(6)

In this case u_R are the output singletons of the height one. This formula contains implicitly that the defuzzyfication is done by computing the centre of gravity of several truncated singletons. Note that a maximum number of four rules can be fired simultaneously.

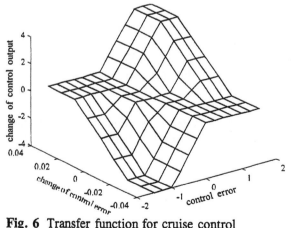

Fig. 6 Transfer function for cruise control

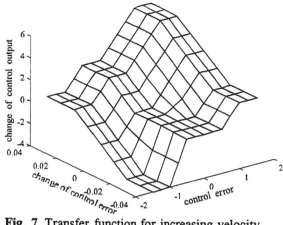

Fig. 7 Transfer function for increasing velocity

The control output value u(k) is obtained by adding the change of control output $\Delta u(k)$ to u(k-1). This calculation is done outside the FKBC and is not reflected in the rules.

5. CLOSED LOOP WITH DRIVER SIMULATION

After designing the FKBC for all modes of operation the fuzzy controlled driver simulation is tested on the engine test stand. The tested engine is a 1.6l turbocharged diesel-engine which is built into a VW Rabbit. Fig. 8 shows the closed loop behavior.

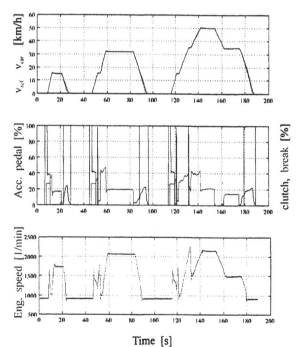

Fig. 8 Closed loop with FKBC

The european community exhaust emission test cycle (ECE-cycle) is chosen as reference velocity. This cycle includes all modes of operation. The range of tolerance of this test cycle is

$$v_+(t) = v_{ref}(t+0.5sec) + 1.0\,km/h$$
$$v_-(t) = v_{ref}(t-0.5sec) - 1.0\,km/h \qquad (7)$$

The closed loop behavior of the FBKC is satisfying. The driver simulation follows the reference velocity in the range of tolerance. The acceleration pedal as manipulated variable shows a smooth trend and the main problems (e.g. pumping acceleration pedal) are avoided.

6. CLOSED LOOP WITH HUMAN DRIVER

Fig. 9 shows the closed loop behavior when a skilled human driver drives the ECE test cycle.

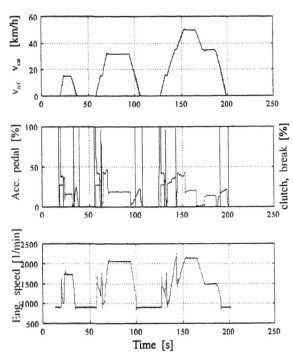

Fig. 9 Closed loop with a human driver

7. CONCLUSIONS

Unfortunately the exhaust semission are not measurable at the moment. Therefore the FKBC can only be compared with a human driver by analysing the quadratic effectiveness criterion on the change of control output and the control error.

$$I_u = \sqrt{\frac{\sum_{i=1}^{n}\Delta u_i^2}{n}}$$

$$I_e = \sqrt{\frac{\sum_{i=1}^{n}e_i^2}{n}} \qquad (8)$$

Table 4   Comparison of FKBC and human driver

Criterion	FKBC	Human driver
I_e	0.640	0.650
I_u	0.654	0.651

The fuzzy controlled driver simulation is a proper tool for driver-simulation during driving exhaust emission tests, because the effectiveness criterions I_u and I_e deliver approximately the same results. The FKBC perform the imposition of the stringent test pattern as well as skilled human drivers.

8. REFERENCES

Böhme, G. (1993). *Fuzzy logic* (in German). Springer-Verlag. Berlin

Driankov, D., Hellendorn, H. and Rheinfrank, M. (1993). *An Introduction to Fuzzy Control*. Springer-Verlag. Berlin

Gebauer, W. (1988). *An Engine Test Stand with High Dynamic Response with Simulation of Driver, Vehicle and Road Resistance*. VDI-Berichte 681. VDI-Verlag. Düsseldorf

Isermann, R. (1989,1990). *Digital Control Systems, Vol. 1 and 2)*. Springer-Verlag. Berlin

Lee, C.C. (1990). Fuzzy logic in control systems: fuzzy logic controller (part 1 and 2). *IEEE transactions on systems, man and cybernetics*. **Vol. 20**. pp 404-435

Pfeiffer, B.M., Isermann, R. (1993). Criteria for Successful Applications of Fuzzy Control. EUFIT '93. Aachen, Germany. Sept. 7-10

Pfeiffer, K., Isermann, R. (1993). Driver Simulation used in Dynamical Engine Test Stands for Exhaust Emission Test Cycles. ACC '93. San Francisco, USA. June 2-4

Preuß, H.P. (1992). Fuzzy Control - heuristische Regelung mittels unscharfer Logik. *atp 34 (Automatisierungstechnische Praxis)*. pp 177-184 and pp 239-246

Thun, von H.J. (1987). A New Dynamic Combustion Engine Test Stand with Real-time Simulation of the Vehicle Driveline. SAE-Paper No. 870085

Voigt, K.U. (1991). A Control Scheme for a Dynamical Engine Test Stand. IEE Control '91. Edinburgh, U.K., March 25-28

Zadeh, L.A. (1973). Outline of a new approach to the analysis of complex systems and decision processes. *IEEE Transactions on Systems, Man, and Cybernetics.* **Vol. 3**, pp. 28-44

Zimmermann, H.J. (1991). *Fuzzy set theory and its applications*. Kluwer. Boston

ANTICIPATION INFLUENCED BY DIFFERENT INTERFACES IN SIMULATED BUS TRAFFIC CONTROL TASKS.

Mailles S., Mariné C., Cellier J. M..

Laboratoire Travail et Cognition
CNRS - URA 1840 - Toulouse - France

Abstract: The type of interfaces used in bus traffic control tasks may influence anticipatory behavior. Former studies showed that there is a very wide variety of interfaces in work situations. In order to study anticipatory behavior under different coding conditions, a bus traffic simulator was developed. The experiment conducted used three different interfaces presenting three degrees of schematization. Results found for three different tasks performed with these interfaces were examined and analyzed in terms of reaction time and nature of responses. They did not reveal significant differences between the interface types.

Keywords: Traffic control, Interfaces, Simulation.

1. INTRODUCTION

Bus traffic control tasks belong to the class of dynamic environments whose complexity has already been demonstrated (see Bainbridge, 1978; Woods, 1988; De Keyser, 1988). Anticipatory behavior, which is needed for prediction of process evolution and effects of action represents a major difficulty. In order to assist the operator in this task, a better understanding of how an operator anticipates is needed, and about the information necessary to facilitate this behavior. A large number of interfaces are displayed in real bus traffic control tasks, but a former study (Mailles, *et al.*, 1993) showed that only a few of them are used by operators. The present study systematically analyzes the accuracy of anticipation performed on a bus traffic control task with various interfaces. First, a presentation of requirements for the tasks is described, then the experiment is presented.

2. THE BUS TRAFFIC CONTROL TASK AND ITS REQUIREMENTS

2.1. The Bus Traffic Control Task

The goal of computer assisted traffic control tasks is to control the regularity of all buses running in a town. These tasks are performed by operators from a central control post. Each bus gets a theoretical schedule for its route. The real present position of each bus is recorded and computed in order to determine discrepancies between the actual and theoretical positions. On this basis, various kinds of information are displayed for the operators.

This task is characterized by the dynamic dimension of the control process. Non-detection or non-processing of a disturbance may involve rapid deterioration of the situation. Therefore, the continuous evolution of the process is assigned substantial temporal constraints. Thus, the operator must anticipate the system's evolution in order to avoid disturbances. Also, the evolution of this process is neither constant nor uniform since not only do speeds vary from one bus to another, but also may vary for a single bus during its run.

These changes in process dynamics depend on several parameters which an operator has to manage in order to evaluate the process state. These parameters include: bus position in real space, number of passengers, the time of day, and driver behavior, among others. If some parameters are static but uncertain, such as the number passengers; others may be affected by the procedural choices of the operators, like the relative positions of buses.
The operator's control task can be divided into several objectives. The first is process monitoring,

especially if an operator detects a deteriorating situation. Because an operator ha number of lines to control, he may have to change and redirect his monitoring. Second, he has to diagnose the causes of disturbances and foresee their immediate or delayed consequences. Finally, the operator has to resolve problems by selecting specific control procedures, deciding the location and time a solution should be applied, and evaluate the direct, immediate, or delayed effects of the procedure applied. These three objectives clearly establish that anticipatory behavior is necessary to perform the task (Cellier and Mariné, 1991).

Anticipatory behavior in bus traffic control tasks may be divided into three assessment categories. First, the operator has to assess the future process state. This assessment involves estimation of the effects of disturbances, assesses the future situation based on the frequency of the buses, and disparities between theoretical and real schedules. Second, anticipatory behavior involves predictive assessment of the effects of a chosen course of action, for the same problem might well be solved by another choice. This demonstrates the need to assess the future consequences of any course of action. Third, the operator has to assess what resources will be available in the future in order to plan his actions (Eyrolle, *et al.*, 1994).

2.2. Variety Of Interfaces Found In Real Situations

The variety of interfaces found in work situations is very large given that the choices often were largely empirical. An initial study led to noticing the diversity of codes used to design bus traffic control task systems. It also showed how it is possibile to identify the uses of interfaces depending on the location studied and on the activity carried out. These results suggested that in this spatio-temporal task, it is necessary to maintain the spatial information (Mailles et al., 1993). It was also found that information presented might be classified in two categories: states (route, stops, etc.) and system dysfunction (advances, delays, etc.). The first information category is always spatial in nature and can be coded analogically or numerically. For example, bus position might be expressed by the position of a square on the computer screen (analogical code), or the position in meters bound to the bus number in a table (numerical code). Information in the system dysfunction category is a little more difficult to encode because of its spatio-temporal nature. Indeed, it can be coded as either spatial or temporal data. Those data can then be coded by analogical or numerical codes. For example, a delay which is spatio-temporal might be a delay in the position of the bus (it should have covered 1 km but it only made 0.6 km in x time so, it is 0.4 km late), or it might be a delay in time (it should have driven 5 minutes to reach a given point but it took 7 minutes so, it is 2 minutes late). Either a delay in position or a delay in time could be coded by analogical or numerical codes. Generally, in empirical situations, when information on states is coded analogically, information on system dysfunction takes on a spatial "second nature" and is also coded analogically. Likewise, when information on states is coded numerically , information on system dysfunction takes on a temporal "second nature" and is also coded numerically .

Bainbridge (1989) has shown that analogical displays gave better results than digital displays for pattern recognition and load on working memory when the goal is to assist operator performance. It remains to be shown whether anticipatory behavior is more accurate according to whether analogical or digital displays are used. Moreover, different quality levels for schematization were found in the analogical interfaces. As noted by Bainbridge (1988), "people do not need a photographically accurate pattern representation" (p.73). Therefore, only relevant features of the environments are necessary in analogical displays. His study tested this second point on interface functionality.

The diversity of the natures of information and the ways of coding it in different work situations led to conducting a systematic study related to anticipatory behavior. A simulated traffic control task system was designed in order to verify and deepen investigation of former results.

3. EXPERIMENTATION

In this experiment, the effect of the schematized interfaces on anticipatory behavior was tested. As mentioned above, interface functionality may favor this behavior.

3.1. Subjects

48 students participated in the experiment. They all knew how to use a computer mouse and had already taken the bus line displayed .

3.2. Material

A simulator was created to test some possible factors which cause anticipatory behavior to vary when bus traffic is being regulated. Therefore, various factors were considered that could be manipulated and adjusted in the simulator. This experiment required setting up three types of factors.

The bus line: The simulator created allows the chosen bus line to be set up with its spatial and temporal features. It uses a real situation as a reference. In this experiment, bus line number 2 of the Toulouse bus network was designed with its length, its stops, its timing during the day, bus speeds (from one point to another), and the frequency of buses .

The disturbances: A disturbance is an algorithm of a

modification (reduction or increase) of bus speeds during a parametered time and over a parametered distance. This provides opportunities to vary the situation's complexity. These parametered disturbances might be interpreted as traffic jams or other kinds of perturbations. Consistent with real situations, a model in the simulator computes the disturbance effects on traffic. For example, if the experimenter introduces a speed reduction lasting 10 minutes over 2 kilometers, the buses delayed by this parametered perturbation will induce an advance for the following buses. This advance is computed by the simulator.

<u>The interfaces</u>: The same information can be represented on three different interfaces. They were designed from those found in real work situations. Some of the frames have been kept and the types of information needed for this task have been analyzed (see above, Chap.2.2). These interfaces have been classified into three degrees of schematization. They all code the buses by isosceles triangles. The delays and advances are presented by bound straight lines to the buses. The length of this line depends on the delay or advance, and its extremity shows the theoretical position of the bus.

In the first degree of schematization (see Figure 1), the interface shows the bus line with its real topography including spatial reference points.

In the second degree of schematization (see Figure 2), the interface shows the bus line in two parts: the outbound direction on the top part of the computer screen, and the inbound direction on the bottom. These two directions are represented by two thick black lines. In this interface, all the non-relevant details were eliminated.

In the third degree of schematization (see Figure 3),

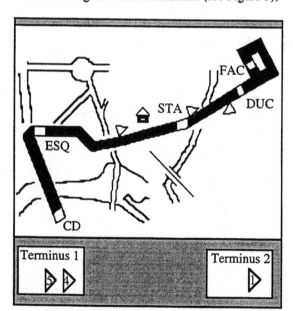

Fig. 1: First degree of schematized Interface. Inscriptions (ESQ, STA, DUC and FAC) are the bus line control stops.

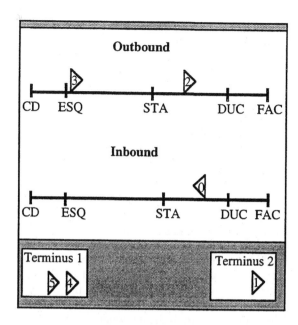

Fig. 2: Second degree of schematized Interface. Inscriptions (ESQ, STA, DUC and FAC) are the bus line control stops.

the interface also shows the bus line in two parts: the outbound direction on the top part of the computer screen, and the inbound direction on the bottom. These two directions are represented by several lines. Each line corresponds to one bus route. Thus, the number of lines is the same as the number of buses running. This interface underscores the relative positions of buses and has more functions.

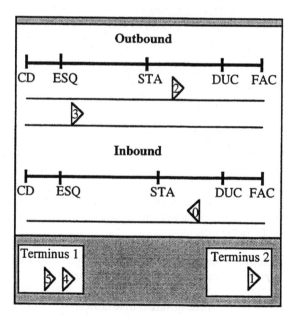

Fig. 3: Third degree of schematized Interface. Inscriptions (ESQ, STA, DUC and FAC) are the bus line control stops.

3.3. Procedure

The subjects were divided into three groups: those for the first degree schematized interface, those for the second degree schematized interface, and those for the third degree schematized interface. They performed the task individually.

In each group, each subject was first invited to familiarize himself with the simulator, then the experiment started. All subjects then performed four detection tasks, four identification tasks, and four memorization tasks. In the first part of all of the tasks, the subjects were shown a real bus traffic situation that they were to observe. In the second part, their task was to imagine the future of the situation they had just seen by answering different questions.

In the Detection task, subjects were asked to mentally imagine how the situation they observed would have progressed in x minutes. Then, they were shown a picture which corresponded to a certain degree of evolution in the future of the observed situation. Subjects had to answer the question by "Yes" or "No" to indicate whether the picture corresponded to the one they had mentally imagined or not.

In the Identification task, subjects were asked to mentally imagine how the situation they had observed would have progressed in x minutes. Then, they were shown three pictures which corresponded to three different degrees of evolution in the future of the situation they had observed. Subjects had to answer the question by selecting the picture which corresponded to the one they had mentally imagined.

In the Dragging task, subjects were asked to draw a given degree of evolution of the observed situation by dragging buses with the computer mouse .

These twelve tasks were counterbalanced taking into account the sought after degree of evolution. Three time lapses were used of 2, 3, and 5 minutes.

3.4. Results

The results for the three types of tasks were analyzed: identification, detection, and dragging tasks. Two kinds of data were examined: the time the subjects took to answer the questions, and the nature of the responses.

Reaction Time: The difference between the three interfaces was found to be insignificant, $p > .05$ for all three tasks. There was no overall difference in times for the three different time lapses.

Nature of Responses: The responses were divided into three types for the detection and the identification tasks: correct responses, overestimated responses, and underestimated responses. An overestimated response represented those occasions for which subjects selected a picture representing a higher degree of evolution of the situation than was sought. Selection of a lesser degree of evolution was considered to be an underestimated response. The mean of the correct responses (0.6) was significantly

higher ($p < 0.0002$) than the mean of the mistaken estimations (0.3). No significant difference between the nature of responses and the type of interfaces was found. The same results were found about the interactive nature of responses and time lapses. Regarding the modification tasks, the discrepancy between the simulated traffic situation and the one designed by the subjects was recorded. It did not show a difference between interfaces, but the time lapses showed disparate results depending on their values. This difference was found significant, $p < 0.0002$ (see Figure 4). The precision in drawing future situations was better for the 2 minute time lapse (0.3), intermediate for the 3 minute time lapse (0.7), and low for the 5 minute time lapse (0.9).

Discrepancies between simulated and drawn traffic situations

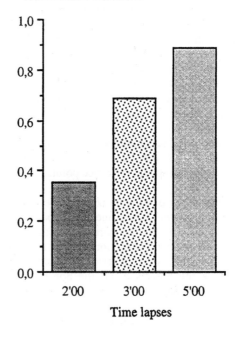

Fig. 4: Effect of the time lapses on the Precision of Anticipation.

4. DISCUSSION

The lack of significant distinction between interfaces and the large number of correct responses showed that subjects performed anticipatory behavior. Moreover, whatever the degree of schematization, all the analogical interfaces allowed prediction of process evolution. Therefore, the degree of schematization neither disturbed nor favored anticipatory behavior.

Results also showed that performance accuracy in the modification task was correlated to the values of the time lapses. The precision of anticipation decreased when the time lapse increased. These results had already been demonstrated in real situations (Cellier and Mariné, 1991). No such difference was found for identification and detection tasks. This might be due to the fact that operators do not have to judge a future state, but instead mentally build one in real work situations.

5. CONCLUSION

In this study, three tasks gave results on the accuracy of anticipation of process evolution. In work situationsas well, operators must anticipate the effects of their actions on the process evolution. The interface functionality might favor this other aspect of the anticipatory behavior involved in bus traffic control tasks. The bus traffic simulator designed for this experiment, can also control task settings. This will be the subject of a future study.

Detection and identification tasks do not seem to disturb anticipatory behavior. Therefore, operators could be helped by the presentation of future process states allowing them to adjust their mental representation of the process evolution.

This research was the first to use the Bus Traffic Control Simulator. It showed us that it can be used easily, and more experiments are planned. This study did not show overall differences between several degrees of functionality for information presentation. The next study will focus on the comparison between the effects of analogical vs. numerical interfaces on anticipatory behavior. Finally, the possibilities offered by the simulator allow comparison of single and double coded interfaces. This study will describe the significance of information redundancy.

REFERENCES

Bainbridge L. (1978). The process controller. In: *The study of real skills, Analysis of Practical skills* (W. T. Singleton, (Ed.)), **1**, pp. 236-263. MTP Press, St. Leonargate.

Bainbridge L. (1988). Types of Representation. In: *Tasks, errors and mental models* (L. P.,Goodstein, H. B.Andersen, S. E.Olsen, (Eds.)), pp. 70-91. Taylor & Francis, Londres.

Bainbridge L. (1989). Development of Skills, reduction of workload. In: *Developing skills with information technology* (L. Bainbridge, S. A. Ruiz Quintanilla (Eds.)), pp. 87-116. John Wiley & Sons, New York.

Cellier J.M. and C. Mariné (1991). Anticipatory knowledge in a bus regulation task. In: *Designing for Everyone* (Y. Queinnec, F. Daniellou (Eds.)), Proceedings of the 11th Congress of the International Ergonomics Association, Vol. 1, pp. 510-512. Taylor & Francis, Londres.

De Keyser V. (1988). De la contingence à la complexité: l'évolution des idées dans l'étude des processus continus, L'ergonomie des processus continus, *Le Travail Humain*, **51**, 1, pp. 1-18.

Eyrolle H., N. Boudes., J. M. Cellier and F. Decortis. (1994). Anticipation aids in traffic management activities, *Temporal management in dynamic situations*, 23rd International Congress of Applied Psychology, Madrid, Spain, July 1994.

Mailles S., C. Mariné and J. M. Cellier (1993). Interface design for dynamic process: an analysis on a bus traffic regulation task, *Cognitive Science Approaches to Process Control*, Copenhagen, August 1993.

Woods D. D. and E. M. Roth (1988). Aiding human performance II: From cognitive analysis to support systems, L'ergonomie des processus continus, *Le Travail Humain*, **51**, 2, pp. 139-172.

HUMAN INTERVENTION INTO AUTOMATIC DECISION-MAKING AND AUTOMATIC CONTROL

S. Kim* and T. B. Sheridan*

*Human-Machine Systems Laboratory, Massachusetts Institute of Technology, Cambridge MA 02139, USA

Abstract. This paper presents a paradigm of human intervention into automatic decision-making and/or automatic control and illustrates four types of intervention. The cause and the prevention of intervention are described from the perspectives of the controlled process, the operating environment, the automatic decision-maker/automatic controller, and the human operator. As an example, this paradigm is applied to the traffic incident management system of the Boston Central Artery/Tunnel Project.

Key Words. Automatic operation, Automation, Human factors, Human-machine interface, Human supervisory control, Integrated vehicle highway systems (IVHS)

1. INTRODUCTION

As the computer has become embedded in various machines and the machines have become automated, the role of the human operator has changed from that of direct controller to that of supervisor. Of the five supervisory roles, namely planning the task, teaching the computer, monitoring the automatic action, intervening in the automatic operation as necessary, and learning from experience, it is the intervention role which currently seems to be most lacking in theoretical or design discipline.

One reason for this, the authors believe, is that in many real applications human intervention into automatic decision and/or automatic control is regarded as a temporary compromise for a bad automatic system, and therefore is tantamount to failure. It is common practice to design automatic systems to avoid human intervention as much as possible, and to regard such systems as less good if they call for any significant degree of human intervention. While in some ideal sense and for very simple tasks that may be true, for tasks of any complexity it is practically impossible to build a system in which some human intervention is not necessary. Therefore, in our opinion, automatic systems should be designed so that the operator can easily intervene when necessary, but at the same time such intervention must not become the cause of de-stabilizing transients or otherwise make system performance worse. To do all of this it is necessary to understand human intervention into automatic operation as a process.

With a continuous control system (a pasteurization plant simulator), Lee and Moray (1994) presented a quantitative model to predict an operator's intervention in terms of trust in automatic control system and self-confidence of the operator. They showed that operators used an automatic control system if trust exceeded self-confidence. Kirlik (1993) used a light helicopter simulation and showed a Markov decision process modeling when a pilot programmed, engaged, and disengaged an autopilot under various engaging and disengaging costs. He assumed that a pilot managed the autopilot system to keep workload and performance at acceptable levels. Riley (1994) shows a qualitative intervention model in terms of risk, trust in automation, self-confidence, and task complexity.

In this research, a paradigm of intervention is developed for a general supervisory control system and to explain why, when and how an operator does or does not engage a decision aid, initiate automatic control, or allow automatic control to be completed. This theory provides guidance in designing decision support systems, manual response procedures, human-computer interaction and operator training.

This paper is organized as follows: In Section 2, the definition and the causes of intervention are introduced. Also, the relationship between the prevention of intervention and the structure of a supervisory control system is discussed. Section 3 describes the Boston Central Artery/Tunnel project and the operator-in-the-loop simulator used to investigate the human op-

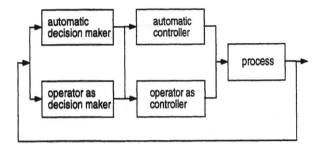

Fig. 1. Structure of Supervisory Control System

erator's intervention. This is followed by the conclusion.

2. INTERVENTION

2.1. *Supervisory Control System*

A supervisory control system can be a decision-making system (one which generates a decision including a decision aid, an expert system, or a decision support system), a control system (one which can execute a decision and bring system performance to some desired state, or in any case one whose expected action can be anticipated relative to the advice of the decision aid), or a combination of decision-making and control system under the supervision of a human operator. For generality, a supervisory control system is assumed to include both a decision-making system and a control system, which are embedded in a computer. Figure 1 depicts the structure of such a supervisory control system. In this structure, both a computer and an operator can make decisions for and control the process.

2.2. *Definition of Intervention*

A supervisory control system can embody many control tasks as well as many control modes. Each task may need a different control mode. The control mode can be an automatic control mode, a manual control mode, or a mixed control mode (Sheridan, 1992).

Since a supervisory control system normally include many control tasks as well as many control modes, a question is raised how to change back and forth among control tasks and control modes. Very seldom are the criteria for the change of control task or for the change of control mode available or preprogrammed. This decision activity typically requires a human operator, and is called intervention.

The intervention can be described as a human operator's action to modify or to reject a decision generated by an automatic decision-maker, or to stop an ongoing automatic control activity, to ad-

just some control parameters, to initiate another automatic control task, or to take over the automatic control and do a control task manually.

2.3. *Causes of Intervention*

An intervention can be initiated by many factors. It can be the controlled process, the working environment, the instrumentation, or the human operator.

Suppose one component of process fails. Unless a system has an automatic back-up component to take over immediately for a failed component, the failure will propagate throughout the system. An operator will notice the deviation from the desired states and eventually take over the automatic decision-making and/or the automatic control to compensate for the deviation.

A disturbance or an unexpected operating environment may bring about the change of priorities of goals to be achieved in a supervisory control task. This priority change can cause an intervention. For example, consider air traffic control. In normal weather conditions, the first priority of traffic management is to maintain smooth traffic flow. In severe weather conditions such as a snow storm, the safety of the system has higher priority than the smoothness of traffic flow. The environmental change often forces an operator to intervene and close an airport.

Poor performance of a previous control task can precipitate an intervention in a subsequent control task. When an automatic decision-maker fails to produce a right decision or an automatic controller fails to perform a given task, it may affect an operator's ability to perceive that another decision-making task or another control task may not be performed well, either. For example, if the false alarm rate is high, the operator tends to ignore alarms, and furthermore may not accept any other decisions made by the computer.

Finally, an intervention can be caused by a human operator. From an operator's own experience, he or she builds a mental model for corresponding control tasks. Whenever the computer makes a decision or controls the process, the operator compares it with his or her mental model. If the output of the computer does not match the operator's mental model, the operator may not accept it and may intervene.

Changes in reward or punishment for the outcome of task may also cause or prevent an intervention. From decision theory, an operator is willing to intervene when sufficient reward is given. With a sufficient punishment, on the other hand, an op-

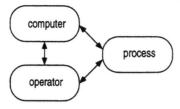

Fig. 2. Parallel-Structured Supervisory Control System

Fig. 3. Serial-Structured Supervisory Control System

erator is reluctant to intervene, because of the liability in case the intervention results in a failure.

2.4. Prevention of Intervention

There are two factors to prevent or limit an intervention: an operator's motivation by reward or punishment as described in the previous section, and the structure of the supervisory control system.

A supervisory control system can be structured with an operator and a computer acting in parallel or in series, determining whether the operator can access a controlled process without any help of computer. In a parallel-structured supervisory control system (Figure 2), the operator can observe and control the process without any interpretation of a computer. There is a direct control channel available between the operator and the process; the operator can always intervene. A conventional (mechanically controlled) airplane belongs to this category.

For a serial-structured system (Figure 3), all control commands of the operator have to go through a computer and be executed by the computer, for example, a 'fly-by-wire' airplane. With a serial-structured system, the operator may or may not intervene in the automatic system; that depends on the degree of automation. In a highly automated serial-structured system, a computer can be programmed to decide what to do and execute it without any approval of or any communication with an operator, and the system designed so that an operator cannot intervene in the automatic process even when he or she wants to. Also in special situations, an operator's actions are constrained in some ways, so called 'envelope of safety', and cannot intervene.

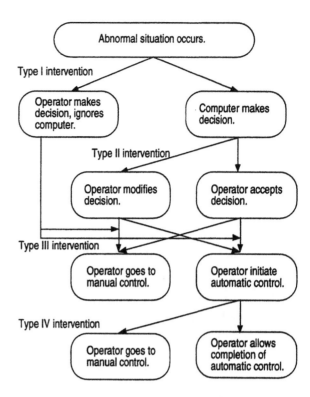

Fig. 4. Paradigm of Intervention

2.5. Types of Intervention

Four types of intervention and their interactions are shown in Figure 4. Suppose an abnormal situation occurs and an operator detects and evaluates it. The operator can employ the decision aid for its advice; or make his or her own decision (Type I intervention) based on his or her knowledge and experience. If the decision aid is employed but the operator does not wish to implement its resulting decision (or automatic action considering the advice received), he or she then can modify the advice partially or fully (Type II intervention). With the decision (whether the decision was made by the computer or the operator), the operator may initiate automatic control; or he or she may decide to go directly to manual control (Type III intervention). In the middle of or at the end of the automatic control the operator may feel that manual control takeover is necessary, and may then control the process manually (Type IV intervention).

Type I intervention and Type II intervention are related to decision-making, and Type III intervention and Type IV intervention to control. Type I intervention happens before a computer generates a decision. Type II intervention occurs after a computer suggests a decision. Type III intervention happens before initiating automatic control and Type IV intervention is in the middle of or at the end of automatic control.

3. CENTRAL ARTERY/TUNNEL PROJECT IN BOSTON

3.1. *Overview*

The above theoretical framework is applicable to many human-machine systems such as a pilot's takeover from an aircraft autopilot, a driver's takeover from the cruise control in a vehicle, or a traffic operator's takeover from the automatic traffic incident management system. Our example is of the latter type.

The Boston Central Artery/Tunnel (CA/T) project is an eight billion dollar project involving about 120 lane-miles of tunnel and two major interstate highways through Boston. In this system once an "incident" (for example, collision, fire, spill of fish, power outage in tunnel, or excess CO buildup in tunnel) occurs, the incident is "managed" interactively by a team of human operators in a central traffic management center and by computers. They are augmented by sensors (several hundred video monitors, as well as optical and electromagnetic sensors) and a communication network (to further interrogate sensors and equipment beyond what is detected and communicated automatically, to dispatch tow trucks, ambulances, fire fighters, police, or repair crews, to regulate traffic by means of variable message signs and lights, and to modify flow and direction of tunnel ventilation air, etc.).

3.2. *Task Analysis*

On the basis of a current design document (Massachusetts Highway Department, 1994), tasks of a traffic operator were analyzed intensively in the Human-Machine Systems Laboratory at Massachusetts Institute of Technology. From the task analysis, there are five tasks for incident management: to monitor the CA/T facilities as well as the traffic, to detect any abnormal events, to classify the detected event, to make response plans for the event if it is an incident, and to clear the incident in time (Figure 5).

When a traffic incident occurs, the incident detection system will detect and generate an alarm. Once an operator accepts an alarm (Detection), the operator has to classify the incident by looking at video monitors and informing the computer about the characteristics of the incident, including the number of vehicles involved the incident, the possibility of fire, any personal injuries, and any hazardous material (Classification). Based on the given information, the computer generates an incident response plan. Then the operator evaluates it, and accepts it or modifies it partially or fully, and sends the response plan back to the computer

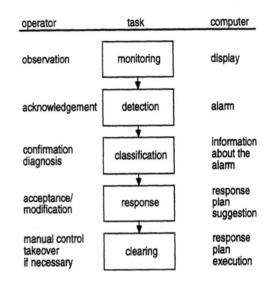

Fig. 5. Tasks of Traffic Operator in the CA/T project

(Response). The resulting response plan will be automatically executed by the computer. The operator will further observe the incident and modify the response plan if necessary (Clearing).

3.3. *Operator-in-the-Loop Simulator*

Our paradigm of intervention can be applied to the CA/T project. To observe and research when and how an operator does or should modify a suggested response plan from the computer or override the automatic traffic control, we developed an operator-in-the-loop simulator on a personal computer equipped with the Intel 486 microprocessor and the Microsoft Windows operating system. The simulator is real-time and consists of six modules: a traffic flow module, a traffic control device module, an incident module, an agency module, a response plan module, and a graphic user interface module.

The traffic flow module based on a macroscopic traffic flow model calculates the average velocity, the average density, and the volume for each segment of highway. The traffic control device module describes LUS (lane use signal), VMS (variable message sign), VSLS (variable speed limit sign), and HAR (highway advisory radio). The incident module represents the dynamic behavior of an incident, which is provided in an input data file. The agency module shows the movements and the incident management activities of various agencies including emergency response teams, the police, the fire department, and the emergency medical service teams. The response plan module generates a response plan for each incident. It determines which agency to dispatch, which traffic control device to be changed, and how to close lanes and ramps. The graphic user interface module shows windows, dialogs, and graphics in order to

310

provide information to the operator and to receive commands from the operator.

Experiments with the simulator are on going, and are to evaluate and verify the means (the layouts of windows, the sequences of windows, etc.) of control, to measure operators' response times with various incidents, and to observe operators' behavior, especially the intervention, to various scenarios.

4. CONCLUSION

In this paper, we present a paradigm of intervention, and relate it to the structure of a supervisory control system, for example parallel-structure vs. serial-structure. We discuss the causes, the prevention, and the types of intervention. An intervention can be caused by the controlled process itself, by changes in operating environment, by the previous performance of the automation, or by changes of an operator's mental model. An operator can be prevented from intervening by the structure of the supervisory system or by a high degree of its automation. There are four types of intervention: Type I intervention is when an operator does not use an automatic decision-making system and makes his or her own decision. Type II intervention is to modify fully or partially the decision generated by the computer. Type III intervention is where the operator controls manually from the beginning, not using the automatic control system. Type IV intervention is to override the automatic control in the middle of or at the end of the automatic control process. This intervention theory is being developed in conjunction with design of an advanced traffic management system for the Boston Central Artery/Tunnel project. For the evaluation of an operator's responses, an operator-in-the-loop simulator has been built and experiments are being carried out.

5. REFERENCES

Kirlik, A. (1993). Modeling strategic behavior in human-automation interaction: why an "aid" can (and should) go unused. *Human Factors*, **35**, 221–242

Lee, J.D. and N. Moray (1994). Trust, self-confidence, and operator's adaptation to automation. *International Journal of Human-Computer Studies*, **40**, 153–184

Massachusetts Highway Department (1994). *Central Artery/Tunnel Project, Integrated Project Control System*. Boston.

Riley, V.A. (1994). *Human Use Of Automation*. Ph.D. Thesis, University of Minnesota.

Sheridan, T.B. (1992). *Telerobotics, Automation, and Human Supervisory Control*. MIT Press, Cambridge.

FLOWER - A System for Domain and Constraint Visualization in Computer-Aided Design

E. RÓZSA*, J.C. DILL**

*MPR Teltech, Department of Broadband Communications, Burnaby, BC, Canada

**Simon Fraser University, School of Engineering Science, Burnaby, BC, Canada

Abstract. This paper explores visualization aids to an interactive intelligent design application. Visualization helps a designer explore data and information in order to gain greater understanding and insight into the design process. Several existing techniques taken from object-oriented expert systems, computer graphics and user interface methodology were combined in order to achieve this goal.

Key Words. Design Systems; CAD; Artificial Intelligence; Intelligent knowledge-based systems; Computer Graphics, Constraint satisfaction problems

1. INTRODUCTION

The current computer graphics technology in computer-aided design systems lacks interpretive ability. Lines are simply lines without a specific meaning such as "these lines represent a room", with the properties of a room, such as walls, neighbors and relations to other rooms. Without such relations or constraints, interpretation of the drawing is completely in the mind of the designer. An expert system can help supply needed interpretation through an appropriately structured knowledge base, and can keep track of many design relationships. However, an expert system on its own, i.e. without visualization tools and support, cannot supply needed visual feedback to the designer. For instance, without appropriate graphic indications from the system, consequences of poor or incorrect design choices are difficult for the designer to see.

Even those systems with a graphics environment and an expert system suffer from a lack of interactiveness; their approach is to request a design goal from the designer and generate a solution with no further input (e.g. (Henry and Hudson, 1988), (Hudson and Yeatts, 1991) or (Seligmann and Feiner, 1991) If not satisfied with the solution, the designer would have to restart from scratch, discarding the entire solution, including acceptable elements. This suggests the need to support an interactive, mixed-initiative approach. The FLOWER (*Floor LayOuts With Expert Recommendations*) system described in this paper represents such an approach: an expert assistant working *with* the user in a *mixed-initiative*

style. The system provides feedback about designer choices, "approving" acceptable choices and indicating errors when they occur. For some errors, the designer may proceed but the error is marked. More severe errors require correction immediately. The system also gives suggestions to the designer about the outcome of certain design choices. For example, the system shows acceptable areas of placement when the designer chooses to add a new room of some type, taking the current layout and constraints into consideration.

Thus our research goals were to create a system capable of

- visualization of the domains of design variables and design constraints to let users *see* how their design space changes as a result of their actions
- letting the user design in an interactive manner rather than simply accepting or rejecting an automated design
- providing visual suggestions or guidelines on how to proceed with the design
- checking design choices and making corrective suggestions or marking problem steps

FLOWER was created as a first step toward fulfilling the above requirements. We were unable to find a description of an existing system that addressed all of these goals, though the need for such system is recognized by many. The system closest to meeting our goals simply produced a long textual explanation for the user about the system's decision (Alpert, 1993).

The specific task chosen was an abstract version

of building floor layout, a tedious task when considering many initial requirements.

2. RELATED WORK

This section briefly summarizes several papers relevant to portions of our work. The papers are sorted into two groups. First systems that were fully automated are described, then those systems where the user plays a role in the design.

2.1. *Automated Design Systems*

The work of Seligmann and Feiner on the use of expert system in designing illustrations (Seligmann and Feiner, 1991) shows that design is a goal-driven process within a system of constraints. When analyzing a partially completed design, their system backtracks for generating a better solution so previous mistakes or off-track solutions can be avoided.

The research of Hudson and Yeatts in (Hudson and Yeatts, 1991) describe a technique for integrating rule-based inference methods into a direct manipulation interface builder. Though they refer to the desirability of the designer control of the process, their system followed an automated approach.

Baykan and Fox in (Baykan and Fox, 1987), (Baykan and Fox, 1988*a*) and (Baykan and Fox, 1988*b*) investigated constraint-directed heuristic search as means of performing design. Their application was very similar to ours: they were designing layouts of kitchens. They were also emphasizing on using constraints throughout the design process. They however, did not deal with feedback from the user.

2.2. *Interactive Design Systems*

The work of Kochar in (Kochar, 1990) was closely related to ours as he is intended to provide help throughout a design process. Kochar's system, *FLATS* is a prototype for design automation via browsing and was constructed to demonstrate the paradigm of cooperation between the user and the computer in CAD applied to the design of small architectural floor plans. A structuring mechanism helps the user explore design alternatives in a systematic way, by varying those properties of the design that are of primary interest.

The work of Kamada et.al. on visualization of abstract data in (Takahashi *et al.*, 1991) is important for the use of feedback from the user throughout the design process. To lessen the need for continuously varying the mapping rules between the

infinite number of possible representations, they used two intermediate, universal representations, and developed a set of rules for mapping one to the other. The intermediate representations can be left unchanged, even if the application changes. The mapping process handles the following representations: Application's Data Representation (AR); Abstract Structure Representation (ASR); Visual Structure Representation (VSR); Pictorial Representation (PR).

$$AR \longleftrightarrow ASR \longleftrightarrow VSR \longleftrightarrow PR$$

Kamada's work represents an important contribution to the area, since it is one of the very few that acknowledge that user and system must work together: that the user must not be left out of the design process.

Finally, the work of Dill et.al ((Calvert *et al.*, 1991) and (Dill *et al.*, 1993)) on intelligent computer aided design was the predecessor of FLOWER of our research group.

3. VISUALIZATION OF CONSTRAINTS

An important part of our work is that FLOWER actually help the user in the process of design. This help is threefold:

- the system acts as a *helper*, showing acceptable areas for a new room, or how to proceed when encountering a design step that contradicts a user-specified relationship
- the system indicates whether a certain step in the design was successful or not

    a) if a constraint representing a physical law was violated (e.g. overlapping rooms), the system does not allow the placement of a new room until the problem is corrected

    b) if a constraint representing a user preference was violated, the placement is allowed but the violation is shown and the offending relationship is indicated

FLOWER uses color-encoded *visual clues* to assist the designer. Each constraint type has a distinct color with a legend as a memory aid. There is also an array of buttons, one for each room type, also color-encoded (as are the corresponding rooms). The designer places rooms by selecting a button and dragging the room with a mouse. The room is shown in its assigned color during this process.

The first visual design suggestion is the valid area of the available space for the placement of that room, shaded to correspond to the color of the room. The shaded area uses the room hue but with lower saturation (the shaded area for a red room is pinkish). If, however, the designer had specified preferences (e.g. *beside* constraint) for the room to be placed, the shaded area for room placement is the color of that constraint.

If the designer ignores the suggestion, and places a room in an unshaded and hence unacceptable area, the system will respond immediately. The problem will be evaluated and if the placement of that room violated a physical law then the user must remove that room from the floor plan in order to continue with the design. However if the room violated a user-specified relationship then the system so indicates. First, the room is shown outlined only indicating a problem. Second, a line will be shown between offending rooms, with the color of the violated constraint, "explaining" what went wrong; if more than one constraint is violated, additional lines are shown). Finally there will be an indication for proper placement of that room in the form of two arrows below the constraint indicator line; the arrows suggest whether the designer should move the room closer ($><$) or further ($<>$) away; the arrows will only appear if the correction is possible (if the user specifies contradictory constraints, there is no way to satisfy them, and no arrow appears).

If a room is in several different relationships to existing rooms, e.g. beside some but far away from others, the available area will still be calculated, but the color of it will be specific to the room to be placed and not to the constraints. An example is shown in Figure 1.

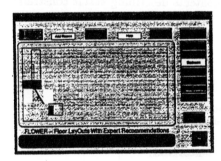

Fig. 1. Visualization Aids for Placing a Master Bedroom with Several User-Specified Constraint

4. SYSTEM ARCHITECTURE

Many researchers (e.g. (Baykan and Fox, 1987), (Calvert et al., 1991), (Dill et al., 1993), (Henry and Hudson, 1988), (Kochar, 1990)) suggest that constraints are a natural way to express design goals. However, because of the lack of readily available constraint processing systems, much work has focused on implementation of such constraints. Kamada et.al in (Takahashi et al., 1991) emphasized the importance of a bi-directional translation between the data representation of an application and the pictorial representations of the user interface as a way to involve the user in the design process. In our case, both the constraint processing system and a bi- directional translation (outlined below) were already available, allowing us to concentrate on further steps.

The structure of FLOWER can be described in Kamada's terms. In FLOWER, rooms form the Pictorial Representation. The VSR data are *draw_house* and *draw_room(new_room, list_of_oldrooms[])* etc. The *draw_room()* describe graphical relations between the *room* graphical objects as it incorporates the representation of minimum and maximum sizes and color information. The ASR data are the room instantiations (*room R2 isa room*), and the constraints, physical (*R2:no_overlap(R1)* and user-specified (*beside(bedroom, bathroom)*). For example, *no_overlap* is an abstract relation between the abstract objects *R1* and *R2*. AR corresponds to the user's request for placement of the next room by means of an input device (mouse).

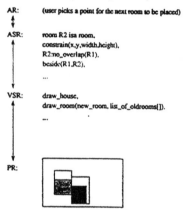

Fig. 2. Structure of FLOWER Described in Kamada's Terms

Again, while Kamada's group had to put a significant effort in developing the communication channel, in our case, the mapping between ASR and VSR was already available. In our case both AR and PR reside in the graphics side of our application. When the user indicates the placement of a room, AR is recognized by the graphics modules and the mapping between AR and ASR is done through knowledge base update routines and the mediator. The knowledge base update routines are responsible for the application dependent part and the mediator is responsible for the link. (Sections 4.4 and 4.3 give more details.) ASR is represented in the knowledge base and becomes accessible through this link. VSR is represented by the graphics database. When our reasoning engine evaluates the user's design goal, the result is sent back to the graphics again through the mediator and the graphics database update routines, thus implementing the mapping from ASR to VSR. The updated graphics database is then mapped back to the graphics module, (VSR to PR) where the picture of the room is created.

Constraints	Meaning
beside	$d = 0$
close	$1 \leq d < 2Dim$
near	$2Dim \leq d < 4Dim$
far	$d \geq 4Dim$
not beside	$d > 0$

d is vertical or horizontal distance between rooms; Dim is corresponding room dimension.

Table 4.1. Preference Constraints

Figure 3 shows a block diagram of our system, the components of which are described in more detail in the following sections.

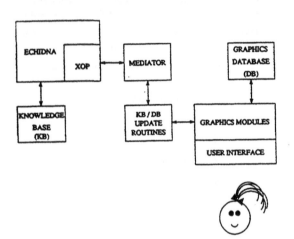

Fig. 3. System Architecture

4.1. *The Knowledge Base*

The following were our initial set of rules for a simple system:

1. houses are rectangular
2. rooms are the smallest element of the design, i.e walls, doors, etc. are not considered
3. rooms are rectangular, and their edges are parallel to those of the house
4. size constraints: rooms have a minimum and maximum size
5. restrictive constraints: rooms must be *inside* the house
6. topological/geometrical constraints, i.e. placement of rooms with respect to each other:

    a) rooms cannot overlap

    b) a given type of room can be beside or not beside, close, near or far from another specific room type.

Table 4.1 briefly describes the meaning of these constraints.

4.2. *Constraint Propagation*

The expert system used in FLOWER is the Echidna model-based reasoning engine. FLOWER's knowledge base is implemented in the Echidna object- oriented constraint logic programming language (Sidebottom *et al.*, 1992).

Constraints represent relationships between variables. A constraint network is constructed during the design session, where the variables are the nodes and the constraints are the arcs between them. In Echidna, the internal propagation of constraints narrows the domains of the variables involved, enabling a solution to be found more efficiently. A constraint is activated whenever the domain of one of its arguments is refined or bound to a particular value. This process can propagate among those variables that share constraints on their parameters.

4.3. *Knowledge Base / Database Update Routines*

The database update routines communicate design goals to Echidna (e.g. getting parameter domains for a specified room, constraining the values of some parameter, etc.) and to accept information from Echidna, such as the domain for a specified room location.

As an example, Figure 4 illustrates knowledge base contents describing *beside* constraints.

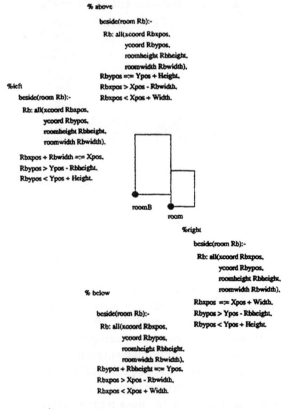

Fig. 4. Beside Constraints: "room" should be beside existing "room B"

316

When the designer places a room, an AR is created. We need the mapping from AR to ASR, or in other words, we have to issue the corresponding design goals. Formulation of these goals is done here, in the knowledge base update module. Then, the system issues these goals to Echidna through the mediator module. After Echidna evaluates the design goal, results are returned through the mediator. Then the interface update routines form a mapping between ASR and VSR. The graphics module will be notified by the interface update module about such changes, and will modify the pictorial representation (mapping from VSR to PR is formed).

4.4. *Mediator*

As described earlier, the abstract design goals (Kamada's ASR) are expressed in the knowledge base. However, a link had to be created through which a connection can be established between the abstract and the visual representations. This link is established through Echidna's External Object Protocol (Sidebottom *et al.*, 1992). To send information from the application to Echidna, queries are issued over this link. Likewise, Echidna terms can be unified with the terms constructed in the application. The mediator supports communication between the reasoner and the graphics part of FLOWER and maintains the communication flow.

4.5. *Graphics Module*

Following our earlier thread of describing our work in Kamada's terms, the graphics module is responsible for generating the pictorial representation (PR) of the design objects. It also supports the user interface and visualization functions. The visualization however, is not part of the general mapping; it is calculated here entirely.

FORMS (Overmars, 1993) was used for the user interface and gl (SGI's Graphics Library) was used to manipulate the graphical objects and show the results of the visualization. Rooms are represented in the same way in the graphics database as they are in the knowledge base, by location and size. Each room type (kitchen, bathroom, etc.) has a minimum and maximum width and height. Rooms are represented visually as rectangles. The minimum size is shown filled and the maximum size is shown outlined. When Echidna notifies the graphics module that the domain of a variable has changed, the representation is changed accordingly.

5. DISCUSSION

5.1. *Complexity*

Use of different colors for different constraints could be a problem if the total complexity of the system is high. The complexity of the system can be described by the number of rooms and the number of relationships between them. One solution could be that the user is presented with a subset of constraints at any given time. This is a restriction but it avoids the confusion of many colors. Another way of dealing with this would be to support some other visual representation. For example, a flag (something attached to an object) could show some of the constraints. Again, this could also represent a problem if we have too many constraints. Finally, we could also use a combination of colors and flags. We could show the available areas of placement in the constraint's color, but indicate a violated constraint with a flag.

5.2. *Rooms with Several Relationships*

Naturally, when designing a floor layout, rooms are expected to appear in more than one relationship. As long as the designer is placing them properly there will be no problem. However, (again naturally), the user cannot be expected to do so. Our system will then give a visual representation of the problem. When the newly added room is part of several constraints, the room will be connected by lines representing all failed constraints to all other rooms in the failed relationships. If many rooms are involved, a badly placed room may result in many "unsatisfied-constraint" lines. This is an issue to be explored. One way to reduce the number, for example, would be to put a priority on the severity of the violation, and only show those above a designer-set threshold.

5.3. *Suggestion for Failure Correction*

FLOWER currently provides suggestions for correcting failed constraints, consisting of arrows on each of the failed-constraints lines. The correct direction might be unclear if there were several such failed constraints. A better solution would be to have FLOWER move the offending room to a place where the relations are satisfied if that exists. This corrective action does not have to be accepted by the user, (s)he could overrule it by removing the room from the workspace. The difficulty of this method lies in the fact that there will be many ways to correctly place the room, and the system would have to try to match the placement to the one that the user specified earlier (but failed). Matching could for example be based on the closest distance from the user picked placement point to an available point.

6. FUTURE WORK

Only a limited number of constraints were implemented in this first version of FLOWER. We implemented dimensional, restrictive, topological and functional constraints. However, many other constraint types would be useful additions: *accessibility constraints* (e.g. room or window placement); *practicality constraints* (e.g. window placement based on location); *interconnection constraints* (e.g. stairs and stairway placement); *aesthetic constraints* (criteria still need to be identified, along with methods for implementing these constraints.)

Another extension is to allow the user to "change his/her mind" and move an already inserted room. In this way he/she would get complete design freedom. A relaxed version of this is to let the user move the room that was placed last, as long as (s)he did not indicate further placement.

Rooms are rectangular in the current version. A required extension would allow arbitrary (but still Manhattan) polygons. Formulating constraints in the knowledge base (close/near/far constraints) would be straightforward as the distance between rooms can be measured between some defined "midpoints" of the polygons. However, beside constraints would be more difficult.

We might free the designer from additional repetitive tasks by allowing partial automation at some stages of the design, such as (initial) placement of passages in placed rooms. Clearly the user still has to accept the placement. Another initiative might be placement of a room based on a design specification.

7. SUMMARY AND CONCLUSION

The objective of this research was to explore interactive intelligent design with visual aids to the designer. Maintaining constraints is managed by the system and is not the responsibility of the user. The user is not eliminated from the design process but rather is integrated into it. Our system also provides some explanations of choices via visualization of design variable domains and constraints on them. The system works in a mixed-initiative style, supporting *interactive* intelligent design by providing visual suggestions or guidelines, checking design decisions, suggesting corrective actions and marking problem steps. We have shown that visualization of variable domains and design constraints aids the design process and have demonstrated intelligent help for designing simple floor layouts.

8. REFERENCES

Alpert, S.R. (1993). Graceful interaction with graphical constraints. *IEEE Computer Graphics and Applications* 13(2), 82–91.

Baykan, C.A. and M.S. Fox (1987). An investigation of opportunistic constraint satisfaction in space planning. In: *Tenth International Joint Conference on Artificial Intelligence*. pp. 1035–1038.

Baykan, C.A. and M.S. Fox (1988a). Constraint satisfaction techniques for spatial planning. Extended Abstract.

Baykan, C.A. and M.S. Fox (1988b). Opportunistic constraint-directed search in space planning. In: *IFIP Working Group 5.2 Workshop on Intelligent CAD*. pp. 1–6.

Calvert, T., J. Dickinson, J. Dill, W. Havens, J. Jones and L. Bartram (1991). An intelligent basis for design. In: *IEEE Pacific Rim Conference on Computers, Communications and Signal Processing*. Victoria, BC. pp. 371–375.

Dill, J., J. Jones and Stefan W. Joseph (1993). Intelligent computer aided design. *International Journal of CADCAM and Computer Graphics* 8(2), 175–184.

Henry, T.R. and S.E. Hudson (1988). Using active data in a UIMS. In: *ACM SIGGRAPH Symposium on User Interface Software*. Banff, AL, Canada. pp. 167–178.

Hudson, S.E. and A.K. Yeatts (1991). Smoothly integrating rule-based techniques into a direct manipulation interface builder. In: *UIST'91*. Hilton Head, SC. pp. 145–153.

Kochar, S. (1990). A prototype system for design automation via the browsing paradigm. In: *Graphics Interface '90*. Halifax, NS. pp. 156–166.

Overmars, M.H. (1993). Forms library; a graphical user interface toolkit for silicon graphics workstations version 2.2. Department of Computer Science, Utrect University.

Seligmann, D.D. and S. Feiner (1991). Automated generation of intent-based 3D illustrations. *ACM SIGGRAPH Computer Graphics* 25(4), 123–132.

Sidebottom, S., W. Havens and S. Kindersley (1992). Echidna constraint reasoning system (version 1): Programming manual. Expert Systems Laboratory, Centre for Systems Science.

Takahashi, S., S. Matsuoka, A. Yonezawa and T. Kamada (1991). A general framework for bi-directional translation between abstract and pictorial data. In: *UIST'91*. Hilton Head, SC. pp. 165–174.

TOWARD THE APPLICATION OF MULTIAGENT TECHNIQUES TO THE DESIGN OF HUMAN-MACHINE SYSTEMS ORGANIZATIONS

Emmanuelle LE STRUGEON, Martial GRISLIN and Patrick MILLOT

LAMIH, URA CNRS 1775, University of Valenciennes
BP 311 - 59304 Valenciennes, FRANCE
e-mail: {strugeon,grislin}@univ-valenciennes.fr

Abstract: This paper proposes to address the problem of work organization design in human-machine systems with the help of multiagent systems techniques. The organization, defined as the structure supporting tasks attribution, coordination and communication, is the subject of many studies in Distributed Artificial Intelligence (DAI). Three organization modes are presented. These modes are seen as foundations on which a global organization including humans and intelligent machines can be built, with some restrictions due to the characteristics of humans. Althought its simplifying aspect, this could be a step toward the application of multiagent concepts to address some of the humans-machines cooperation problems.

Keywords: Human-Machine system, Work organization, Distributed artificial intelligence.

1. INTRODUCTION

Current supervision systems include both computing modules dedicated to specific tasks (graphical displays, data processing, management of alarms, etc.) and human operators. When such a system leaves its normal functionning state, the dysfunction is generally solved by a team, and not by a single solver. Human and artificial agents compose the solving team, in which all the members have to cooperate. How can this heterogenerous team succeed in obtaining a joint solution? They have to exchange knowledge and results to contribute to a global coherent result. This is an organizational problem. We propose to address the problem of work organization design in human-machine systems, with the help of multiagent systems researchs.

This paper is composed of three parts. The first one shows the interest of the multiagent approach for humans-machines interactions study. The second part deals with our experience in the field of the multiagent organizations. In the third part, some applications perspectives for humans-machines organizations are presented.

2. INTERACTIONS IN THE HUMANS-MACHINES TEAMS

According to recent definitions (see, for instance, Grant, 1991), agents are Artificial Intelligence systems or humans. Thus, human-machine systems can be considered as multiagent systems, providing that the involved machine is enough "intelligent", if it includes an expert system, for example. The recent assistance systems are based on artificial intelligence concepts and techniques; they take decisions and act on the supervised process. So, it is possible to see new supervision teams as multiagent systems including two subsystems: a "multi-human" system and a multi-artificial agents system. These subsystems work together to solve problems and act on the process (fig. 1).

The human group acts on the process through a human-machine interface and interacts with the multiagent system through another interface. This latter interface is generally used to request some assistance from the AI agents. AI agents give explanations for the actions and solutions they recommend or directly perform on the process. Between the multiagent system and the process another interface is required. It is a smaller one because it takes place between two machines (it is not a HCI problem). This interface exists —as a filter— because these machines do not speak the same language: the agents exchange messages using a high level communication protocol, whereas the process uses logic states between sensors and control data.

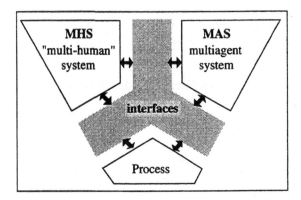

Fig. 1. The supervision team

In the human and artificial agents (MHS and MAS) work team, one can distinguish five interaction types corresponding to different research approaches:

1. *the interactions among the humans of the system,*
2. *the interactions among artificial agents,*
3. *the interactions between one human and one artificial agent,*
4. *the interactions of one artificial agent with several humans,*
5. *the interaction of several humans with the same agent.*

The first interaction type is studied by sociologists and psychologists (i.e., managing and team work, see CACM, 1994). The Distributed Artificial Intelligence (DAI) deals with the second interaction type. The second part of this paper develops this point.

The interactions between one human and one artificial agent (3) has been studied for a long time by human-factor specialists. New types of "intelligent" agents appeared recently. These new agents take more and more the place of personnalised and autonomous sotware assistants. For example, Maes (1994) proposes a metaphor of "the personnal assistant that collaborates with the user": the agent learns how he can assist — and even stand in for — the user in his interaction with the process, by observation and imitation. The interaction between a human and several intelligent systems seems to be quite the same as this interaction type. It can be reduced to the case of a human faced to one machine when he has the possibility to manage himself the communication. An example is given by recent interfaces with many windows opened on the same screen, with in each of them, informations relative to a different and concurrent assistance systems.

The interactions of one artificial agent with several humans (4) is rather complex. The idea is to assist cooperative work, as it is studied by CSCW (Computer Supported Cooperative Work) or HCCW (Human Computer Cooperative Work) researchers (see, for example, Jones *et al.*, 1994; Li *et al.*, 1994). Current systems of this type are often design systems. "The cooperative design is a complex group activity including participants of heterogeneous experiences" (Edmonds *et al.*, 1994). It is a typical communication problem among agents speaking different languages, in the aim to achieve a common task. Existing systems propose a hardware and software support for joint thoughts. Such systems unify the different formalisms used by the group's members, integrate and centralise data from various origins.

The *interaction between several humans with the same agent (5)* is complex too. What is complex is the management of more than one action at the same time on the same piece of data. An AI machine that is connected to a network with several terminals is faced to this type of problem. This is generally solved by queues mechanisms.

The combination of these five interaction types can be found in the interaction between a multiagent system and a human group. Such interactions produce difficulties to share work (Debernard *et al.*, 1992). Especially, it is difficult to attribute tasks to persons or machines and to coordinate the attributed tasks performing. Such problems have been studied, in the more restricted context of the artificial agents, in the DAI field.

3. MULTIAGENT ARTIFICIAL ORGANIZATIONS

Traditionnally, DAI researchers saw the interaction from the individual agent's point of view. Three cooperation modes were distinguished (Hautin, 1986):

— in the command mode, one agent (the "master") breaks down the problem and allocates the sub-tasks among the other agents (the "slaves");

— in the bidding mode, one agent (the manager) invites all the others to tender for a task performance. Some agents answer by the offer of their capacities. The manager selects some of them, that contract to perform the requisite task;

— in the competition mode, all the agents propose solutions for all the sub-tasks. A master agent makes the final selection among them.

In the current works in this field, the "organizations" term becomes more used than the "cooperation modes": "It has been clear for some time in multiagent circles that organization is a powerful concept for thinking about how to structure the interactions of collections of problem solvers" (Gasser, 1994). Actually, it is the same problem studied at a larger and more external level. The cooperation mode, which is observable between two agents is an internal view of the organization observable at the global system level. The multiagent organization can be defined as the structure supporting the communication and control links among the agents.

Multiagent system organizations vary according to the type of agents that compose them. Cognitive agents, that act on their environment, usually form organizations with a structure that is fixed from the system's design. Reactive agents, that are subjected to their environment, form various organizations which are called "emergent" organizations. This study is more focus on cognitive agents' organizations.

Relations between cognitive agents are regulated by some rules. These rules vary with the importance of the control and knowledge decentralisation. Gasser (1992) classified distributed organizations in four categories: centralised, market-like, pluralistic community and community with behavior rules.

— The *centralised organization* is easy to understand because it is an usual structure in human groups. It is a hierarchic configuration, the links between the agents are of the master-slave type. At each level of the hierarchy, masters centralise the decision and control powers. It is a generalization of the command mode of cooperation.

— The *market-like organizations* are based on the principle of contract, like the bidding mode of cooperation. The reference example in DAI is the Contract-Net of Smith (1980).

— The *pluralistic communities* are made of independant agents. They prepare solutions and communicate them to the other members of the community. Each of them verifies, validates or improves the propositions of the others (see, for example, Kornfeld and Hewitt, 1981; Lesser and Corkill, 1983).

— The *community with behavior rules* is a community of specialists. They interact according to defined and known protocols. Multiexpert and multiknowledge bases systems (Gleizes *et al.*, 1990) belong to this category. Control and communication among the specialists is often done by blackboard (BB) techniques (the most known BB systems and tools are Hearsay, GBB and BB1). Some hierarchy and priority rules manage the access to the BB levels.

These categories can be summed up by the four following organization forms:

1. the *hierarchy*, based on master-slave relations
2. the *market*, based on contracts
3. the *community*, based on peers relations
4. the *society*, that allow the coexistence of the three previous control modes.

As a matter of fact, the society form is the more realistic one to design or describe complex

organizations. It includes the three other categories, which conversely appear as basic organization forms. Thus, it seems more interesting to consider the organization classification including the three categories: hierarchy, market and community. Based on this classification, a multiagent system has been designed. It is able to adopt one of these organization modes according to its needs (Le Strugeon, 1995).

In the aim to implement these different types of organization, a representation model must be found. Adapted from a representation described by Pattison *et al.* in 1987, an organization model is proposed. It defines an organizational structure as being made of functional components and of relations between these components (see fig. 2).

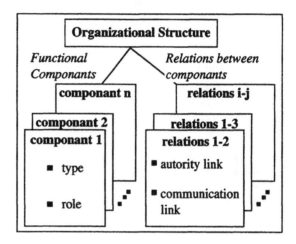

Fig. 2. Organizational structure model

A *component* is described by:
— *a type*, that allows to distinguish a single component from a group component,
— *a role*, that allows to represent the component's position in the organization.

Three roles are defined: coordination, performance and expertise. The coordinator's responsability consists in managing the coordination among the components, that means to organize the actions to perform in a chronological order. The underling makes no decision, it performs actions. The expert provides information to the others.

There are two types of *relations* between components: autority and communication relations. The *autority* links state the relations existing between a component that has a role i with a component with a role j. The *communication* links state the communication direction that can occure between two specific components. Thus, relations among components are determined by their respective roles in the organization.

This description can be used at the different levels of a complex organization structure. If the component is a group, the same method can be used to describe its own organization (its internal components and the links among them). Thus, an organization including many sub-organizations can be described in a recursive way.

The three organization types of our classification are modeled on this pattern. Among the relations between components, only existing (autorized) ones are represented.

The hierarchy model (fig. 3) is composed of two components types: coordinators and underlings. The coordinator is superior to the underling in the hierarchy. Coordinator gives orders to underling components, that perform the tasks and send back results to coordinators.

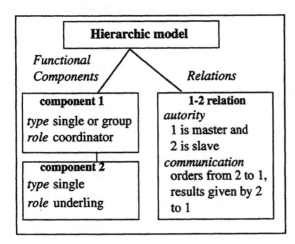

Fig. 3. The hierarchy model

Initially, in the market organization (fig. 4), all the components are peers. Then, contracts create several couples among them. The invitation to tender is made by a coordinator. When the contract exists, some components that had no role become underling ones.

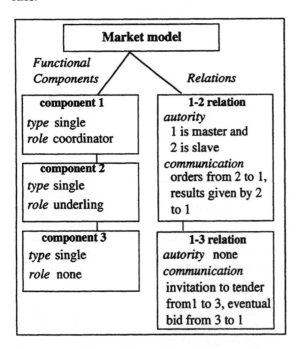

Fig. 4. The market model

The community model (fig. 5) includes only one component role. The organization relies mainly on negotiation and partial plans exchanges. Each component takes alternately expert, coordinator or underling roles.

Each artificial agent of the system has social knowledge about the group and some of the other agents. It knows what the system's current organization structure is (hierarchy, market or community). It knows the roles of the agents it is in contact with and is able to decide how it must behave towards them.

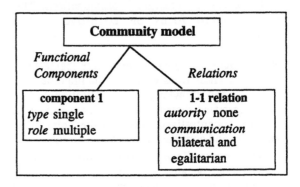

Fig.5. The community model

An agent's behavior is different according to its role and the other agent's role. For example, when receiving an order from an agent B, an agent A has various reactions according to their respective roles, as it is described on table 1. In a hierarchy, an agent which receives an order to perform a task, can agree or not: if its role is "underling", it refuses the orders that come from another "underling", but agrees those from its "coordinator".

Table 1. The agent A's behavior towards an agent B in a hierarchy. A has received an order to perform a task from B (the letter 'X' takes the place of an undifferentiated value).

Does B know A?	A's role	B's role (from A's point of view)	Is A able to perform the task?	A's behavior
yes	underling	underling	X	refusal to perform the task
yes	underling	coordinator	yes	agreement
yes	underling	coordinator	no	agreement & looking for an agent able to perform the task
yes	coordinator	X	X	refusal
no	X	X	X	refusal

322

4. MULTI-ARTIFICIAL AND HUMAN AGENTS ORGANIZATIONS

The organizational techniques used in artificial agents systems are not directly applicable for the teams that include humans. Indeed, human agents have very particular characteristics:

— The autority link between a human and an artificial agent is specific. Even if, in some cases, both of them take decisions and act directly on the process, in most cases the human is the main decision-maker who has (and must have) the last word.

— A human is generally less quick than an artificial agent in routine operations, but it is the opposite in cognitive and complex strategies. Humans are better than artificial agents for the resolution of unexpected problems.

— Human is not always reliable. Human acts are sometimes irrational and his attention may decrease during long time monotonous tasks performing. On the opposite, artificial agents are obviously always rational, but they are not reliable either. Wrong decisions come from erroneous or incomplete knowledge bases.

These features must be taken into account in designing multi-human and artificial agents systems. Because of the kind of experience we have in the multiagent systems field, we focus on decision-making, tasks allocating and coordinating problems. The three organizational structures models described above could help to address the problem of the interactions in humans-machines systems, with restrictions due to the humans characteristics. This application of multiagent techniques to work organization pose some problems. Especially, two problems can be noticed:

1. The representation of the human group in the artificial agents' knowledge bases. More specifically, the pertinent information must be determined: what the artificial agents have to know about their human partners?

2. Which kinds of organizations can be realized in a "multihuman" and multiagent system?

As a response to the first problem, we think that the artificial agents must have a mean to know if the agent they contact is an artificial or a human one. This is necessary to distinguish different behaviors depending on the nature —artifical or not— of the interlocutor. This piece of information about the component's nature can be easily added to components' description and implementation. Some human characteristics can be handled by the implementation of some privileges, that allow, for example, to consider a human as being able to take the role of a coordinator, and forbid to let him only in underling roles.

By reason of the human and artificial agents compared characteristics, the global organization must be designed in letting as much as possible coordinator roles to humans and underling roles to artificial agents. This leads to create an organization based on the hierarchic model. It can be used in usual and normal functionning states of the system.

In dysfunctionning situations, it can be useful to adopt another structure. For example, if there is a problem entailing many concurrent actions performance, the community organizational model is better than the hierarchic one. This organization enables to address these situations in an efficient way. In situations that require to group together a great amount of knowledge of different kind, the market organization can be used. The bidding mode allows the tasks to be handled by the appropriate agents, which have the information that are necessary to achieve them.

Thus, in response to the second problem, one can notice that:

— the hierarchy, with machines performing what humans have decided, is the most used and natural organization;

— the community organization should be used to perform efficiently concurrent actions on the process;

— the market organization enables an efficient use of distributed knowledge among many agents.

The characteristics of different organizations have been assessed by some researchers in the DAI, sociology and economy fields (a synthesis of some of these works is given in Le Strugeon, 1995). With such knowledge, it could be possible to determine the situations in which each organization is appropriate more precisely. However, this is a very simplifying way to see humans-AI agents interactions. In most cases, different kinds of societies —depending on the activity and ressources context— are the organization modes that should be used to address these problems. But this reflection could help to clarify complex and multiagent teams' work.

5. CONCLUSION

This paper is an attempt to contribute to the research about coordination and task allocation in groups that include both human and artificial agents decision-makers. On the base of works in DAI, three organization modes have been presented. These modes are seen as fundations on which a global organization including humans and intelligent machines can be built. Althought its simplifying aspect, this could be a first step toward the application of multiagent concepts to address some of the humans-machines cooperation problems. Multiagent organizations are strongly based on human societies models. That is why we consider this work as a return of multiagent techniques to their origins, by their application to human-machine systems.

REFERENCES

CACM (1993). Special issue about Participatory Design. *Communication of the ACM*, **36** (4).

Debernard S., F. Vanderhagen, P. Millot (1992). *An experimental investigation of dynamic allocation of tasks between air traffic controller and A.I. system*. 5th IFAC/IFORS/IEA Symposium on Analysis, design and Evaluation of Man-Machines Systems, The Hague, The Netherlands, June 9-11.

Edmonds E., L. Candy, R. Jones, B. Soufi (1994). Supports for collaborative design: agents and emergence. *Communications of the ACM*, **37**, n°7, July, pp.41-47.

Gasser L. (1992). An Overview of DAI. In: *Distributed Artificial Intelligence: Theory and Praxis* (Avouris & Gasser (Ed.)), pp. 9-30. Eurocourses, Kluwer Academic Publishers, The Netherlands.

Gasser L. (1994). *Agent organizations for information retrieval and electronic commerce: the next frontier*. Int. Symp. on 5th generation computer systems, Workshop on heterogeneous cooperative knowledge bases, 15-16 Dec., pp. 49-63.

Gleizes M.P., P. Glize, S. Trouilhet (1990). *La résolution distribuée de problèmes dans un environnement multi-agent*. Convention IA'90, Proceed. of the "2ème conférence européenne sur les techniques et applications IA en milieu industriel et de service", Hermes, Paris, Janu., pp.121-135.

Grant T.J. (1991). A review of multi-agent systems techniques, with application to Colombus user support organisation. *Future generation computer systems*. **7**. North-Holland, pp. 413-437.

Hautin F., A. Vailly (1986). *La coopération entre systèmes experts*. Proceedings of the "Journées Nationales du PRC-IA", Cepadues, Nov., pp. 187-204.

Jones P.M., S.M. Kaplan, P.M. Sanderson, S.L. Star (1994). *ALCHEMIST: Support for Emergent Models of Work Practices in Collaborative Systems*. IEEE Int. Conf. on SMC, San Antonio, Texas, 2-5 Oct., pp. 373-378.

Kornfeld W., C. Hewitt (1981). The scientific community metaphor. *IEEE Transactions on Systems, Man and Cybernetics*. SMC-**11**, n°1, pp. 24-33.

Lesser V.R., D.D. Corkill (1983). The distributed vehicle monitoring testbed. *The A.I. Magazine*, Fall 1983, pp.15-33.

Le Strugeon E. (1995). *Une methodologie d'auto-adaptation d'un système multi-agents cognitifs*. Ph. D. thesis, University of Valenciennes, Jan., France.

Li X.M., M. De, K.W. Hipel (1994). *A DSS Architecture for Multiple Participant Decision Making*. IEEE Int. Conf. on SMC, San Antonio, Texas, 2-5 Oct., pp. 1208-1214.

Maes P. (1994). Agents that reduce work and information overload. *Communications of the ACM*, vol. **37**, n°7, pp.31-40.

Pattison H., D. Corkill, V. Lesser (1987). Instantiating descriptions of organizational structures. In: *Distributed artificial intelligence* (Huhns (Ed.)), tome 1, pp. 59-96. Research notes in artificial intelligence, Pitman, London.

Smith R.G. (1980). The contract net protocol: high-level communications and control in a distributed problem solver. *IEEE Transactions on computers*, n°12, vol. C-**29**, pp.1104-1113.

DESIGNING SUPPORT CONTEXTS:
HELPING OPERATORS TO GENERATE AND USE KNOWLEDGE

Gunilla A. Sundström

GTE Laboratories Incorporated
40 Sylvan Road
Waltham, MA 02254, U.S.A.

Abstract: Support system design requires designers to make decisions about what type of information operators need to achieve task goals. In the literature, various operator models have been proposed to help designers make such decisions. However, to use multiple operator modeling approaches, designers need to find a way of integrating support types generated based on these different models. In the present paper it is proposed to use the notion of user support contexts to integrate operator support derived from various operator modeling approaches. The notion of user support context is described and illustrated using an example from the telecommunications domain. In particular, this example demonstrates how knowledge from various modeling approaches can be integrated and seamlessly made available to operators.

Keywords: System Design, Modeling, User Interfaces, Telecommunications

1. INTRODUCTION

In the past decade, a variety of approaches to operator modeling have been proposed. Such models have primarily been used to design and structure operator support in complex systems such as power plants and satellite ground control centers. Examples of such operator models are Rasmussen's skill-, rule-, and knowledge-based taxonomy (Rasmussen, 1983, 1986); Mitchell's operator function model (Mitchell, 1987); and Lind's multilevel flow modeling approach (Lind, 1981). All three approaches, as well as others described in the literature, provide the support system designer with concepts for analysis, and design of human interaction with complex systems. For example, Mitchell's approach is based on discrete modeling techniques and requires that the designer, at some point in the design process, structures his/her analysis using concepts such as "nodes" and "transitions." On the other hand, a designer using Lind's multilevel flow model will focus on generating a representation of the technical system using concepts such as "function," "sink," and "source." The two examples illustrate the diverse nature of concepts provided by various operator models. Because of this conceptual diversity, it is difficult for a system designer to use multiple models during a design process. However, using a diverse set of models might be very desirable when each type of

modeling abstraction provides operators with a different view of the "world." For example, one model could be used to help operators to generate a "functional" view, whereas another model could support operators' choices of the most appropriate actions in a particular situation. However, for an operator to be able to use the various types of knowledge provided by different models, these knowledge types need to be integrated into the same "processing" context. If such an integration is possible, an operator support system would be able to offer operators access to a more diverse knowledge set. If this knowledge set can be adequately organized and managed by the support system, the support system should be able to help operators acquire and use the various types of knowledge they need to perform their tasks adequately. However, very few approaches have been proposed that provide designers with concepts for integrating various knowledge types into one integrated "processing context."

In the present paper, one such "integration" approach is described. It is proposed to focus the design process on the creation of "user support contexts." This contrasts with "piecemeal" support types identified by using one particular operator modeling approach. The described design approach is based on the assumption that user support context design must be focused on capturing the overall context for possible user actions. In the paper, some basic characteristics of user support contexts are described as well as a methodology for specification of aspects of such contexts. The approach is illustrated by providing an example from the telecommunications domain.

2. USER SUPPORT CONTEXTS: SOME ASSUMPTIONS

Focusing the design process on user support contexts rather than on support of, for example, a particular procedural flow, defined by one particular operator model, reflects some basic assumptions about human cognitive processes in complex systems. The most important assumption is that a human in a complex system "constructs" a mental representation of the situation and that all judgments and decisions are made in the *context* of this dynamic mental representation. In recent years, such a contextual view of human cognition in complex systems has been contrasted with a procedural view of cognition (e.g., Bainbridge, 1993; Hollnagel, 1994). Contextual models differ from procedural operator models by their focus on understanding the events and knowledge required to achieve task goals rather than only on the particular steps associated with procedures related to these task goals. A second important assumption when adopting the "contextual" view is that the primary function of an operator support system is to help operators to construct a representation of the current "context" such that they are enabled to acquire and assimilate a diverse set of knowledge, thus enabling operators to consider many alternative courses of action in any particular situation. Consequently, the operator support system should not simply attempt to replace (or automate) operators' cognitive processes. Rather, the design goal should be for the operator support system to help operators utilize their cognitive abilities to construct a rich representation of the current situation and, therefore, enabling operators to consider a variety of options and procedures.

2.1 What Is a "User Support Context?"

A user support context is defined as any situation in which the user interacts with the user support system to achieve a task goal. Thus, the context is defined by an event which makes the user interact with the system in a goal-oriented way. Examples of events triggering operators to use an operator support system are incoming alarms or requests for information by other human operators. From the designer's perspective, a user support context defines situations in which the user needs support from the system to achieve task goals. In any particular design process, the designer's task is to identify user support contexts and design the user support based on this knowledge. Thus, the key elements of a design process based on the user support context notion are (1) a methodology to identify and represent user support context; and (2) A process for deciding what tools are required to help operators to generate and use knowledge in each user support context.

From the designer's perspective, a user support context consists of information and knowledge required in the context and graphical user interface elements designed to enable the operator to acquire, use, and "navigate" in the context. Fig. 1 illustrates the basic components of a user support context.

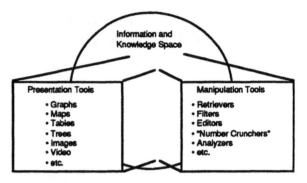

Fig. 1. The "elements" of a user support context

The information and knowledge associated with a user support context corresponds to the information and knowledge (I/K) space associated with the user's task (cf. Sundström, 1993b; Sundström and Ayub, 1994). An I/K space differs from other types of task representations through the focus on *what* information a user needs and *how* this information is used to generate knowledge types required to achieve task goals. It is important to note that the I/K space does not represent a user's "mental model" rather, it represents what the designer has learned about users' tasks specifically with respect to a task's information and knowledge requirements. As Fig. 1 illustrates, the user support context enables the designer to integrate various knowledge types. Thus, the designer can use various operator modeling notions to capture a diverse set of knowledge without having to constrain the actual support design to a set of concepts associated with any particular operator model.

Three types of user support contexts are distinguished based on the assumption that operators' reasons for interacting with a particular operator support system can be classified into one of three high-level categories: (1) *Situation assessment user support contexts* in which the primary intent is to provide operators with access to information, knowledge, and tools to support situation assessment; (2) *Choice of action user support contexts* which are designed to support selection of an appropriate action as well as support to carry out a single action or a sequence of actions; (3) *Evaluation of impact of action user support contexts* which are designed to help the operator assess impact of actions. Two basic types of "resources" are provided in each user support context: First, "information and knowledge resources," i.e., the user support context will define an information and knowledge set which should help the operator to acquire and generate knowledge required by the particular task. Second, "tool resources," i.e., the user support context will provide operators with tools to support both task-related information and knowledge acquisition behavior as well as domain specific actions.

A design process focused on the design of user support contexts needs to be structured in the following phases: (1) Identify user tasks which need to be supported in each type of user support context. (2) Identify knowledge types which need to be manipulated in the user support context, i.e., specify I/K spaces associated with user support contexts. (3) Design graphical user interface tools to enable users to create and use their mental representation of the current context. To achieve the goals in the first two phases, the designer needs to use a design methodology which supports identification of user tasks and information and knowledge required to achieve the task goals. To achieve the design goals associated with the third phase, the designer needs to know how to map information and knowledge elements identified during Phases 1 and 2 into a graphical user interface design. These graphical user interface elements should enable users to manipulate and navigate the information and knowledge represented in the user support context.

The process of designing user support contexts is described in the following sections using an example from the telecommunications domain. In this example, the designer's task was to develop interactive user support for network operators responsible for monitoring incoming alarms and messages in an emerging centralized network control center. In such a centralized monitoring center, network operators base their decisions on whatever information is available from network elements which are distributed across a wide geographic area. Example of information types available from such remote network elements are alarms, informational messages, and results from diagnostic test.

3. PHASES 1 AND 2: IDENTIFYING USER TASKS AND I/K SPACES ASSOCIATED WITH USER SUPPORT CONTEXTS.

Both the identification of user tasks and the I/K spaces associated with these tasks require a task analysis methodology. This methodology must help the designer to structure task analysis results so that they can be used to describe user support contexts. Consequently, the task analysis approach

should support analysis of the three support contexts: situation assessment, choice of action, and evaluation of impact of actions. Moreover, the task analysis needs to be focused on capturing the I/K spaces associated with users' tasks rather than any particular procedures used to achieve task goals. In the design example described in the following sections, the Functional Information and Knowledge Acquisition (FIKA-) modeling approach was used to capture the "contextual" knowledge associated with user tasks. Unlike most task analysis approaches (e.g., GOMS described by Card, Moran & Newell, 1983), the FIKA approach provides designers with a mechanism to focus on information and knowledge types used across a diverse set of procedures.

3.1 Using FIKA to Capture Network Operator Tasks

The complete task analysis approach has been described in detail in Sundström and Salvador (1995). In what follows, the major steps are described with focus on illustrating how the FIKA approach supports design of user support contexts.

An overview of network operators' tasks was obtained by conducting field visits and observations at field sites, i.e., existing network control centers. As in other complex dynamic systems, network operators' tasks fall into one of the categories of monitoring, failure detection, failure diagnosis, failure compensation, failure correction, and maintenance. The user support design process focused on support for the first three tasks.

The FIKA approach was used to specify the I/K spaces associated with monitoring, failure detection, and failure diagnosis. (See Sundström 1991, 1993a,b for detailed description of the FIKA approach.) During field visits, FIKA-based forms were used to help the designer to structure observations and results from observations in "FIKA" categories. In Fig. 2, such a FIKA-based form is illustrated. A detailed description of the forms and how they are used is available in Sundström and Salvador (1995).

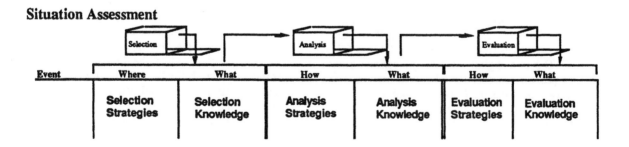

Fig.2. A FIKA based form

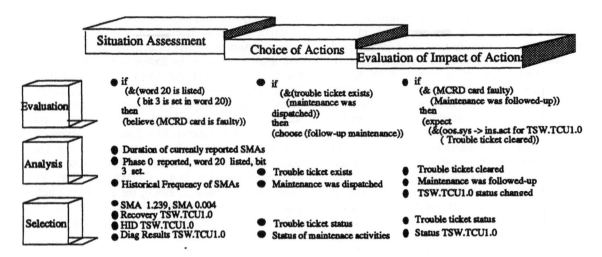

Fig. 3. Information and knowledge space for MCRD card failure diagnosis

The FIKA approach enables the designer to structure results from a task analysis in the following way: The FIKA model's notion of *information processing goal* (IPG) structures task analysis results into the three relevant types of user support contexts, i.e., situation assessment, choice of action, and evaluation of impact of action. The designer can specify each I/K space associated with a support context using the FIKA modeling concepts of *means* and *requirements*. Both concepts specify a specific knowledge type within the user support context. *Means* knowledge describes knowledge about *how* to generate various types of knowledge (e.g., information processing strategies), whereas *requirements* knowledge refers to knowledge required by means to generate other types of knowledge. FIKA distinguishes three classes of means, each associated with a requirement. These categories are: *selection* of relevant information, *analysis* of relevant information, and *evaluation* of analyzed information. Consequently, the designer can use these categories to specify what knowledge needs to be available for the user, what information processing needs to take place to generate the various types of knowledge, and how various knowledge types depend on each other.

3.2 Example of I/K Space Related to Failure Diagnosis

The example I/K space in Fig. 3 illustrates knowledge types required to diagnose a failure in a digital switch. In the failure scenario, the abnormal network behavior is caused by a failure of a Master Clock Receiver Distribution card (MCRD) in a GTD-5. A GTD-5 is a digital telecommunications switch used to process requests for telephone services. The MCRD card provides the Network Clock Unit (NCU) with all clock synchronization signals required by the switch's Time Switch (TSW) and Facility Interface Units (FIUs). The primary function of the TSW is to manage time slots assigned to phone calls, whereas the primary function of FIUs is to provide an interface between switching and transmission equipment. The MCRD card is associated with the

Bus Interface Controller Time card (BICT), which reports all alarms associated with cards in the TSW. In this example, the MCRD card fails in copy zero of the first Time Switch and Peripheral Control Unit in the Time Switch (TSW.TCU1.0). When such an MCRD card failure occurs, the network operator sees the following sequence of messages (words in square brackets typically appear on a new line on the CRT): (1) SMA 1.239 "Network Maskable Interrupt reported by the BICT," [Additional Data: 0000 0000 0000 0023 0011 0000], [TSW.TCU1.0], (2) SMA 0.004 "Base Unit Network Time Switch Fault," [Additional Data 0000 0012 0000 1000 0000 0000], [TSW.TCU1.0], (3) Message: "Recover TSW.TCU1.0," (4) Message: "Diag Results TSW.TCU1.0, Phase 0: 0000 0001 0000" and (5) Message: "Status Change TSW.TCU1.0," [from ins.act to oos.sys]. Trouble shooting of a GTD-5 is done using either system malfunction analysis based on an interpretation of bit settings in the SMA alarms (1 and 2 above) and/or phase analysis based on the output of diagnostic procedures (4 above). The network operator observes the pattern of alarms and messages and concludes that the trouble is in TSW.TCU1.0. In the case of an MCRD card failure, the diagnostic result indicates that diagnostic Phase 0 was completed and that bit 3 was set in word 20. An experienced network operator immediately knows that this indicates a faulty MCRD card.

Fig. 3 illustrates a portion of the I/K space identified for the MCRD failure situations, i.e., the required knowledge for achieving task goals. The I/K space is structured using the previously described FIKA notions. Note that, while the information types are specific to the MCRD failure, the information categories related to Selection, Analysis, and Evaluation are more general, i.e., the same categories would be used to diagnose a variety of card failures.

The situation assessment I/K space identifies the information on which the network operator needs to focus attention (information selection), how it needs to be analyzed, and how the analyzed information

328

needs to be evaluated to infer the correct failure diagnosis. The *Selection* category identifies the information required to generate the failure diagnosis, whereas *Analysis* and *Evaluation* illustrate knowledge types the network operator must generate to actually generate the fault diagnosis. Note that the designer can add any type of knowledge into the structure provided by the FIKA concepts. For example, if a functional representation of a switch is used to enhance the failure diagnostic process, the *Selection* category associated with the situation assessment I/K space would include a list of functions and their status in the pertinent information set. Similarly, if the operator support system to be designed were to include a component which provides operators with intelligent advice with respect to what actions to consider (e.g., based on Mitchell's operator function model), the I/K space associated with the choice of action support context could be augmented with information which supports the operator to integrate the advice given by the system.

4. PHASE 3: DESIGN GRAPHICAL USER INTERFACE TOOLS ASSOCIATED WITH THE USER SUPPORT CONTEXT

The final task of the designer is to provide the user with tools to view and manipulate knowledge in the I/K space associated with the failure diagnosis process. Thus, the designer needs to decide the following: (1) How to indicate that a "new" support context is available; (2) How to provide access to the various types of knowledge provided in the support context; and (3) What tools to provide the user with to manipulate and use knowledge in a particular support context. How these decisions were made in the case of the support design for the network operators is illustrated focusing on the designed support for situation assessment. The two user interface elements described below are part of the Integrated Graphical Support System (IGSS), designed and developed at GTE Laboratories for telecommunications network monitoring (e.g., Sundström, 1993c). The two user interface elements are "generic" in the sense that the concepts underlying them can be used in any domain and their graphical layout can be tailored to the specific requirements of the particular domain.

The first user interface element, the Task Manager Window (TMW), indicates to users that a new task exists, i.e., in the case of the network operators, the user interface elements indicate that a network event has occurred that requires attention. These new tasks are identified by automatic alarm pattern analysis and are indicated by an icon in the New Task Section of the window (See Fig. 4). The network operator acknowledges tasks by clicking on task icons. Acknowledged task icons are moved to the Acknowledged Tasks section. The middle TMW section contains the Monitoring Task icon providing access to the monitoring user support context. This icon is always present because network operators have to continue monitoring the network while they are performing other types of tasks, such as failure diagnosis.

Each task icon indicates the trouble type, e.g., a problem in a switch might be presented as a task of type *switch*. Task icon labels are used to present a brief descriptive summary of the task represented by the icon. The label includes specific task knowledge, e.g., a card failure in a switch. This label can also include the severity level, the status of the equipment, the time the task was originally presented, and whether responsibility for the task is shared with another user or with an automated system. Pop-up menus associated with task icons enable users to clear, ship, and share tasks. The four menus in the menu bar provide users with the ability to take specific actions, such as shipping a job to another location or sorting task icons by task type.

The primary function of the Task Manager Window is to provide users with structured access to user support contexts, but it does not provide the tools to access and use information and knowledge in the particular support context. The user interface element providing this functionality is the Task Support Window. As with the Task Manager Window, the functionality provided by the Task Support Window is generic, whereas the specific graphical layout can be changed as required by the domain.

To open a Task Support Window (TSW), the network operator double clicks on the task icon in the Task Manager Window. The TSW opens and provides the network operator with access to all information

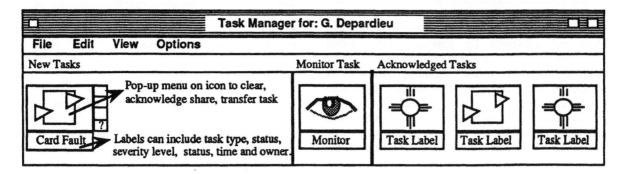

Fig. 4. The Task Manager

directly related to three user support contexts associated with the occurring network event (See Fig. 5). These three support contexts are situation assessment, choice of action, and evaluation of impact of action.

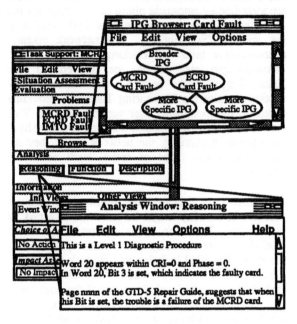

Fig. 5. The Task Support Window

All TSWs have three partitions, each providing access to information and tools related to one of the three mentioned user support contexts. Thus, each TSW partition is structured to provide access to support related to information selection, analysis, and evaluation. To illustrate the structure, the Situation Assessment partition is described in more detail.

The *Information* segment provides access to Selection information (cf. Fig. 5) required in the current context. The user can access this information by selecting from the *Inf. Views* list. For example, the user can access an Event Window showing currently active events related to this particular task from the Inf. Views list. Using the example above, in the case of an MCRD failure, the Event Window would contain a representation of SMA 1.239 "Network Maskable Interrupt reported by the BICT." In addition to the Selection information, users can access any other information they choose to view from the *Other Views* list. The *Analysis* segment provides tools and assistance for analyzing the selected information. In the MCRD failure scenario, the Analysis support would primarily consist of identifying which bits were set in the diagnostic tests run by the switch. In general, Analysis tools could include reasoning support as well as support to create alternative graphical representation for the same data. The reasoning support is activated by clicking on the *Reasoning* button, which opens a window where an explanation of key information is provided. Additional reasoning support can be accessed by clicking on the *Function* and *Description* buttons

respectively. The *functional* reasoning support indicates the network functions affected by the abnormal events. If the system design incorporates reasoning based on models providing functional representations such as Lind's Multilevel Flow Model, these would be accessed through the *Function* button. The *Description* reasoning support provides the physical description of network components. The *Evaluation* segment provides access to the fault diagnosis. This diagnosis could be generated by IGSS or by any other system. The fault diagnosis indicates the most likely cause for the abnormal events. If the network operator enters or selects another fault diagnosis, the information contained in the Inf. Views and Analysis segments will be re-defined accordingly. Finally, the *Browse* button allows network operators to traverse a network of problem types, represented as situation assessment user support contexts. The browser's top node represents the most likely failure diagnosis category; sub-nodes represent specialization of this category. The network operator can select any other diagnosis category from this network to override the system generated diagnosis. The selected category is then used by the support system to re-define the I/K space(s) used to populate the TSW segments.

If the designer wants to add additional information types as defined by a diverse set of operator models, these can be added to the TSW without fundamentally changing it. For example, if a new information source provides historical information about telephone switches, this information becomes a part of FIKA's *Selection* category. The user could access this information from the *Inf. View* list in the TSW *Information* segment. New types of graphical representations for information already available to users can also be added. For example, instead of textually describing physical network components, these components could be shown using available video resources and other forms of multimedia presentation. The user could access these new resources by simply clicking on the *Description* button in the TSW *Analysis* segment.

The two described user interface elements illustrate how the notion of user support contexts provides the designer with a structure which not only enables the designer to add new information but also to incorporate new presentation formats without fundamentally changing the user support design. Consequently, an operator support system based on the notion of user support contexts rather than on a specific operator support model can migrate as system requirements evolve over time.

REFERENCES

Bainbridge, L. (1993). The change in concepts needed to account for human behavior in complex dynamic tasks. *Proc. of the 1993 IEEE International Conference on Systems, Man, and Cybernetics*, 1, pp. 126-131.

Card, S., Moran, T., and Newell (1983). A. *The Psychology of Human-Computer Interaction*, Erlbaum, Hillsdale, NJ.

Hollnagel, E.(1994). *Reliability of Cognition: Foundations of Human Reliability Analysis.* Academic Press, London, United Kingdom.

Lind, M. (1981). The use of flow models for automated plant diagnosis," in *Human Detection and Diagnosis of System Failures,* J. Rasmussen & W.B. Rouse (Eds.), pp. 411-432, London, Plenum Press.

Mitchell, C.M. (1987). GT-MSOCC: A domain for research on human-computer interaction and decision aiding in supervisory and control systems, *IEEE Transactions on Systems, Man, and Cybernetics,* 17, pp. 553-572.

Rasmussen, J. (1983). Skills, rules, and knowledge; signals, signs, and symbols, and other distinctions in human performance models. *IEEE Transactions on Systems, Man, and Cybernetics,* 13, 257-266.

Rasmussen, J. (1986). *Information Processing and Human-Machine Interaction. An Approach to Cognitive Engineering.* Amsterdam: North Holland.

Sundström, G.A. (1993a). User modeling for graphical design in complex dynamic environments: Concepts and prototype implementations. *International Journal of Man-Machine Studies,* 38, 567– 586.

Sundström, G.A. (1993b). Towards Models of Tasks and Task Complexity in Supervisory Control Applications. *Ergonomics,* 36, 1413-1423.

Sundström, G.A. (1993c) Model-Based User Support: Design Principles and an Example. *Proceedings of the IEEE International Conference on Systems, Man, and Cybernetics,* 1, pp. 355–360.

Sundström, G.A. and Ayub, S. (1994). Graphical User Support in Distributed Environments. *Proceedings of the 1994 IEEE International Conference on Systems, Man, and Cybernetics,* 1, pp. 379-385.

Sundström, G.A., and Salvador, A.C. (1995). Integrating Field Work in System Design: A Methodology and Two Case Studies. *IEEE Transactions on Systems, Man, and Cybernetics,* 25.

BUILDING NATURAL LANGUAGE INTERFACE
- ITS METHODOLOGY AND TOOLS

Hiroshi Tsuji, Yasuharu Namba, Hisao Mase,
Yukiko Morimoto and Hiroshi Kinukawa

Systems Development Laboratory, Hitachi, Ltd.
3-6-1 Bakuroh-Machi, Chuo, Osaka, 541 Japan.
e-mail : tsuji@sdl.hitachi.co.jp

Abstract: This paper presents methodology and tools for building natural language interface between users and application systems. The goal of this research is to allow any system designer to implement a natural language interface. A basic premise of this research is that there are command languages for application systems. The methodology proposes ten steps, including acquiring knowledge for semantic analysis and usability enhancement. The tools consists of seven components which require domain knowledge and a dictionary for the relation between natural language and command language.

keywords :   Natural languages, Software tools, Human-machine interface,
Knowledge representation, Systems methodology

1. INTRODUCTION

Although direct manipulation is an effective interface technology for some classes of problem, it is limited in many ways. In particular, it provides little support for identifying and operating on large sets of entities. On the other hand, these are precisely the strengths of natural language (Shneilderman, 1987, Cohen, 1991). A natural language interface (NLI) might be defined as the operation of computers by people using a familiar language to give instructions and receive responses. Users do not have to learn a command syntax or to select an operation from menus.

Early attempts at general-purpose machine translation have faded, but there are continuing efforts to provide many ways for computers to deal with natural language, such as for travel consulting, command operation (Wilensky, *et al.*, 1984), and database queries (Kinukawa, 1988). However, only the natural language professional can build the NLI for the

applications. It is not easy for general system designers who are not familiar with natural language processing technique to construct a NLI. To build natural language interfaces in a more illuminating, and hopefully more efficient, manner than just by trial-and-error requires some guiding principles.

This paper presents methodology and tools for building natural language interfaces between users and application systems. Because the NLI part and the rest of an application (the functionality part) are modeled conceptually as separate parts, the separation should be clearly manifested in a physical implementation. A basic premise of this research is that there are command languages for application systems. An NLI should facilitate information exchange in both directions: from the user to the application part and from the application part to the user. The purpose of NLI software (tools + domain knowledge) is thus to transform natural language into command language, (a transformation including functional chaining and resolving ambiguities) and to show what the user can or should do. This paper also presents the experimental building of a natural language interface for a full text search system.

should do next, or shows what the problem is. With the discourse knowledge, the NLI should guide the users.

2. NATURAL LANGUAGE : POSITION AND REQUIREMENTs

The NLI component receives a user message, translates it into command language, and executes application systems as shown in Fig. 1. It also transmits the application's return message to the user in natural language. Thus the requirements for the NLI in this paper are as follows:

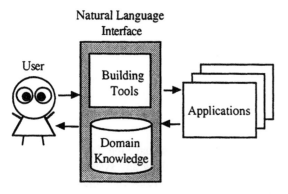

Fig. 1. Relation between NLI and Applications.

NLI receives key-in messages. Ideally, NLI should recognize voice and handwritten language. However, even in Japan, where Kanji characters are used, this restriction is currently acceptable because most people are familiar with word processors.

Each application system has a command language.
The NLI designer can assign a relevant verb for each command. For example, "move" may be the relevant verb for command "cd". Consider the difference between "cd ABC" and "move to the directory whose name is ABC". If "ABC" is recorded as a directory in the domain dictionary, the expression "move to ABC" might be also acceptable for the NLI.

The domain should be restricted. Both the designer and the user should understand that the acceptable vocabulary is restricted. Note that any NLI cannot work as a human translator. The NLI stores domain knowledge, including a dictionary and semantic analysis knowledge. If the NLI cannot translate a user sentence into command language, then it should explain which word in the input sentence cannot be recognized and/or which parts (operation, condition, or object) is not expressed.

NLI knows the standard dialogue procedures.
If the NLI knows the standard dialogue for users to achieve the goal, it prompts the users with what he

3 . NATURAL LANGUAGE INTERFACE BUILDING METHODOLOGY

To develop computer application systems, the traditional methodology requires realworld modeling strategy. For example, to build an expert system, most methodologies propose (1) build an abstract problem solving model, (2) map this abstract model into a knowledge representation such as rule and frame, for example see (Shoen, *et al.*, 1987). On the other hand, to build an object-oriented system, the methodology (Rumbaugh, *et al.*, 1990) requires an object identification stage, a functional model building stage, and a state-transition model analysis stage.

However, these system building methodologies cannot be applied to develop a natural language interface. The methodology for building a NLI should provide ways to analyze acceptable expression, to implement the software, and so on. The proposed methodology has ten steps, including acquiring knowledge for semantic analysis, as follows:

Task domain and application selection.
First of all, the range in which a user can interact with a system in natural language should be specified. At the same time, which applications are appropriate for the domain should also be clarified where each application formulates not only command syntax but also the environmental objects names (for example, the data base names and the field names in the query domain). In general, when the application is identified, the kind of sentences that should be accepted is also clarified. Note that the NLI generates commands for a variety of applications, if needed, and generates them not only for software but also for hardware such as a voice synthesizer or a VTR player which have command interface.

Example NL expression collection.
To analyze what kind of sentence should be accepted, the example sentences are collected. It is desirable that the sentence are expressed by a variety of personnel including the system designers, the end-users, and also natural language researchers. Note that it is meaningless to accept an input sentence that directs the execution of a function not supported by the applications. Then, the methodology proposes the following: If the sentence cannot be accepted by the NLI, the designer records the reason that it cannot; otherwise the corresponding instantiated application command should be written down.

Generalization.
To generalize the acceptable sentences, clarify the variable part in the example sentences. How to generalize depends strongly on the selected domain. Then, the corresponding abstract commands for the generalized sentences should be written down.

Designing normal NL expression.
For the generalized expression, the designer should write down the normal expression. The normal expression form includes the function, conditions, objects, and all of the relations between them. The object should be expressed by its type and the name. Although the normal expression may be redundant for the person who has vocabulary knowledge, it is essential for the computer software which does not have common knowledge. Semantic analysis in the NLI can then decide the unique command for the normal NL expression. The following is an example normal expression: "retrieve the document which owner is Tsuji, from the file which name is newsDB.

Knowledge formalization.
To express the semantic structure of natural language, a diagram, called chained function structure (CFS), should be clarified. The chained function structure which concerns a natural language analysis technique is discussed elsewhere in detail (Namba, *et al.*, 1993). The verbs are assigned for the functions, and the nouns are assigned for the objects and the conditions. To express the normal expression, the following special function is presented: (1) *is_a relation* expresses the slot name and the slot value, (2) *has_p relation* expresses the condition, and (3) *equ relation* expresses the connection between objects. The example CFS is drawn in Fig. 2, where a hexagon represents a function and a square represents an object or a condition.

Knowledge Instantiation.
The semantic analysis knowledge is represented in a machine-readable formula. The corresponding command template that should be filled by the NLI is also represented. The relations between the CFS and command parameters are also specified. The framework of the knowledge representation is provided by the building tools discussed later.

Domain dictionary generation.
An environmental name vocabulary in the selected domain should be registered in the dictionary. If the special words are registered with its usage, the NLI has a chance to supplement an elliptic expression, and the user need not always express the strict normal expression. Note that human can infer that "newsDB" is a file name and Tsuji is person's name. Then, if the dictionary stores the meaning that word "newsDB" is file name and the "Tsuji" is persons name, then the NLI can recognize the sentence: "retrieve the

document of Tsuji from newDB. Another example was shown earlier for the "cd" command.

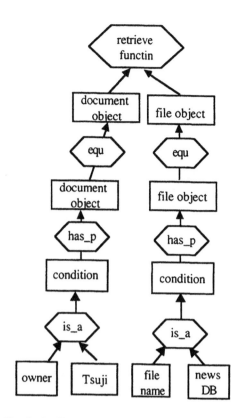

Fig. 2. An Example of Meaning Representation in a Chained Function Structure.

Test & Debug.
Whether or not all of the designed normal expression are analyzed exactly should be examined. The success of the normal expression analysis is requirement for the NLI at least. In this phase, batch mode examination is desirable.

Dialogue robustness analysis.
How to analyze the example sentences collected before is examined here. In general, human conversation includes anaphoric expression and elliptic expression, for example, see (Hayes, *et al.*, 1979). Whether these expression are acceptable or not is an optional problem in the presented natural language interface building. However, the challenge should be examined in this phase. If there are alternatives for the meaning analysis, the NLI shows them and confirms the user's intention.

Discourse design.
In a specific domain, there may be usual procedures in which one function would often be used after another function is executed. Then, the NLI can suggest possible actions and/or prompts the preceding actions in specific situations. An example discourse can be drawn as shown in Fig. 3. Furthermore, the

usage for conjunctions and adverbs should also be considered. Sometimes they are useful to express the special commands of applications (for example, "further", and "on the other hand", et al.).

Utility function for usability.

As mentioned before, the basic NLI is a translator between natural language and command language. Then, the window for dialogue should be designed to provide the following for the usability:

Example illustration menu. This consists of examples of acceptable sentences. The user refers to the menu, selects one, and modifies it for the use. The cut & paste function is useful for rapid operation. It not only allows the NLI to raise the rate of correct meaning analysis, but also allows the users to find out what they can say and how to express what they would like to do.

Animation for facial expression. The user may want an NLI to be as intelligent as a human. However, it is difficult for an NLI to express all the internal states of the system at once. Because a multi-modal interface technique (Cohen, 1991) is expected, several kinds of channel will be useful for supplementing natural language conversation. A typical example is expression animation.

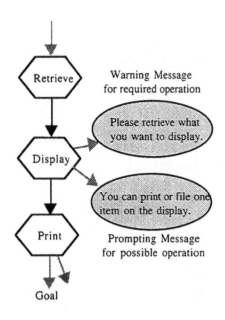

Fig. 3. Discourse Representation for User Guide Message.

4. NATURAL LANGUAGE INTERFACE BUILDING TOOLS

Even with a building methodology, a general system designer cannot implement the natural language interface. Note that most natural language interfaces have the same components, including semantic analyzer, command generator, word dictionary, and so on. Therefore, a tool that interprets the domain knowledge is also necessary for building an NLI.

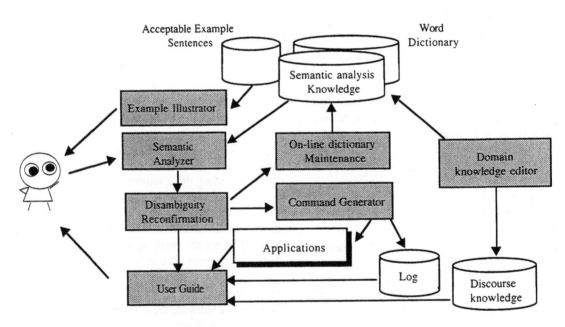

Fig. 4. Constitute of Natural Language Interface Building Tool NLI Builder.

'NLI builder' has been developed as the partner to the methodology. It is implemented for Japanese, but its basic idea is not limited that language. The components of NLI builder are shown in Fig. 4. Using NLI builder, the system designer needs only to define domain knowledge, including a word dictionary according to the methodology. NLI builder provides a variety of built-in functions for specifying the semantic analysis knowledge and consists of the following seven components:

Semantic analyzer. A natural language offers the ability to avoid extensive redescription through the use of pronouns and other anaphoric expressions. Such expressions are usually intended to denote the same entities as earlier ones, and NLI builder is intended to infer these connections. Thus the use of anaphora provides an economical benefit to the user at the expense NLI builder's having to draw inferences (Namba, *et al.*, 1993). Finally, NLI builder transforms the natural language expression into a meaning representation.

Disambiguity reconfirmation function. NLI builder engages in some form of clarification or confirmation subdialogue in order to ascertain whether its interpretation is in fact the intended one. NLI builder recognizes utterance and displays the recognition. If the recognition is correct, the corresponding command is executed. An example of this technique in the database query domain is discussed elsewhere (Mase, *et al.*, 1994).

On-line dictionary maintenance function. If, on the other hand, the recognized meaning is not correct, the user disconfirms it and corrects the confirmation sentence. Thus, there are chances to maintain the domain knowledge base.

Command generator. The meaning representation is transformed into the command language. Again, because an order in natural language is transformed into an application command, the user cannot order to execute a function not supported by the application.

User guide. According to the discourse knowledge, NLI builder recognizes the sequential order in which to execute commands. A typical example is that an object should be retrieved before it is printed or filed. Interpreting the standard discourse knowledge, the NLI shows the user in natural language what the user should do or notify what the user can do.

Example illustrator. The tool provides a framework for showing the full range of possible example operations in the menu. Users can thereby avoid the frustration of probing the boundaries of system functionality. Example sentences guarantee semantically and syntactically correct operation.

Domain Knowledge editor. A special editor is devoted to maintaining the domain knowledge, including the word dictionary, semantic analysis knowledge, and discourse knowledge. And a fundamental dictionary storing fifty thousands words is provided.

5. NLI BUILDING EXPERIMENTATION

The methodology and tools have been applied to a full text search system. Basically, the full text search system had eight commands, each having a variety of operands that allow a user to specify the options in detail. The retrieval command generated retrieval sets and the user could execute logical operations on these sets. The experimentation is summarized as follows:

The domain is articles search from a newspaper database, so the database name and fields names are fixed. The original text search system has no printing or filing function, so the NLI executes OS commands for such functions. The news text skimming system is also utilized for an application.

More than one hundred example sentences are collected. After generalization for the examples, seventeen normal NL expressions are designed. Each NL expression is written down for NLI builder, taking about thirty minutes for each one. About fifty words, some corresponding to commands and others that are environment names, are registered in the dictionary.

For robust communication, it is designed that the current retrieval set and the current displayed article number are supplemented by NLI builder so that they can be omitted in the sentence. In this domain, NLI builder interprets the user's intention as retrieval if the verb is omitted. The CFS in the semantic analysis knowledge is powerful enough to realize such robustness.

In this domain, the following discourse is stored in the knowledge base : (1) If over hundred articles are retrieved, the system prompts "add the condition to the current retrieval set" and shows the example sentence, (2) After retrieval and article display, the system prompts "how about text skimming then?", (3) If the user tries to skim the text without article display, the system warns "why do not you confirm the article content?", and so on.

For usability, a couple of idea for a multi-modal interface are implemented :

(1) If the user's sentence cannot be analyzed semantically, the facial expression animation shows the embarrassment.

(2) The messages are generated by a voice synthesizer.

(3) If there is no response within a specified period, the systems shows what the user can do in that situation and prompts the next operation.

The man-machine interface are shown in Fig. 5.

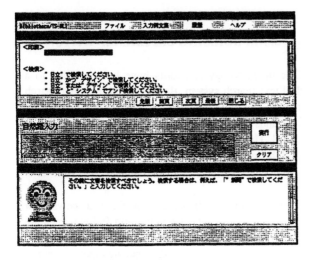

Fig. 5. Man-Machine Interface for
a Full Text Search System.

Twelve persons have used the application in a usability test. To achieve the goal in the test, a user had to key in five sentences in fifteen minutes. Six persons used the system with the user guide and example illustrator, and others used the system without them. Four of the former achieved the goal, although only one of the latter did. Therefore, the utility functions for the usability is indispensable for the natural language interface.

6. CONCLUSION

This paper has presented a methodology for building natural language interface, and has introduced building tools, called NLI builder that allow non-professionals in natural language technique to build NLIs between users and application systems. The methodology and tools have been applied to a full text search system.

The experiment showed that engineers who are not acquainted with natural language processing but familiar with application systems could build a NLI in two months. They define only the domain knowledge and are not required to develop a semantic analyzer or dictionary.

If building an NLI is easy for any designers, the chances are there that any application system may

have a natural language interface. This is the goal of this research. The usability problems will be manifested by the end-users, whereas today these problems are pointed out only by the NLI researchers.

The proposed methodology and tools are not restricted to building natural language interfaces. Their more general concerns are how to represent the meaning of the user's operation command, and how to analyze the man-machine communication. The prospects for multi-modal interfaces (Blattner, 1992, Cohen, 1991), the next-generation interfaces, will also be discussed according to this research. Conversely, if multi-modal interfaces can not be built by general system designers, they will not be in common use even in next-generation computer systems.

REFERENCES

Blattner, M. et. al. (Ed.) (1992). *Multimedia Interface Design*, ACM Press.

Cohen, P. (1991). The Role of Natural Language in a Multi-modal Interface, In : *Proc. of International Symposium on Next Generation Human Interface*, pp. 1-p8.

Hayes, P. et al. (1979). An Anatomy of Graceful Interaction in Spoken and Written Man-Machine Communication, *CMU-CS-79-144*.

Kinukawa, H. (1988). A Natural Language Interface Processor based on the Hierarchical-Tree Structure Model of Relation Tables, In: *Journal of Information Processing*, **Vol. 11, No. 2**, pp. 83-91.

Mase, H, and H. Kinukawa (1994). Confirmation Sentence Generation for Natural Language Interface in Database Retrieval Domain and its Evaluation (in Japanese). In : *Transaction of Information Processing*, **Vol. 35, No. 8**, pp. 1579-1590.

Namba, Y. and H. Kinukawa (1993). Chained Functions Structure : Meaning Representation for Natural Language Interface, In: *Proc. of NLPRS '93 (at Fukuoka)*, pp. 238-247.

Rumbaught, J., M. Blaha, W. Premerlani, F. Eddy and W. Lorensen (1990). *Object-Oriented Modeling and Design*, Prentice-Hall.

Shneilderman, B. (1987). *Designing the User Interface*, Addison-Wesley.

Shoen, S. et al. (1987). *Putting Artificial Intelligence into Work*, John Wiley & Sons, Inc..

Wilensky, R., Y. Arens, and D. Chin (1984). Talking to UNIX in English : An Overview of UC, In : *Comm. ACM*, **Vol. 27, No. 6**, pp. 574-593.

EVALUATION OF INTELLIGENT ON-BOARD PILOT ASSISTANCE IN IN-FLIGHT FIELD TRIALS

T.Prévôt[+], M.Gerlach[+], W. Ruckdeschel[+], T.Wittig[*], R.Onken[+]

[+]*Universität der Bundeswehr München, Institut für Systemdynamik und Flugmechanik D-85577 Neubiberg*
[*] *Daimler-Benz Aerospace AG, Navigation and Flight Guidance Sensors , D-89070 Ulm*

Abstract: In this paper the in-flight evaluation of the Cockpit Assistant System CASSY is presented. The design principles, functionalities and the evolution of the intelligent assistant are briefly described. The evaluation program in simulation and flight is pointed out. Interfaces to the pilot, especially speech input, and to ATC are investigated. Results are presented of the situation assessment capabilities concerning pilot behaviour and the complex planning and decision aids, CASSY provides. It is the first system of this kind, which was integrated into an aircraft to be tested in the real aviation environment.

Keywords: automation, human-centered design, man-machine interaction, intelligent knowledge based systems, aircraft operations

1. INTRODUCTION

Cockpit automation is permanently under discussion. Research is done to understand the impact of cockpit automation on crew behaviour in modern glass cockpits (Wiener, 1989). Obviously, recent cockpit automation efforts have not always achieved what they were designed for. Accident rates do not decrease and the attribution of accidents to human error stays at a constant level of some 75 %. It is generally recognized that there are some problems with the current state of cockpit automation. As Norman (1990) pointed out, the current problems may arise from an inappropriate design and application of these systems and a lack of intelligence and conversation possibilities they obtain.

To bypass the current insufficient intermediate level of automation new generation cockpit systems should be integrated into the aircrafts. These cockpit systems should carefully reflect the experiences gained with current automation and the guidelines and requirements resulting from these experiences (Wiener and Curry, 1980; Billings, 1991; Onken, 1993).

The design of the Cockpit Assistant System CASSY is based on the fundamentals of human centered-automation. During the development activities at the University of the German Armed Forces, Munich, with support by DASA the system has undergone different simulation

experiments to be evaluated and improved. Flight-testing the resulting system in the real aviation environment was to demonstrate the feasibility and the operational value of intelligent human-centered pilot support. This paper is focused on the flight experiments and their results. It starts with design principles and implemented functions of the intelligent assistant, though.

2. THE INTELLIGENT ASSISTANT

The intelligent assistant to be evaluated is CASSY (Onken and Prévôt, 1994). CASSY is a knowledge based on-board assistant system with interfaces to the flight crew, to air traffic control and to aircraft systems. It is primarily aimed at increasing flight safety under instrument flight conditions. In addition, there is the objective of improving economical efficiency in short and medium range transport. CASSY's design is the result of research and development activities over seven years and based on the following principles.

2.1 Design Principles

A major impact on the design of the intelligent assistant has been the research on the cognitive behaviour of humans denoted by Rasmussen (1983; 1986) and on Parallel Distributed Processing as described in the work of Rummelhart and McClelland (1986). In the aviation field principles and guidelines for human-centered aircraft automation have been established by Billings (1991). Regarding these findings and the results of extensive simulation experiments with a first generation crew assistant (see chapter 3), two basic requirements for machine support specifications in flight guidance have been formulated by Onken (1993) (Onken and Prévôt, 1994).

The decisive point of these requirements is that any overtaxing of the crew must be avoided by technical means. In basic requirement (1) it is expressed that the crew must always be aware of the current situation and its most urgent task or subtask. The intelligent assistant is responsible for presenting the situation in an appropriate way, which assumes that the machine assesses and understands the situation completely on its own. But even when requirement (1) is accomplished, still the situation may overtax the crew with respect to planning and plan execution tasks. In this case requirement (2) states that this situation must be transformed by technical means into another one back to normal.

To comply with these requirements the tasks cannot be rigidly allocated to the human pilot and the machine. Different situations require different levels of autonomy and interaction. The extent of task sharing has to be situation-dependent cooperative. Therefore, each of the partners - human and computer- must know the current state and the intention of each other.

2.2 Realization of CASSY

All situational elements of the entire flight situation, including mission, aircraft, systems, environment, and crew aspects, are stored in a central object-oriented representation. A specific communication module, the Dialogue Manager (Gerlach and Onken, 1993) is responsible for extracting the decisive patterns and coordinating their output to the crew via speech and/or display.

Vice versa the Dialogue Manager picks up the inputs of the crew and directs them to the respective module of the assistant. This is done via speech recognition. Since the pilot should not permanently tell his electronic partner about his state or his intentions, this information must be gained in a different manner. The Automatic Flight Planner (Prévôt and Onken, 1993) generates a complete 3-D/4-D flight plan. This is done autonomously or interactively with the crew, which depends on the situation. According to the generated flight plan, the Piloting Expert (Ruckdeschel and Onken, 1994) elaborates the expected pilot action patterns on the basis of pilot modeling such that the Pilot Intent and Error Recognition (Wittig, 1994) can compare this expected behaviour to the actual behaviour of the crew. Thereby, it identfies discrepancies in behaviour and their reasons. There are three possible reasons for this:

a) Pilot error:
   The pilot deviates from the objectively correct expectation of behaviour, derived by CASSY.

b) Temporary discrepancy of pilot intent:
   Events, which CASSY does not know yet, cause the pilot to deviate from the behaviour expected by CASSY.

c) Machine error:
   An inappropriate or erroneous modeling or information processing within the machine (including CASSY) leads to an objectively wrong expectation of pilot behaviour by CASSY.

In case of a pilot error, a warning or hint is given to the pilot to correct the error. In order to cope with the temporary discrepancy of pilot intent, CASSY tries to figure out the intention, modify the flight plan, accordingly, and elaborate the consistent expected behaviour. Machine error should not appear, but realistically it sometimes does and must be considered. The errors are less serious, when they can be detected easily, be recovered with very few commands and have no safety critical consequences. Therefore, capabilities for in-flight restart of the system or recovery from erroneous states have to be provided as an important functionality. In the following the main features and functions of CASSY are summed up.

2.3 Assistance Functions

In addition to the situation assessment concerning pilot behaviour, conflicts with the flight plan are detected autonomously and conflict hints are given or replanning is initiated to solve the problem. In the conflict case CASSY decides whether to initiate an interactive replanning or a completely autonomous replanning, depending on the available time and human resources. If the pilot decides to initiate planning, the amount of inputs he gives is up to him. The assessed situation is permanently shown with respect to the current flight plan on the display in heading-up or plan-mode. When no problems occur and everything is working properly, the crew does not become aware of CASSY actvities other than the appropriate presentation of the flight situation.

The planning and decision making assistance includes:
- autonomous or interactive generation and evaluation of routings or routing alternatives and trajectory profiles for the complete flight or local portions of the flight
- evaluation and selection of alternate airports and emergency fields
- prediction of the remaining flight portions, when ATC redirects the aircraft or the pilot intentionally deviates from the plan.

The monitoring capabilities include
- monitoring of the pilot actions with regard to nominal flight plan values, i. e. altitude, heading/track, vertical velocity, speeds and configuration management, e.g. flaps, gear, spoiler and radio navigation settings
- monitoring of violations of specific danger boundaries, including minimum safe altitudes, stall and maximum operating speeds and thrust limits.

Basic services are provided for
- configuration management by speech input, approach briefings, departure, approach and profile charts generated from the actual flight plan on request,
performance and navigation calculations on request.

The range and kind of provided functionalities and the way, they are realized, result from extensive simulation experiments and consultations of pilots at each decisive development step. A brief review of the different evaluation phases on the way to the flight experiments is given next.

3. REVIEW OF SIMULATION EXPERIMENTS AND CONSISTENT SYSTEM IMPROVEMENTS

The first significant results have been gained in extensive simulation experiments with the CASSY predecessor system ASPIO (Assistant for Single Pilot IFR Operation) in 1990 at the German Armed Forces University in Munich (Dudek, 1990). This system had been prototyped primarily for the approach and landing phases of flights in the Munich area. The experimental environment consisted of a very heterogeneous hardware from Personal Computers to VAX-stations. For speech recognition a single word recognizer was used. With respect to this architecture the results were astonishingly promising. The design philosophy proved to be well accepted by the pilots. The significant improvement of flight accuracy and acceleration of planning and decision making tasks were further factors, which gave way for continuing projects and industrial support (Onken, 1992).

The following activities were aimed at the two men cockpit of regional aircrafts. One major part of system improvement was extending the support functions for all flight portions. Integrating a dialogue management module was stressed as well as exploiting crew modeling techniques. Consistently, the pilot intent recognition became part of the system, too. After two more years in 1992 a first CASSY prototype could be demonstrated in the one-seat fixed base flight simulator at the university. In the meantime the computer hardware had been changed to a homogeneous workstation architecture and continuous speech recognition systems. Again, airline pilots were consulted for further improvements. The flight procedures were adapted to major airline procedures and specific details of the man-machine cooperation were discussed and improved.

In 1993 the system was tested in the DO 328 flight simulator at DASA, Friedrichshafen. Flights in the Frankfurt area have been simulated with different pilots. Good pilot acceptance was for the planning and the monitoring functions (Onken and Prévôt, 1994). The speech input was marginally accepted because of technical shortcomings. Therefore, the speech input part of the Dialogue Manager has been improved and is still under improvement.

Before finally integrating the system into the test aircraft, flights have been simulated with the hardware in the loop in the simulation facilities of the German Aerospace Research Establishment (DLR) in Braunschweig. In these simulator runs the software was checked and the test pilots for the flight were introduced to the functionalities of CASSY and the experimental environment. After the successful completion of these simulator trials the flight experiments were performed in June 1994.

4. FLIGHT EXPERIMENTS

The flight experiments were aimed at evaluating the CASSY performance in the real aviation environment. The system was integrated into the experimental cockpit of the Advanced Technologies Testing Aircraft System ATTAS of the DLR and typical regional flights in high traffic areas were performed. In the following the experimental environment and the flight scenarios are presented.

4.1 Experimental Environment

The flying simulator ATTAS, an especially developed modification of the 44-seat cummuter jet VFW 614, is equipped with an experimental fly-by-wire flight control system and a versatile computer and sensor system. Beyond many other test programs it is used as the airborne segment in DLR's air traffic management demonstration programme (Adam, et al, 1993) and is equipped with very good facilities for testing complex on-board systems in instrument flight scenarios. In addition to the two safety pilots seated in the front cockpit, the ATTAS aircraft can be flown by the test pilot in an experimental cockpit, which is installed in the rear cabin directly behind the front cockpit. The experimental cockpit is a generic flight deck (one seat) with side-stick, airbus display and autopilot techniques and ARINC control panels. Therefore, it represents a realistic pilot working environment for IFR operation. The CASSY hard- and software has been integrated into the experimental cockpit.

The hardware of the assistant system, consisted of
- an off-the-shelf Silicon Graphics Indigo (R 4000) workstation to run the core modules of the assistant system connected to the ATTAS experimental system via ethernet
- a PC/QT equipped with a Marconi MR8 PC-cart providing speaker dependent continuous speech recognition with a speech button on the side stick
- a DECtalk speech synthesizer with various voices for speech output connected to the ATTAS intercommunication facilities
- a BARCO monitor (about 25cm) connected to the graphics channel of the SGI-Indigo and built into the experimental cockpit.

The computers were located in the rear of the main cabin. There are several experimenter work stations in the aircraft. One was equipped with a laptop for starting and maintaining CASSY.

4.2 Knowledge Acquisition

During the flight tests CASSY has been running throughout the complete flights from taxi-out to taxi-in. All data, which CASSY received via the avionics data bus, have been recorded with a frequency of 10 Hertz. All in- and output messages have also been recorded and every time, the flight plan had changed because of a major planning activity or when a checkpoint has been passed, the whole situation representation has been stored. These data enable a replay of all flights and a reproduction of all situations.

The presented results have been gained by observing the behaviour of the pilot and the intelligent assistant during the flights on-line, by off-line evaluating the collected data and in debriefings immediately after the flights. Two professional pilots served as experimental pilots and additional pilots from Lufthansa German Airlines were participating as observers.

4.3 Flight Scenarios

A total amount of about 10 flight hours has been performed, comprising eight flights from the regional airport Braunschweig (EDVE) to the international airports of Frankfurt (EDDF), Hamburg (EDDH) and Hannover (EDVV) at which a missed approach procedure was conducted before returning back to Braunschweig.

Table 1 Flight test scenarios

Flight	T/O T/D	time airb.	G/A in	after	ATC instr.	Pilot
1	EDVE	1:03	EDDH	0:33	26	1
2	EDDF	0:50	*inflight*	*simul.*	13	1
3	EDVE	1:27	EDDF	0:43	27	1
4	EDVE	0:50	EDVV	0:09	24	2
5	EDVE	1:32	EDDF	0:41	32	1
6	EDVE	0:57	EDDH	0:32	27	2
7	EDVE	0:57	EDDH	0:31	24	1
8	EDVE	0:58	EDDH	0:31	21	1
9	EDVE	1:31	EDVV	1:14	42	1

Flight no. 2 has been an in flight simulation of departure and approach to Frankfurt, which was necessary to investigate certain incidents, which would have been safety critical in the real Frankfurt area, e.g. descending below the minimum safe altitude. In all other flights nothing has been simulated and no special situations have been provoked, since the system should be evaluated in the real environment, which includes coping with all events, which occur during an IFR flight in a high density area.

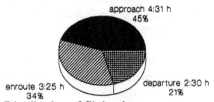

Fig. 1. Distribution of flight phases

5. EXPERIMENTAL RESULTS

One important result held true throughout the complete test program: There was no significant difference in system performance between the flight tests and the simulation trials. Consistently, the following discussion of results concentrate on the major questions concerning the real environment rather than system performance.

5.1 Operationality of the Interfaces

To evaluate the speech recognition performance three different speakers made the speech input during the flight tests, summarized in table 2.

Table 2: Speech Inputs

Speaker	Time	Inputs
Pilot 1	8:18	324
Pilot 2	0:50	36
Experimenter	0:57	56

In their first flight pilot 1 and pilot 2 were not very familiar with the speech recognition system and the specific syntax to be used. In flight no. 6 a CASSY experimenter made the complete speech input for the pilot. He was familiar with the syntax and the speech recognizer from simulation experiments. The results are shown with regard to recognition performance.

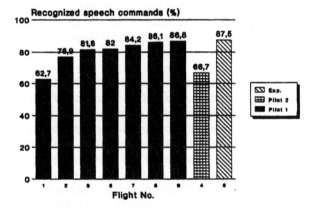

Fig. 2. Percentage of recognized speech commands

Obviously, speech recognition inside the noisy aircraft is possible. It takes some time for the pilot to become familiar with the recognizer and the syntax to be used. This learning process can also be done in the simulator, as the flight with the experimenter has shown. The achieved percentage of recognized speech commands is almost of the same level as could be achieved in simulator runs with the same recognition system.

For entering the ATC commands into the system, two different experiments have been made throughout the flights. 92 ATC commands of a total of 236 have been fed into the system by the pilot using speech. The remaining 144 commands have been keyed into the system by one of the experimenters onboard the aircraft immediately after receiving the message, to simulate a data link from the ground into the aircraft. This took some seconds. The pilot reacted to the commands at the same moment he received the ATC message, but acknowledged the command with some time lag. This time lag resulted in delayed reaction of CASSY, which sometimes led to unnecessary warnings and hints. The percentage of occurence of these incidents compared to the respective number of ATC messages and the mean phase lag is illustrated in figure 3:

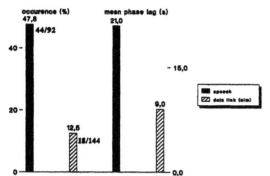

Fig. 3. Unnecessary warnings or hints resulting from time lags in entering ATC commands

This effect was typical for the one-pilot configuration of the experimental cockpit. The figure illustrates the importance of a fast and powerful ATC interface. Optimal system performance can only be achieved with a digital data link.

5.2 Situation Assessment with Respect to Pilot Behaviour

The basic requirements described in chapter 2.1 point out the necessity for a complete understanding of the global flight situation. To get an impression of the situation assessment capabilities the duration of discrepancies between the actual and the expected pilot behaviour has been related to the total flight time. This has been done on the basis of the stored data for the six flights 2, 3, 5, 6, 7 and 8.

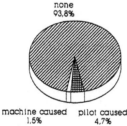

Fig. 4 . Dicrepancies in actual and expected pilot behaviour

Figure 4 indicates that for almost 94 % of the flight time in a high density environment the pilot and the machine assessed the situation equally, because otherwise they would not expect or perform the same action patterns.

A total amount of 100 incidents leading to warnings have been evaluated to find out the reasons for the warnings and the consequences they had. All incidents have been related to one of the three categories ex-

plained in chapter 2.2: pilot error, pilot intent and machine error (i.e CASSY errors in this case).

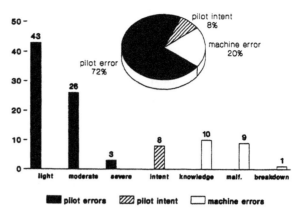

Fig. 5. Error classification

In five cases of the intentional deviations from the flight plan the intention was autonomously figured out by the assistant system and the flight plan has been adapted, accordingly. In three cases the pilot had to inform CASSY about his intention.

Half of the machine errors were caused by an incomplete knowledge base, e. g. insufficient modeling of the aircraft performance and the other half by malfunctions of CASSY, i.e. software implementation errors due to less rigorous application of software development procedures. In one case such a malfunction led to a complete breakdown of the assistant system. In all machine error cases the pilot realized that a wrong warning was issued by CASSY. No negative influence on the pilot's situation assessment could be observed. In the one breakdown case, the complete CASSY system had to be restarted in flight, which took about 15 seconds. The only pilot input needed for such a recovery procedure is the flight destination. In all other machine error cases the warnings disappeared autonomously, when the incorrect assessed maneuver had been completed by the pilot.

Concerning the pilot errors the light errors are considered to result in an inaccurate or uneconomical, but safe maneuver. Moderate errors, probably would lead to a safety critical situation, and severe errors surely would lead to a dangerous safety hazard unless an immediate correction is made. All pilot errors, which occured during the flight tests, were detected by CASSY. All moderate and severe errors as well as about 70% of the light errors were immediately corrected by the pilot after having received the warning or hint.

5.3 Flight Planning and Decision Aiding

CASSY's flight planning capabilities have been stated by the experimental pilots and the observers as very impressive. As a matter of fact, all planning proposals have been accepted and none of the autonomous radar vectoring predictions has been modified or caused any doubt from the pilot. The time needed for planning a complete flight from one airport to the other is illustrated in figure 6.

Before every flight the flight destination and the departure runway were entered into CASSY and an autonomous planning of the complete flight was initiated. After the go around procedure at this destination the pilot initiated an interactive planning to return to the airport, from which he had departed, by entering its name. CASSY elaborated and presented two routing proposals in parallel, which the pilot could select from or modify. After the selection the trajectory profile was planned in detail and recommended speeds, times of overflight, radio aids etc. were inserted.

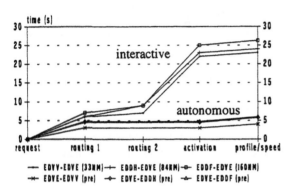

Fig. 6. Duration of flight planning actvities

The distance to the destination had only little impact on the duration of planning. The autonomous planning took between 4 and 6 seconds, the interactive replanning up to 26 seconds, of which the pilot needed about 16 seconds to decide for a proposal. This confirms the approach to replan autonomously, when the flight plan must be generated very fast. When there is more time available, replanning can be done interactively, too, in order to keep the pilot more involved (see chapter 2.3).

5.4 Pilot Acceptance

The acceptance of the planning and monitoring functions was at least as good as in the previous simulator trials (Onken and Prévôt, 1994). All participants in the evaluation attested CASSY a nearly operational performance and a very promising concept. It was noted that situation assessment, monitoring and good planning capabilities are completely in line with human-centered design.

6. CONCLUSION

The successful flight tests of the Cockpit Assistant System CASSY in real IFR flights have demonstrated that it is possible to integrate intelligent on-board systems into modern aircraft. Speech recognition proved to be a powerful device as pilot interface, but other devices should also be considered. Optimal performance can be achieved with a digital data link between ground and airborne segments. The amount of detected and avoided pilot errors, the availability of features like pilot intent recognition as well as the power demonstrated in complex planning indicate the performance level of CASSY. This kind of system can be considered as the solution to overcome existing criticism with respect to current flight management systems.

REFERENCES

Adam, V., Klostermann, E. and Schubert, M. (1993). *DLR's ATM Demonstration Programme*. AGARD Meeting on 'Machine Intelligence in Air Traffic Management'

Billings, C. E. (1990). *Human-Centered Aircraft Automation: a Concept and Guidelines*, NASA Technical Memorandum 103885, Moffett Field, CA.

Dudek, H.-L. (1990). *Wissensbasierte Pilotenunterstützung im Ein-Mann Cockpit bei Instrumentenflug*. Dissertation, UniBw, Munich

Gerlach, M. and Onken, R. (1993). A Dialogue Manager as Interface for Communication between Pilot and Pilot's Assistant System. In: *Human Computer Interaction* (G. Salvendy and M. J. Smith, (Ed.)) pp 98-103. Elesvier, Amsterdam.

Norman, D.A. (1990). The 'problem' with automation: inappropriate feedback and interaction, not 'overautomation'. In: *Human Factors in Hazardous Situations* (D.E. Broadbent, A. Badderly and J. T. Reason (Ed.)) pp. 585-593. Clarendon Press, Oxford.

Onken, R. (1993). *Funktionsverteilung Pilot-Maschine: Umsetzung von Grundforderungen im Cockpitassistenzsystem CASSY*. DGLR FA Anthropotechnik, Berlin

Onken, R. (1992). *New Developments in Aerospace Guidance and Control: Knowledge-Based Pilot Assistance*. 12th IFAC Symposium on Automatic Control in Aerospace, Ottobrunn.

Onken, R. and Prévôt, T. (1994). *CASSY-Cockpit Assistant System for IFR Operation*. ICAS 94-8.7.2 Proceedings **Vol.2**, Anaheim, CA

Prévôt, T. and Onken, R. (1993). *On-board Interactive Flight Planning and Decision Making with the Cockpit Assistant System CASSY*. In: 4th HMI-AI-AS, Toulouse.

Rasmussen, J. (1983). *Skills, Rules, and Knowledge: Signals, Signs, and Symbols, and other Distinctions in Human Performance Models*. IEEE SMC-**13** (3), pp 257-267.

Rasmussen, J. (1986). *Information Processing and Human Machine Interaction An Approach to Cognitive Engineering*. Elsevier, Amsterdam

Ruckdeschel, W. and Onken, R. (1994). *modeling of Pilot Behaviour Using Petri-Nets*. 15th Int. Conf. on Application and Theory of Petri Nets, Zaragoza.

Rummelhart, D.E and McClelland, J. L. (1987). *Parallel Distributed Processing. Explorations in the Microstructure of Cognition*. MIT Press, Cambridge, MA.

Wiener, E. L. (1989). *Human Factors of Advanced Technology ("Glass Cockpit")Transport Aircraft*. NASA Report 177528, Moffett Field, CA.

Wiener, E. L. and Curry, R. (1980). *Flight Deck Automation: Promises and Problems*. Ergonomics 23, pp 995-1011.

Wittig, T. (1994). *Maschinelle Erkennung von Pilotenabsichten und Pilotenfehlern über heuristische Klassifikation*, Dissertation, UniBw, Munich.

MODE USAGE IN AUTOMATED COCKPITS:
SOME INITIAL OBSERVATIONS

Asaf Degani[1], **Michael Shafto**[2], **Alex Kirlik**[3]

San Jose State University Foundation, CA[1]. NASA Ames Research Center,
Mountain View, CA[2]. Georgia Institute of Technology, Atlanta, GA[3].

Abstract: Mode confusion is increasingly becoming a significant contributor to accidents and incidents involving highly automated airliners; in the last seven years there have been four airline accidents in which mode problems were present. This paper attempts to provide some initial observations about modes and how pilots use them. The authors define the terms "mode," "mode transitions," "mode configurations," and propose a framework for describing and classifying modes. Preliminary results from a field study that documented mode usage in "Glass Cockpit" aircraft are presented. The data were collected during 30 flights onboard Boeing 757/767 type aircraft. Summary of the data depicts the paths that pilots use in transitioning from one mode to another. Analysis of the data suggest that these mode transitions are influenced by changes in aircraft altitude as well as by two factors in the operational environment: the type of air traffic control facility supervising the flight, and the type of instruction (clearance) issued.

Keywords: Automation, Aircraft operations, Man/machine systems, Mode structure.

1. INTRODUCTION

Modes are found in almost any supervisory control system. Yet, it appears that in some highly automated systems, mode confusion is a trigger for many accidents and incidents. Modes, as a method for human-automation interaction, are now recognized as an operational problem by both operators and manufacturers of these systems (*Aviation Week and Space Technology*, 1995a, 1995b). But just what are modes? How are they used by operators in supervisory control systems? This paper attempts to provide some initial insights into these two questions.

The first part of the paper discusses mode usage from several viewpoints: (1) an historical perspective, (2) symptoms of mode problems, (3) definitions, and (4) a framework for describing and classifying modes. In particular, the discussion focuses on mode transitions—a critical aspect of user interaction with a modal system. The second part of the paper discusses preliminary results from a field study documenting how operators transition between the various modes of operation, and what factors prompt these transitions. The discussion is set in the context of pilots using the automatic flight control system of a modern "glass cockpit" aircraft.

1.1 *Historical perspective*

Historically, the issue of modes in human computer interaction emerged as more and more functions were added to early word processors, and yet the size of the interface (e.g., number of function keys, screen area, etc.) stayed constant. One solution was to use the same key to engage several commands; this was implemented by providing the user with some mechanism to switch the application from one mode to another. Depending on the mode, hitting the same key would execute different commands. In this paper the term *format/data-entry* modes is used to describe this type of mode implementation. For example, the *vi* text editor has two modes of operation: "Command" and "Insert." In "Command" mode, pressing the "x" key will delete a character; in "Insert" mode this action will write the letter "x" on the screen.

Users of these early applications, however, were not always happy with such mode implementations: errors, or *mode-errors*, as these were termed by Norman (1981), caused confusion and frustration (Lewis, and Norman, 1983). Tesler (1981) captured this growing frustration in his influential article in *Byte* magazine and his pointed cry: "don't mode me in." Research on modes in the human computer

345

interaction literature has mostly focused on various implementations for the mode switching mechanism (Monk, 1986; Sellen, Kurtenbach, and Buxton, 1992; Thimbleby, 1982). The problem, nevertheless, has not disappeared: designing efficient modes and switching mechanisms continues to be part of any human-computer interface.

The same growing pains are now shared by designers and operators of supervisory control systems (*Aviation Week and Space Technology*, 1995a; Woods, Johannesen, Cook, and Sarter, 1993). Since most supervisory control systems are managed via a computer, *format/data-entry* modes for input of information and display switching are heavily used. But in most supervisory control systems there is also another type of mode: one that is used for controlling the process. This unique mode is the method used for engaging various control behaviors (e.g., reverse/drive gears in a car). In this paper, the term *control* modes is used to describe this type of implementation.

1.2 Symptoms of mode problems

In the last seven years, there have been four major airline accidents in which mode problems were cited. In the first, an Air France Airbus A-320 crashed in Habersheim-Mullhouse Airport, France, following a low altitude fly-by (Ministry of Transport, 1990). The crew, flying close to the ground, engaged a pitch mode that provides relatively slow thrust response to throttle movement. In the second accident, an Indian Airlines A-320 crashed during a visual approach to Banglore Airport, India (Gopal and Rao, 1991). The crew, intentionally or unintentionally, engaged a pitch mode in a way that provided no speed or altitude protection. In the third accident, an Air Inter A-320 crashed during a nighttime approach into Strasbourg-Entzheim Airport, France. The accident report suggests that the crew may have mistakenly engaged the wrong mode for the situation at hand (*Aviation Week and Space Technology*, 1994a). In the fourth accident, a China Airlines A-300/600 crashed during an approach into Nagoya International Airport, Japan. The crew, unintentionally or intentionally, engaged a mode that commanded climb with full thrust, and at the same time *manually* pushed the control wheel down in order to prevent the aircraft from climbing. In a conflict between manual versus autopilot commands, the aircraft achieved an extreme pitch attitude of 36 degrees with decaying airspeed, rolled to the right, and crashed (*Aviation Week and Space Technology*, 1994b).

2. MODES

Before studying mode usage, it seems important to describe what are modes and what types of human-machine interaction they foster. Unfortunately, in the context of human-machine systems, no common terminology for describing modes is available. The following discussion suggests a terminology and proposes a framework for classifying different types of modes.

2.1 Terminology and definitions

A mode is defined here as *a manner of behaving*. This general definition satisfies the use of the term within any system, may it be behavioral, social, organizational, or a hardware/software system (Ashby, 1956; Goldberg and Goldberg, 1991; Nadler, 1989; Perrow, 1986). A system can have several ways of behaving; but at any point in time only a single mode can be active. If each mode behavior can be captured as a vector of several operands (e.g., c, d, d, b), then the transition table in Figure 1 can describe this modal system.

$\downarrow$	a	b	c	d
M_1	c	d	d	b
M_2	b	a	d	c

Figure 1. Modal system.

For a given system, M_1 corresponds to a mode-switch set to position 1, and M_2 to position 2. Mode transition, or the change of M's subscript from 1 to 2, is a transformation from one manner of behaving to another (Ashby, 1956). The machine's overall behavior is a combination of its various mode behaviors and transitions.

The human operator interacts with the machine via its modes. Problems in the human-machine interaction, or in particular mode confusion, usually result from misidentification of the machine's behavior—its mode behavior and its mode transitions. Such mode confusion may lead to error, namely mode error. These mode errors occur when the user takes some action (e.g., issues a command) believing that the machine is in one mode, when in fact it is in another (Norman 1983). Since the machine's behavior changes as a result of a mode transitions, it is not surprising that such transitions are a critical ingredient of mode confusion and subsequent mode errors.

2.2 Mode transition

Ashby (1956) describes a system that exhibits various manners of behavior as a machine with input. This input is the determining factor in making the transition from one mode to the next. In the context of modes in human-machine systems, three types of inputs may be used: manual, automatic, and automatic/manual. In a *manual* input, or a manual mode transition, the user directly engages the mode (and consequently disengages another). This is the most commonly used mode type (e.g., modes on an electronic watch, or a text editor's insert/replace modes). In an *automatic* input, or a automatic mode transition, a controller (another machine) initiates the transition. This type of mode transition is mostly used in fully automatic systems (e.g., an automatic braking system in a modern car). In a *automatic/manual* mode transition, either the human or the machine, initiates the transition. This kind of transition is used in quite a few systems and appliances (e.g., a microwave can switch from "Cook" mode to "Idle" mode either automatically or when the user intervenes manually).

2.3 Mode classification

Earlier we distinguished between two primary mode functions: *format/data-entry* and *control*. These two types of functions, combined with the three types of inputs (*manual*, *automatic*, and *automatic/manual*) form a matrix that can be represented in a 2 by 3 table. This table can be used for classifying modes (Figure 2).

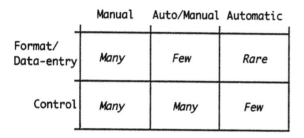

	Manual	Auto/Manual	Automatic
Format/ Data-entry	*Many*	*Few*	*Rare*
Control	*Many*	*Many*	*Few*

Figure 2. Mode classification.

This proposed classification is not always crisp. Some may argue that the term "control" can be applied to both writing a document on a word processor and flying an airplane. The various systems that we surveyed had modes that fell naturally into one of the cells in the table. Only *format/data-entry* modes that transition *automatically* were a rarity. Nevertheless, some do exist—certain ATM machines automatically switch to another format (or mode) once the expected entry is typed.

2.4 Mode configuration

An additional input to any modal system are the parameters, or target values, that the machine has to maintain (Lambergts, 1983). In other words, these target values constrain mode behavior. For example, the pitch component of an automated flight control system has several modes: "Vertical Speed," "Vertical Navigation," and others (Figure 3). Mode transitions, depicted by the arrow on the top, can occur either manually, automatic/manually, or entirely automatically. Once a mode is active, it will operate according to its characteristic behavior while attempting to maintain these target values. A target value, say airspeed, may come from various sources: if the "Vertical Speed" mode is active, the target value is obtained from the mode control panel; if the "Vertical Navigation" mode is active, airspeed target value is obtained from the flight management computer.

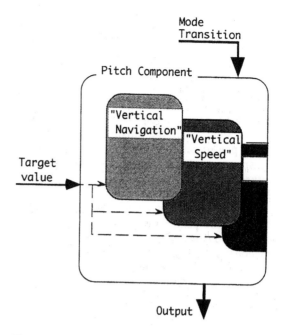

Figure 3. Mode transition, target value, and output.

The originator of the target values can be either the pilot or the machine. Continuing the above example, when the "Vertical Speed" mode is active, the pilot (the originator) enters the desired airspeed into the mode control panel; when the "Vertical Navigation" mode is active, the flight management computer (the originator) calculates the most economical airspeed for the particular flight situation.

The pilot, therefore, has several options to control the aircraft: he or she can change the target values of the current mode, or transition to another mode. The term *mode configuration* is used here to describe the type and value of the various target values entered into the machine. For example, a change in mode configuration occurs when the pilot enters a new rate of descent while the "Vertical Speed" mode is active or when the pilot changes the vertical profile while the "Vertical Navigation" mode is active. Over time, the changes in target values define the *mode configuration trajectory*.

2.5 A system with several modes

In many complex domains a given system is made up of several sub-systems, or components. Each of these components may have its own set of modes. Therefore, unlike a simple system that may exhibit only one mode at a time, the status of a complex system, with respect to its modes, is a vector of all active modes. Furthermore, since by definition some relationship exists between the components of a system, interactions exist between a mode of one component and a mode of another component. Thus we propose here several definitions and terms for describing human-machine interactions via modes. In the following sections, we use these terms to describe how pilots interact with the automated flight control system of a modern airliner.

3. TASK DEMANDS AND MODES

The various accidents mentioned is section 1.2, as well as hundreds of mode-related incidents (ASRS, 1991; *Aviation Week and Space Technology*, 1995a), suggest a link between mode design/usage and operational problems (Sarter and Woods, 1994). The authors of this paper hypothesize that some of these problems stem from the mismatch between the demands placed on the human supervisor and the mode structure of the system. The term *mode structure* is used here to describe the hierarchy of modes in a system, the transitions among modes, and the transformations that occur from one mode to another. In the context of a complex system with several components, mode structure also signifies the interactions between the modes of one component and the modes of another component.

On the one hand, the pilot has formulated a set of goals that he or she attempts to accomplish in a logical, efficient, and safe manner. On the other hand, the system has a predetermined set of methods, or modes, that are available for controlling the system. Various paths exist for transitioning between these modes. If and when task demands do not match the mode structure of the system, mode confusion and unwanted results may ensue. This link is only amplified when the operating environment as well as the system are highly dynamic: frequent changes in environmental demands (e.g., ATC clearances) and aircraft situation (e.g., imminent stall) require frequent mode transitions.

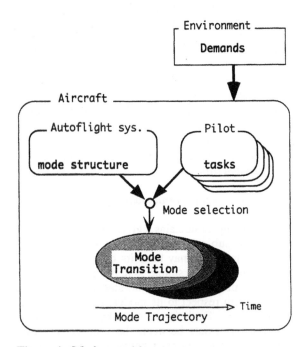

Figure 4. Mode transition.

The mode structure of the system, the task demands, and the pilot's goals all combine to produce mode selection. This can be recorded as a mode transition from the previous mode to the current mode, and over time as a continuum of mode transitions that forms a *mode trajectory* (Figure 4). Problems in identifying

the automatic flight control system's behavior appear to be a critical component in many accidents and incidents. It appears reasonable, therefore, that one approach for studying mode usage is to document and understand mode trajectories.

4. METHOD

Mode transition data was collected by an observer onboard an airliner during the climb-to-cruise and descent-to-land phases of a flight. The observations were conducted during two typical trips, each comprised of three flights. Each of the two trips was observed five times. This design of experiment yielded 30 flights (2*3*5). Subjects were airline pilots from a major US carrier, flying regular revenue flights in either the Boeing B-757 or B-767—both modern "glass cockpit" aircraft equipped with an automatic flight control system (AFCS). The AFCS is composed of three major components: autopilot, autothrottle, and flight management computer (FMC). Sitting in the jumpseat, the observer recorded the following variables: changes in pitch and roll modes, thrust modes, FMC modes, as well as whether the autopilot, flight-director, and autothrottle were "On" or "Off." Other information such as aircraft altitude, distance/bearing from airport, weather, air traffic control (ATC) clearances, and the type of ATC facility supervising the flight were also recorded. Crew information, such as rank (captain, first officer) and duty (pilot-flying, pilot-not-flying) were collected. The dataset analyzed here contained 30 flights which amounted to some 700 records of both mode changes and mode configuration changes.

5. ANALYSIS

The objective of this analysis was twofold: (1) to describe mode transitions and the frequency of occupying a certain mode (mode occupancy), and (2) to identify possible factors that prompt these mode transitions. In particular, the authors hypothesized that one of the strategies that flight crews use to combat complexity of the system (e.g., its mode structure and mode behaviors) is by using a small subset of all possible modes, and that these strategies are influenced by task demands coming from the operational environment. Of the some 700 records in the dataset, only those that documented mode transitions were included (mode configuration were excluded). The reduced dataset contained 291 records.

5.1 *Mode occupancy and transition*

Mode occupancy. The various pitch and roll modes of the automatic flight control system (AFCS) are represented in a 5*8 table (Figure 5). On the horizontal legend (columns) are listed the five modes of the roll component; on the vertical legend (rows) are listed the eight modes of the pitch component. Since the status of the AFCS in this analysis is described as a vector of both pitch and roll modes, each cell in the table indicates such a combination. On the Northwest corner of the table, the combination of "Manual Roll" mode and "Manual Pitch" mode indicates a situation in which the pilot is flying manually: autopilot and autothrottle are disengaged, and he or she is flying *without* reference to the flight director guidance. On the Southeast corner of the table, the combination of "Lateral navigation" mode and "Vertical Navigation" mode indicates a situation in which the aircraft is flown fully automatic. The numerical value in each cell indicates the occupancy frequency.

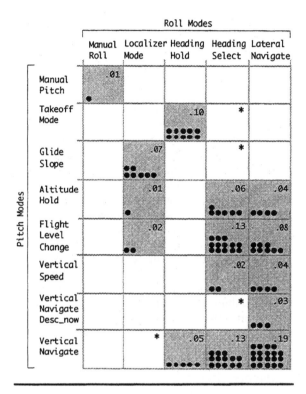

Figure 5. Mode occupancy
(* indicates 0 < occupancy < 0.01).

Two observations can be made from Figure 5:
(1) only some of the pitch/roll mode combination are occupied, and (2) heavy occupancy is either associated with a procedure (e.g., using "Takeoff Mode"/Heading Hold" during takeoff is a standard operating procedure in this airline), or a preferred mode combination (e.g., "Heading Select" and "Flight Level Change").

Mode transitions. Figure 6 depicts mode transitions among the pitch/roll mode combinations (only those that were shaded in Figure 5). The transitions between these mode combinations shows the possible paths that pilots use from takeoff to touch-down. Broken lines shows the initial transition from *start of flight* to "Takeoff Mode"/Heading Hold" mode combination as well as the final transitions from "Flight Level Change"/"Localizer Mode," "Manual Pitch"/"Manual Roll," and "Glide Slope"/"Localizer mode" to *touchdown*.

Dark shading indicates the heavy occupancy of the "Lateral navigation"/"Vertical Navigation" and the

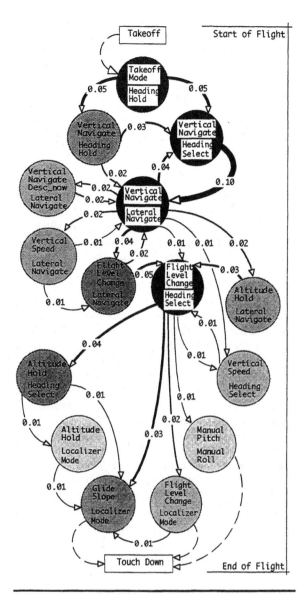

Figure 6. Mode transitions (dark shading indicates high occupancy).

"Flight Level Change"/"Heading Select" mode combinations; both are pivots for transitioning to other modes combinations. As a unit, the diagram shows how pilots traverse within the mode structure of the ACFS.

5.2 *Factors influencing mode transition*

The previous section summarized and depicted the mode transition in the ACFS. The current section attempt to identify some of the factors that prompted such transitions by using two statistical analysis procedures. The data collected during the flights poses some challenges for such analysis, since the values are mostly discrete and the status of the AFCS is a vector of several modes. Using indicator variables, the discrete data were coded numerically. For example, there were 11 types of ATC clearances; this required 10 indicator variables for coding. A similar coding scheme was used for all other discrete variables. Our analysis approach was to employ two types of procedures in order to identify the factors that prompted mode transitions: (1) a multivariate regression analysis, and (2) a categorical canonical correlation test. For both tests, mode transitions from the 30 flights were randomly split to two equal size sets: a model building set (15 flights), and a validation (hold out) set (15 flights).

Regression. The purpose of the regression analysis was to obtain the relationship between the active mode combination and the dependent factors (e.g., crew duty, rank, leg, trip, phase of flight, altitude, distance from airport, type of clearance, type of ATC facility, type of aircraft). In order to build the regression model, the vector containing the pitch and roll modes was combined into a single ordinal value (the dependent variable—"Y"). This was done by assigning high values to a combination of pitch and roll modes that were highly automated and low values to a combination of modes that were manual. The criterion for the value assignments was the precision of the mode combination for tracking a predetermined path. The advantage of the regression is its simplicity; the disadvantages are the limits on the amount of raw information that enters the model due to using this composite "Y" variable (Walker and Catrambone, 1993), and the normality assumptions associated with this type of analysis.

The results indicated that 61% of the variance in mode transitions can be explained via three factors: the aircraft altitude, the type of ATC facility supervising the flight, and the type of clearance issued by ATC ($R^2_{adj.} = 0.61$, $p < 0.001$). Cross-validation of the model on the hold out dataset yielded a comparable fit ($R^2_{adj.} = 0.51$, $p < 0.001$).

Canonical correlation. This procedure is an extension of the multiple regression approach, in that a vector of dependent variables (pitch and roll indicator variables) is used instead of a single dependent variable. Canonical correlation finds the linear combination of independent variables (altitude, ATC clearance, etc.) and the linear combination of dependent variables (pitch and roll mode indicator variables), such that the correlation between the two linear combinations is maximized (Tatsuoka, 1988). Because of the obvious inapplicability of normal-distribution theory to a mostly discrete dataset, a "Monte-Carlo" randomization procedure (Edgington, 1987) was used to test the significance of the canonical correlation, and a "jackknife" method was used to compute an approximate confidence interval (Efron and Tibshirani, 1993).

The preliminary analysis indicates a high canonical correlation between the linear combination of the dependent set and the linear combination of the independent set ($r = 0.95$, $p < 0.01$ by randomization test; approximate 95% confidence interval = [0.90, 0.99]). The analysis showed that ATC facilities ("Departure," "En-route," and "Approach") highly influenced mode transitions. Aircraft altitude had only a moderate influence, and the type of clearance had almost no influence in this analysis. Since canonical correlation allows for a vector of dependent variables ("Y's"), identification of pitch and roll modes that correlate with the dependent variables was performed. On the pitch modes, "Altitude Hold," "Flight Level Change," "Vertical Speed," and "Vertical Navigation" appear to be highly influenced by the independent set. On the roll modes, only "Lateral Navigation" appears to be influenced; the remaining roll modes showed only moderate relation to the independent set.

6. CONCLUSIONS

The preliminary analysis discussed here is the result of an observational study. This methodology poses some limitations for identifying cause-effect relationships—mainly that the factors are not directly manipulated by the experimenter (Cook and Campbell, 1979). Bearing in mind this limitation, the initial results presented here suggest the following:

First, within the possible mode space there are certain mode combinations that are frequently used. Pilots use several standard and preferred paths for mode transitions during the progress of the flight. Second, these mode transitions are influenced by the aircraft altitude and two environmental factors: type of ATC clearance, and the type of ATC facility (Approach Control, En Route Control, etc.) providing these clearances. We offer several possible explanation for this. (1) Altitude is a primary factor with respect to both short term (tactical) and long term (strategic) activity on the flight deck; and therefore, directly or indirectly it influences mode transitions; (2) ATC clearances prompt mode transitions. This comes as no surprise, since modes are a method for executing the tasks directed by ATC; (3) ATC facilities vary in the type and rate of clearances. For example, ATC controllers in an Approach Control facility issue mostly tactical clearances (e.g., maintain heading of 280 degrees, descend to 6000 feet) at a high frequency while demanding a quick response. In contrast, ATC controllers in En Route Control facility issue mostly strategic clearances (e.g., a complete route of flight between several waypoints). Evidence on the influence of both ATC Facility and clearance type on pilots' mode engagement was also found by Casner (in press).

Taken as a whole, these preliminary findings point to the important relationship between the modes structure of the automated system, and the task demands coming from the operational environment. The result of this relationship, or interaction, are the mode transitions in the system (see Figure 4).

Understanding both the automated system and the operating environment, as well as their interaction, appears valuable for designing new automatic flight control systems. This may be particularly important as future aircraft and the next-generation ATC system are likely to be very different from today's.

REFERENCES

Ashby, R. W. (1956). *An introduction to cybernetics.* new York: John Wiley & Sons.

ASRS (1990). *Advanced cockpit autoflight control reports* (Special request No. 1823, [Database search]). Aviation Safety Reporting System office, Mountain View, California: Battelle.

Aviation Week and Space Technology. (1994a). Human factors cited in French A-320 crash. January 3, pp. 30-31.

Aviation Week and Space Technology. (1994b). Autopilot go-around key to China Air Lines crash. May 9, pp. 31-32.

Aviation Week and Space Technology (1995a). Automated cockpits: who's in charge? Part 1. January 30, pp. 52-65.

Aviation Week and Space Technology (1995b). Automated cockpits: who's in charge? Part 2. February 6, pp. 48-57.

Casner, S. M. (in press). Understanding the determinants of problem-solving behavior in a complex environment. *Human Factors.*

Cook, T. D., and Campbell, D. T. (1979). *Quasi-experimentation: Design and analysis issues for field settings.* Boston: Houghton Mifflin Company.

Edgington, E. S. (1987). *Randomization tests* (2nd ed.). New York: Marcel Dekker.

Efron, B., & Tibshirani, R. J. (1993). *An introduction to the bootstrap.* New York: Chapman & Hall.

Goldenberg, I., and Goldenberg, H. (1991). *Family therapy: An overview* (3rd ed.). Belmont, CA: Brooks/Cole Publishing Company.

Gopal, B. S., & Rao, C. R. S. (1991). *Indian Airlines A-320 VT.EP: Report of the technical assessors to the court of enquiry.* Indian Government.

Lambergts, A. A. (1983). *Operational aspects of integrated vertical flight path and speed control system* (Society of Automotive Engineers technical paper 831420). Warrendale, PA: Society of Automotive Engineers.

Lewis, C., and Norman, D. A. (1986). Designing for error. In D. A. Norman and S. W. A. Draper (Eds.), *User centered system design.* Hilsdale, NJ: Lowrence Earlbaum Associates.

Ministry of Transport (1990). *Air France Airbus A-320 F-GFKC, Mulhouse Habsheim, June 26, 1988* (Reprinted in Aviation Week and Space Technology, June 4, 1990). French Ministry of Planning, Housing, Transport and Maritime Affairs Investigation Commission: France.

Monk, A. (1986). Mode errors: a user-centered analysis and some preventative measures using keying-contingent sound. *International journal of man-machine studies, 24,* 313-327.

Nadler, D. A. (1989). Concepts for the management of organization change. In M. L. Tushman, C. O'reilly, and D. A. Nadler (Eds.), *The management of organizations: strategies, tactics, analysis* (pp. 490-504). New York: Harper & Row.

Norman, D. A. (1981). Categorization of action slips. *Psychological review, 1*(88), 1-15.

Norman, D. A. (1983). Design rules based on analysis of human error. *Communication of the ACM, 26*(4), 254-258.

Perrow, C. (1986). *Complex organizations* (3 ed.). New York: Random House.

Sarter , N. B., and Woods, D. D. (1994). Pilot interaction with cockpit automation II: An experimental study of pilot's mental model and awareness of the flight management and guidance system. *International Journal of Aviation Psychology, 4*(1), 1-28.

Sellen, A. J., Kurtenbach, G. P., and Buxton, W. A. (1992). The prevention of mode errors through sensory feedback. *Human-computer interaction, 7,* 141-164.

Tatsuoka, M. M. (1988). *Multivariate analysis: Techniques for educational and psychological research* (2nd ed.). New York: Macmillan Publishing Company.

Tesler, L. (1981). The smalltalk environment. *Byte, 6*(8), 90-147.

Thimbleby, H. (1982). Character level ambiguity: consequences for user interface design. *International Journal of Man-Machine Studies, 16,* 211-225.

Walker, N., & Catrambone, R. (1993). Aggregation bias and the use of regression in evaluating models of performance. *Human Factors, 35*(3), 397-411.

Woods, D. D., Johannesen, L. J., Cook, R. I., & Sarter, N. B. (1993). Behind human error: Cognitive systems, computers and hindsight (Cognitive Systems Engineering Laboratory). Columbus, OH: Ohio State University.

THE EFFECT OF DATA LINK-PROVIDED GRAPHICAL WEATHER IMAGES ON PILOT DECISION MAKING*

A. T. Lind, A. Dershowitz, D. Chandra, S. R. Bussolari

MIT Lincoln Laboratory, Lexington, Massachusetts

*This work was sponsored by the Federal Aviation Administration.

ABSTRACT – This paper summarizes the methodology and results of two human factors evaluations of a data link-provided Graphical Weather Service (GWS) for general aviation pilots. The transmission of these complex images requires far more bandwidth than is available with any practical data link implementation. Transmission is made possible through application of a compression algorithm. However, the use of compression can result in distortion of the weather image. Phase One was conducted to assess the effects of GWS on pilot situational awareness and decision making. Results indicated that GWS had a substantial positive effect on the weather-related decision making of pilots. Phase Two was conducted to determine the maximum level of compression-induced distortion that would be acceptable for transmission of weather images to the cockpit.

Keywords: Graphic display; data compression; image distortion; human factors; performance analysis.

1. INTRODUCTION

Among the most important information that affects the situational awareness of pilots of both commercial transport and general aviation (GA) aircraft is the location and severity of hazardous weather. The flight crews of commercial transport aircraft have a variety of on-board systems to assist them with maintaining awareness of potentially dangerous weather. Many of these aircraft are equipped with airborne weather radar, which detects hazardous weather ahead of the aircraft. Weather information and advisories are provided via VHF voice-radio frequency by company airline dispatchers and staff meteorologists on the ground. In contrast with the airline crew, the GA pilot faces the demanding task of flying the aircraft, navigating, communicating with Air Traffic Control (ATC), and interpreting the verbal weather reports from a Flight Service Station (FSS), often without the benefit of a second crew member to share the workload, nor with any of the supporting technology available to the airline pilot. In addition, GA aircraft are more sensitive than air carriers to the effects of hazardous weather.

MIT Lincoln Laboratory, through the sponsorship of the Federal Aviation Administration (FAA), is developing a data link application that will provide graphical, as well as text, weather information to the GA pilot in the cockpit. The goal is to provide relevant and timely information at an affordable cost to the GA pilot community.

To assess the effects, as well as to aid in the proper design, implementation and certification of use of the Graphical Weather Service (GWS) in aircraft, a series of human factors studies is being conducted. This paper is an overview of the methodology and results of two human factors evaluations conducted thus far: Phase One (Lind, *et al.*, 1994) and Phase Two (Lind, *et al.*, 1995).

2. GRAPHICAL WEATHER SERVICE: A DATA LINK APPLICATION

The initial GWS product is a composite precipitation image derived from a mosaic of ground-based weather radars. The mosaic is a commercial product provided by WSI Corporation, and is a nationwide image of the six National Weather Service precipitation levels with a resolution of 2 kilometer x 2 kilometer (km). In its current implementation, GWS uses three colors to depict intensity levels: green for weak, yellow for moderate, and red for strong to extreme precipitation.

The data link transmission of the raw precipitation mosaic would require far more bandwidth than is available with any practical data link implementation. However, the transmission of these complex images is accomplished through application of a compression algorithm developed at MIT Lincoln Laboratory. The algorithm (Gertz, 1990) is based upon the underlying geometric structure of weather phenomena and operates by coding the graphical image as a set of polygons and ellipses. Figure 1 shows an uncompressed and compressed weather image.

The components of GWS are seen in Figure 2. To use GWS, the aircraft must be equipped with a data link "modem" such as a Mode S transponder or a VHF data radio that transmits and receives the data

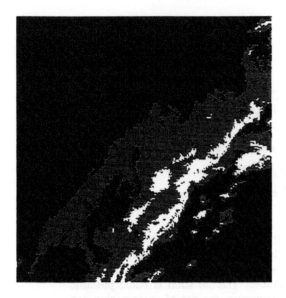

Figure 1. Uncompressed and Compressed Weather Images. Without data compression, the 256x256 km image on the left would require 131,000 bits to transmit. The image on the right has been compressed to 2413 bits using the Polygon-Ellipse algorithms. The compressed image can be transmitted to the aircraft by Mode S data link in approximately 10 seconds.

link messages. An onboard Control and Display Unit is used for the pilot to request services and system to display information. It is estimated that this equipage will cost approximately $5000 to $8000 (Bussolari, 1994). To receive a GWS image: a data link request for a specific image is received from an aircraft, it is passed to a ground-based image compression processor, the processor selects the appropriate image area from a weather data base (based on location, time, and scale specified in the request), the processor compresses the image and encodes it for transmission to the requesting aircraft. When received by the aircraft, an on-board processor decodes the message, decompresses the image and displays it to the pilot.

3. HUMAN FACTORS EVALUATION OF GWS

The availability of near-real-time graphical weather information via data link will significantly affect pilot situational awareness and decision making. Phase One was conducted to assess the overall effect of GWS on pilot decision making. It was seen as a first step in validating the need for GWS and as a proof of concept. Once Phase One findings validated that GWS is useful and effective, then we proceeded to Phase Two testing to determine what amount of compression would be acceptable for transmission of images to an aircraft. Since these complex images need to be compressed due to limited bandwidth, the resulting image is somewhat altered from the original image. Therefore, the key issue in Phase Two was the determination of how much compression, and associated distortion, is considered acceptable for transmission of images to the aircraft and at what point is the level of compression no longer acceptable. Phase Two also addressed the issue of determining whether there is a computational

measure of image quality that could be used to predict subjective acceptability of images.

Both phases were conducted in an office setting. GWS images were displayed on a Macintosh personal computer. All of the weather information and images used were constructed from actual recorded data, made available by WSI Corporation.

In both phases, the subjects were told that the aircraft flown in these hypothetical flights is a light, single engine piston aircraft, such as a Cessna 172. They were told to assume that they have full fuel for the flight and could assume that they are planning to travel with one passenger who is not a pilot.

Twenty volunteer instrument rated pilots participated in each phase. The majority of subjects who participated in Phase One also participated in Phase Two. Subjects had a range of flight time of 500 to over 27,000 hours, and a range of actual instrument time of 35 to over 2,000 hours.

4. PHASE ONE

4.1 Method

Phase One tested the effect of GWS on decision making during hypothetical flights in challenging weather conditions. Each subject participated in four flights (two with GWS and two without GWS). For each flight half of the subjects had access to GWS and half of the subjects did not. This design enabled the testing of the GWS versus No GWS Condition.

Prior to each hypothetical flight, the subject received a prepared flight plan, relevant navigational charts and weather briefing materials. The subject was questioned at each of three decision points within the flight. The first decision point was at departure, prior to starting the aircraft engine. The second decision point was in the cruise portion of the

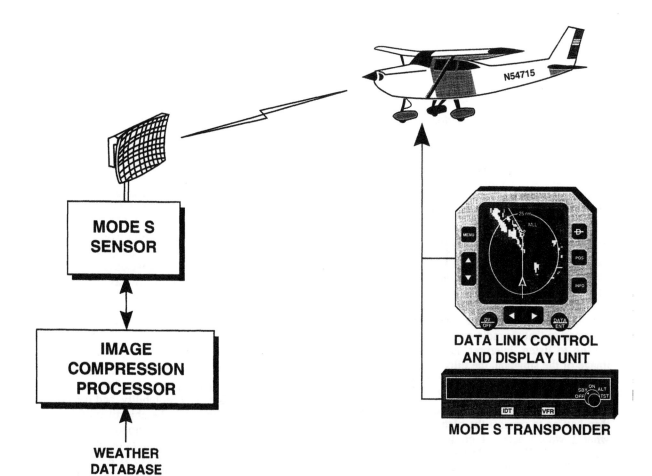

Figure 2. Graphical Weather Service.

flight, and the third was near the destination. Since the subject did not have the benefit of the sensory experience of flight, the experimenter told the subject what the pilot would be experiencing, e.g., ride quality, visibility, and precipitation. The subject was then asked what action he would take. The subject could respond immediately or could seek additional information using GWS (in the GWS Condition) or via queries to Air Traffic Control (ATC) or Flight Watch (FW) (in the GWS and No GWS Condition). An experimenter, sitting in the room with the subject, played the role of ATC and FW personnel.

For each decision point in which the subject had GWS, experimental images could be accessed for four locations (present position, departure, destination, alternate), at four different ranges (25, 50, 100, 200-nmi radius). The route of flight was in the 200-nmi range.

The subject was asked to "think aloud" throughout the experimental session. Verbal requests for information from ATC and FW, choices of GWS images, comments and actions taken at each decision point were recorded. Actions taken included go and no-go decisions and, decisions to deviate or to proceed on course. After selecting the action to be taken, the subject gave two ratings: a rating of confidence in his ability to assess the weather situation, given the information available, and a rating of the level of hazard presented.

4.2 Results

Action Taken

Results indicated that, by using GWS, subjects were able to make decisions based on an improved awareness of the actual weather situation. This was true for go / no-go decisions at the beginning of a simulated flight and also throughout flight when subjects were able to determine the need for course deviations and whether or not to proceed to destination or go to an alternate. For example, in one flight there was forecast to be a chance of imbedded thunderstorms in the area. Subjects with GWS were able to see that none of them were pertinent to the planned route of flight. As a result, all subjects with GWS decided to go on this flight, while half of the subjects without GWS decided not to go. In another flight, the plan called for the pilots to fly parallel to a cold front. The subjects with GWS were able to determine how close they were going to be to the front, while subjects without GWS were less sure of the location of the front and, subsequently, requested deviations. In this flight the presence of GWS resulted in less deviations. In another flight, the presence of GWS resulted in more deviations than if the subject did not have GWS. With GWS, subjects could see that a region of precipitation was localized and that by making a deviation the area could be avoided. Without GWS, subjects could not judge the expanse and severity of the weather and tended to plan to fly through it. The conclusion that follows is that GWS had a pronounced positive effect on situational awareness and subsequent decision making.

Information Requests to ATC and FW

Chi-square analyses were performed to assess the difference in frequency of calls and no calls for weather information in the GWS versus No GWS Conditions. Results indicated significantly fewer calls in the GWS Condition ($p < 0.05$). Results indicated this to be the case at each decision point. This effect was found for both moderately and extensively experienced instrument rated pilots.

When fewer calls for weather information are made there are several benefits. Since GWS provides a graphical presentation of weather, the pilot does not have to query ATC or FW about particulars on the exact location of cells and their intensity. This not only saves time for the pilot but it provides information that the radio cannot provide. Instead of having to listen, interpret, and construct a mental image of the situation, the pilot can see the situation graphically on the GWS cockpit display. Many of the subjects made comments to this effect. For example, "... graphically I can see the picture rather than having to concoct the picture in my brain from what I would hear. I could see how close the weather is." GWS could also provide a benefit to ATC. The ATC workload could be reduced, since pilots may call ATC less often for weather information. In the future, GWS will provide other information, such as surface observations. Therefore, even fewer calls would be made to ATC.

Confidence in Ability to Assess the Weather

Wilcoxon Rank Sum Test results indicated that pilot confidence in ability to assess the weather (rated on a scale of (1) Not at all Confident to (5) Very Confident) was higher at each decision point when GWS was used. This higher confidence was shown to be statistically significant at all decision points ($p < 0.05$). This effect was found for both moderately and extensively experienced instrument rated pilots.

Ratings of Perceived Weather Hazard

A concern in providing graphical weather information in the cockpit is the fear that pilots may have more information about the weather and this might result in underestimating the level of risk. Instead Wilcoxon Rank Sum Test results indicated that pilot perception of hazard (rated on a scale of (1) Not at all Hazardous to (5) Very Hazardous) did not change significantly when GWS was used.

Ratings of GWS Usefulness

GWS was rated as being more than moderately useful to very useful (rated on a scale of (1) Not at all Useful to (5) Very Useful) for each flight in which it was used. This approval of GWS was substantiated by subject comments made throughout the study. In addition, when subjects were asked if they would

purchase the equipage necessary for receipt of data link services, the answer was overwhelmingly positive.

5. PHASE ONE – CONCLUSIONS

Results indicated that GWS had a substantial positive effect on the weather-related decisions made by the subjects. With GWS, subjects could see the weather graphically displayed and could make decisions based on their improved awareness of the situation. This was found for pilots with both moderate and extensive experience in actual instrument flight. It was also found that subjects with GWS made fewer calls for weather information to ATC and FW, thus indicating a potential decrease in workload for both the pilot and ground personnel. Subjects' confidence in their ability to assess the weather situation was markedly increased when GWS was used. Subjects found GWS to be useful and cost-effective.

6. PHASE TWO

6.1 Method

Phase Two tested the effect of various levels of compression of GWS images on pilot perception of distortion, opinion of acceptability, and performance on a route drawing task. The objectives of this phase were to determine: 1.) what amount of compression would be acceptable for transmission of images to an aircraft, and 2.) whether there is a computational measure of image quality that could be used to predict subjective acceptability of images.

Twenty images were compressed to three compression levels, representing high, moderate, and low compression. These images where used in the two parts of this phase which are described below. As previously mentioned, this paper provides an overview of findings. For full documentation of Phase Two, the reader is referred to (Lind, *et al.*, 1995).

Part One – Subjective Ratings of Distortion and Acceptability

In the Distortion Task, the subject saw a series of pairs of GWS weather images. Each pair contained an uncompressed image (original) and compressed image (altered version). The subject judged the degree to which the compressed image had been distorted relative to the uncompressed image by assigning a numerical value to the level of distortion perceived. The rating was based on the quantitative amount of distortion of the compressed image and not on the usefulness and functionality of the compressed image. The functionality of the image was rated later in the Acceptability Task. For the distortion rating a magnitude estimate technique was used (Engen, 1971).

In the Acceptability Task, the subject saw a series of pairs of GWS weather images. Each pair contained an uncompressed image and compressed image. The subject judged the acceptability of the compressed image as a replacement for the uncompressed image. The subject was asked to judge acceptability in terms of the compressed image's functionality for the flight task as compared with the functionality of the uncompressed image for the flight task, regardless of the degree of image distortion. Ratings were operationally defined by the experimenter and a reference sheet given to the subject. Acceptability was rated by assigning a rating of Good or Excellent (for an acceptable image) or Poor or Very Poor (for an unacceptable image).

Part Two – Route Drawing Task

The subject saw a series of single GWS weather images. Each image was either an uncompressed image or a compressed image at high, moderate, or low compression (no designation of compression level was made to the subject.) The subject was asked to draw a route of flight from one designated point to another designated point, indicated on the screen as point "A" and "B". The route was drawn by using the mouse and clicking. In addition to drawing the route, the subject answered two questions regarding willingness to go on the flight and the amount of hazard in the depicted weather.

6.2 Results

Results of the distortion task indicated that subjects were in general agreement in their perception of amount of distortion in the images. However, there was a large amount of between-subjects variance in the acceptability ratings. That is, subjects tended to differ on how many of the most distorted images they were willing to call acceptable. Subject comments indicated the main objection to the highly compressed images was that detail was lacking, and the configuration of the weather was too greatly distorted, i.e., information content had been changed and found to be no longer acceptable. When images are highly distorted the algorithm tends to convert

areas of precipitation to elliptical shapes that may not adequately represent the shape of the actual weather.

The number of bits per image is one computed measure of image quality that can be used to predict whether an image would be acceptable and, therefore, should be transmitted to the cockpit. Figure 3 illustrates an example in which the experimenter set a criteria of 80% acceptability, i.e., if 80% of the subjects rated the image as acceptable, the image was considered acceptable for transmission to the cockpit. In this example, it is seen that images displayed with fewer than 2000 bits are generally considered unacceptable for transmission.

Bits per image, as well as several other computed measures of image quality, were examined to determine the measure that best predicted subjective ratings (both distortion and acceptability). Stepwise regression analyses indicated the best predictor to be a compression ratio of run-length encoding to number of bits per image. This measure was the best predictor within almost every individual subject.

In route drawing, the effects of compression level on route area error, route length, and proximity to levels of precipitation were examined. Results of Analysis of Variance of route area error indicated a significant main effect ($p < 0.05$) of compression level. The results of the Scheffe Procedure indicated that the difference in distance between routes is significant for the high compression group, i.e., with a mean of 19 nmi in contrast to 16 nmi for the low and moderate compression groups. No significant difference in route length was found to be a function of compression level. Regarding how close pilots drew a route to each level of precipitation intensity, Multiple Analysis of Variance and subsequent

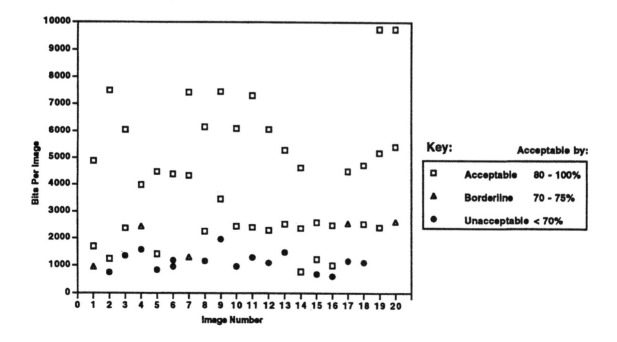

Figure 3. Application of Bits per Image in selecting images to be transmitted to the cockpit.

analysis indicated a significant effect only for Level 1 (weak precipitation intensity) when compression was high. When compression was high, the mean distance of routes from Level 1 was 2 nmi, rather than 1 nmi when compression was moderate or low.

Results indicated that compression did not significantly affect go / no-go decisions. Results indicated that hazard ratings increased significantly as a function of increased compression level.

7. CONCLUSIONS

Phase One validated the positive effects of GWS in increasing situational awareness and aiding pilots in making weather related decisions. In Phase Two images of low and moderate compression were found to be generally acceptable. Computed measures of image quality have been identified to enable the establishment of a criteria for transmitting images to an aircraft. Regarding pilot performance (as measured by the route drawing task), low and moderate compression did not result in significant differences in performance. High compression resulted in significant differences in route area error and proximity to weak precipitation intensity. A very important finding was that the subjects generally found the presence of ellipses to be unacceptable. While the algorithm preserves the fidelity of representation of precipitation intensity levels, the configuration of these levels were considered to be "too distorted" when a high degree of compression was applied. Work is underway on algorithm development to represent highly compressed images with greater fidelity to the shape of the uncompressed version.

REFERENCES

Bussolari, S. R. (1994). Data Link Applications for General Aviation, In *MIT Lincoln Laboratory Journal*, Vol.. 7, No. 2, Lexington, MA,.

Engen, T. (1971) Psychophysics: II. Scaling Methods. In J.W. Kling & L.W. Riggs (Eds.) *Woodworth and Schlosberg's Experimental Psychology* (3rd ed.), pp. 47-86. New York: Holt Rinehart and Winston.

Gertz, J. L., (1990). Weather Map Compression for Ground to Air Data Links, In *Proceedings of the Aeronautical Telecommunications Symposium on Data Link Integration*, pp. 131-40.

Lind, A. T., A. Dershowitz, S. R. Bussolari (1994). *The Influence of Data Link-Provided Graphical Weather on Pilot Decision Making*, MIT Lincoln Laboratory, Lexington, MA, ATC (Air Traffic Control) Project Report–215.

Lind, A. T., A. Dershowitz, D. Chandra, and S. R. Bussolari (1995, In Progress). *The Effects of Compression-Induced Distortion of Graphical Weather Images on Pilot Perception, Acceptance, and Performance*, MIT Lincoln Laboratory, Lexington, MA, ATC Project Report.

INTELLIGENT TUTORING SYSTEMS TO SUPPORT
MODE AWARENESS IN THE 'GLASS COCKPIT'

Alan R. Chappell and Christine M. Mitchell

Center for Human-Machine Systems Research
School of Industrial and Systems Engineering
Georgia Institute of Technology

Abstract: The proliferation of modes in modern glass cockpit aircraft places new demands on the pilot, i.e., mode management. This paper presents two computer-based tutors designed to provide the knowledge and operational skill required to safely and effectively use these modes. The VNAV Tutor, through visualization and context dependent instruction, addresses one mode often cited by pilots as confusing. The Mode Management Tutor, designed on the lessons learned from the VNAV Tutor, uses case-based teaching and intelligent tutoring system concepts to addresses a wider range of modes and mode problems.

Keywords: Aircraft control, Operators, Complex systems, Modes, Training, Computer-aided instruction, Displays.

1. INTRODUCTION

In conjunction with the increased levels of automation in modern glass cockpit aircraft, the number of modes available to the pilot has similarly increased (Sarter and Woods, 1993). Although these modes extend the functionality of the auto flight system, they also introduce new "perceptual and cognitive" demands on the pilot (Billings, 1991), i.e., mode management. Mode management is a multifaceted task. In order to make safe and effective use of the modes available, the pilot must recognize the modes that achieve the current flight objectives, know the procedures for using those modes, and monitor the modes by comparing expected aircraft behavior to actual performance. Even more demanding than the sheer number of modes, is the dynamic nature of these modes in systems such as the Flight Management System (FMS). Not only can these modes change manually at pilots commands, but also automatically at system events (without direct or immediate pilot input). In order to maintain mode awareness in this dynamic environment, the pilot must be continuously vigilant to indications in several locations in the cockpit.

Indeed, pilot confusion over mode management issues has been a dominant factor in at least two recent aircraft accidents (Gopal and Rao, 1991; Sparaco, 1994). Training is one method of addressing this problem. This paper describes two training systems, the VNAV Tutor and a Mode Management Tutor, that help to build pilot's knowledge of mode management concepts and expectations of mode behavior.

2. THE VNAV TUTOR

In a study of human factors of advanced technology ('glass cockpit') aircraft, Wiener (1989) found that pilots frequently identified vertical navigation as a problem with understanding and using auto flight systems. Sarter and Woods (1992) asked pilots to describe instances in which FMS behavior surprised them or to identify FMS modes/features that surprised them. The largest number of reported problems (63 of the 159 problems) concerned vertical navigation modes (VNAV). The reported problems suggest that pilots do not feel that the underlying FMS logic is clear; they can not visualize the intended path, and thus are unable to anticipate or understand VNAV activities. From these research findings it is evident that pilots need additional knowledge about vertical path navigation (VNAV) modes. The VNAV Tutor is a proof-of-concept computer-based training system that teaches VNAV concepts and operations.

The VNAV Tutor presents training in the context of the pilot flying a part-task Boeing 757/767 simulator through four Line-Oriented Flight Training (LOFT)

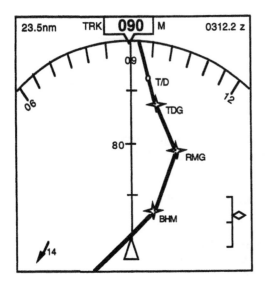

Fig. 1. Horizontal Situation Indicator showing the programmed path and the aircraft's progress along that route.

scenarios. These scenarios guide the student pilot through the FMS vertical profile, FMS execution of that profile through use of the VNAV function, interaction between FMS and other vertical navigation modes, and the use of FMS vertical navigation by the pilot for the completion of various in-flight maneuvers. LOFT scenarios specifically include entire flights: take-off, climb, cruise, descent, and landing. This design makes the scenarios similar to the pilot's operating environment, increasing the realism of the task being performed.

2.1 Vertical Profile Display

One of the primary functions of the flight management system (FMS) is to provide efficient 3-dimensional path navigation. The pilot configures the FMS and enters a set of points in space that defines the desired path of flight. The FMS then controls the aircraft such that it proceeds from point to point on this path using speeds that meet given restrictions where necessary and are fuel efficient where possible.

FMS navigation is divided into two modes: LNAV and VNAV. Lateral path navigation (LNAV) controls the horizontal movement of the aircraft, similar to driving a car along a prescribed route on a map. Pilots use the Horizontal Situation Indicator (HSI) to visualize the LNAV path, i.e., to preview the programmed route and to monitor the aircraft's progress as it flies that route. As depicted in Figure 1, the HSI provides a map-like representation of the programmed path. From the HSI, the pilot can see at a glance what LNAV is currently doing and what changes are upcoming. Based on the lack of complaints in the Wiener (1989) and Sarter and Woods (1992) studies, pilots are relatively comfortable with the LNAV auto flight mode.

Vertical path navigation (VNAV) controls the altitude and speed of the aircraft. However, the pilot is

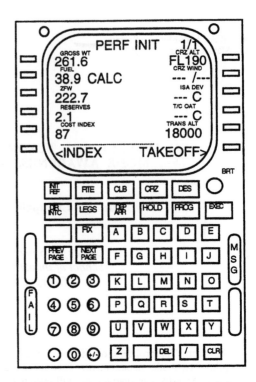

Fig. 2. Control and Display Unit containing pages of information defining the programmed path.

provided no integrated view of VNAV information. In order to ascertain the current state of VNAV and anticipate upcoming changes, the pilot must monitor three widely separated sources of data. The altitude and speed components of the programmed path (the vertical profile) are distributed over four pages of the Control and Display Unit (CDU) like the one depicted in Figure 2. All four pages of alphanumeric data must be viewed, one at a time, to obtain the data defining the vertical profile. Information telling which modes are active is located on the Attitude Direction Indicator (ADI) depicted in Figure 3. Constraints on VNAV control and upcoming mode transitions must be deduced from the Mode Control Panel (MCP) depicted in Figure 4. Given the labor-intensive task of integrating all these data sources, it is not surprising that pilots find VNAV confusing.

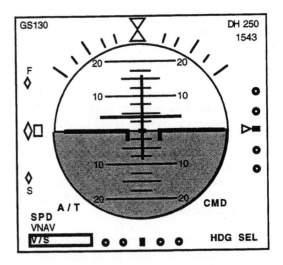

Fig. 3. Attitude Direction Indicator showing currently active modes.

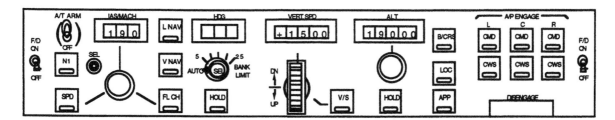

Fig. 4. Mode Control Panel showing constraints on VNAV control and information on the interaction of modes.

The VNAV Tutor uses visualization to represent the current and predicted aircraft behavior in the vertical plane. The primary mechanism is the Vertical Profile Display (Crowther, et al., 1994) depicted in Figure 5. This display is an analog to the HSI, integrating the information from the four CDU pages into a dynamic, pictorial representation of the vertical path. The currently active vertical control mode, normally found on the ADI, is displayed in the upper left corner. Additionally, the MCP constraint on altitude is shown using the MCP ALT line through the center. Highlighting and color coding are used to forecast aircraft behavior and mode transitions.

2.2 Evaluation

An evaluation of the VNAV Tutor was conducted on-site at a major commercial airline's training facility, with subjects drawn from Boeing 757/767 ground school. Subjects were pilots who were B757/767 students with little or no previous experience with FMS equipped aircraft, i.e., pilots transitioning to the glass cockpit. Five subjects participated in the evaluation. A pre-training questionnaire showed that the subjects had little or no prior knowledge of VNAV.

The evaluation used the four LOFT scenarios that comprised the VNAV Tutor. After the four tutorial sessions, the pilot flew a fifth, evaluation session

that did not incorporate the Tutor or the Vertical Profile Display. This session had periodic pauses at predetermined points in time in order to allow the experimenter to ask the pilot specific questions regarding the state of the FMS and other auto flight equipment, specifically focusing on vertical navigation awareness.

Figure 6 is typical of the results of the evaluation. This figure shows the performance of the pilots on a group of eight VNAV tasks. These tasks included the use of VNAV mode, speed intervention, descend now, and an approach change operations. A more complete presentation of the data can be found in Crowther et al. (1994). The data indicate that pilots typically learned to use and understand VNAV. In addition, this new knowledge resulted in a good level of mode awareness. Hence, the VNAV Tutor, through the use of appropriate visualization and situated instruction, was a successful training system.

Nonetheless, pilot performance was not perfect. Examination of the incorrect responses given by the different pilots shows that the subjects made different errors. Thus, mistakes are attributable to individual differences in learning rates and styles. This indicates that repetitions of existing lessons to enhance performance may be of marginal value. Forcing everyone to invest the time needed by only a few students on any particular training objective is an inefficient use of training time. Instead, these errors

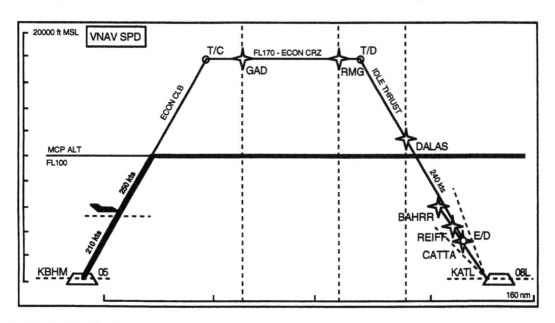

Fig. 5. Vertical Profile Display during a VNAV climb. The current vertical control mode is VNAV speed. Altitude is constrained to 10,000 ft.

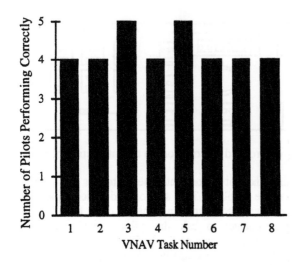

Fig. 6. Number of pilots performing correctly on VNAV tasks.

could more effectively be corrected by detecting and remediating individual problems.

Additionally, observation of the evaluation brought into question the efficiency and necessity of LOFT-based scenarios. When presenting a focused training objective, LOFT scenarios require the student to fly through relatively long segments containing little training activity. A tutor that employs short duration, focused scenarios could provide training in a realistic environment (situated instruction) and maintain training efficiency.

3. THE MODE MANAGEMENT TUTOR

The VNAV Tutor provides the system knowledge necessary to enhance the pilot's mode awareness when using VNAV functions. Thus, the VNAV Tutor addresses one mode management problem. Pilots, however, have also indicated concern over managing the large number of modes available throughout glass cockpit aircraft. Possible combinations of modes and

sub-modes provide a wide range of control capabilities. However, the pilot must learn the purpose, capabilities, and limitations of each of those individual modes, and more importantly, mode combinations, to select, monitor, and manage an appropriate mode for a given situation. The aviation literature is replete with instances where such knowledge was found to be lacking (Sarter and Woods, 1992; Wiener, 1989).

The Mode Management Tutor is an *intelligent* tutoring system (ITS) that focuses on the more general problem of mode management. The design of the Mode Management Tutor takes advantage of the capabilities of an ITS to tailor instruction to individual students. Case-based scenarios focus instruction on known mode management problems. Hence, the Mode Management Tutor builds on the lessons learned from the VNAV Tutor and ongoing educational systems research.

3.1 Intelligent Tutoring Systems

An Intelligent Tutoring Systems (ITS) is *intelligent* because it adapts training material to the needs of the individual student. An ITS performs this adaptation through the application of models of expert performance and of the student's knowledge.

An ITS is generally composed of four basic components as depicted in Figure 7: a pedagogy, an expert model, a student model, and an interface (Wenger, 1987). The pedagogy is the system's knowledge about how to teach and usually controls the progress of the tutor. The domain expert is a model that provides the declarative and procedural knowledge that characterizes expert performance in the domain being taught. The student model represents the current and evolving knowledge of the student. Finally, the interface provides communication between the tutor and the student.

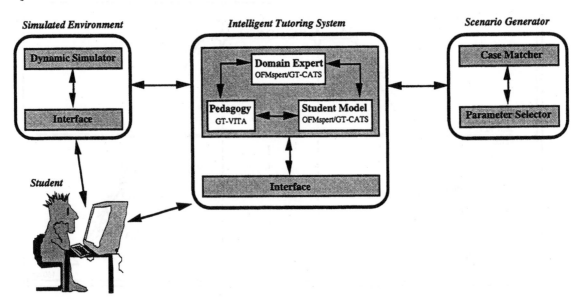

Fig. 7. Architectural design of a Case-Based Intelligent Tutoring System.

A fifth component, a simulation environment, is sometimes included in an ITS design. When teaching operators to control complex, dynamic systems, the process of becoming an expert involves more than learning declarative and procedural knowledge. A simulation is essential for teaching operational skill, such as situational cues indicating when to use specific procedures (Chu and Mitchell, 1995).

3.2 Case-Based Scenarios

Another emerging area of educational systems research is Case-Based Teaching. Case-based teaching is based on the idea that students learn on a "need-to-know basis" (Schank, 1991). The student is placed in a situation that creates interest in or a need for a specific piece of information. A case that teaches that information is then presented. Aviation training appears well suited to case-based teaching. Ample cases exist in the literature in the form of incident and accident reports. Using situations that have caused other pilots problems is likely to produce high levels of motivation. To illustrate how these documented problems can be used, an example case and its use are presented.

An Example Case. The aircraft is performing a programmed climb using the VNAV mode. An upper altitude limit of 11,000 feet is set on the Mode Control Panel (MCP). Around 10,000 ft. air traffic control (ATC) commands the pilot to slow the plane to 240 knots. Upon reaching 11,000 ft., the aircraft levels out according to the MCP limit. ATC then gives a clearance for the climb to be continued to 14,000 ft.

In this situation the pilot originally performed two sets of actions. When commanded to slow to 240 knots, a VNAV sub-mode called speed intervention was used to slow the speed without interfering with the programmed climb. When the clearance to a higher altitude is given after the level-off at 11,00 ft. the MCP altitude limit was raised to the new cleared limit and VNAV was re-engaged to restart the climb. However, since VNAV was completely disengaged at the level-off, re-engaging VNAV does not re-activate the speed intervention sub-mode. Hence, the aircraft will accelerate to the programmed speed which is usually well above the 240 knot restriction. In addition to the actions taken the pilot should have re-engaged speed intervention.

Using the Example Case. In the process of teaching VNAV speed intervention, one of the many modes a mode management tutor must cover, this case can be used to highlight a limitation, or common 'gotcha', of the mode. The given scenario can be used in several ways: as an example, to demonstrate the original error and its consequences, as a coaching opportunity, allowing the student to fly the scenario with the tutor providing hints or suggestions, and in a testing situation, allowing the student to fly without intervention and evaluating performance at the end. Multiple scenarios can be generated from one case by changing, in a coordinated and appropriate manner, the locations, altitudes, speeds, and phase of flight. These multiple scenarios are used to provide less repetitive practice with the same basic case.

3.3 Conceptual Design

Figure 7 depicts the architectural design of a case-based intelligent tutoring system for mode management. Like the VNAV Tutor, the Mode Management Tutor design is built upon a Boeing 757/767 part-task simulator. This allows the pilot to gain operational experience in the concepts presented. The scenarios used in the simulator are short duration, focused scenarios generated from the cases of past incidents and accidents. Cases are selected to exemplify the concepts selected for instruction by an intelligent tutoring system (ITS).

The design of this ITS is based primarily on the Georgia Tech Visual Inspectable Tutor and Aide (GT-VITA). GT-VITA (Chu and Mitchell, 1995) provides the framework for building training systems that teach declarative and procedural knowledge, and the operational skill needed to control effectively a complex dynamic system. This framework includes a structured succession of pedagogical strategies that become less intrusive as the student's knowledge state progresses.

The domain expert model and the student model designs are based on an adaptation of the Operator Function Model (OFM) (Mitchell, 1987), the computational implementation of OFM, OFMspert (Rubin, et al. 1988), and the Georgia Tech Crew Activity Tracking System (GT-CATS). The OFM provides a hierarchical and heterarchical description of the operator's job. The hierarchy provides for decomposition of the functions into successively smaller activities, i.e., functions, subfunctions, tasks, subtasks, down to the level of individual actions. The heterarchy models the concurrent nature of many tasks performed by expert operators.

GT-CATS (Calentine and Mitchell, 1994) contains an implementation of the OFM and OFMspert modified to understand pilot actions in glass cockpit aircraft. Figure 8 shows a representation of how activities are decomposed by GT-CATS including an additional mode selection level. By tracing the model down from the desired function or mode of interest, a normative list of requisite actions can be predicted. Hence, GT-CATS acts as a domain expert model. In addition, tracing branches up from detected actions leads to the classification of those actions as expected, non-normative but valid, or unexplainable. Detection of expected actions represents operational activities that the student knows. Detection of non-normative, unexplained, or missed actions (as compared to the expectation of the model expert) leads to remediation.

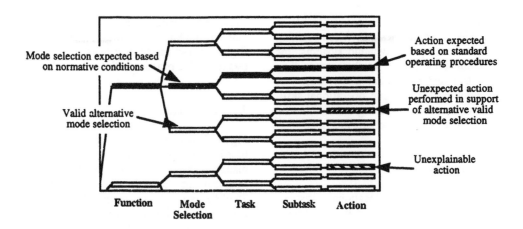

Fig. 8. GT-CATS' decomposition of part of the pilots task.

4. SUMMARY

The VNAV Tutor provides training for one problematic mode of the auto flight system in a modern glass cockpit aircraft. Evaluation of the VNAV Tutor with line pilots showed that a computer-based tutor that uses appropriate visualization and context dependent instruction can be successful teaching mode management. Lessons learned from the VNAV Tutor guide the design of a Mode Management Tutor. This design takes advantage of state-of-the-art educational systems research, including case-based teaching and intelligent tutoring systems for operators of complex systems.

5. ACKNOWLEDGMENTS

This research was supported by the NASA Ames Research Center grant NCC2-824 (Everett Palmer, Technical Monitor). The authors would also like to thank the pilots and ground training personnel at Delta Air Lines who provided extensive training and documentation on the Boeing 757/767.

REFERENCES

Billings, C.E. (1991). *Human-Centered Aircraft Automation: A Concept and Guidelines* (NASA Technical Memorandum 103885). NASA Ames Research Center, Moffett Field, CA.

Calentine, T. and C.M. Mitchell (1994). A methodology for understanding how operators select and use modes of automation to control complex dynamic systems. In: *Proceedings of the 1994 IEEE International Conference on Systems, Man, and Cybernetics*, pp. 1751-1756. San Antonio, TX.

Chu, R., and C.M. Mitchell (1995). Using the operator function model and OFMspert as the basis for an intelligent tutoring system: Towards a Tutor/Aide paradigm for operators of supervisory control systems. To appear in: *IEEE Transactions on Systems, Man, and Cybernetics*.

Crowther, E.G., A.R. Chappell and C.M. Mitchell (1994). VNAV Tutor: System knowledge training for improving pilots' modes awareness. In: *Proceedings of the 1994 IEEE International Conference on Systems, Man, and Cybernetics*, pp. 747-752. San Antonio, TX.

Gopal, B.S. and C.R.S. Rao (1991). *Indian Airlines A-320 VT EP: Report of the technical assessors to the court of inquiry*. Indian Government.

Mitchell, C.M. (1987). GT-MSOCC: A domain for modeling human-computer interaction and aiding decision making in supervisory control systems. *IEEE Transactions on Systems, Man, and Cybernetics*, **SMC-17**(4), 553-572.

Rubin, K.S., P.M. Jones and C.M. Mitchell (1988). OFMspert: Inference of operator intentions in supervisory control using a blackboard architecture. *IEEE Transactions on Systems, Man, and Cybernetics*, **SMC-18**(4), 618-637.

Sarter, N.B. and D.D. Woods (1992). Pilot interaction with cockpit automation: Operational experiences with the flight management system. *International Journal of Aviation Psychology*, 2(4), 303-321.

Sarter, N.B. and D.D. Woods (1993). "How in the world did I ever get into that mode?" Mode error and awareness in supervisory control. In: *Cognitive Engineering in Aerospace Application: Pilot Interaction with Cockpit Automation* (NASA Contractor Report 177528), pp. 51-64. NASA Ames Research Center, Moffett Field, CA.

Schank, R.C. (1991). *Case-Based Teaching: Four Experiences in Educational Software Design*. The Institute for the Learning Sciences, Northwestern University, Evanston, IL.

Sparaco, P. (1994). Human factors cited in French A320 crash. *Aviation Week & Space Technology*, January, 30-31.

Wenger, E. (1987). *Artificial Intelligence and Tutoring Systems: Computational and Cognitive Approaches to the Communication of Knowledge*. Morgan Kaufmann, Los Altos, CA.

Wiener, E.L. (1989). *The Human Factors of Advanced Technology ("Glass Cockpit") Transport Aircraft* (NASA Contractor Report 177528). NASA Ames Research Center, Moffett Field, CA.

TOWARDS A NEW PARADIGM FOR AUTOMATION: DESIGNING FOR SITUATION AWARENESS

Mica R. Endsley

Department of Industrial Engineering
Texas Tech University
Lubbock, TX 79409

Abstract: Automation is being implemented in a variety of systems in an effort to improve performance and overcome high operator workload. In examining accidents with automated systems, it becomes apparent that current automation approaches may underlie these problems by reducing operator situation awareness. Furthermore, evidence suggests that in many ways current automation approaches fail to achieve the desired reduction in workload, yet the prevailing approach to system design is still to automate to reduce workload. An alternate design approach is presented that focuses on enhancing situation awareness.

Keywords: automation, failure detection, human error, aircraft, human centered design

1. INTRODUCTION

For the past several decades automation has been heralded as a panacea for the ills of complex systems. Its goal has been to improve system reliability and performance by removing error prone humans from direct control and by aiding humans with difficult tasks. In an effort to improve system performance, some form of automation has become embedded in almost every type of system from process control to manufacturing systems, from aircraft to automotive systems.

1.1 Advantages of automation

In many cases, automation additions to systems have performed admirably, significantly improving system performance. For instance, Norman and Abbott (1988) show decreased crew-caused accident rates associated with aircraft containing higher levels of automation. Although this data may be highly confounded (with age of the aircraft for instance), improvements in system performance can be found with many specific automation implementations. The incidence of controlled flight into terrain accidents, for example, has significantly decreased since the introduction of ground proximity warning systems in transport aircraft (Billings, 1991).

1.2 Problems with automated systems

Despite the advantages of automated systems, significant problems remain in the integration of automated systems and the human operator. With many systems, automation has only been applied piece-meal, to tasks for which problems are identified and technological solutions available. This has been dubbed the technology-centered approach to automation. In addition to performing tasks not easily automated, the human operator usually remains in the system as a monitor to insure that the automated systems perform properly and to detect the occurrence of aberrant conditions.

Unfortunately, this places human operators into a role they are ill-suited for — that of monitor — and sets up a situation in which different *types* of errors are likely to occur. An increase in the complexity level of the system and, correspondingly, an increased propensity for catastrophic failures has been associated with the incorporation of automation (Wickens, 1992; Wiener, 1985). The occurrence of more frequent small errors in a non-automated system is replaced by an automated system with fewer large errors with significant consequences (Billings, 1988; Wiener, 1985).

The out-of-the-loop performance problem is a major issue associated with automation. Human operators acting as monitors have problems in detecting system errors and performing tasks manually in the event of automation failures (Billings, 1988; Wickens, 1992; Wiener and Curry, 1980). In addition, they have a more complex system to monitor. In a review of automation problems, Billings (1988) noted six major aircraft accidents that could be traced directly to failures to monitor automated system or the parameters controlled by the automated systems.

In addition to delays in detecting that a problem has occurred necessitating intervention, operators may require a significant period of time to develop a sufficient understanding of the state of the system to be able to act appropriately. This delay may prohibit operators from carrying out the very tasks they are there to perform or diminish the effectiveness of actions taken. In 1989, a US Air flight failed at take-off at LaGuardia Airport landing in the river and killing two passengers when an autothrottle was accidentally disarmed (National Transportation Safety Board, 1990). The time taken for the crew members to attempt to gain control of the aircraft without understanding the problem resulted in delaying the decision to abort the take-off until too late.

1.3 Impact on situation awareness

These highly reported problems with automated systems can be directly attributed to lower levels of operator *situation awareness* (SA) that can occur with automation approaches that place people in the role of passive monitor. Situation awareness, a person's mental model of the state of a dynamic system, is central to effective decision making and control, and is one of the most challenging portions of many operator's jobs. In a study of aircraft accidents involving major air carriers over a five year period, it was found that 88% of accidents citing human error could be attributed to problems with SA (Endsley, 1994). Hartel, et al. (1991) reported poor SA to be the leading causal factor in military aviation mishaps.

Situation awareness is formally defined as *"the perception of the elements in the environment within a volume of time and space, the comprehension of their meaning, and the projection of their status in the near future"* (Endsley, 1988). The development of SA first involves perceiving critical data, (e.g. the state of flight parameters, system status, location of other aircraft and ground features, etc....) comprising Level 1 SA – *perception*. Shortcomings in perceiving the status of the automated system and the parameters it controls are classic problems with SA at this level.

Data perceived also must be integrated and compared to operational goals to provide an understanding of what it really means, forming Level 2 SA - *comprehension*. At the highest level, operators have a well developed enough understanding of the situation to be able to predict the future actions of the system (Level 3 SA - *projection*), allowing them

to behave proactively instead of reactively in dealing with the environment to achieve their goals.

With many automated systems, forming the higher levels of SA can pose a significant difficulty. Empirical evidence of lower Level 2 SA under automated conditions has been found in experimental studies (Carmody and Gluckman, 1993; Endsley and Kiris, 1994). The accident at Three Mile Island is a classic example of a situation where the operators received needed data, but did not correctly understand the meaning of that data. Nor do operators appear to have a high level of understanding regarding the future behavior of automated systems.

Problems with situation awareness under automation have been directly linked to several major factors (Endsley and Kiris, in press): 1.) Vigilance decrements associated with monitoring, complacency due to over-reliance on automation, or a lack of trust in automation can all significantly reduce SA as people may neglect monitoring tasks, attempt to monitor but do so poorly, or be aware of indicated problems, but neglect them due to high false alarm rates. 2.) Passive processing of information under automation (as opposed to active manual processing) can make the dynamic update and integration of system information more difficult. 3.) Changes in form or a complete loss of feedback frequently occur either intentionally or inadvertently with many automated systems.

In addition, problems with Levels 2 and 3 SA can be partially attributed to the increased level of complexity associated with many automated systems, poor interface design and inadequate training (Endsley, in press-a). As system complexity increases, achieving a good mental model of the system can be more difficult. There are more parameters that may interact in more complex ways, and many combinations of circumstances may rarely be seen. Pilots have reported significant difficulties in understanding what their flight management systems are doing and why (Wiener, 1989). Although this understanding tends to increase with experience (McClumpha and James, 1994), there is also an indication that the interfaces of these systems are frequently not well designed to meet operator information needs (Billings, 1991; Norman, 1989) yielding significant difficulties in achieving the higher levels of SA with many automated systems.

1.4 Summary

Significant decrements in the SA of operators of automated systems exist creating many problems in controlling and monitoring these systems; Yet, automation is felt to be necessary to provide operators with manageable workload levels and reduced error rates. At present, it appears that this state of affairs has been largely accepted as something to be tolerated until further automation can be added to solve each new human failure. Successive levels of automation only add to the level of complexity and remove the operator further from active control of the system, however, exacerbating the very problem these efforts attempt to solve. The answer to this conundrum lies in examining the underlying assumptions of automation and devising a different approach to the problems automation attempts to solve.

2. THE TRADITIONAL AUTOMATION PARADIGM

Traditional human factors design has focused on function allocation — the division of tasks between man and machine. In the era of automation, the focus has been on determining high workload tasks that operators need assistance with. These tasks are then allocated to automation in order to reduce the human's tasks to more manageable levels.

2.1 Reduce workload through automation

This strategy is well ensconced in the aircraft design process and underlies much of the thinking within the human factors community. The Air Force Studies Board of the National Research Council (1982 , pp. 36-39) advocated automation as a desirable means of dealing with the high workload brought on by previous piecemeal automation and increases in system complexity. The underlying assumption of workload reduction has been implicit in automation projects. The Boeing 767, like many newer aircraft, was designed with the specific objective of "increasing cockpit automation and decreasing cockpit workload" (Ropelewski, 1982).

Current automation programs, featuring advanced artificial intelligence or expert systems, are also frequently aimed at dealing with workload problems. Even programs that recognize the need for automation that is better integrated with the needs of the human operator recommend basing this integration around workload, dynamically allocating tasks in real-time during the course of a mission based on the workload level of the pilot (Emerson and Reising, 1992; Morrison and Gluckman, 1994). In attempting to implement this approach, however, Pilot's Associate noted significant difficulties in implementing a workload based strategy for adaptively allocating functions (Hammer and Small, in press).

2.2 Fallacy of the workload assumption

Evidence indicates that the under-riding principle involved in this automation strategy is flawed. Human workload does not always respond to automation as predicted. Hart and Sheridan (1984) noted that automation often replaces workload involving physical activity with workload involving cognitive and perceptual activity. "Pilots recognize that the new aircraft call for more programming, planning, sequencing alternative selection and more thinking" (Wiener, 1988b). Furthermore, pilots complained that automation requires constant

scanning adding to workload. Wiener's (1985) studies in commercial aviation found a significant number of pilots reporting that automation does not necessarily reduce their workload, but actually may increase it during critical portions of the flight. Bainbridge (1983) called it the irony of automation that when workload is highest automation is often of the least assistance.

Many recent studies are beginning to confirm a lack of correspondence between workload and automation or people's use of automation. In studying adaptive automation techniques, Harris, et al. (1994) found that when operators were required to initiate task automation in response to an unanticipated increase in workload, it was accompanied by a significant increase in performance error on other manual tasks. This confirmed work by Parasuraman, et al. (1992) indicating that operator initiation of automation was likely to increase demands when they were already high. Riley (1994) investigated factors that influenced when people would choose to initiate automation. He found that a subject's choice to use automation in a task was not related to the workload level of the task, but rather to factors such as reliability, trust and risk.

Furthermore, it appears that monitoring itself may induce high workload. A series of studies have shown that in tasks where people must provide sustained attention as monitors over a period of time, it induces considerable fatigue (Galinsky, et al., 1993) and perceived workload is rated as fairly high (e.g. Becker, et al., 1991; Dittmar, et al., 1993; Scerbo, et al., 1993), contrary to characterizations of monitoring activities as boring but non-demanding.

The evidence is building that automation fails to decrease and may even increase workload. Operators indicate that under high workload they frequently turn the automation off, which Wiener (1988a) calls the paradox of automation. Even when task load is not high, the requirement to vigilantly monitor automated systems imposes its own workload.

2.3 Summary

Despite significant problems with human performance and evidence that workload does not necessarily decrease with automation, workload remains the fundamental human factors consideration in automation decisions. The overriding approach for dealing with workload is to automate. This schizophrenic mind set is firmly entrenched in today's automation projects, perhaps because no other alternative is readily apparent. As automation may not reduce workload and indications are that it can compromise SA, a new approach is needed.

3. A NEW APPROACH TO AUTOMATION

A solution to the problem posed here lies in examining when things go well in system operations. This is when operators are involved in their tasks, aware of what is going on, but not overloaded. Researchers have made the mistake of focusing on the overload problem with out taking into account the issues of involvement and awareness.

Design solutions that decrease SA increase the probability that errors will occur. Traditional automation approaches have done just that — decrease operators' SA by removing them from involvement in system operation. Using a fairly ingenious feedback paradigm, Pope et al. (1994) found an index based on EEG signals that corresponds to the degree to which subjects are "engaged" in performing a task. This index responded negatively to automation and positively to manual control, demonstrating this effect.

As a lack of SA appears to be at the heart of a large majority of human errors, it makes sense to focus on SA in the design process. Although the design community has focused increased attention on SA, this is currently being done in conjunction with the existing automation paradigm. These two approaches are not additive or even fundamentally compatible. An alternate approach focuses on system design and automation strategies that enhance SA by keeping operator involvement.

3.1 Design for situation awareness

A structured approach is required to incorporate SA considerations into the design process, including a determination of SA requirements, designing for SA enhancement, and measurement of SA in design evaluation.

SA Requirements Analysis. Designing interfaces that provide SA depends on domain specifics that determine the critical features of the situation that are relevant to a given operator. A goal-directed task analysis methodology (Endsley, 1993) has been used successfully for determining SA requirements in several different domains, including aircraft, air traffic control and remote maintenance control centers. This methodology focuses on the basic goals of operators (which may change dynamically), the major decisions they need to make relevant to these goals, and the SA requirements for each decision. SA requirements are established in terms of the basic data that is needed (Level 1 SA), required integration of the data for a comprehension of system state in light of goals (Level 2 SA), and projection of future trends and events (Level 3 SA).

The method is significantly different than traditional task analyses in that: 1.) it is not pinned to a fixed timeline, a feature which is not compatible with the work flow in dynamic systems, 2.) it is technology independent, not tied to how tasks are done with a given system, but to what information is really, ideally needed, and 3.) the focus is not just on what data is needed, but on how that data needs to be

combined and integrated to support decision making and goal attainment. This last feature, defining comprehension and projection needs, is critical for creating designs that support SA instead of overload the operator with data as many current systems do.

SA-Oriented Design. The development of a system design for successfully providing the multitude of SA requirements that exist in complex systems is a significant challenge. A set of design principles have been developed based on a theoretical model of the mechanisms and processes involved in acquiring and maintaining SA in dynamic complex systems (Endsley, in press-c). These guidelines are focused on a model of human cognition involving dynamic switching between goal-driven and data-driven processing and feature support for limited operator resources, including: 1.) direct presentation of higher level SA needs (comprehension and projection) instead of low level data, 2.) goal-oriented information display, 3.) support for global SA, providing an overview of the situation across the operator's goals at all times (with detailed information for goals of current interest), enabling efficient and timely goal switching and projection, 4.) use of salient features to trigger goal switching, 5.) reduction of extraneous information not related to SA needs, and 6.) support for parallel processing. To date, an SA-oriented design has been successfully applied as a design philosophy for systems involving remote maintenance operations and flexible manufacturing cells.

Evaluation. Many concepts and technologies are currently being developed and touted as enhancing SA. Prototyping and simulation of new technologies, new displays and new automation concepts is extremely important for evaluating the actual effects of proposed concepts with in the context of the task domain and using domain knowledgeable subjects. If SA is to be a design objective, then it is critical that it be specifically evaluated during the design process. Without this it will be impossible to tell if a proposed concept actually helps SA, does not effect it, or inadvertently compromises it in some way. The Situation Awareness Global Assessment Technique (SAGAT) has been sucessfully used to provide this information by directly and objectively measuring operator SA in evaluating avionics concepts, display designs, and interface technologies (Endsley, in press-b).

3.2 New roles for automation

In addition to supporting SA through system design, automation strategies must maintain or enhance SA. Many of the negative impacts of automation on SA and human performance may be attributable to the way that automation has traditionally been implemented. New approaches are currently being explored that challenge the relegation of the operator to passive monitor. These approaches redefine the assignment of functions to people and automation in terms of an integrated team approach that maintains operator involvement. One approach seeks to optimize the assignment of control between the human and automated system by keeping both involved in system operation. The other recognizes that control must pass back and forth between the human and the automation over time, and seeks to use this factor to increase human performance.

Adaptive Automation. Adaptive automation has been found to aid in overcoming the out-of-the-loop performance problem (Parasuraman, 1993). Adaptive automation attempts to optimize a dynamic allocation of tasks by creating a mechanism for determining in real-time when tasks need to become automated (or manually controlled) (Morrison and Gluckman, 1994). In direct contrast to historical efforts which have featured fixed task allocations, adaptive automation provides the potential for improving operator performance by keeping them in the loop. Carmody and Gluckman (1993) found SA to be impacted by adaptive automation of certain tasks. A concern associated with this technique, however, is that it may impose an additional task management load, requiring operators to keep up with who is doing what as the allocation changes. Research is needed to examine the full consequences of this approach for operator SA.

Level of Control. A second, complementary approach maintains operator involvement by identifying an optimal level of automation that keeps the operator in the loop, thus maintaining SA. In automating cognitive tasks via expert systems, five levels of automation (or control) can be considered. Decisions can be made: (a) manually with no assistance from the system, (b) by the operator with input in the form of recommendations provided by the system, (c) by the system, with the consent of the operator required to carry out the action, (d) by the system, to be automatically implemented unless vetoed by the operator, or (e) fully automatically, with no operator interaction.

Endsley and Kiris (in press) implemented automation of an automobile navigation task at each of these five levels. They found that the out-of-the-loop performance problem was significantly greater under full automation than under intermediate levels of automation. This corresponded with a greater decrement in SA under full automation than under intermediate levels, as compared to manual control. By implementing functions at a lower level of automation, keeping the operator in the decision loop, SA remained at a higher level and subjects were more able to assume manual control when needed.

Thus, even though full automation of a task may be technically possible, it may not be desirable if the performance of the joint human-machine system is to be optimized. Intermediate levels of automation may be preferable for certain tasks, in order to keep human operators' SA at a higher level and allow them to perform critical functions.

3.3 Summary

Following a structured approach from analysis to design to testing, SA can be incorporated as a significant and attainable design goal. It is important that such procedures are applied in the design process and automation strategies that maintain operator involvement exploited if the SA and performance of operators in complex settings are to be improved.

REFERENCES

Air Force Studies Board, Committee on Automation in Combat Aircraft, National Research Council. (1982). *Automation in combat aircraft*. National Academy Press, Washington D.C.

Bainbridge, L. (1983). Ironies of automation. *Automatica*, **19**, 775-779.

Becker, A. B., J. S. Warm and W. N. Dember (1991). Effects of feedback on perceived workload in vigilance performance. In: *Proceedings of the Human Factors Society 35th Annual Meeting*, pp. 1491-1494. Human Factors Society, Santa Monica, CA.

Billings, C. E. (1988). Toward human centered automation. In: *Flight deck automation: Promises and realities* (S. D. Norman and H. W. Orlady (Eds.)), pp. 167-190. NASA-Ames Research Center, Moffet Field, CA.

Billings, C. E. (1991). *Human-centered aircraft automation: A concept and guidelines* (NASA Technical Memorandum 103885). NASA Ames Research Center, Moffet Field, CA.

Carmody, M. A. and J. P. Gluckman (1993). Task specific effects of automation and automation failure on performance, workload and situational awareness. In: *Proceedings of the Seventh International Symposium on Aviation Psychology* (R. S. Jensen and D. Neumeister (Eds.)), pp. 167-171. Department of Aviation, The Ohio State University, Columbus, OH.

Dittmar, M. L., J. S. Warm, W. N. Dember and D. F. Ricks (1993). Sex differences in vigilance performance and perceived workload. *J. of Gen. Psych.*, **120**(3), 309-322.

Emerson, T. J. and J. M. Reising (1992). The effect of an artificially intelligent computer on the cockpit paradigm. In: *The human-electronic crew: Is the team maturing?* (T. Emerson, M. Reinecke, J. Reising and R. Taylor (Eds.)), pp. 1-5. Wright Laboratory, Air Force Materiel Command, Wright-Patterson AFB, OH.

Endsley, M. R. (1988). Design and evaluation for situation awareness enhancement. In: *Proceedings of the Human Factors Society 32nd Annual Meeting*, pp. 97-101. Human Factors Society, Santa Monica, CA.

Endsley, M. R. (1993). A survey of situation awareness requirements in air-to-air combat fighters. *Int. J. of Av. Psych.*, **3**(2), 157-168.

Endsley, M. R. (1994, March). *A taxonomy of situation awareness errors*. Paper presented at the Western European Association of Aviation Psychology 21st Conference, Dublin, Ireland.

Endsley, M. R. (in press-a). Automation and situation awareness. In: *Automation and human performance: Theory and applications* (R. Parasuraman and M. Mouloua (Eds.)). Lawrence Erlbaum, Hillsdale, NJ.

Endsley, M. R. (in press-b). Measurement of situation awareness in dynamic systems. *Human Factors*.

Endsley, M. R. (in press-c). Towards a theory of situation awareness. *Human Factors*.

Endsley, M. R. and E. O. Kiris (1994). The out-of-the-loop performance problem: Impact of level of control and situation awareness. In: *Human performance in automated systems: Current research and trends* (M. Mouloua and R. Parasuraman (Eds.)), pp. 50-56. Lawrence Erlbaum, Hillsdale, NJ.

Endsley, M. R. and E. O. Kiris (in press). The Out-of-the-Loop Performance Problem and Level of Control in Automation. *Human Factors*.

Galinsky, T. L., R. R. Rosa, J. S. Warm and W. N. Dember (1993). Psychophysical determinants of stress in sustained attention. *Human Factors, 35(4)*, 603-614.

Hammer, J. M. and R. L. Small (in press). An intelligent interface in an associate system. In: *Human/technology interaction in complex systems* (W. B. Rouse (Ed.)). JAI Press, Greenwich, CT.

Harris, W. C., P. N. Goernert, P. A. Hancock and E. Arthur (1994). The comparative effectiveness of adaptive automation and operator initiated automation during anticipated and unanticipated taskload increases. In: *Human performance in automated systems: Current research and trends* (M. Mouloua and R. Parasuraman (Eds.)), pp. 40-44. LEA, Hillsdale, NJ.

Hart, S. G. and T. B. Sheridan (1984). Pilot workload, performance, and aircraft control automation. In: *Human factors considerations in high performance aircraft (AGARD-CP-371)*, pp. 18/1-18/12. NATO-AGARD, Neuilly Sur Seine, France.

Hartel, C. E., K. Smith and C. Prince (1991, April). *Defining aircrew coordination: Searching mishaps for meaning*. Paper presented at the Sixth International Symposium on Aviation Psychology, Columbus, OH.

McClumpha, A. and M. James (1994). Understanding automated aircraft. In: *Human performance in automated systems: Current research and trends* (M. Mouloua and R. Parasuraman (Eds.)), pp. 183-190. Lawrence Erlbaum, Hillsdale, NJ.

Morrison, J. G. and J. P. Gluckman (1994). Definitions and prospective guidelines for the application of adaptive automation. In: *Human performance in automated systems: current research and trends* (M. Mouloua and R. Parasuraman (Eds.)), pp. 256-263. Lawrence Erlbaum, Hillsdale, NJ.

National Transportation Safety Board (1990). *Aircraft accidents report: USAIR, Inc., Boeing 737-400, LaGuardia Airport, Flushing New York, September 20, 1989* (NTSB/AAR-90-03). Author, Washington, D.C.

Norman, D. A. (1989). *The problem of automation: Inappropriate feedback and interaction not overautomation* (ICS Report 8904). Institute for Cognitive Science, U. C. San Diego, La Jolla, CA.

Norman, S. and K. Abbott (1988). Current flight deck automation: Airframe manufacturing and FAA certification. In: *Flight deck automation: Promises and realities* (S. D. Norman and H. W. Orlady (Eds.). NASA-Ames Research Center, Moffett Field, CA.

Parasuraman, R. (1993). Effects of Adaptive Function Allocation on Human Performance. In: *Human factors and advanced aviation technologies* (D. J. Garland and J. A. Wise (Eds.)), pp. 147-158. Embry-Riddle Aeronautical University Press, Daytona Beach, FL.

Parasuraman, R., T. Bahri, J. E. Deaton, J. G. Morrison and M. Barnes (1992). *Theory and design of adaptive automation in aviation systems* (AWCADWAR-92033-60). Naval Air Warfare Center, Aircraft Division, Warminster, PA.

Pope, A. T., R. J. Comstock, D. S. Bartolome, E. H. Bogart and D. W. Burdette (1994). Biocybernetic system validates index of operator engagement in automated task. In: *Human performance in automated systems: Current research and trends* (M. Mouloua and R. Parasuraman (Eds.)), pp. 300-306. Lawrence Erlbaum, Hillsdale, NJ.

Riley, V. (1994). A theory of operator reliance on automation. In: *Human performance in automated systems: Current research and trends* (M. Mouloua and R. Parasuraman (Eds.)), pp. 8-14. Lawrence Erlbaum, Hillsdale, NJ.

Ropelewski, R. R. (1982). Boeing's new 767 eases crew workload. *Aviation Week & Space Technology*, **August 23**, 41-53.

Scerbo, M. W., C. Q. Greenwald and D. A. Sawin (1993). The effect of subject-controlled pacing and task type on sustained attention and subjective workload. *J. of Gen. Psych.*, **120(3)**, 293-307.

Wickens, C. D. (1992). *Engineering Psychology and Human Performance* (2nd ed.). Harper Collins, New York.

Wiener, E. (1988a). Field studies in automation. In: *Flight deck automation: Promises and realities* (S. D. Norman and H. W. Orlady (Eds.)), pp. 37-55. NASA-Ames Research Center, Moffett Field, CA.

Wiener, E. L. (1985). Cockpit automation: In need of a philosophy. In: *Proceedings of the 1985 Behavioral Engineering Conference*, pp. 369-375. Society of Automotive Engineers, Warrendale, PA.

Wiener, E. L. (1988b). Cockpit automation. In: *Human Factors in Aviation* (E. L. Wiener and D. C. Nagel (Eds.)), pp. 433-461. Academic Press, San Diego.

Wiener, E. L. (1989). *Human factors of advanced technology ("glass cockpit") transport aircraft* (NASA Contractor Report No. 177528). NASA-Ames Research Center, Moffett Field, CA.

Wiener, E. L. and R. E. Curry (1980). Flight deck automation: Promises and problems. *Ergonomics, 23(10)*, 995-1011.

MODE AWARENESS IN ADVANCED AUTOFLIGHT SYSTEMS

Sanjay S. Vakil
R. John Hansman, Jr.
Alan H. Midkiff
Thomas Vaneck

Aeronautical Systems Laboratory,
Massachusetts Institute of Technology
Cambridge, MA 02139

Abstract: An examination of autoflight systems in modern aircraft was made, with empasis on the complex mode structure which is suspect in several recent accidents. Aviation Safety Reporting System reports from 1990-94 were examined. Flight Mode Annunciator conventions were inventoried. Focussed interviews with pilots and check airmen were conducted. Principal results identified the lack of a consistent global model of the AutoFlight System architecture and identified the vertical channel as requiring enhanced feedback. An Electronic Vertical Situation Display to help mitigate the identified problems was prototyped.

Keywords: Aircraft Contol, Mode Structure, Complex Systems, Displays, Human Supervisory Control, Human Factors, Human-Machine Interface, Mode Analysis

1.0 INTRODUCTION

Current advanced commercial transport aircraft rely on AutoFlight Systems (AFS) for flight management, navigation and inner loop control. These systems have evolved into multiple computers which are capable of sophisticated tasks including automatic flight planning, navigation, and automated landings. These systems also provide envelope protection, which prevents pilots from committing obvious mistakes, such as stalling or lowering flaps at a high speed.

In the simplest form, these AutoFlight Systems switch between different modes (states of aircraft automation), stringing together modes to create complicated flight trajectories. As the AFS becomes more complicated, it is speculated that the proliferation of modes and the specific characteristics of each mode become more difficult to model mentally, leading to problems in mode awareness.

1.1 AutoFlight System Overview

The AFS in a modern aircraft typically separates the guidance into uncoupled horizontal and vertical components. Examples of horizontal guidance modes include flying a preprogrammed trajectory using a Lateral Navigation mode (LNAV), or flying on a selected heading. For horizontal flight, the path is controlled through roll. Typical vertical guidance modes include flying level (Altitude Hold), or maintaining a selected Vertical Speed (V/S). The vertical flight path coupled with the speed of the aircraft is controlled by a combination of thrust and elevator.

Unfortunately, the increasing complexity of AutoFlight Systems has caused an increase in problems associated with the management of the system. In flight mode awareness problems, there can be confusion between the pilots' expectations of the AFS and what it is actually doing. In this research, the working definition of a mode awareness problem is one in which the aircraft AFS executes an action, or fails to execute an action that is anticipated or expected by one or more of the pilots.

1.2 Motivation

Mode awareness problems are suspected in a number of aircraft incidents, specifically vertical path control confusion on an Airbus A320 at Strasbourg (Sparaco, 1994) and pitch control in go-around mode on an A300 at Nagoya (Mecham, 1994). Additional incidents include events such as overspeed during takeoff or go-around on Boeing 757/767s and inconsistent envelope protection (loss of the alpha protection floor) on the A320.

1.3 Approach

The approach involved two parts: an investigation to identify the problems and issues involved in mode awareness, and an investigation of potential mechanisms to mitigate mode awareness problems, including enhanced mode feedback

2.0 ISSUE IDENTIFICATION METHODOLOGY

Several mechanisms were used to explore the issues surrounding mode awareness problems. These included a review of Aviation Safety Reporting System (ASRS) reports, an investigation of the structure of current AFSs, including the Flight Mode Annunciator, as well as flight observations and focussed interviews.

2.1 Aviation Safety Reporting System

The Aviation Safety Reporting System allows pilots to detail safety problems or incidents with a degree of amnesty. A search was performed on the ASRS database over the years 1990-94 with a set of keywords designed to elicit problems related to mode awareness. The keywords consisted of the following: annunciation, annunciator, FMC, flight management computer, FMS, flight management system, CDU, mode, capture, arm, automatic flight system, vertical, horizontal, and program.

2.2 Inventory of Current AutoFlight Systems

The structure of the AutoFlight Systems in current aircraft was examined, including an inventory of flight mode annunciation schemes, via training manuals, and a review of the open literature. The

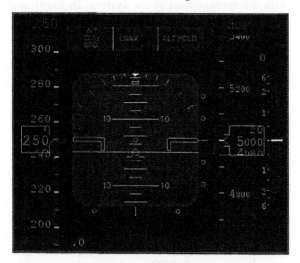

Figure 1. Primary Flight Display on 747-400

aircraft examined included the Boeing 757/767, the MD-11 and the Airbus 300-600R

The Flight Mode Annunciator (FMA) on the Primary Flight Display (PFD) of modern aircraft is normally the primary location of mode status information. FMAs typically display the current mode configuration of the aircraft in text. A PFD from the Boeing 747-400 is shown in Figure 1. The FMA is located in the middle of the display directly above the artificial horizon. The FMA shows the aircraft's AutoThrottle is engaged, that it is in LNAV Mode, and in Altitude Hold Mode. The display conventions used in the FMA were compared across a set of aircraft, including the Boeing 737-500/600, 757, and 767, MD-11, MD-80, Fokker 100 and Airbus 300-600R.

2.3 Focussed Interviews

Information was gathered from a variety of informal sources, including direct flight deck observations, discussions with flight crews, and discussions with simulator check airmen about their observations of crews during recurrent training.

3.0 PRELIMINARY OBSERVATIONS

Two main observations emerged from the investigation. First the results indicated that vertical modes appear to be more prone to serious mode related problems. Second was a lack of underlying structure to the automation, making it difficult for pilots to develop consistent mental models. Tertiary observations included a lack of commonality between Flight Mode Annunciators and a reduction in vestibular cueing.

3.1 Vertical Flight Path

Three hundred ASRS reports (from 1990-94) were returned by the keyword search and analyzed. Of these 184 were categorized as mode awareness problems using the working definition detailed earlier.

As shown in Figure 2, these reports were then categorized by the perceived cause of the problem and by the flight path (vertical/speed, horizontal or both) that was impacted. Since the vertical flight path and the speed are implicitly coupled, problems with either were grouped together. In instances where the problems spanned multiple causal categories, the reports were counted in each relevent causal category.

In Figure 2, it can be seen that vertical/speed problems dominate many of the categories. A total of 62.7% of the categorized reports were vertical/speed related. In particular, the Mode Transition Problems causal category is dominated by vertical/speed

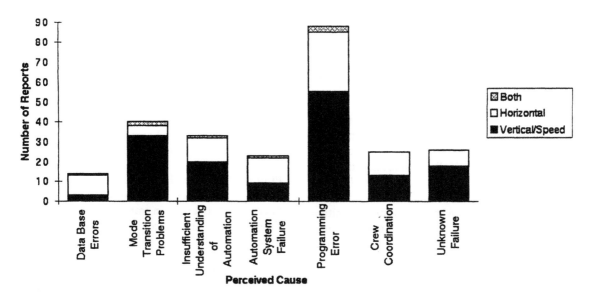

Figure 2.   Breakdown of ASRS reports into Perceived Causes and Flight Domain.

problems. The data in the Insufficient Understanding of Automation causal category also suggests a deficiency in knowledge of the vertical domain automation.

It should be noted that there exists a potential for over-reporting vertical deviations. Since Air Traffic Control radar has very precise surveillance of altitude versus position, there may be more cause to report vertical/speed incidents due to the amnesty clause in the ASRS.

3.2   Structure of Current Autoflight Systems

AutoFlight Systems have evolved from more basic autopilots as manufacturers incorporated additional functionality, including fuel optimal descents, flight path angle controlled vertical maneuvers and speed based vertical maneuvers. The evolutionary development of these systems and their increased capabilities have resulted in a complex structure of operating modes. Since each generation of AutoFlight System incorporates much of the functionality and modes of the previous generation, without reducing the number of features, there has been an entropic growth of complexity.

The other effect of this entropic growth has been to create a system that appears to lack a simple, consistent, global model. This lack requires pilots to create their own ad-hoc models.

Mode Definitions. A mode can be defined as a specific state of the aircraft automation. A mode includes the set of targets (heading, speed, vertical speed, pitch etc.) to which the aircraft is to be controlled as well as the actuators that are to be used (elevators, thrust,

ailerons). Current AutoFlight Systems switch between two basic types of operating modes which are termed base-modes and macro-modes. Base-modes are used in quasi-steady-state conditions and have an invariant set of targets. A base-mode example would be a Vertical Speed Mode where the aircraft attempts to maintain a specific vertical speed target by controlling pitch (via the elevators) and airspeed with thrust.

Macro-modes consist of a linked sequence of base-modes. Each base-mode in the macro-mode has its own set of targets, implying a set of targets which vary over the course of the macro-mode. Transitions between the base-modes are made based on specific state criteria, such as altitude or indicated air speed. An example of a macro-mode is the Autoland sequence, which transistions (in the vertical channel) between Flight Level Change, Glideslope Capture, Flare, and Rollout with a different set of targets in each base-mode.

Mode Transitions. There are three types of transitions between modes. A *commanded transition* is active as soon as selection is made. An example is pressing the vertical speed button. An *uncommanded transition* is one that is not directly activated by the pilot. These transitions are usually some type of envelope protection. An example is a transition caused by overspeed protection. Finally, *automatic/conditional transitions* occur when, after arming, a mode engagement occurs at a target state. An example is the use of Glide Slope Capture to transition to a descent mode after the aircraft intersects with the glide slope trajectory.

AFS Input-Output Relationships. Two basic types of AFS input-output relationships can exist. The simpler is a quasi-steady-state model where each output variable is controlled by a single input (SISO). A typical Vertical Speed mode engages two independent SISO controllers: the aircraft's pitch controls the vertical speed target and the thrust controls the air speed target. SISO models appear to be functionally adequate for simple modes.

Some mode transitions appear to utilize Multi-Input Multi-Output (MIMO) controllers, where each output variable can be controlled by more than one input. These transitions are typically of short duration and they do not appear to be modelled in detail by the crew. An example of a complex mode transition is the 0.05g capture used in an Altitude Capture transition of the MD-11: when the aircraft is approaching a selected flight altitude, the intercept maneuver limits the normal acceleration to 0.05g.

3.3 Flight Mode Annunciators

Preliminary analysis of Current Flight Mode Annunciators showed a limited commonality between displays. There are differences in the number, type and location of windows, in the usage of color, in the layout of text, and in the conventions used to indicate arming. More recently designed FMAs include Target State Values. An example of a target state value would be the commanded vertical speed in Vertical Speed Mode.

The Fokker 100 and the MD-11 FMAs also displayed the *control allocation* in each mode. This information explicitly identifies which actuator is controlling a particular output. Knowledge of control allocation is particularly important in the vertical domain, since pitch and thrust can be used interchangeably to control the vertical path and speed. Note that the availability of control allocation in these FMAs implies a set of parallel SISO (Single Input, Single Output) controllers being used.

3.4 Pilot models of AutoFlight Systems

Based on the analysis of the structure of current AutoFlight Systems, there does not appear to be a simple, consistent, global model of current AFS. Such a model is not available in flight manuals, which focus on crew interface and procedures. In the absence of a simple consistent model, pilots appear develop their own ad hoc models of the AFS.

The observation that pilots are constructing their own ad hoc models in the absence being presented with one has serious implications. Any empirically derived model of the AFS is going to be largely based on nominal operation. Without explicit knowledge of the system structure during non-nominal situations, pilots may find their models inadequate when they are most

critical. The implication for the training of new pilots is that a consistent model of the automation would provide a solid basis upon which to understand aircraft automation.

Based on interviews and observations, pilot's appear to model the AFS modes as independent SISO control loops. They do not appear to model MIMO transitions in detail, relying instead on an understanding of the final target criteria and some smoothness criteria to monitor the AFS performance.

Pilot models were observed to be of a SISO variety. As such, they may not accurately represent AFS operation, especially in transitions between modes. Transitions to and from envelope protection modes appear to be particularly troublesome.

Since these models are constructed empirically, individual pilot models may vary significantly. In future aircraft, this empirical model construction may cause additional difficulties, especially as control designers migrate to more common MIMO control designs.

3.5 Vestibular Cueing Concerns

An additional factor in mode awareness is the reduction of vestibular cues of mode transitions in some aircraft. For the purpose of improving ride quality, some AutoFlight Systems impose nominal vertical acceleration limits close to the human vestibular detection threshold. This provides a smooth ride for passengers, but may reduce mode awareness.

The vestibular thresholds of blindfolded subject in the Z body-axis is between 0.1Hz to 1Hz measured under laboratory conditions is between 0.001G and 0.015G (Gundry, 1978). The nominal goal of the control system on the Airbus A320 and the McDonald-Douglas MD-11 during nominal maneuvering is thought to be 0.05G.

Under operational conditions, 0.05G may be below the detection threshold of individual pilots. More work on human detection capabilities in operating environments is required to address this issue.

4.0 INVESTIGATION OF FEEDBACK MECHANISMS

In order to improve vertical mode awareness, a set of crew information requirements was developed to help mitigate some identified Mode Awareness Problems. An Electronic Vertical Situation display was designed based on these requirements. The display is currently being prototyped on the Aeronautical Systems Laboratory's Advanced Part Task Simulator.

4.1 Information Requirements for Enhanced Vertical Mode Feedback

Information requirements were based on operator requirements and on known mode awareness problems.

As shown in Table 1, the information requirements have four components: information regarding the current mode, the anticipated modes, the mode transitions and the consequences based on no further inputs.

Table 1. Crew Information Requirements

Current Mode
Mode Identification
Target States
Control Allocation

Anticipated Mode*
Mode Identification
Target States
Control Allocation

Mode Transitions*
Envelope Limits
Retroactive Cueing

Consequences*

*if predictable

For the *current mode*, the mode identification and the specific attributes of the mode, namely target state values and control allocation, are displayed in text. Where possible, graphical depiction is given of the current state and targets.

Anticipated modes have the same set of information as the current mode when this information can be determined. While this determination is straightforward for many preprogrammed macro-modes, it may be inaccurate for uncommanded mode changes. For example, the prediction of entering an envelope protection mode far in the future would be difficult.

Mode transition alerting requires anticipation of mode transitions. The depiction of envelope protection limits alerts the pilot when envelope protection is not available. In certain situations, it may not be possible to predict transitions. Retroactive cueing may be necessary to show the mode that was previously engaged. This is important in situations where an uncommanded mode change creates a confusing set of transitions.

The *consequences* of a mode change are based on an extrapolation of the current state of the aircraft automation. A predictive profile based on the current automation state is shown to display the locations of anticipated mode transitions and their consequences on the flight path.

4.2 Prototype Electronic Vertical Situation Display

It is hypothesized that an Electronic Vertical Situation Display (EVSD) which incoporates the information elements discussed earlier will provide enhanced vertical mode feedback. This conclusion has been drawn by other researchers both to improve mode awareness by Hutchins and Palmer and to accurately control in the vertical domain by Fadden, Braune, and Wiedemann. A simple EVSD is currently flying on the GulfStream Corporation's G4 cockpit.

An EVSD is envisioned to provide an analog to the Electronic Horizontal Situation Indicator (EHSI), or "map" display currently available in glass cockpits. The display shows the programmed vertical path of the aircraft and the associated modes referenced to that path. The functional requirements for a prototypical EVSD were based on a functional model of the AFS and were to be consistent with pilots' mental models and the control allocation displays on newer aircraft. Basing the requirements on a SISO model maintains this consistency.

A preliminary prototype of the EVSD has been implemented on the Aeronautical Systems Laboratory's Part Task simulator. An example of this display is shown in Figure 3. This prototype has four major areas. At the top of the display is the mode display window, showing the current and anticipated modes, control allocations and target states. At the left is a scaleable altitude tape. The bottom window can either display the path distance (if in a lateral navigation mode), or the range directly ahead of the aircraft. Finally, the main window shows the aircraft in vertical relation to the upcoming waypoints and mode transition points. It should be noted that because of the prototypical nature of this EVSD, certain informational requirements (retroactive cueing, predictive profile, some graphical elements) have yet to be implemented.

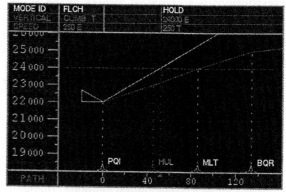

Figure 3.   Electronic Vertical Situation Display during Flight Level Change Climb

This display is implemented in color conventions consistent with the 747-400 Electronic Horizontal Situation Indicator (EHSI) as shown in Table 2.

Table 2.   Color Conventions

Information Element	Graphical Representation
MCP) Input	Dashed Magenta Line
FMS Path	Magenta Line
Programmed Waypoint	White Symbol
Current Waypoint	Magenta Symbol
Calculated Altitude Intercept Location	Green Arc
Top of Descent	Green Circle
Extrapolation of Aircraft State	White Line

The *current mode* is identified in this prototype on the top window, in white directly above the aircraft symbol. Underneath are the control allocations for the mode, and the target states. In this example, the aircraft is in Flight Level Change (FLCH) Mode with the vertical path controlled by throttle (T) and the speed controlled by the elevator (E).

The *anticipated mode* is shown in the top window above the point where it is calculated to be engaged. The anticipated target state and control allocations are depicted in a manner similar to the current mode. Note that both the target states and the control allocation changes when the new altitude is captured.

Mode transition alerting currently consists of the anticipated mode sliding across the top window and into the current mode slot when engaged. In this case, an altitude capture arc highlights where the transition is calculated to occur.

An example of the *target state* is the dashed magenta line at 24000ft, which is the altitude currently dialed into the MCP. The target state is also shown in the mode identification window, as a vertical path at 24000ft.

The solid magenta line connecting waypoint crossing restrictions is a graphical display of the current vertical path programmed into the FMS. On the bottom of the screen are the lateral waypoints programmed into the FMS. Altitude crossings are shown by the height of the dashed lines which intersect the relevent waypoint. The path distance between each waypoint defines the horizontal scale.

The example shown in Figure 4, is a Vertical Navigation (VNAV) macro-mode descent. While still in VNAV macro-mode, note that the aircraft changes the target state of the vertical path at the Top of Descent point, a predetermined transition criteria.

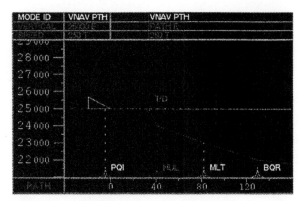

Figure 4.   EVSD during VNAV descent

In the mode identification window is the current target state, an altitude of 25000 ft. At the anticipated Top of Descent point, the target state becomes a path based descent.

ACKNOWLEDGEMENTS

This work was supported by the National Aeronautics and Space Administration under grant NAG1-1581. The authors would like to thank the following individuals for their suggestions and contributions: William Corwin, Honeywell; Peter Polson, Jim Irving, Sharon Irving, University of Colorado; Michael Palmer, Kathy Abbot, Terrence Abbot, Everett Palmer, NASA.

REFERENCES

Mecham, M. (1994), Autopilot Go-Around Key to CAL Crash. *Aviation Week & Space Technology*. May 9, 1994.

Sparaco, P. (1994), Human Factors Cited in French A320 Crash, *Aviation Week & Space Technology*. January 3, 1994.

Ministry of Planning, Housing, Transportation and Space (1992). Preliminary Report Relative to the Accident which Occurred on Jan 20, 1992 near Mon Sainte-Odile (BAS-Rhine) to the Airbus A320.

Fadden, Braune, and Wiedemann, 1988. *Spatial Displays as a Means to Increase Pilot Situational Awareness*, Boeing Commercial Aircraft Company, Seattle, Washington.

Hughes, D. (1995), Incidents Reveal Mode Confusion. *Aviation Week & Space Technology*. January 30, 1995.

Printed and bound by CPI Group (UK) Ltd, Croydon, CR0 4YY

03/10/2024

01040322-0017